Analysis and Design of Heating, Ventilating, and Air-Conditioning Systems

Analysis and Design of Heating, Ventilating, and Air-Conditioning Systems

Second Edition

Herbert W. Stanford III and
Adam F. Spach

CRC Press
Taylor & Francis Group
Boca Raton London New York

CRC Press is an imprint of the
Taylor & Francis Group, an **informa** business

CRC Press
Taylor & Francis Group
6000 Broken Sound Parkway NW, Suite 300
Boca Raton, FL 33487-2742

Printed on acid-free paper

International Standard Book Number-13: 978-1-138-60241-0 (Hardback)

Visit the Taylor & Francis Web site at
http://www.taylorandfrancis.com

and the CRC Press Web site at
http://www.crcpress.com

Contents

Preface xiii
Authors xv

Section I Basic Concepts

1. HVAC Basics 3
1.1 Overview of HVAC 3
1.1.1 Definition of Air-Conditioning 3
1.1.2 Energy Impact of HVAC Systems 4
1.2 Buildings as Thermodynamic Systems 5
1.2.1 Systems Concepts 5
1.2.2 Energy Gains and Losses in Building Spaces 6
1.2.3 Building Envelope Design 7
1.3 Common HVAC Elements 9
1.3.1 Structural Support for HVAC Components 9
1.3.2 Electrical Aspects of HVAC Systems 12
1.4 Definitions of HVAC Terms 25
Bibliography 31

2. The Air-Conditioning Process 33
2.1 Introduction 33
2.2 Air and Its Properties 34
2.2.1 Atmospheric Air 34
2.2.2 Water Vapor 35
2.2.3 Terminology 35
2.2.4 Basic Psychrometrics 36
2.3 Methods of Measurement and Analysis 36
2.3.1 Thermodynamic Wet-Bulb Temperature 36
2.3.2 The Psychrometer 37
2.3.3 The Psychrometric Chart 38
2.4 Typical Air-Conditioning Processes 41
2.4.1 Sensible Heating or Cooling 41
2.4.2 Combined Sensible and Latent Cooling (Dehumidification) 42
2.4.3 Heating and Humidification 43
2.4.4 Evaporative Cooling 44
2.4.5 Air Mixing 45
2.5 Psychrometric Analysis of Complete Systems 46
2.5.1 Space Heating and Cooling Loads 46
2.5.2 Sensible Cooling Load 47
2.5.3 Latent Cooling Loads 48
2.5.4 Psychrometric Chart Representation of Space Conditions 49
2.5.5 The Coil Line 50
2.5.6 Coil Contact and Bypass Factors 51

2.5.7 Psychrometric Analysis of Complete Systems 52
Bibliography 52

3. HVAC Systems Concepts 53
3.1 Introduction 53
3.2 All-Air Systems 53
3.2.1 Single-Zone Systems 56
3.2.2 VAV Systems 58
3.2.3 Dual–Duct Systems 59
3.2.4 Multizone Systems 62
3.2.5 VAV/Variable Temperature Systems 62
3.3 Air–Water Secondary Systems 65
3.3.1 Active and Passive Chilled Beam Systems 65
3.3.2 FCUs and Unit Ventilators 67
3.4 Unitary Systems 68
3.4.1 Incremental Units 68
3.4.2 Packaged Units 69
3.4.3 Split Systems 69
3.4.4 Variable Refrigerant Flow Systems 70
3.5 Heat Pumps 72
3.5.1 The Heat Pump Concept 72
3.5.2 Air-Source Heat Pumps 74
3.5.3 Water-Source Heat Pumps 76
3.5.3.1 Closed Circuit Heat Pump Systems 76
3.5.3.2 Closed Circuit Geothermal Heat Pump Systems 77
3.5.3.3 Open Circuit Geothermal Heat Pump Systems 80
3.5.3.4 Gas-Fired Engine-Driven Heat Pumps 80
3.5.3.5 Heat Recovery Chiller/Heat Pump System 80
3.5.4 Advanced Technology Heat Pumps 81
Bibliography 82

Section II The Design Method

4. HVAC Systems Design 85
4.1 Introduction 85
4.2 Criteria for HVAC Design 87
4.3 Compliance with Building Codes and Standards 88
4.3.1 International Mechanical Code 88
4.3.2 International Energy Conservation Code 88
4.3.3 NFPA Standards and Guidelines 90
4.3.4 ASHRAE Standards and Guidelines 90
4.4 Designing for Maintainability 91
4.4.1 HVAC Equipment Location and Access 92
4.4.2 Mechanical Equipment Rooms for Air-Handling Equipment 93
4.4.3 Mechanical Equipment Rooms for Boilers and Chillers 93
4.4.4 Ceiling Cavity Space 94
4.5 Designing for Energy Conservation 95

4.5.1 Introduction to ASHRAE Standard 90.1 ... 95
4.5.2 Simplified Energy Analysis ... 96
4.5.3 Building Energy Modeling ... 97
4.5.4 Exhaust Air Heat Recovery ... 99
4.5.5 Internal Source Heat Recovery ... 102
4.5.6 Thermal Energy Storage ... 103
4.6 Designing for Sustainability ... 105
4.6.1 High-Performance Building Elements ... 105
4.6.2 Introduction to ASHRAE Standard 189.1 ... 108
4.6.3 Net Zero/Sum Zero-Energy Buildings ... 108
4.7 Construction Documents Quality Control ... 110
Bibliography ... 112

5. Heating/Cooling Load Calculation ... 113
5.1 Introduction ... 113
5.1.1 Building Thermal Balance ... 113
5.1.2 Purpose of Load Calculations ... 113
5.2 Mechanics of Building Heat Loss and Gain ... 113
5.2.1 Design Conditions ... 114
5.2.2 Heat Transfer through Walls and Roofs ... 115
5.2.3 Heat Transfer through Fenestration ... 118
5.2.4 Heat Loss through Floors and Basements ... 122
5.2.5 Internal Heat Gains ... 124
5.2.6 Infiltration Loads ... 125
5.3 Load Computation Methodologies ... 128
5.3.1 Heat Balance and Radiant Time Series ... 128
5.3.2 Total Equivalent Temperature Difference with Time Averaging ... 130
5.3.3 Transfer Function ... 131
5.3.4 Computer-Based Calculation Tools ... 131
5.5 Benchmark Heating/Cooling Loads ... 132
Bibliography ... 134

6. Air Tempering and Distribution ... 135
6.1 Heating and Cooling Coils ... 135
6.1.1 Direct Expansion Refrigerant Cooling Coils ... 138
6.1.2 Chilled Water Cooling Coils ... 140
6.1.3 Hot Water Heating Coils ... 141
6.1.4 Steam Heating Coils ... 141
6.1.5 Electric Resistance Heating Coils ... 141
6.1.6 Preheating and Precooling Coils ... 143
6.1.7 Desiccant Cooling and Dehumidification ... 146
6.2 Air Distribution Design ... 147
6.2.1 Room Air Diffusion ... 147
6.2.2 Typical Air Distribution Patterns ... 148
6.2.3 Large Space Air Distribution ... 154
6.2.4 Displacement Ventilation Air Distribution ... 155
6.3 Terminal Units ... 157
6.3.1 VAV Single-Duct TUs ... 157

6.3.2 VAV Single-Duct Fan-Powered TUs 160
6.3.3 VAV Dual Duct TUs 160
Bibliography 161

7. Duct Design 163
7.1 Introduction 163
7.1.1 Steps in Duct System Design 163
7.1.2 Duct Air Balancing Rationale 164
7.2 Energy and Pressure Relationships 164
7.3 Friction Losses 166
7.3.1 Calculation of Friction Losses in Ducts 166
7.3.2 Calculation of Friction Losses in Fittings 170
7.4 Methods of Duct Design 171
7.4.1 Constant Velocity Method 171
7.4.2 Velocity Reduction Method 171
7.4.3 Static Regain Method 171
7.4.4 Equal Friction Method 172
7.4.5 Duct Construction Criteria 173
7.4.6 Duct Design Guidelines 174
7.5 Special Exhaust Systems Design 176
7.5.1 Laboratory Exhaust Systems 176
7.5.2 Industrial Exhaust Systems 179
7.6 Duct Insulation 181
7.7 Fire and Smoke Control in Duct Systems 182
Bibliography 183

8. Piping Design 185
8.1 Introduction 185
8.1.1 Viscosity 185
8.1.2 Weight Density, Specific Volume, and Specific Gravity 185
8.1.3 Mean Velocity of Flow 186
8.1.4 Piping Basics 187
8.2 Hydronic Piping 191
8.2.1 Pipe Routing and Sizing 192
8.2.2 Pressure Loss Calculation 196
8.2.3 Expansion and Air Removal 198
8.2.4 Freeze Protection 201
8.3 Steam and Condensate Piping 202
8.3.1 Pipe Routing and Sizing 203
8.3.2 Steam Trap Application and Sizing 206
8.3.3 Control of Steam Piping Expansion 209
8.4 Refrigerant Piping 210
8.4.1 Design Considerations 210
8.4.2 Suction Lines 213
8.4.3 Liquid Lines 215
8.5 Piping Insulation 215
8.5.1 Hot Piping Insulation 215
8.5.2 Cold Piping Insulation 215
Bibliography 218

Section III Systems and Components

9. Pumps and Fans 221
9.1 Introduction 221
9.2 Pumps and Applications 221
9.2.1 Pump Types and Characteristics 221
9.2.2 Pump Arrangements 225
9.2.3 Water Flow Volume Control 230
9.3 Fans 231
9.3.1 Fan Energy Relations 231
9.3.2 Fan Types and Characteristics 233
9.3.3 Fan and System Characteristics 237
9.3.4 System Effects 238
9.3.5 Variable Air Volume Fan Control 242
9.4 Pump and Fan Affinity Laws 243
Bibliography 243

10. Terminal Systems and Components 245
10.1 Systems Types and Applications 245
10.1.1 Selecting the Right System for the Application 245
10.1.2 Space Needs and Other Requirements for Terminal Systems 245
10.2 AHUs and Components 249
10.2.1 AHUs 249
10.2.2 Particulate Air Filtration 253
10.2.3 Gas-Phase Filtration 255
10.2.4 Ultraviolet Air and Surface Treatment 256
10.2.5 Airside Economizer Cycle 257
10.2.6 Cooling Coils Condensate Drainage 261
10.3 Humidification 262
10.4 Ventilation and Building Pressurization 264
10.4.1 Outdoor Air Intakes 265
10.4.2 Understanding and Applying ASHRAE Standard 62.1 267
10.4.3 Demand Control Ventilation 271
10.4.4 Dedicated Outdoor Air Systems 274
Bibliography 275

11. Refrigeration Systems and Components 277
11.1 Refrigeration Basics 277
11.1.1 Refrigerant Issues 277
11.1.2 Vapor Compression Refrigeration Cycle 278
11.1.3 Vapor Absorption Refrigeration Cycle 282
11.2 Direct Expansion Refrigeration Systems 285
11.3 Chilled Water Refrigeration Systems 288
11.3.1 Determining Chilled Water Supply Temperature 289
11.3.2 Establishing Chilled Water Temperature Range 289
11.3.3 Vapor Compression Cycle Water Chillers 290
11.3.4 Scroll Compressor Water Chillers 292
11.3.5 Rotary Screw Compressor Water Chillers 297
11.3.6 Centrifugal Compressor Water Chillers 297

11.3.7 Absorption Water Chillers ... 299
11.3.8 Chilled Water System Configurations ... 300
11.3.9 Chilled Water Buffer Tanks ... 307
11.4 Condensers and Cooling Towers ... 308
11.4.1 Air-Cooled Condensers ... 308
11.4.2 Cooling Tower Fundamentals ... 309
11.4.3 Tower Configuration and Application ... 314
11.4.4 Closed Circuit Evaporative Cooler ... 319
11.5 Evaporative Air Cooling ... 320
11.6 Waterside Economizer Cycle ... 320
Bibliography ... 323

12. Heating Systems and Components ... 325
12.1 Firing Fossil Fuels ... 325
12.1.1 Oil-Fired Systems ... 325
12.1.2 Gas-Fired Systems ... 328
12.2 Furnaces ... 329
12.3 Boilers ... 330
12.3.1 Boiler Types, Ratings, and Efficiency ... 330
12.3.2 Application Considerations ... 331
12.3.3 Boiler/Furnace Venting ... 332
12.4 Hydronic Heating Systems ... 333
12.5 Steam Heating Systems ... 335
12.5.1 Steam Quality ... 335
12.5.2 Steam Heat Transfer ... 338
12.5.3 Steam-to-Water Heat Exchangers ... 338
12.5.4 Feedwater System ... 339
12.5.5 Steam Pressure Reducing Valves ... 339
12.5.6 Steam Condensate Receivers and Pumps ... 343
Bibliography ... 344

13. HVAC Controls ... 345
13.1 DDC Fundamentals ... 346
13.2 DDC Input/Output Points ... 351
13.2.1 Input/Output Basics ... 351
13.2.2 DI Devices ... 352
13.2.3 AI Devices ... 353
13.2.4 DO Devices ... 363
13.2.5 AO Devices ... 364
13.3 Final Control Elements ... 366
13.3.1 Control Valves ... 366
13.3.2 Control Dampers ... 368
13.4 Direct Digital Controllers and Systems ... 370
13.4.1 DDC System Architecture ... 370
13.4.2 Controllers and Control Loops ... 371
13.4.3 Networks and Communication ... 372
13.5 HVAC Sequences of Operation ... 375
13.6 DDC System Security ... 377

13.7 DDC System Design Checklist 379
Bibliography 380

Section IV Special Considerations

14. Special HVAC Design Considerations 383
14.1 Indoor Air Quality 383
14.2 Antiterrorism Design for HVAC Systems 385
14.2.1 Introduction 385
14.2.2 Terror Threats against Buildings 386
14.2.3 HVAC Design to Minimize Explosive Threats 387
14.2.4 HVAC Design to Minimize CBR Threats 388
14.3 HVAC Water Treatment 390
14.3.1 Hot and Chilled Water Systems Chemical Treatment 390
14.3.2 Evaporative Cooling Water Systems Chemical Treatment 390
14.3.3 Evaporative Cooling Water Systems Nonchemical Treatment 394
14.3.4 Steam and Condensate Systems Chemical Treatment 396
14.3.5 *Legionella* Risk Management in HVAC Water Systems 399
14.4 Vibration and Noise Mitigation 400
14.4.1 Noise Definition and Design Criteria 400
14.4.2 Indoor Noise Control in HVAC Systems 401
14.4.3 Outdoor Noise Issues in HVAC Systems 404
14.5 HVAC Systems Start-Up and Commissioning 407
14.5.1 HVAC Systems Operation Prior to Commissioning 407
14.5.2 The Commissioning Process 408
14.6 Correcting HVAC Performance for Altitude and Temperature 410
Bibliography 411

15. Engineering Economics and Design Decision-Making 413
15.1 Introduction 413
15.2 Defining HVAC Design Alternatives 415
15.3 Estimating Capital Requirements 417
15.3.1 Capital Cost Estimating 417
15.3.2 Time/Money Relationships 418
15.4 Recurring Costs Determination 420
15.4.1 Utilities Costs 420
15.4.2 Maintenance Costs 421
15.5 Component Service Life and Replacement 422
15.6 Comparing Alternatives and Dealing with Uncertainty 424
15.6.1 Alternatives with Different Economic Lives 427
15.6.2 Sensitivity Analysis 427
15.6.3 Selecting Alternatives within the Project Budget 428
15.7 Overview of LCCA 429
15.7.1 Basic Practices for LCCA 429
15.7.2 Calculations and Analysis Tools 429
Bibliography 430

16. Building Information Modeling 431
16.1 Introduction 431
16.2 BIM Advantages 432
16.3 Applying BIM 434
16.3.1 BIM Use by HVAC Designers 435
16.3.2 BIM Use by Contractors/Subcontractors 436
16.3.3 BIM Use by Owners 436
16.4 AIA Contracts 436
Bibliography 437

17. Construction Contract Administration 439
17.1 Introduction 439
17.2 Design Period Considerations 440
17.3 Preconstruction Procedures and Practices 442
17.4 Construction Period Procedures and Practices 444
17.5 Completion and Closeout Procedures and Practices 448
17.6 Avoiding Litigation Pitfalls 449
17.7 Construction Administration Checklist 452
17.7.1 Construction Administration Practices and Procedures Checklist 453

Index 457

Preface

The first edition of this book came about as the outgrowth of a conversation between a group of practicing heating, ventilating, and air-conditioning (HVAC) design engineers in 1983 concerning their dissatisfaction with both the textbooks used in HVAC college courses (what few there were) and the design handbooks then available. Thus, I set out to write a text that I would have liked to have had as a student and one that I would have retained as a practicing engineer. I feel that the first edition had some success in meeting the following objectives:

1. To build a text around a strong "systems" point of view with emphasis on the energy consumption ramifications of design decisions
2. To primarily address HVAC applications for institutional, commercial, and industrial buildings
3. To make this text valuable as a resource tool for the practicing engineer
4. To use the methodologies of the American Society of Heating, Refrigerating, and Air-Conditioning Engineers (*ASHRAE*) *Handbook* as the primary reference source.

These objectives continue to apply in this second edition.

Another major impetus to the writing of this book was the fact that for many years I made a very good living investigating and solving HVAC system operating problems in buildings, many of which were the result of inadequate or incorrect design. Even now, a recent survey conducted by one major industry magazine indicates that the majority of all building owners queried responded that there were problems and ongoing dissatisfaction with the HVAC systems in their buildings. These experiences indicate that HVAC design engineers are still too often doing a less-than-adequate job in the delivery of the final product: an HVAC system that "satisfies."

The "science" of HVAC design is well developed. One has only to read the *ASHRAE Handbook* and other texts currently in use in the academic environment to realize that the basics of psychometrics, heat gain/loss calculations, fluid flow, fan and pump operations, and so on are well defined and readily available to the practicing engineer. Thus, while this text also addresses these fundamental topics, it is, rather, the "art" of successful HVAC design, including proper system and component application, which was incorporated into the first edition and significantly expanded in this second edition.

Design goes beyond load calculations and duct and pipe sizing. In fact, these are the smallest parts of the design effort. It is more important for the design engineer to have a proper grasp of systems applications, controls, and the basics of "fitting the right solution to solve the problem." In this text, significant emphasis is placed on "systems" considerations of each basic topic.

A word about units: In 1981, ASHRAE published their *Handbook-Fundamentals* in which the primary unit designations were in accordance with the *Système International* (SI). The result was a handbook that was useless to the practicing professional in the United States, and there subsequently was a "revolt" within the membership of ASHRAE between the academic and practicing sides of the organization. The compromise was the publishing of the *1985 Handbook-Fundamentals* (and all subsequent volumes) in two separate editions, one using the inch-pound (IP) system and the other using SI.

The first edition of this book, published in 1988, utilized only IP units. Since then, though, the HVAC market has become more global in nature and now ASHRAE requires dual unit publications. Thus, this edition uses IP units and equations as primary, with SI units and equations as secondary, in accordance with the *SI Guide for HVAC&R*, American Society of Heating, Refrigerating, and Air-Conditioning Engineers, Inc., Atlanta, Georgia, 2013.

Now, 30 years since the first edition of this book was published, a number of HVAC topics have changed: heating/cooling load calculation methodologies, indoor air-quality considerations, control systems and sequences of operation, increased emphasis on energy efficiency and sustainability, etc. Since my semiretirement in 1998, I have written and taught on a number of these developing issues in HVAC design and this new text incorporates much of that material.

However, the HVAC design business has also changed over the last 30 years… production methods (the transition from hand drafting to computer-aided drafting to building information modeling), the general reduction in contractor skills and capabilities (resulting in a need for the detail and coordination within design documents), the impact of greatly expanded building codes and industry standards, etc., are all markedly different that they were in 1988.

This created the need for a coauthor who had the expertise and experience to address these areas in this edition. Therefore, I asked Adam F. Spach, P.E., a friend and colleague with whom I've worked over the last 15 years and consider to be one of the brightest younger engineers I've ever met, to join me in the writing of this new edition.

Finally, I rededicate this book to Guy Hammer Cheek, P.E. (1927–1987), my friend, my mentor, and the man who introduced me to the idea of "HVAC systems concepts" and who diligently encouraged me to pick the right one(s) for each application.

Herbert W. Stanford III, P.E.
North Carolina
2019

Authors

Herbert W. Stanford III, P.E, is a North Carolina native and 1966 graduate of North Carolina State University with a B.S. in mechanical engineering. He is a registered professional engineer in North Carolina. In 1977, he founded Stanford White, Inc., an engineering consulting firm located in Raleigh, North Carolina and semiretired in 1998. Currently, Mr. Stanford is actively engaged in investigative and forensic engineering, teaching, and writing within a broad range of topics relative to buildings, especially HVAC systems, indoor environmental quality, and building operations and maintenance. Mr. Stanford developed the Facilities Condition Assessment Program for the State of North Carolina and the Life Cycle Cost Methodology used by North Carolina to evaluate building design decisions. He has taught a series of short courses on current building topics at the University of Toledo (Ohio), North Carolina State University, and the University of North Carolina at Charlotte and is a life member of the ASHRAE. Mr. Stanford is the author of HVAC Water Chillers and Cooling Towers: Fundamentals, Application, and Operation (Second Edition, CRC Press, 2011); The Health Care HVAC Technician (MGI Systems, Inc., 2008), a program and training manual for hospital HVAC maintenance personnel; and Effective Building Maintenance (Fairmont Press, 2010).

Adam F. Spach was born in upstate New York and graduated from Alfred University with a B.S. in mechanical engineering in 2000. After graduation, he relocated to North Carolina and has called it his home since. Mr. Spach has worked as an HVAC consulting engineer since 2001 and is licensed as a professional engineer in North Carolina and several other states. He joined Stanford White, Inc., Raleigh, NC in 2005 and is now an associate with that firm. He currently focuses on providing engineering services for educational (K-12 and university), commercial, recreational, research, and health care facilities. Mr. Spach has a passion for sustainable design in the built environment. He was appointed to the Wake County Citizen's Energy Advisory Committee, serving as the vice chair in 2010–2014, and currently co-chairs the High Performance Building Task Force sponsored by North Carolina Chapter of the American Institute of Architects. He has made presentations at the annual N.C. State Construction Conference on three occasions and at the annual North Carolina Sustainable Energy Conference on five occasions. Mr. Spach routinely lectures at the North Carolina State Design School.

Section I

Basic Concepts

1

HVAC Basics

1.1 Overview of HVAC

1.1.1 Definition of Air-Conditioning

Modern-day *air-conditioning* was created in the early 20th century based on the vision and works of Hermann Rietschel, Alfred Wolff, Stuart Cramer, and Willis Carrier. Cramer, a textile engineer in Charlotte, North Carolina, is credited with coining the phrase "air-conditioning" in 1906. Willis Carrier, though he did not actually invent air-conditioning or take the first documented scientific approach to applying it, is credited with integrating the scientific method, engineering, and business of this developing technology and creating the industry we know today as *heating, ventilating, and air-conditioning* (HVAC).

By 1929, the American Society of Heating and Ventilating Engineers defined *comfort air-conditioning* as "… the process of treating the air so as to control simultaneously its temperature, humidity, cleanliness, and distribution." Today the organization incorporates that term in its current name: The American Society of Heating, Refrigerating and Air-Conditioning Engineers, Inc. (ASHRAE). The same definition of air-conditioning applies today, but, of course, there have been major refinements in the art since 1929.

Comfort air-conditioning means the maintenance of those indoor atmospheric factors affecting comfort:

1. The desired range of air temperatures
2. An acceptable humidity
3. Minimal atmospheric particulates, including pollens and bacteria
4. An acceptable odor level
5. An acceptable degree and pattern of air motion

Air-conditioning, in its broadest sense, is the engineering science of designing means of controlling the air in the human environment for comfort and health. This broad area includes control of humidity and heating, as well as cooling. It is distinguished from *refrigeration,* in that refrigeration is one of the processes typically involved in an air-conditioning system (i.e., refrigeration systems cool the air, either directly or indirectly). However, it is still true that the term air-conditioning is taken by the lay public to refer only to cooling and dehumidification of air. For this reason, the acronym HVAC is still the most common term used when referring to the multipurpose systems found in modern buildings.

This book is concerned with the engineering design involved in application, sizing, component selection, layout, and control of HVAC systems in buildings. These tasks are generally performed by consulting engineers and designers, working with the building owner (and, as applicable, the architect for the building). These designers are not concerned with the detailed design of HVAC components themselves (fans, pumps, chillers, and so on), although HVAC engineers are expected to know the general operating characteristics and respective advantages of, for example, centrifugal-electric refrigeration machines versus steam absorption water chillers, they need not necessarily be familiar with the detailed heat transfer characteristics or the internal control mechanisms of the machines. HVAC designers are best referred to as "assemblers" who bring together multiple components under common control to satisfy the performance requirements listed previously.

But, long gone are the days when an HVAC designer had to know only about HVAC systems and components. Today, there are a myriad of issues that impact HVAC design and have to be included as part of the design process, including general construction elements, energy efficiency and consumption, indoor air quality, anti-terrorism, and others addressed by this book.

1.1.2 Energy Impact of HVAC Systems

Before any building is constructed, the space that the building will occupy consumes no energy. However, the moment the building is created, there arises a need to provide an internal environment that is different from the surrounding environment (i.e., the building must be either warmer or cooler and more or less humid than outdoors as outdoor conditions change over time). And, since walls and roofs block natural light, artificial lighting must usually be provided even during daylight hours. Thus, an energy-use burden is created by almost every building in satisfying these environmental requirements.

Although there were Roman hot air heating systems and ice-based cooling was applied by the Egyptians, HVAC is basically a modern technology dating, realistically, only from the early 20th century. Significant scientific and practical development did not occur until after World War I and the widespread use of building cooling did not take hold until after World War II.

The design and application of HVAC systems can be divided into five historical periods:

1. *Pre-1960.* Prior to the late 1960s, HVAC systems tended to have noncritical design criteria and were basically simple in scope. Heating systems were designed for 65°F [18°C] to 70°F [21°C] indoor temperature, while the indoor temperature for cooling was generally selected to be 15°F [8°C] below the outside peak temperature (typically, about 80°F [27°C]). Systems tended to be single zone, multi-zone, or, in very large buildings, dual-duct or induction types. Emphasis in this period was on low first cost, and since energy was cheap, little or no attention was paid to operating efficiency.
2. *1960–1975.* This period was the "heyday" of complex and very inefficient HVAC systems. No longer was 80°F [27°C] interior design and multi-zone air distribution acceptable. Summer indoor design temperatures fell as low as 72°F [22°C] and *every* space had to have a thermostat to maintain temperature control within ±1.5°F [0.8°C]. Ventilation air quantities equivalent to 25% of the system supply airflow were very common. To satisfy these more critical design criteria while keeping first costs reasonable, reheat systems became very popular.

3. *1975–1984*. The energy cost upheavals of 1973–1974 and 1977–1978 forced designers to reevaluate design criteria and system concepts to reduce energy consumption and cost. During this period, design temperatures were 76°F–78°F [24°C–26°C] for cooling and 68°F–72°F [20°C–22°C] for heating. Design ventilation airflow rates were reduced to about 20% of the 1960–1975 values. Non-reheat systems, particularly those incorporating variable air volume concepts, gained wide acceptance. The first energy codes came into being during this period.
4. *1984–1998*. In the early 1980s, the "energy crisis" had disappeared. But, many of the lessons learned worked to help offset the rapid inflation of energy costs during that period. Variable air volume systems became more common. However, design indoor temperatures returned to values that resulted in real comfort conditions and, by 1989, ventilation rates had returned to about 75% of their pre-1975 values due to widespread complaints about indoor air quality and the common occurrence of "sick building syndrome."
5. *1999-Today*. In addition to ever more stringent limits on building energy consumption imposed by building codes, in the mid-1990s, the term "sustainability" began to enter the HVAC designer's lexicon. Some building owners wanted their buildings and systems to be not only energy efficient but also to reduce their negative imprint on the environment in general. While the definition of what constitutes "sustainable building design" is still evolving, modern HVAC designers are often required by their clients to focus their designs on sustainability, without the function and performance of HVAC systems being impaired.

Today, while the primary goal for any HVAC system remains its satisfactory function and performance, there are two other important secondary goals for the HVAC systems designer, as discussed in Chapter 14.

1. Minimize the energy burden imposed by the system. ASHRAE standards, energy conservation building codes, and simply good design stewardship are all imposed to meet this goal.
2. Minimize the system's overall negative impact on the environment, not limited to just energy consumption, making it more "sustainable."

1.2 Buildings as Thermodynamic Systems

1.2.1 Systems Concepts

Buildings may be viewed as thermodynamic systems. The advantage of such an approach is that it allows an overall view of the processes of heat gain and loss that are operating continually to change the environmental conditions felt by the occupants. Figure 1-1 shows the general concept of a system as applied to a building. The system boundary is a real or imaginary separation between the area "inside" and the external environment.

We are concerned with maintaining some desired conditions of temperature and humidity within the boundary. However, the energy inputs and outputs act to change these conditions unless we compensate for them. According to the first law of thermodynamics, all

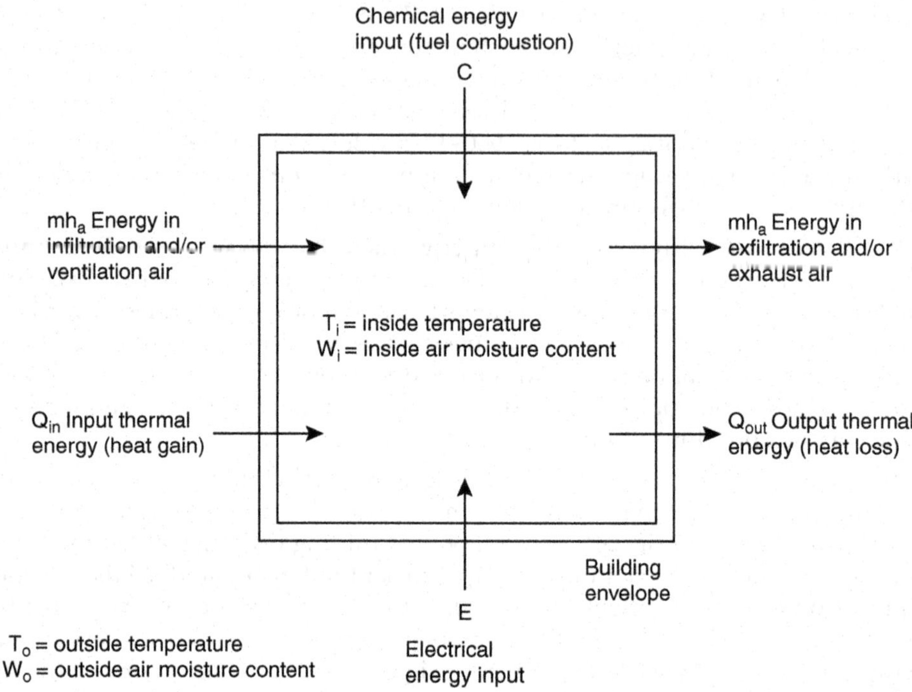

FIGURE 1-1
Building viewed as a thermodynamic system.

of the energy entering the system boundaries can be accounted for … it is either stored in the system or it is returned back through the system boundaries.

Virtually all of the energy that enters the boundaries of a building is transformed eventually into thermal energy, and it is typically felt as an increase in indoor temperature (internal energy) of the building. Some energy enters or leaves the building directly in the form of thermal energy. The energy that enters or leaves as a result of a differing inside and outside temperature is referred to as "heat" (although heat is also loosely used to refer to any form of thermal energy).

1.2.2 Energy Gains and Losses in Building Spaces

Figure 1-2 shows the typical methods by which energy enters and leaves building spaces.

From this figure, it is quite clear that energy typically enters in one form and leaves in another. For example, on a hot summer day, the building is heated by the solar radiation on the walls (which results in conduction through the walls) and by direct solar radiation through glass (which results in heating of the interior surface). Conduction through the walls and glass due to indoor/outdoor temperature differences also occurs. The electrical energy that enters the building is eventually converted to heat, either mechanically by fans or other motor-operated devices, or indirectly, such as by lighting.

The important concept is that all of this energy must eventually begin to leave the building at roughly the same daily rate that it enters or else the building will just become increasingly warmer. The primary mechanism for relieving the buildup of thermal energy is conduction (and convection) through the envelope, which will occur when the outside temperature falls below the inside temperature (such as during the evening). This is, in the

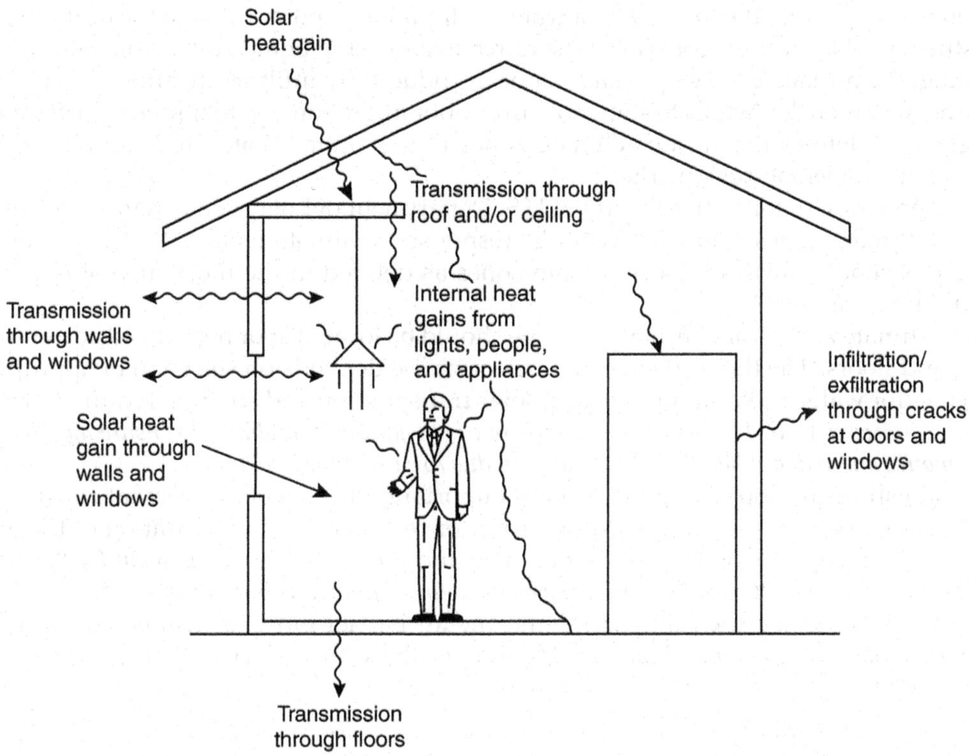

FIGURE 1-2
Basic elements of heat transfer to and from a building space.

absence of mechanical cooling, a self-limiting process; the building would lose heat until only the indoor temperature becomes nearly equal to the outside temperature, at which point the conduction heat transfer would cease. If there were significant internal heat generation. The temperature inside would continue to climb and conduction from inside to outside would resume. At some point, however, the temperature may be just high enough to maintain a conduction rate that would balance the internal heat generation rate and a steady-state condition would have been achieved. This sort of dynamic temperature fluctuation happens continuously in all buildings, although it may be diminished by mechanical heating or cooling.

This is a simplified look at the process, and there are several complicating factors. For example, the walls of the building have mass, which has a thermal storage capacity. The effect of thermal mass is to retard the flow of thermal energy (since a portion of it must be stored in the walls along the way). That is, the wall materials are heated or cooled as the "wave" of energy makes its way from the hotter to the cooler side of the wall.

1.2.3 Building Envelope Design

Every building must have both a thermal and a moisture envelope, and it is an important part of the HVAC designer's role to ensure that the envelope is correctly designed and constructed, even though that is not normally part of his or her direct design responsibility.

The role of the thermal envelope is to reduce or eliminate heat transfer across envelope boundaries driven by temperature differences and solar radiation. The role of the moisture envelope is to prevent the migration of water, in liquid or vapor form, into the building. The moisture envelope always consists of the rain barrier to stop liquid water from entering the building, the air barrier to stop water vapor introduced by infiltration, and the vapor barrier (roofs) or retarder (walls) to stop moisture vapor migration due to humidity differences.

Figure 1-3 defines the various climate zones in the United States and can be used for establishing envelope design criteria.

Note, however, that ASHRAE and the U.S. Department of Energy may revise the boundaries of climate zones every 3–5 years in response to climate change that is underway. Designers should always review climate zones as defined in the most current edition of ASHRAE *Standard 90.1*.

In all climate zones, the thermal envelope should be located at or near the inside of walls, roofs, and floors. The thermal envelope performance depends on selection of appropriate U-factors for walls, roofs, and floors and solar transmission and shading requirements for building fenestration. *The HVAC engineer's role is to advise the architect to ensure high thermal performance and reduced heat loss and heat gain due to each of these factors.*

The moisture envelope design is equally important. This envelope element consists of a rain barrier, an air barrier, and a vapor barrier or retarder. The configuration of the moisture envelope depends on the outdoor climate and *it is a primary requirement for the HVAC designer to ensure that the wall design(s) correctly define the correct configuration.*

In all climate zones, the rain barrier and the air barrier must be installed at or immediately inboard of the wall cladding. The role of these two barriers is to prevent liquid

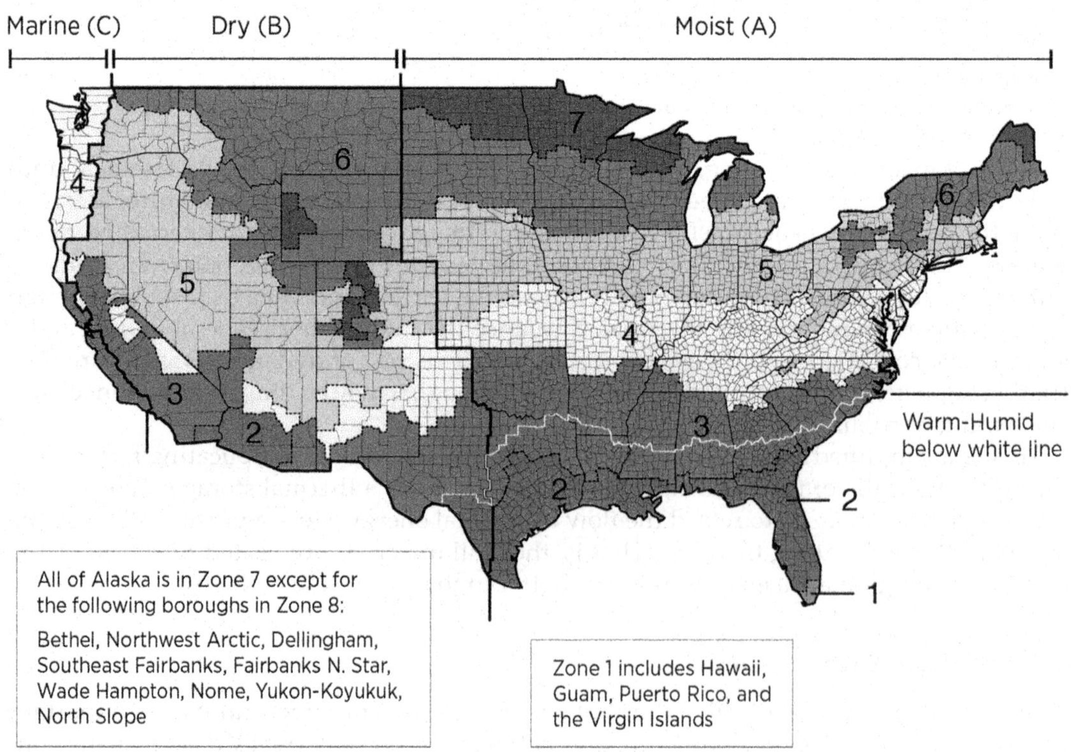

FIGURE 1-3
Climate zones in the United States. (Public Domain (U.S. Department of Energy) https://basc.pnnl.gov/system/files_force/images/IECCmap_Revised.jpg?download=1)

moisture from entering the exterior wall and, ultimately, into the air-conditioned building. But, the location of the vapor barrier or retarder, which is designed to prevent water vapor from entering the exterior wall, is more complex and varies by the type of climate as defined in the following:

1. In *hot, humid climate zones 1, 2A, 2B, and 3A,* the vapor retarder must never be placed on the indoor wall since the outdoor moisture will be trapped in the wall. Vinyl wallcovering is an effective vapor barrier that should never be used in hot, humid climates. Rather, the vapor retarder must be located immediately inboard (or as part of) the air barrier, near the outer boundary of the wall construction.
2. In a *mixed, humid climate* such as zone 4A and in a *mixed, dry climate* such as zone 4B, the placement of the vapor retarder requires more analysis by the designer. In this type of climate, the vapor retarder should be installed roughly in the "thermal middle" of the wall assembly.

   An interior vapor barrier would be detrimental, as it would prevent the wall assembly from drying toward the interior during cooling periods. The wall assembly is more forgiving without the interior vapor barrier than if one were installed.

   For hospitals, laboratories, and so on, located in mixed, humid climates, for which humidification is provided during the winter and the indoor dew point temperature remains above 40°F [4°C], detailed analysis of the vapor retarder placement is required to establish the correct vapor retarder location.
3. In *cold climate zones 5, 6, and 7,* indoor moisture levels tend to be higher than outdoor levels during winter, and thus, the vapor retarder must be located at or near the inside surface of the wall.
4. In *hot, dry climates,* such as zone 3B, no vapor retarder is required since there is little moisture vapor outdoors or indoors to negatively impact wall construction.

1.3 Common HVAC Elements

There certain elements that apply to almost all HVAC systems that are generally referred to as "common work" or "common elements." *Again, while these elements may not be part of the HVAC designer's direct design responsibility, he or she is charged with insuring that they are addressed correctly.*

1.3.1 Structural Support for HVAC Components

HVAC systems typically consist of many different components that must be installed, supported, and anchored individually.

HVAC equipment must be supported in accordance with the equipment manufacturer's requirements, coordinated with the equipment location within the building and building structural system.

1. *Floor and Housekeeping Pad Installation.* Floor-mounted equipment must always be installed on concrete housekeeping pad (or equivalent). The pads for air-handling units should be 6+" [150+ mm] high to allow for the installation of drain pan traps,

while pads for all other equipment should be 4″ [100 mm] high. The equipment should be anchored and grouted to housekeeping pads.

2. *Structural Slab, Pier, or Foundation Installation.* Cooling towers, tanks, chillers, and (sometimes) boilers are typically heavy enough to require a structural slab, pier(s), or foundations for support. *Review each component with the project structural designer to determine requirements.*
3. *Suspended Installation.* For suspended equipment, structural steel framing to distribute the imposed operating loads without stressing building structural elements or causing damage to the building substrate is normally required. This frame, which is then suspended from the building structure with vibration isolators, an appropriately sized all-thread rod, serves an equipment platform.
4. *Rooftop Installation.* The equipment may be mounted on structural steel supports (the preferred method) or on continuous roof curb:
   a. Structural steel supports must be as detailed on the drawings, and their design requires the input of a structural engineer. The supports must allow clear space under the equipment for roof maintenance and replacement (at least 36″ [1,000 mm] under equipment up to 60″ [1,500 mm] wide and at least 48″ [1,200 mm] under equipment over 60″ [1,500 mm] wide) and include appropriate service walkways and handrails, steps, ladders, and so on.
   b. Continuous roof curb tops are required to be at least 17″ [430 mm] above the roof surface to prevent leakage into the building in the event of roof flooding due to blocked roof drains. The top of all roof curbs must be level, with pitch built into curb when deck slopes are 2% or greater. *Often, structural steel reinforcement below the roof is required to avoid stressing building structural elements and must be reviewed with the project structural designer.*

Ductwork support requirements are well defined in *HVAC Duct Construction Standards—Metal and Flexible*, Sheet Metal and Contractor's National Association's (SMACNA), with which every HVAC designer should become familiar.

1. *Horizontal Suspended Installation.* Ductwork hangers may be fabricated with sheet metal straps or all-thread rod. SMACNA *HVAC Duct Construction Standards* provides numerous details for individual duct hanger methods. Where multiple ducts are routed together, they may be supported by a common "trapeze" hanger, essentially a single cross member with two or more vertical hangers.
2. *Vertical Installation.* Vertical ducts must be anchored and supported at each floor.
3. *Rooftop Installation.* Horizontal ductwork installed above a roof must be supported by roof support rails integrated with the building roof and structural members to both support the weight of the duct and to provide vertical and horizontal anchoring against wind forces. *Consult with the project structural designer to determine requirements and provide details as necessary.*

Piping systems are generally classified on the basis of their service temperature, which impacts strength, potential thermal expansion/contraction, and insulation requirements (see Chapter 6), as summarized in Table 1-1.

Based on these classifications, hangers and supports for piping systems are typically selected in accordance with the requirements of *Standard Practice-58, Pipe Hangers and Supports*, Manufacturers Standardization Society (MSS).

TABLE 1-1

Piping Systems Classification

Classification	Temperature Range	Typical HVAC Applications
Type 1: Hot systems	Type 1A: 120°F–250°F [50–120°C]	Hot water, low pressure steam (≤15 psig [100 kPa]), steam condensate, low pressure boiler feedwater
	Type 1B: >250°F [120°C]	High pressure steam (>15 psig [100 kPa]), high pressure boiler feedwater, high temperature hot water
Type 2: Ambient systems	71°F–120°F [22–50°C]	Condenser water (indoor), oil, fuel gas
Type 3: Cold systems	Type 3A: 32°F–70°F [0°C–49°C]	Chilled water, cooling coil condensate, condenser water (outdoor), condenser water (all) with waterside economizer cycle
	Type 3B: <32°F [0°C]	Liquid and cold gas refrigerant, chilled water with ice thermal storage system

1. *Suspended Horizontal Installation.* Horizontal piping is typically supported by factory-fabricated horizontal-piping hangers complying with MSS 58. Where multiple horizontal pipes are routed together, "trapeze" hangers may be field fabricated from structural steel members or from preformed channel members and suspended by two or more all-thread hanger rods. Each pipe on a trapeze hanger must be individually supported, and if the piping is insulated, the insulation must be protected by using MSS Type 39 pipe saddles for Classification Type 1B piping and MSS Type 40 insulation shields for Classification Types 1A, 2, and 3 piping at each pipe support. Install supports with maximum spacing and all-thread hanger rods sized in accordance with Table 1-2, where NPS defines *nominal pipe size.*
2. *Vertical Installation.* Vertical piping requires the use of factory-fabricated riser clamps complying with MSS Type 8 for support. Vertical piping must be anchored and supported at each floor. In tall, concrete structure buildings, it is typically necessary to provide expansion devices on every other floor or so to account for long-term "creep" or shrinkage of the structure.
3. *Rooftop Installation.* Horizontal rooftop piping requires the use of factory-fabricated roof pipe rails with individual pipe supports (as for trapeze hangers).

TABLE 1-2

Pipe Hanger Spacing and Hanger Rod Sizes

NPS (in.) [DN]	Max. Span for Copper Tubing (ft) [m]	Max. Span for Steel Pipe (ft) [m]	Min. All-Thread Hanger Rod Size (in.) [mm]
<1 [25]	5 [1.5]	7 [2.1]	3/8 [10]
1 to 1-1/4 [25–32]	6 [1.8]	7 [2.1]	3/8 [10]
1-1/2 [40]	8 [2.4]	9 [2.7]	3/8 [10]
2 [50]	8 [2.4]	10 [3.0]	3/8 [10]
2-1/2 [65]	9 [2.7]	10 [3.0]	1/2 [13]
3 [80]	10 [3.0]	12 [3.7]	1/2 [13]
4 [100]	10 [3.0]	12 [3.7]	5/8 [16]
6 [150]	10 [3.0]	12 [3.7]	3/4 [19]
8–12 [200–300]	10 [3.0]	12 [3.7]	7/8 [22]
14–16 [350–400]	10 [3.0]	12 [3.7]	1 [25]
18–20 [450–500]	10 [3.0]	12 [3.7]	1-1/4 [32]
24 [600]	10 [3.0]	12 [3.7]	1-1/2 [40]

The HVAC designer must review seismic requirements with the project architect/structural engineer at the beginning of the design process. The responsibility for the design of manufactured equipment and/or field-fabricated components installed to withstand seismic loading, including comprehensive engineering analysis by a qualified professional engineer using performance requirements and design criteria specified by the HVAC designer, should be delegated to the contractor by the project specifications. *This is a very specialized design process and is typically not within the skill set of most HVAC designers.* Note that seismic design criteria may trigger building code requirements for *special inspections* by a third party for this portion of the work during the construction period.

As with any design, there is certain basic information that is required before seismic restraints can be selected and placed. The building owner, architect, and structural engineer must make the decisions that form the basis for the information required to select the seismic restraints for the building. This is the information that is included in the project specifications. The following parameters must be defined by the design professionals having responsibility for HVAC systems in a building and should be determined by the structural engineer of record.

1. *Occupancy Category or Seismic Use Group.* This is defined by the applicable building code on the basis of building use and specifies which buildings are required for emergency response or disaster recovery.
2. *Seismic Design Category.* This determines whether or not seismic restraint is required.
3. *Short Period Design Response Acceleration Parameter.* This value is used to compute the horizontal seismic force used to design and/or select the seismic restraints required.

The responsibility for the design of manufactured equipment and/or field-fabricated components installed outdoors to withstand wind loading, including comprehensive engineering analysis by a qualified professional engineer, using performance requirements and design criteria specified by the HVAC designer, should be delegated to the equipment suppliers and/or the project structural designer by the project specifications. Again, this is a very specialized design process and is typically not within the skill set of most HVAC designers.

Equipment and/or field-fabricated components installed outdoors must be fabricated and anchored to the ground or building structure, as applicable, to withstand a wind load imposed on the largest vertical or projected surface area at the maximum wind speed dictated by the applicable building code based on the project location. Anchoring by using structural frames, straps, or other hold-down devices attached to foundations, structural supports, or roof curbs, as applicable, is required.

1.3.2 Electrical Aspects of HVAC Systems

For most HVAC designers, electrical engineering, in general, and Article 430 of the *National Electrical Code* (NEC), "Motors, Motor Circuits, and Controllers," in particular, are not his or her strong suit. But, every HVAC designer must have a working knowledge of basic electrical terms and relationships and the electrical requirements of various HVAC components, particularly motors, and be aware of the information required by the project electrical designer to ensure that electrical service to each component is adequate.

Building power systems utilize alternating current (AC) voltage levels, roughly defined as LV (low voltage, ≤600 V) and HV (high voltage, >600 V), with a frequency of 50 or 60 Hz. An electrical circuit has the following three basic components:

1. *Voltage* (V) is defined as the electrical potential difference that causes electrons to flow.
2. *Current* (I) is defined as the flow of electrons and is measured in amperes (amps).
3. *Resistance* (R) is defined as the opposition to the flow of electrons and is measured in ohms.

These components are defined by Ohm's law in accordance with Eq. (1-1).

$$V = I \times R \tag{1-1}$$

In order to flow, electricity must have a continuous, closed path. The word *circuit* refers to the entire course an electric current travels, from the source of power, through an electrical device, and back to the source. Every circuit is comprised of three major components:

1. A conductive "medium," such as a wire.
2. A "source" of electrical power.
3. A "load" that needs electrical power to operate.

The current flows to the loads through a "hot" wire and returns via a "neutral" wire, under normal conditions, maintained at zero volts.

There are also two optional components that can be included in any electrical circuit: control devices and protective devices. Control and protective devices, however, are not required for a circuit to function. A power circuit is defined as any circuit that carries power to electrical loads. A control circuit is a special type of circuit that uses control devices to determine when loads are energized or de-energized by controlling the current flow.

In *direct current* (DC) circuits, power (P) is simply a product of voltage and current (amps), as shown by Eq. (1-2).

$$P = V \times I \tag{1-2}$$

where

P = Power (W or kW)

For AC circuits, *power factor* (PF) must be considered. PF is defined as the ratio of the "real" power flowing to the load to the "apparent" power in the circuit. A PF of less than one means that the voltage and current waveforms are not in phase, and real power is the capacity of the circuit for performing work under specific loading, while apparent power is the product of the current and voltage of the circuit.

For resistive loads, such as electric heating coils, and so on, PF = 1.0. For inductive or capacitive loads, such as motors, the PF can be estimated from Table 1-3.

For single-phase power circuits, power is defined by Eq. (1-3).

$$P = V \times I \times \text{PF} \tag{1-3}$$

TABLE 1-3
Typical Motor PF

		PF		
Motor Nameplate (HP)	**Speed (rpm)**	**50% Load**	**75% Load**	**100% Load**
0–5	1,800	0.72	0.82	0.84
5–20	1,800	0.74	0.84	0.86
20–100	1,800	0.79	0.86	0.89
100–250	1,800	0.81	0.88	0.91

For three-phase power circuits, there are three conductors supplying the load rather than only two as for single-phase loads. The current in one conductor supplying the three-phase load is 120° shifted in phase from the current flowing in each of the other wires. A factor that takes all of this into account is the number 1.73, the square root of three. Thus, polyphase power is defined by Eq. (1-4).

$$P = 1.73 \times V \times I \times \text{PF} \tag{1-4}$$

For motors in the United States, power is typically defined by the actual load imposed on the motor in terms of *brake horsepower* (bHP), while the motor nameplate rating is given as *horsepower* (HP). In either case, the conversion from W to bHP or HP is defined by Eq. (1-5).

$$\text{bHP or HP} = P \times \text{EFF}/746 \tag{1-5}$$

Where EFF is the motor efficiency at the given load percentage.

The total amount of energy consumed by the load in a power circuit is computed by multiplying the P by the length of time the load is "on" (hours). This is most commonly expressed in "Kilowatt Hours" (or kWh), where a kW is equal to 1,000 W.

While the design of the electrical service to HVAC equipment is typically performed by the project electrical engineer, the HVAC designer must carefully coordinate HVAC electrical loads, voltage requirements, and safety and control devices required by the NEC to ensure that (1) electrical power is provided as needed, (2) in each case it is the right type of power (voltage and phase), and (3) procedures of connection of electrical power to each HVAC component is clearly defined in the project specifications.

The general term *conductor* applies to anything that conducts the flow of electricity. In the United States, electrical wires are conductors that are sized using two different systems: the *American Wire Gauge System* (AWG) and the *Thousand Circular Mill System*. Both systems designate wire size based on diameter or cross-sectional areas. In the AWG system, every six-gauge *decrease* corresponds to a doubling of the wire diameter and every three-gauge *decrease* doubles the wire's cross-sectional area. In the SI system, wire gauge is defined as 10 times the wire diameter in millimeters, so a 50-gauge metric wire would be 5 mm in diameter.

Note that in AWG, the diameter goes up as the gauge goes down, but for SI gauges, it is the opposite. To avoid confusion, in SI units, wire size is more commonly specified in millimeters rather than in gauges.

The current carrying capacity of a particular wire is dictated by its "capacity," it can be allowed to conduct, as limited to the maximum permitted by the NEC, based on the type of wire (copper or aluminum) and the maximum temperature rating of the wire's insulation (60°C, 75°C, or 90°C).

As electrical energy is generated, it is transformed and transported through a network of wires, substations, and transformers to the consumer. Typically, electric energy leaves the utility distribution substation and is distributed via three-phase distribution lines to switchboards within the building. Switchboards take a large block of power and break it down into separate circuits, each of which is controlled and protected by the fuses or switchgear of the switchboard.

A *panel board* is an enclosed assembly with circuit breakers. Branch circuits feed power to receptacles, switches, and equipment in the building. Likewise, *motor control centers*, with integral switches and/or motor starters, may be supplied as part of the electrical distribution system ready to connect field wiring to the electrical motors served.

Overcurrent protection is installed to provide automatic means for interrupting ("opening") a circuit in which the current rises above their rating due to a fault or short circuit. Two types of over-current devices are in common use: circuit breakers and fuses, both rated in amperes.

A *circuit breaker* is a switching device capable of protecting the distribution line or feeder connected to it from overloads and faults. If a circuit is overloaded, the mechanism inside the breaker trips the switch and breaks the circuit. The circuit breaker may be reset by simply flipping the switch. A circuit breaker is capable of ignoring short-period overloads (such as the initial current required in the starting of a motor) without "tripping" but protects against prolonged overloads.

A *fuse* is a thermal device used for protecting, typically, switchgear equipment and cables against over-currents. When a fuse element overheats due to an over-current condition and "blows," the circuit breaks.

A *ground fault interrupter* is a device that detects small current leaks and disconnects the hot wire to the circuit. It can also be part of a circuit breaker or part of an electrical outlet.

Relays are small, very fast-acting automatic switches designed to protect an electrical system from faults and overloads. It is usually an electromagnetic device which has a coil. When this coil is supplied with power, a magnetic field is created which will operate a mechanical switch. When a relay senses a problem, it quickly sends a signal to one or many circuit breakers to open, or trip, thus protecting it from damage as well as human life from injury.

A *contactor* is an electrically operated switch, a large relay in effect, which can be made to switch a motor, heater bank, capacitor bank, and so on, on and off directly or by a remote controller, such as a thermostat, humidistat, timer, pilot devices, or any other protective devices. Although it is a switch, a contactor is designed to interrupt an electric current repeatedly and frequently. When a contactor breaks the current, an electrical *arc* is established across the contacts and a good amount of heat energy is generated. This increases when the frequency of breaking the current increases and may result in the "welding" or fusing of the contacts and contactor failure.

The simplest and most common starting device for HVAC electrical motors is the *across-the-line* (ACL) starter, consisting of a main contactor and a thermal or electronic overload relay. The disadvantage of the direct-on-line method is very high starting current (6–10 times the rated motor currents) and high starting torque, causing slipping belts, heavy wear on bearings, and gear boxes, and so on. ACL starting is typically applied to motors only 40 HP [30 kW] or less.

For larger motors, reduced voltage starters are typically applied:

1. *Wye-delta* starting device consists normally of three contactors, an overload relay, and a timer for setting the time in the star position (starting position). The starting current is about 30% of the direct-on-line starting device. The starting torque

is about 25% of the direct-on-line starting torque. The stress on an application is reduced compared to the direct-on-line starting method.

2. *Part-winding* starting uses only a portion (usually one-half but sometimes two-thirds) of the motor winding, increasing the impedance seen by the power system. It is to be used only for voltage recovery and must not be left on the start connection for more than 2–3 s. The motor is not expected to accelerate on the start connection. The advantages of part-winding starting include the following:
   a. Starting current is 60%–75% of normal, depending on the specific winding connection.
   b. Starting torque is very low (may not even turn the shaft).
   c. Winding heating is very high on start connection.
3. An *autotransformer* starter is connected so the motor is on the secondary of an autotransformer during starting. The autotransformer has taps, to limit the voltage, applied to the motor at 50%, 65%, or 80% of full voltage. Because the line current varies as the square of the impressed voltage, these same taps equate to 25%, 42%, and 64% of the full-voltage value of line current. The autotransformer starter was historically the most flexible of reduced voltage starters until the advent of the solid-state starter.
4. *Solid-state* starters use back-to-back *thermistors* for each line to the motor. These six thermistors control power to the motor. The power adjusts by not completely turning on the thermistors during starting. In other words, only a portion of the three-phase sinusoidal wave is supplied to the motor during start.

   Because of these control features, the big advantage of the solid-state starter is the large number of starting characteristics. The standard soft-start mode simply ramps the voltage from a preset initial torque value to 100% during a user-selected time of 0–30 s.

   Another available control mode is a start based on current limitation. In this mode, the current is limited (between 50% and 600%), as is the duration (between 0 and 30 s). Other available operating modes include kick start, soft stop, and pump control options. The last option starts a pump motor on a curve rather than a straight line ramp. This causes the hydraulic system to react as if there were a closed discharge valve behind the pump, opening during starting.
5. A variable frequency drive (*VFD*) is a type of motor controller that includes an advanced solid-state power controller. Instead of simply opening and closing the power circuit such as an ACL starter or ramping the motor voltage up or down such as a soft start to turn the motor on and off, a VFD controls motor speed and, thus, power.

The distance between the VFD and the motor it controls can create an opportunity for motor winding failure due to reflected wave HVs caused by locating the motor too distant from the VFD. Several drive manufacturers have Internet-based calculators that will tell designers if the distance between the motor and the drive is too far, but the common recommendation is to keep the VFD within 50 ft. [15 m] and within the line of sight.

VFDs utilized for most HVAC applications are the *integrated gate bipolar transistor pulse width modulation* type in accordance with National Electrical Manufacturers Association (NEMA) *Publication ICS 2*, listed and labeled as a complete unit and arranged to provide variable speed of an inverter duty induction motor by adjusting output voltage and frequency.

As building electrical systems have incorporated more and more electronic devices that are "nonlinear load," including electronic lighting ballasts, computers, and VFDs, problems with harmonic interferences have arisen. Consequently, some designers require the VFD manufacturer to demonstrate compliance with IEEE *Standard 519*, which is written to manage harmonic contributions from facilities into the primary electrical distribution system operated by an electric utility. The context of the standard is to provide a basis for determining if a customer reflects an excessive amount of harmonic content, with respect to the electrical demand load and the size of electrical service infrastructure, into the utility distribution system.

However, applying IEEE *Standard 519* to a VFD makes no sense! This is the equivalent of an HVAC designer requiring an air handler manufacturer to demonstrate that his products do not compromise the ability to achieve a 30% energy reduction in the building. *Thus, project specifications for VFDs should not include a reference to IEEE Standard 519.*

Typical HVAC VFDs are six pulse drives that are adequate for most applications *if the electrical designer is competent in configuring the electrical distribution system for the imposed HVAC equipment loads and potential harmonics.* For electrical motors exceeding 75 HP [56 kW] (such as for large fans, pumps, or chillers), 12 or even 18 pulse drives may be specified as needed.

Basic VFD requirements for HVAC applications include the following:

1. *Input AC Voltage Tolerance.* Plus or minus 10%
2. *Input Frequency Tolerance.* Plus or minus 2 Hz
3. *Minimum Efficiency.* Ninety-six percent at design frequency and full load
4. *Minimum Harmonics.* VFDs should be equipped with internal 5% impedance DC link reactors to minimize power line harmonics and to provide near unity PF. VFDs without DC link reactors should include 5% impedance line side reactors.

   Total harmonic distortion should not exceed 10% for motors <10 HP [7.5 kW], 8% for motors $10 \leq HP < 25$ [$7.5 \leq kW \leq 18.6$], and 5% for motors ≥25 HP [18.6 kW]. Maximum individual allowable distortion at any individual harmonic shall not exceed 5% for motors <10 HP [7.5 kW], 5% for motors $10 \leq HP < 25$ [$7.5 \leq kW \leq 18.6$], and 3% for motors ≥25 HP [18.6 kW]. Testing and documentation of total harmonic distortion at the VFD-circuit breaker terminals is required.
5. *Short Circuit Interrupt Capability.* ≥65,000 A
6. *Input Transient Protection.* Integral surge suppressors
7. *Phase Protection.* Loss-of-phase, reverse-phase, and phase imbalance
8. *Motor Overload Relay.* Adjustable and capable of NEMA 250, Class 20 performance, with notch filter to prevent operation of the controller-motor-load combination at a natural frequency of the combination.
9. *Temperature Compensation.* VFD should automatically adjust current fallback based on output frequency for temperature protection of self-cooled fan-ventilated motors at slow speeds.
10. *PF.* VFD should automatically boost PF at lower speeds by including a full-wave diode bridge rectifier to maintain displacement PF at near unity regardless of speed or load.
11. *Manual Bypass.* VFD should include magnetic contactor to safely transfer motor between controller output and bypass controller circuit when motor is at zero speed. The motor shall operate at full speed when in bypass control.

Controller-off-bypass selector switch sets the mode, and indicator lights give indication of the mode selected. The unit must be capable of stable operation (starting, stopping, and running), with motor completely disconnected from controller (no load).

12. *Isolating Switch.* A non-load-break switch is required for service/isolation of the VFD and permits safe troubleshooting and testing, both energized and de-energized, while the motor is operating in bypass mode.
13. *Integral Disconnecting Means.* Each VFD must include factory-assembled, molded case circuit breakers with permanent instantaneous magnetic trips in each pole, with fault current interrupting the rating suitable for application and ampere ratings as indicated.
14. *Unit Controls.* Manufacturer-furnished, factory-installed VFD controls, including final control elements, must interface seamlessly with the HVAC system direct digital control system.

For fractional HP [kW] motors and smaller packaged HVAC equipment (≤5 tons [18 kW]), single-phase power is typically sufficient. Single-phase circuits may also be required to provide power for control systems and "interlock wiring" (see Chapter 13). For motors larger than 1 HP [0.75 kW] and for larger HVAC packaged equipment, three-phase (typically referred to as "polyphase") electrical service is normally required.

The NEMA defines 20 types of motor enclosures, which fall into two broad categories: "open" and "totally enclosed." Open motors have ventilation openings allowing for air cooling of the motor windings, while totally enclosed motors do not.

The most common open motor is the *open drip-proof* (ODP) type in which ventilation openings are positioned to keep particles and water from falling into the motor. Most motors found in commercial buildings are ODP motors. The water resistance of ODP motors can be improved by requiring "splash-proof" motors that add protection from material that may enter the motor from below, while "guarded" motors use screens or baffles to protect the motor from particle entry.

Totally enclosed motors are designed to prevent free exchange of air between the inside and the outside of the motor. The most common totally enclosed motor is the *totally enclosed fan-cooled* (TEFC) in which a fan on the opposite end of the motor from the load draws air over the case to provide cooling.

Pumps, fans, cooling towers, and so on utilize general purpose motors. But, for some applications, a "definite purpose motor" may be provided by equipment manufacturers. A definite purpose motor is any motor designed, listed, and offered in standard ratings with standard operating characteristics and mechanical construction, for use under service conditions other than usual or for a particular application. The definite purpose family of motors includes those used for refrigeration compressors, oil burners, unit heaters, packaged terminal air-conditioning units and heat pumps, computer room air conditioning units, steam boiler feedwater pumps, and many more.

All electrical motors must be rated for continuous duty at ambient temperature of 40°C and at an altitude of 3,300 ft [1,000 m] above sea level. Motor capacity and torque characteristics must be sufficient for the motor to start, accelerate, and operate its connected load at designated speeds, at installed altitude and environment, with indicated operating sequence, and without exceeding nameplate ratings or service factor. (Note: The need for starting torque when large fans that are powered by low horsepower motors, common for return air applications, can be a critical factor in motor selection.)

Table 1-4 summarizes the typical motor standard ("nameplate") horsepower (HP) ratings utilized for HVAC applications.

Designers should select motors on the following basis to minimize oversizing and loss of efficiency, while maintaining overall motor reliability:

1. If the imposed load brake horsepower is <6.0 [4.5 kW], select the first available motor with a nameplate rating *exceeding* the required brake horsepower. If the first available motor has a nameplate horsepower that exceeds the required brake horsepower by *less than 50%*, select the next available motor. For example, a 5.9 bHP [4.4 kW] load should be met with a 10 HP [7.5 kW] motor since a 7.5 HP [6 kW] motor provides only 27% more horsepower than the load. But, a 5.0 bHP [3.7 kW] load may be met with a 7.5 HP [5.6 kW] motor.
2. If the required brake horsepower is ≥6.0 [4.5 kW], select the first available motor with a nameplate rating exceeding the required brake horsepower [load kW]. If the first available motor has a nameplate horsepower that exceeds the required brake horsepower by *less than 30%*, select the next available motor.

Synchronous motor speed is its speed under "no load" conditions, and the actual speed under normal load conditions will typically be approximately 50 rpm lower. The synchronous speed of an induction motor depends on the frequency of the power supply and on the number of poles for which the motor is wound. The higher the frequency, the faster a motor runs. The more poles the motor has, the slower it runs. Generally, power supply frequency in the western hemisphere is 60 Hz, while Europe and the rest of the eastern hemisphere utilize 50 Hz power. Thus, motor synchronous speeds can be summarized as shown in Table 1-5.

Motors for HVAC applications are normally selected for a 1,800 rpm rotational speed at 60 Hz [1,500 rpm at 50 Hz]. Motors rated at higher rpm may be required for pumps with high pressure requirements, and motors rated at lower rpm may occasionally be needed for pumps or fans with low pressure requirements.

TABLE 1-4

Standard Electrical Motor Ratings (HP [kW])

Single Phase	Polyphase				
1/8 [0.1]	1 [0.75]	5 [3.7]	20 [15]	50 [37]	125 [90]
1/4 [0.18]	1.5 [1.1]	7.5 [5.5]	25 [18]	60 [45]	150 [110]
1/2 [0.37]	2 [1.5]	10 [7.5]	30 [22]	75 [55]	200 [150]
3/4 [0.56]	3 [2.2]	15 [11]	40 [30]	100 [75]	250 [190]

TABLE 1-5

Induction Motor Synchronous Speeds

Poles	Rpm at 60 Hz	Rpm at 50 Hz
2	3,600	3,000
4	1,800	1,500
6	1,200	1,000
8	900	750
10	720	600
12	600	500

Single-phase, constant speed motors may be any of the following types: *permanent-split capacitor; split phase; capacitor start, inductor run;* or *capacitor start, capacitor run.* Constant speed motors 1/12 HP [0.07 kW] and smaller shall be *shaded-pole type.* Two-speed motors should be *variable-torque, permanent-split-capacitor type.*

Variable speed single-phase motors should be *electronic commutation motors* (ECMs), i.e., brushless DC motors using a built-in inverter to convert AC power supplied to DC power and a magnet rotor needed to operate the motor. This type of motor is speed-controllable via internal circuitry down to about 20% of full speed (i.e., 80% or 5:1 turndown) via a controller output signal. ECMs are able to achieve greater efficiency in airflow systems than conventional *permanent split capacitor* (PSC) motors used in the past. Initially, silicon rectifier controllers were added to PSC motors to provide variable speed control. But, as the PSC motor speed was reduced, efficiency suffered, falling from 65% to 70% to as low as 12%. ECMs, on the other hand, maintain a high level (65%–75%) of efficiency over almost its full speed range. Additionally, unlike PSC motors, ECMs are not prone to overheating and do not require additional measures to offset the generation of heat.

Polyphase motors, located indoors or otherwise weather protected, should be in accordance with NEMA *Publication MG 1,* Design B, medium duty induction open ODP type with 1.15 service factor and Class F insulation, rated as "Premium" efficiency motors, meeting the requirements of DOE in the United States or the requirements of IEC 60034-30 IE3 ratings promulgated by the European Union. The use of premium efficiency motors reduces energy consumption by an average 6% (2%–10% depending on motor rating) over standard efficiency motors.

Multispeed polyphase motors with a 2:1 speed ratio, often used in cooling tower applications, may be *consequent pole, single winding type* in order minimize first cost. Multispeed polyphase motors with other than a 2:1 speed ratio should have separate winding for each speed.

Polyphase motors used with VFDs must comply with additional requirement(s):

1. Early VFDs produced a great deal of electrical "noise" that resulted in transient currents overheating motor windings and causing insulation failure. Additionally, the step frequency pattern produced by a VFD to emulate the AC power sine wave can impose HVs for short periods that can also result in breakdown of the motor insulation. To address these issues, motor windings must be copper magnet wire with moisture-resistant insulation varnish and be designed and tested to resist transient spikes, high frequencies, and short-time rise pulses produced by pulse-width modulation inverters. Motors controlled by VFDs must comply with NEMA *Publication MG 1,* Part 31, to provide windings capable of withstanding up to 1,600 peak volts with a rise time of 0.1 μs.
2. Harmonics produced by VFDs may produce transient currents along the motor shaft that ultimately may result in motor bearing failure. A relatively low cost measure to address this problem is to install a *shaft grounding ring* (SGR) to "bleed" current from the motor shaft to the motor casing, which is grounded. In effect, the SGR creates a low-resistance electrical path from the motor shaft to a ground that bypasses the bearings. It is recommended that the designer specify SGRs on motors that are 10 HP [7.5 kW] or larger.

   For large motors, replacement of bearings becomes hugely expensive, sometimes exceeding the cost of the motor. Therefore, NEMA recommends that, for motors 100 HP [75 kW] rating and larger ceramic bearings or electrical insulated bearings, to provide minimum 50 MΩ resistance at up to 1,000 VDC be incorporated.

With either of these measures, the bearings cease to be part of an electrical path to ground and pitting will not occur.

When using insulated bearings, it is also necessary to use shaft brushes. If insulated bushings are used without the brushes, there is no mechanism to remove voltages that accumulate on the motor rotor/shaft assembly. Eventually, the voltage will increase to the point where an electrical arc will occur around the insulated bearing or, even worse, to the motor windings. This particular problem is much more prevalent where the motor mechanically drives a load through an insulated belt; in this case, the insulated belt allows a static charge to accumulate on the insulated rotor similar to the operation of the old hand-cranked static electricity generators.

3. Polyphase motors located outdoors, including motors within cooling towers or evaporative coolers, should comply with the following:
   a. To provide the *minimum* level of weather protection, the motor must be a TEFC type with weather-resistant motor junctions, terminal box, and motor coating.
   b. To provide the *maximum* level of weather protection, the motor should be TEFC, "Cast Iron Construction Only" complying with the requirements of IEEE *Standard 45, Recommended Practice for Electric Installations on Shipboard,* and be classified as a "Marine Motor."
   c. For any motor installed outdoors that does not run for significant periods of time, condensation within the motor windings must be prevented by equipping the motor with internal cartridge or strip heaters that are energized during nonoperating periods to maintain a temperature inside the motor 5°F–10°F [3°C–6°C] warmer than the ambient air.

V-belt drives, consisting of belt and pulley assemblies, are used in HVAC applications (most commonly for fans) to perform two functions: transfer motor power to fans and to reduce the fan speed from motor speed to the required load speed. The power transfer requirement dictates the belt(s)' required load rating, while the sheaves (one on the motor and one on the load) dictate the speed change in direct ratio to their diameters. For example, a motor sheave half the diameter of the fan sheave will reduce the fan's speed to half the motor's speed.

V-belt drives should be specified to utilize American National Standards Institute (ANSI)/Rubber Manufacturers Association (RMA) "cogged" V-belts with properly selected motor pulley and drive sheaves. Generally, drives should be selected for the motor nameplate horsepower, *plus the recommended ANSI/RMA service factor (but, not less than 20%) to account for drive mechanical losses and potential load fluctuations,* in addition to meeting the ANSI/RMA required allowances for pitch diameter, center distance, and arc of contact.

1. V-belt drives rated as 1 HP [0.75 kW] and less may be provided with single *standard* V-belt.
2. Drives 1.5 HP [1.1 kW] and larger should be provided with *raw edge cogged* V belts, the number of belts necessary to transmit the required power with 95% minimum efficiency but in no case less than 2.
3. Belt drives for fans utilized as part of smoke control and/or smoke venting system must be rated for the motor nameplate horsepower, *plus 50% additional service factor,* in addition to the ANSI/RMA allowances for pitch diameter, center distance, and arc of contact and must have at least two belts.

Sheaves and pulleys should be *fixed pitch* type, statically and dynamically balanced. While *adjustable pulleys* are available, they are expensive, prone to early failure, accelerate belt wear, and are not recommended.

For personnel safety, *belt drive guards* are required for any exposed belt drive. Belt drive guards are also required for any belt drive installed within an air-handling unit that is designed to be personnel-accessible when in operation. Belt drive guards must comply with U.S. Occupational Health and Safety Administration (OSHA) and SMACNA requirements. Each guard must be designed with provisions to allow routine adjustment of belt tension, lubrication, and use of tachometer to measure the load rotational speed with the guard in place.

The following should be considered with regard to belt drives:

1. Drives should be installed with adequate provisions for center distance adjustment to accommodate belt stretch.
2. Centers should not exceed 2-1/2 to 3 times the sum of the sheave diameters nor be less than the diameter of the larger sheave.
3. The *angle of wrap* (arc) created by belt contact on the smaller sheave should not be less than 120°, computed by Eqs. (1-6) and (1-7).

$$\alpha 1 = 180 - 2\ \sin - 1\left[(R-r)/C\right] \tag{1-6}$$

$$\alpha 2 = 180 + 2\ \sin - 1\left[(R-r)/C\right] \tag{1-7}$$

where

$\alpha 1$ = Angle of wrap for the small pulley (degrees)

$\alpha 2$ = Angle of wrap for the large pulley (degrees)

R = Radius of large pulley (inches) [cm]

r = Radius of small pulley (inches) [cm]

C = Center-to-center distance of pulleys (inches) [cm]

4. Sheave ratios should not exceed 8:1. To determine pulley diameters to drive a load, use Eq. (1-8).

$$D_l = D_m (N_m / N_l) \tag{1-8}$$

where

D_l = Diameter of load sheave (inches) [cm]

D_m = Diameter of motor sheave (inches) [cm]

N_m = Speed of motor (rpm)

N_l = Speed of load (rpm)

5. The belt speed should not exceed 5,000 ft/min [1,500 m/min] nor be less than 1,000 ft/min [300 m/min]. A speed of 4,000 ft/min [1,200 m/min] is considered best practice. Belt speed can be calculated by Eq. (1-9).

$$\mathrm{Sb} = \pi dN/12 \ \text{(IP units)} \tag{1-9}$$

$$\mathrm{Sb} = \pi dN/100 \ \text{(SI units)} \tag{1-10}$$

where

Sb = belt speed (ft/min) [m/min]

d = pitch diameter of pulley (inches) [cm]

N = rotating speed of pulley (rpm)

6. Sheaves should be dynamically balanced when rim speeds in excess of 5,000 ft/min [25 m/s] occur.
7. A service factor of 1.2 is normal for belts operating between 16 and 24 h/day.

For pumps and fans using variable speed motors, *direct drive* motor connections are preferred. In this case, the pump or fan shaft is aligned with the motor shaft and the two are connected by use of a *shaft coupling*. There are several advantages to direct drive over the use of belt drives:

1. Lower first cost
2. Reduced maintenance cost by eliminating the need for periodic belt and sheave replacement
3. Reduced energy consumption due to eliminating the 3%–5% mechanical losses associated with belts and sheaves

Shaft couplings directly connect a motor shaft to the shaft of a fan or pump so that both the motor and the driven load rotate at the same speed. Couplings connecting to polyphase motors should be a flexible type capable of absorbing vibration and be rated for motor nameplate horsepower [kW] plus an additional 50% service factor. Couplings should be "drop-out" type to allow disassembly and removal without removing equipment shaft or motor.

Shaft coupling *guards* are required for each directed connected motor assembly and must comply with ANSI B15.1 and OSHA 1910.219. Guards should be easily removable for access and service and provided with openings for speed checks, and so on without removal.

Enclosures for disconnect switches, starters, VFDs, control panels, and any other panel enclosures housing electrical equipment needed by HVAC components are rated based on NEMA standard ratings. Panel enclosures must be suitable for the environment in which they will be installed, and HVAC designers should require NEMA-rated enclosures based on Table 1-6.

Finally, HVAC designers must be aware that coordination between them and electrical designers is required to prevent interference with electrical equipment by HVAC pipes, ducts, and so on, as dictated by code. Figure 1-4 provides a graphical summary of the separations required in installations.

TABLE 1-6

NEMA Enclosure Ratings for Electrical Equipment

NEMA Type	Environment Condition
1	Indoors only, dry, low dust, and noncorrosive environment
3R	Outdoors, weatherproof, and rainproof
4	Outdoors, watertight, and rainproof
4X	Same as 4 plus corrosion resistant (recommended for coastal environments)
7	Hazardous locations Class I, Groups A, B, C, or D
9	Hazardous locations Class II, Groups E, F, or G
12	Indoors subject to circulating nonhazardous dust or dripping noncorrosive liquids

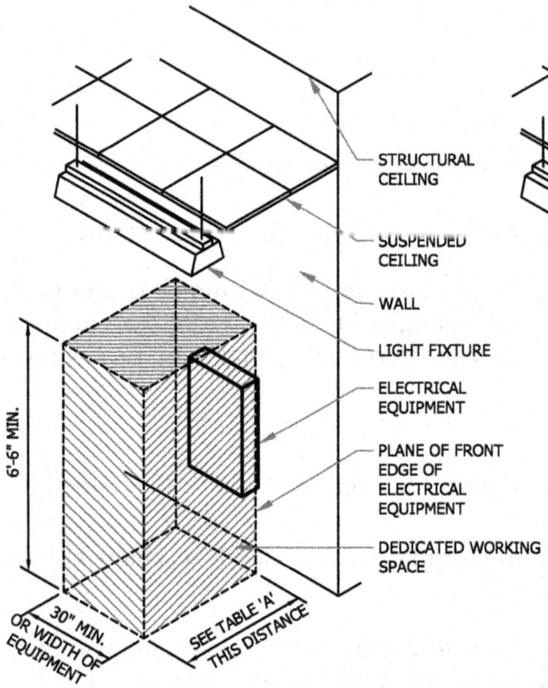

FIGURE 1

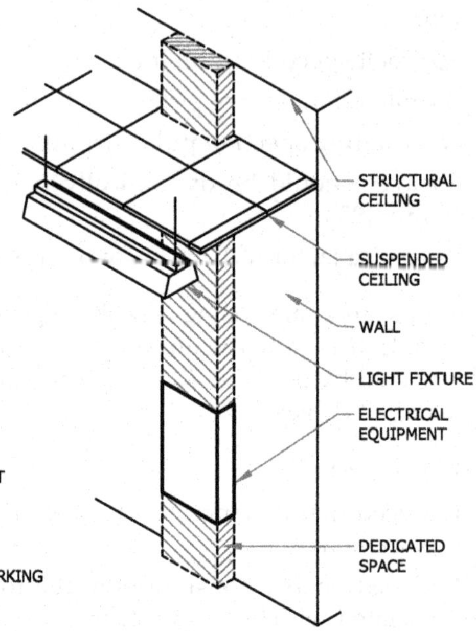

FIGURE 2

TABLE A - WORKING CLEARANCES				
VOLTAGE TO GROUND NOMINAL	CONDITION	MINIMUM CLEAR DISTANCE (FEET) 1	2	3
0-150		3	3	3
151-600		3	3 1/2	4

WHERE THE CONDITIONS ARE AS FOLLOWS :

1. EXPOSED LIVE PARTS ON ONE SIDE AND NO LIVE OR UNGROUNDED PARTS ON THE OTHER SIDE OF THE WORKING SPACE, OR EXPOSED LIVE PARTS ON BOTH SIDES EFFECTIVELY GUARDED BY SUITABLE WOOD OR OTHER INSULATING MATERIALS. INSULATED WIRE OR INSULATED BUSBARS OPERATING AT NOT OVER 300V SHALL NOT BE CONSIDERED LIVE PARTS.
2. EXPOSED LIVE PARTS ON ONE SIDE AND GROUNDED PARTS ON THE OTHER SIDE.
3. EXPOSED LIVE PARTS ON BOTH SIDES OF THE WORK SPACE (NOT GUARDED AS PROVIDED IN CONDITION 1), WITH THE OPERATOR BETWEEN.

NOTES:

1. THESE FIGURES ILLUSTRATE THE WORKING CLEARANCE AND DEDICATED SPACE AROUND ELECTRICAL EQUIPMENT AS REQUIRED BY NEC SECTION 110-26.
2. DEDICATED SPACE RUNS TO A HEIGHT OF 6'-0" ABOVE EQUIPMENT. DEDICATED SPACE CONTINUES THROUGH SUSPENDED CEILING OR UP TO STRUCTURAL CEILING. ANY FOREIGN SYSTEMS TO THE ELECTRICAL EQUIPMENT SHALL NOT RUN WITHIN THIS SPACE. (FIGURE 2)

FIGURE 1-4
Clearances for electrical equipment. (SW Standard Detail MR03.)

1.4 Definitions of HVAC Terms

The HVAC business is rift with specialized engineering terms, archaic terms from historical usages, acronyms, and so on. The following are some common HVAC terms and abbreviations used in the industry and within this text.

Absorption: The process by which one material extracts other substances from a mixture of gases or liquids.

Access Door: A door or panel provided in any structure, as in a duct, wall, and so on, or in a cooling or heating unit, to permit inspection and adjustment of the inside components.

Air, Ambient: The air surrounding any object.

Air Barrier: Those parts of the building envelope designed to minimize the level of infiltration of outdoor air into the building.

Air Change: The amount of air required to completely replace the air in a room or building, not to be confused with recirculated air.

Air Circulation: Natural or forced movement of air.

Air Cleaner: A mechanical, electrical, or chemical device (usually a filter) for removing dust, gas vapor, fumes, smoke, and other impurities from air.

Air Conditioner: A machine that controls the temperature, moisture, cleanliness, and distribution of air.

Air Conditioner, Unitary: An air conditioner consisting of several factory-made components within an insulated casing, normally including an evaporator (cooling) coil, a condenser and compressor combination, and controls and fans for distributing conditioned air; some also contain filters, heaters, and dampers.

Air Conditioning System: A system that controls the total indoor environment during all seasons under all comfort requirements.

Air Diffuser: An outlet which discharges supply air in a spreading pattern.

Air Handler: A fan or blower for moving air within a distribution system.

Air Handling Unit: An air handler, heating element and cooling coil, and other components in a cabinet or casing.

Air, Recirculated: Air drawn from a space and passed through the conditioner and discharged again into the conditioned space.

Air, Ventilation: The quantity of supply air drawn from outdoors needed to maintain the desired amount of oxygen and the quality of air within a designated space.

Anemometer: An instrument for measuring the velocity of air.

Atomize: To break up a liquid into a fine spray, as in oil burner nozzles and atomizing water humidifiers.

Baffle: A surface used to deflect or direct airflow, usually in the form of a plate or wall.

Barometer: A device for measuring atmospheric pressure.

Blow (Throw): In air distribution, the distance an airstream travels from an outlet to a point where air motion is reduced to a specific flow measured in feet per minute (terminal velocity).

Blower (Fan): An air-handling device for moving air in a distribution system.

British Thermal Unit (BTU): A unit for measuring heat quantity equal approximately to the amount of heat produced by the burning of an ordinary wooden kitchen match. More specifically, it is the amount of heat required to raise the temperature of one pound of water 1°F.

Building Core: That portion of the building interior space which is essentially unaffected by the outside conditions when the perimeter areas are maintained at desired conditions.

Building Perimeter: That portion of the building adjacent to the building envelope and affected by the outside conditions.

Capillary Tube: A small-diameter tube (metering device), whose bore and length are designed to permit the passage of a specific amount of liquid refrigerant or other liquid at a specific pressure drop.

Change of Air: Introduction of new air to a conditioned space, measured by the number of complete enclosed-space changes per unit time, usually per hour.

Chill: To reduce temperature moderately without freezing.

Coil: A heating or cooling element made of pipe or tubing, with or without interior or exterior surface extensions or fins.

Coil, Cooling: A commonly used term meaning the evaporator.

Comfort Zone: The range of temperature, humidity, and air velocity at which the greatest percentage of people feel "comfortable."

Compression: In a compression refrigeration system, a process by which the pressure of the refrigerant is increased.

Compression Ratio: The ratio of absolute pressures after and before compression.

Compressor: A device for increasing the pressure of heat-laden refrigerant vapors thus increasing the heat level (super-heating) within those vapors so that the heat contained can be released into the outside atmosphere.

Condensate: The liquid formed by the condensation of a vapor.

Condensation: The process whereby a vapor is changed to a liquid by removal of heat after its dew point (condensation temperature) is reached.

Condenser: A vessel, or arrangement of pipes or tubing, in which vapor is liquified by removal of heat.

Condensing Unit: A refrigerating machine consisting of one or more power-driven compressors, condensers, liquid receivers, and other components.

Conduction: The process of transferring heat along the elements of a substance, as from a tube to a fin.

Controller: A device for regulating a system, or part of a system, in normal operation. Such a device can be either manually operated or automatic.

Convection: The transfer of heat by movement of fluid or air.

Cooling Load (Design): Maximum probable net rate of heat gain in the space which would have to be extracted by the space cooling system to maintain some desired temperature condition, expressed in Btu/h.

Cycle, Refrigeration (Absorption): The complete cycle of a refrigerating agent through a system whereby the agent absorbs heat, transports it to a point where a second, unrelated, substance extracts the heat from the initial agent which is returned through the system to absorb more heat, and repeat the process. Energy source for such a cycle is heat.

Cycle, Refrigeration (Compression): Complete cycle of a refrigerant from a condensing unit (compressor-condenser) through a system of lines or tubing to the evaporator coil and back into the condensing unit; a cycle in which heat is absorbed in the evaporator and given up in the condenser.

Damper: A device for adjusting the amount of air flowing through an outlet, inlet, or duct.

Defrosting: The removal of accumulated ice or frost from a cooling element or coil.

Dehumidification: The reduction of water vapor in air by cooling the air below the dew point or by removal of water vapor from air by chemical means.

Dew Point: The temperature at which water vapor turns to liquid at a given pressure.

Draft: A current caused by the movement of air from an area of high pressure to an area of low pressure; usually considered objectionable.

Dryer: A device placed in a refrigeration system to attract and collect unwanted moisture which may be in the system.

Dry: To separate or remove liquid or vapor from another substance, such as moisture from a refrigerant.

Duct: A pipe or closed conduit made of sheet metal or other suitable material used for conducting air to and from an air-handling unit.

Economizer Cycle: Operation of a system during a cooling requirement without refrigeration due to low loading or favorable weather conditions.

Entrainment: The induction of room air into an airstream from an outlet. (Also, see Induction.)

Evaporator: That part (heat exchanger) of a cooling system in which refrigerant is vaporized.

Exfiltration: Airflow outward from an enclosed space through a wall, leak, membrane, and so on.

Expansion Coil: An evaporator (heat exchanger) constructed of bare or finned pipe or tubing in which direct expansion of liquid refrigerant occurs. It is usually known as an evaporator or an evaporator coil.

Expansion Valve: A thermostatically controlled valve, or metering device, for controlling the boiling point and flow of refrigerant to the cooling element (evaporator).

Fan (Blower): A device for moving air.

Fan Coil Unit: An air-handling unit especially designed to condition small spaces.

Filter: A device for removing dust particles from air or unwanted elements from liquids.

Filter Dryer: A combination of a liquid refrigerant line strainer and dehydrator.

Fluid: Any liquid or gas. In HVAC applications, typically, air, water, or steam.

Freezing Point: The temperature at which liquids will solidify or freeze upon removal of heat (e.g., the freezing point of water is +32°F [0°C]).

Furnace: That part of an environmental system which converts gas, oil, electricity, or other fuel into heat for distribution within a structure.

Gage (Gauge): An instrument for measuring temperatures, pressures, or liquid levels. Also, an arbitrary scale of measurement for sheet metal thicknesses, wire and drill diameters, and so on, liquid levels within boiler pipes, and tanks and other enclosures.

Gas, Fuel: An expanded hydrocarbon fluid used to provide heat in environmental systems.

Gas, Inert: A gas that neither undergoes nor causes chemical reaction nor change of form.

Gravity, Specific: Density measured on the basis of the known density of a given substance, usually air or water.

Grille: A covering for an opening through which air passes.

Heat Addition Rate: Rate at which the heating system adds heat to the conditioned space, expressed in Btu/h [W]. This would be approximately equal to the heating load only during design heating conditions, and then only at steady state.

Heat Exchanger: Usually considered to be that part of a furnace that transfers heat from burning fuel or electricity into air or other medium. Also, it is any device in which heat is absorbed by one fluid from another fluid, such as a condenser, evaporator, or boiler.

Heat Extraction Rate: Rate at which the cooling system removes heat from the conditioned space, expressed in Btu/h [W]. This would be approximately equal to the cooling load only during design cooling conditions and then only at steady state.

Heat Gain: Rate at which the conditioned space is heated by one or more sources, such as conduction through walls, solar radiation through glass, internal appliances, and so on. Expressed in Btu/h [W].

Heat, Latent: Heat energy which changes the form of a substance (e.g., ice to water) without changing its temperature. Expressed in Btu/h [W].

Heat Loss: Rate at which the conditioned space loses heat, such as through walls, glass, infiltration, and so on. Expressed in Btu/h [W].

Heat Pump: A reverse-cycle refrigeration system designed to perform both heating and cooling operations.

Heat, Sensible: Heat which changes the temperature of a substance without changing its form.

Heat Transmission: Any time-rate flow of heat; usually refers to conduction, convection, and radiation.

Heating Load (Design): Maximum probable net rate of heat loss from the space which would have to be made up by addition of heat from the space heating system to maintain some desired temperature condition, expressed in Btu/h [W].

Heating System, Warm Air: A heating system whereby warmed air is delivered from a heating unit through ducts and is discharged into a room from diffusers or grilles.

Heating System, Warm Air, Perimeter: A heating system characterized by diffusers or registers situated along outside walls particularly under windows or other areas having high rates of heat loss.

High Side: That part of a refrigeration system where refrigerant is under the greatest amount of pressure where heat is rejected (condenser section). This is the area of highest temperatures and highest pressures.

Humidifier: A device that adds moisture to warm air being circulated or directed into a space.

Humidistat: A device designed to regulate humidity input by reacting to changes in the moisture content of the air.

Humidity, Relative: The percentage of moisture in the air measured against the amount of moisture the air could hold at a given temperature. (e.g., since cold air is capable of holding less moisture than warm air, if the temperature drops and the moisture volume remains constant, the relative humidity will increase. Conversely, if the temperature rises and the moisture volume remains constant, the relative humidity will decrease.)

Induction: The entrainment of room air into an airstream from an air outlet.

Infiltration: Airflow inward into a space through walls, leaks around doors and windows, or through the building material used in the structure.

Insulation, Thermal: A material having a relatively high resistance to heat flow, used primarily to reduce heat loss and heat gain.

Liquid line: The tubing or piping carrying liquid refrigerant from the condenser or receiver or a refrigeration system to a pressure-reducing device.

Louver: A series of vanes that permit directional adjustment of airflow. They are usually installed on outside grilles or intake openings to impede water entry into ducts, elbows, and to minimize turbulence; they may also include indoor grilles to eliminate light penetration.

Low Side: Those parts of a refrigeration system where heat is absorbed at evaporator pressure; the cooling coil or evaporator with associated components; also commonly referred to as the suction side; this is the area of lowest temperature and lowest pressure.

Main: Pipe or duct for distributing fluids such as air, water, or steam to various branch ducts and collecting fluids from various branches.

Make-Up Air Unit: Unitary equipment which introduces outside air into a space and conditions it to offset air exhausted from that space.

Manometer: An instrument for measuring pressures by the difference in liquid volume.

Meter: An instrument for measuring rates of flow of energy or fuel over a given period.

Motor (Operator), Air (Pneumatic): An air-operated device normally used for opening and closing dampers and valves.

Multi-Zone System: An air-conditioning system designed to serve one or more areas having different heating/cooling/humidification requirements. Such a system can consist of more than one heating/cooling/humidification unit or can consist of a single unit controlled by dampers, bypasses, and thermostats.

Muffler, Noise Reducing: A device for the prevention or reduction of sound or vibration being transmitted from one space to another via air conditioning distribution systems or equipment.

Oil Separator: A device for separating oil or oil vapor from refrigerant.

Plenum Chamber: An air compartment connecting one or more ducts with the heating or cooling unit; any enlarged section within an air duct with several duct, grille, or diffuser connections.

Pneumatic: Any device actuated by air pressure is said to be "pneumatic."

Power roof ventilator: A motor-driven exhaust fan mounted above the roof on a roof curb.

Pressure, Absolute: That pressure which will register on a pressure gauge, plus atmospheric pressure present, for example, a gauge pressure of 68.5 psi [472 kPa] at sea level would represent an absolute pressure of 83.2 psi [574 kPa] (68.5 plus 14.7 equals 83.2). Absolute pressure is expressed as psia [kPa absolute].

Pressure, Atmospheric: Pressure resulting from the weight of the atmosphere at a given temperature; or, that pressure in the outdoors or within a space that is present as a result of the forces of nature with no artificially induced pressure change.

Pressure Drop: That pressure lost between any two points of a piping or duct system due to friction, leakage, or other reasons.

Pressure Static: Force (per unit area) expended against the walls of a container such as an air duct. Commonly, force is measured in heating and air-conditioning psi [kPa] or inches of water pressure [Pa].

Pressure, Total: A combination of both static and velocity pressures.

Pressure, Velocity: Force of air as it moves in an air duct.

Psychrometer: A thermometer-like instrument for measuring wet- and dry-bulb temperatures simultaneously.

Radiation: Transmission of heat or energy by electromagnetic waves, such as heat transmitted by an electric heater and absorbed by a disjoined substance.

Radiator: A mechanical device that transmits heat by the process of radiation.

Refrigerant: The fluid in a refrigeration cycle that absorbs heat at low temperatures and rejects heat at higher temperatures.

Register: A combination grille and damper assembly covering an air-supply outlet, designed to distribute air into a room.

Return Air: Air recirculated through a return air system to an air-handling unit or furnace.

Saturation: A condition which occurs when air contains all the moisture that can possibly hold at a given temperature.

Solar Load: Cooling load imposed by direct solar radiation through glass windows, doors, skylights, and so on, which is absorbed by the space contents. Although winter solar radiation will offset annual energy consumption for heating, it is not normally taken as a credit against (peak) winter design heating loads expressed in Btu/h [W].

Steam: Water that has turned to vapor as a result of the application of heat at a given pressure or the reduction of pressure at a given temperature.

Strainer: A device for withholding foreign matter from a flowing liquid or gas.

System, Duct: A series of tubular or rectangular sections, elbows, and connectors fabricated as a channel to carry air from one point to another.

Temperature: The thermal level a substance, °F [°C].

Temperature Range: A measurement of the difference in heat between two points of a system.

Temperature, Dry-Bulb: The temperature of the air at any given location indicated by an accurate thermometer and not influenced by outside interferences such as radiation or water.

Temperature, Wet-Bulb: The temperature of the air in a space or outdoors as measured by a thermometer whose bulb is covered with a wet tubular wick or cloth.

Thermometer, Dry-Bulb: An instrument for measuring the temperature of sensible heat in air and other substances.

Thermometer, Wet-Bulb: A dry-bulb thermometer with a wet wick; used for measuring the difference in air temperature due to the evaporation of moisture.

Thermostat: An instrument which reacts to changes in temperature, normally used to control heating and cooling equipment. A thermal-operated switch.

Throw: The horizontal or vertical distance that an airstream travels after leaving an air supply outlet before its force is reduced to a specified terminal velocity.

Ton (of Refrigeration): Removal of 12,000 Btu/h [3,517 W] from a given area.

Unit: An assembly for heating, cooling, dehumidifying, and/or ventilating.

Unit Heater: A direct-heating, factory-made assembly including a heating element, fan and motor, and directional outlets.

Vapor Barrier: A moisture-proof layer of material used to reduce moisture migration.

Ventilation: The process of supplying or removing air to or from a room or space either by natural or mechanical means.

Ventilation Load: Heating or cooling load imposed by introducing outside air into the building. Since ventilation air is normally taken in by the HVAC systems directly, these loads are generally considered to be on the system rather than on the conditioned space, expressed in Btu/h.

Water Barrier: Those parts of the building envelope designed to eliminate the direct entry of water into the building.

Wet-Bulb Depression: The difference between dry- and wet-bulb temperatures.

Zone: Group of rooms or building spaces which have sufficiently similar control and load dynamics as to be capable of control from a single thermostat location.

Bibliography

ASHRAE Handbook-Fundamentals, American Society of Heating, Refrigerating, and Air-Conditioning Engineers, Inc. (ASHRAE), Atlanta, GA, 2017.

Standard Practice-58, Pipe Hangers and Supports—Materials, Design, Manufacture, Selection, Application, and Installation, Manufacturers Standardization Society (MSS) of the Valve and Fittings Industry, Inc., Vienna VA, 2009.
HVAC Duct Construction Standards—Metal and Flexible, Sheet Metal and Contractor's National Association's (SMACNA), Chantilly, VA, 2006.
Publication MG 1, National Electrical Manufacturers Association (NEMA), Arlington, VA, 2016.
National Electrical Code, Standard 70, National Fire Protection Association (NFPA), Quincy, MA, 2017.

2

The Air-Conditioning Process

2.1 Introduction

Effective design and analysis of heating, ventilating, and air-conditioning (HVAC) systems to accomplish air-conditioning processes depends on knowledge of the properties of moist air and familiarity with the tools of psychrometrics. The industry now depends heavily on the analytical techniques outlined by Willis Carrier in 1911 when he published relationships for moist air properties and his psychrometric chart. In this chapter, we define certain air properties fundamental to HVAC design, present methods of determining these properties, and illustrate accepted methods of analyzing air-conditioning processes in HVAC systems.

The variables used throughout this text for the properties of air are as follows:

Gas Constant	
R_a	Gas constant for dry air, 53.352 ft-lbf/(lbm-°R) [287.042 J/kg-°K]
R_w	Gas constant for water vapor, 85.778 ft-lbf/(lbm-°R) [461.524 J/kg-°K]
R^*	Universal gas constant, 1,545.32 ft-lbf/(lbm-°R) [8,314.472 J/kmol-°K]
Temperature	
°R	Degrees absolute (Rankin) °R=°F+459.67
°K	Degrees absolute (Kelvin) °K=°C+273.15
°F	Degrees Fahrenheit
°C	Degrees Celsius
T	Absolute temperature, °R [°K]
T_{db}	Dry-bulb temperature, °F [°C]
T_{wb}	Wet-bulb temperature, °F [°C]
T_d	Dew-point temperature, °F [°C]
T_{adp}	Apparatus dew-point temperature, °F [°C]
t^*	Thermodynamic wet-bulb temperature, °F [°C]
Volume, Density, and Mass	
cf	Cubic feet [m³]
V	Volume, cf [m³]
V_a	Volume of dry air, cf/lb of dry air [m³/kg]
V_s	Volume of moist air at saturation, cf/lb of dry air [m³/kg]
D	Density of moist air, lb/cf of dry air [kg/m³]
m	Mass, lbm [kg]
m_a	Mass of dry air, lbm [kg]

(*Continued*)

m_w	Mass of water vapor, lbm [kg]
M	Molecular weight
Enthalpy	
h	Enthalpy of moist air, Btu/lb of dry air [kJ/kg of dry air]
h_a	Enthalpy of dry air, Btu/lb of dry air [kJ/kg of dry air]
h_s	Enthalpy of moist air at saturation, Btu/lb of dry air [kJ/kg of dry air]
h_w	Enthalpy of water vapor, Btu/lb of water[kJ/kg of water]
Pressure	
P	Total barometric pressure of moist air, psia [kPa]
P_w	Vapor pressure of water in moist air, psia [kPa]
P_a	Pressure of dry air, psia [kPa]
P_{ws}	Vapor pressure of water in saturated moist air, psia [kPa]
Humidity	
W	Humidity ratio of moist air, lb of water/lb of dry air [kg of water/kg of dry air]
W_s	Humidity ratio of saturated air, lb of water/lb of dry air [kg of water/kg of dry air]
u	Specific humidity, lb of water/lb of moist air [kg of water/kg of moist air]
RH	Relative humidity, %
Moles	
N	Number of moles of moist air
N_a	Number of moles of dry air
N_w	Number of moles of water vapor
X_a	Mole fraction of dry air
X_w	Mole fraction of water vapor

2.2 Air and Its Properties

2.2.1 Atmospheric Air

The atmosphere is composed of a mixture of many gases, contaminants, and water vapor. To provide a basis for system analysis and design, Ref. [1] gives the following approximate composition, by mass, of dry air, excluding contaminants:

Nitrogen	78.084%
Oxygen	20.948%
Argon	0.934%
Carbon dioxide	0.031%
Minor gases	*0.003%*
	100.00%

Standard atmospheric data are used by air-conditioning engineers to estimate the properties of air at various altitudes. The temperature of the air is assumed to decrease linearly with increasing altitude in the lower atmosphere, which is assumed to consist of dry air, which behaves as a perfect gas. At sea level, the standard temperature is taken to be 59°F [15°C] and the standard barometric pressure is 29.921 in. Hg or 14.696 psia [101.325 kPa]. Reference [1] lists standard atmospheric properties for altitudes up to 30,000 ft [10,000 m]. But, for any altitude, the barometric pressure may be computed by Eq. (2-1) or (2-2).

$$\text{IP units:} \quad P = 14.696\left(1 - 6.8754 \times 10^{-6} Z\right)^{5.2559} \text{ psia} \tag{2-1}$$

$$\text{SI units:} \quad P = 101.325\left(1 - 2.25577 \times 10^{-5} Z\right)^{5.2559} \text{ kPa} \tag{2-2}$$

where Z is the altitude in feet (meters).

2.2.2 Water Vapor

The properties of water vapor in air are important to air-conditioning processes. The maximum possible amount of water vapor in an air–water mixture depends on the temperature of the mixture at any given pressure. If the amount of water vapor in air is a maximum for the existing temperature and pressure, the condition of the air is referred to as *saturated.*

2.2.3 Terminology

The following terms are fundamental to the analysis of air-conditioning processes and systems:

Air density, D: density of a moist air sample of volume *V.*

Specific volume, V: reciprocal of density:

$$V = 1/D \tag{2-3}$$

Atmospheric pressure, P: pressure of the atmosphere at a particular elevation and temperature.

Humidity ratio, W: ratio of the mass of water vapor to the mass of dry air in a sample of moist air:

$$W = m_w / m_a \tag{2-4}$$

Specific humidity, u: ratio of the mass of water vapor to the total mass of the moist air sample

$$u = m_w / (m_w + m_a) \tag{2-5}$$

or

$$u = W/(1 + W) \tag{2-6}$$

Relative humidity, RH: ratio of the mole fraction of water vapor, in a given moist air sample, to the mole fraction, *of water vapor in saturated air at the same pressure and temperature:*

$$\text{RH} = (X_w / X_{ws})_{t,p} \tag{2-7}$$

Dew-point temperature, T_d: temperature at which the air with a given humidity ratio and pressure becomes saturated (i.e., at which "dew" or condensation occurs).

Wet-bulb temperature, T_{wb}: equilibrium temperature achieved by a thermometer with a wetted wick over which the sample airflows. It is an approximation of the "thermodynamic wet-bulb temperature," which is used to determine moist air properties. This parameter is discussed in detail later in the chapter.

Enthalpy, h: total heat content of a sample of moist air at a given temperature and pressure.

2.2.4 Basic Psychrometrics

Psychrometrics is the quantitative study of the thermodynamic properties of moist air and the use of these properties in analyzing air-conditioning processes. A useful relationship between relative humidity and humidity ratio can be derived from the perfect gas law:

$$PV = NRT \tag{2-8}$$

or

$$MPV = MR^*T \tag{2-9}$$

where R^* is the universal gas constant, V is the total mixture volume, and M is the molecular weight. The mass of air in a sample, then,

$$m_a = P_a V M_a / R^*T \tag{2-10}$$

and the mass of the water vapor in the air sample is

$$m_w = P_w V M_w / R^*T \tag{2-11}$$

and, therefore, utilizing Eq. (2-4), we have

$$W = m_w / m_a = P_w M_w / P_a M_a \tag{2-12}$$

Substituting the actual molecular weights for M_w and M_a yields

$$W = 0.622\ (P_w / P_a) \tag{2-13}$$

Since the barometric pressure, P, is the sum of the partial pressures of the dry air, P_a, and the water, P_w, the humidity ratio can be expressed in Eq. (2-13) as

$$W = 0.622[P_w / (P - P_w)] \tag{2-14}$$

2.3 Methods of Measurement and Analysis

2.3.1 Thermodynamic Wet-Bulb Temperature

To fully determine the thermodynamic state of moist air, one property in addition to the temperature and pressure must be known. Any of the typical properties discussed previously, such as specific volume, specific enthalpy, relative humidity, or humidity ratio, could be used, but as a practical matter, these properties are not easily measurable. Instead, the idealized process of *adiabatic saturation* is used to infer the state of moist air from a property called the *wet-bulb temperature.*

The adiabatic saturation process can be visualized conceptually as shown in Figure 2-1. Sample air with some humidity ratio W and at temperature T is introduced into an insulated chamber filled with water with a large surface contact area. If the air is allowed to reach saturation in its flow through the chamber, eventually the water in the reservoir

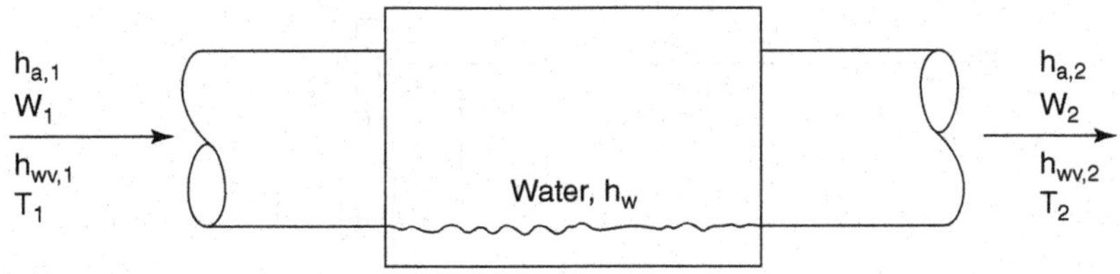

FIGURE 2-1
Adiabatic saturation chamber.

will reach this saturation temperature as well. The air is said to leave at its *temperature of adiabatic saturation* or its *thermodynamic wet-bulb temperature*. Thus, the thermodynamic temperature can be defined as "that temperature at which water may be evaporated into the air to bring it to saturation at that temperature."

We may show that the humidity ratio of moist air is a function of this wet-bulb temperature and other known properties of air and water by applying the first law of thermodynamics to the system illustrated in Figure 2-1. Since this is an open system at steady state, the rate of energy input equals the rate of energy output, or

$$h_{a1} + W_1 h_{w1} + (W_2 - W_1) h_1 = h_{a2} + W_2 h_{w2} \tag{2-15}$$

where h_1 is the enthalpy of water at its entering temperature, or

$$(\text{dry air in}) + (\text{water vapor in}) + (\text{water vapor added}) = (\text{dry air out}) + (\text{water vapor out})$$

Using this relation gives us

$$W_1 = \left[C_p (T_{wb} - T_1) + W_2 H_{fg}\right] / (h_{w1} - h_1) \tag{2-16}$$

where

h_{fg} = enthalpy difference between liquid water and saturated vapor at the temperature T_{wb}

h_{w1} = enthalpy of pure water at the temperature T_1

W_2 = humidity ratio of saturated air at temperature T_{wb}

C_p = specific heat of the saturated air

Thus, the humidity ratio of a given sample of air may be determined from the measured thermodynamic wet-bulb temperature T_{wb} and the tabulated properties of moist air and water vapor at T_{wb} and T_1.

2.3.2 The Psychrometer

The adiabatic saturation chamber is an impractical device for determining the thermodynamic wet-bulb temperature (it requires infinite wetted surface area in contact with the air). Instead, the thermodynamic wet-bulb temperature is approximated by the use of a device called a *psychrometer*, shown in Figure 2-2, which consists of two thermometers, one that measures the dry-bulb temperature and another, with a wetted wick at the bulb, which

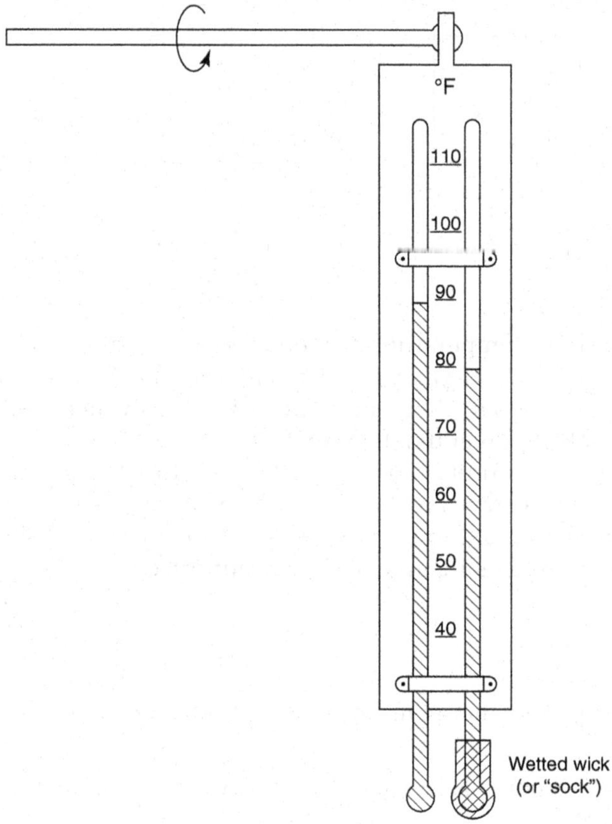

FIGURE 2-2
Typical "sling" psychrometer.

measures the wet-bulb temperature. Although the process is not adiabatic saturation, but is actually simultaneous heat and mass transfer from the wet bulb, the resulting difference between true thermodynamic wet-bulb temperature and measured wet-bulb temperature with a psychrometer is very small and may be ignored. In 1911, Dr. Willis Carrier presented formulas that make corrections for these differences, and the plots of these formulas are the basis for psychrometric charts used extensively in the HVAC industry to this day.

If the air is saturated, the wet-bulb, dry-bulb, and dew-point temperatures all have the same value. For non-saturated air, however, the dry-bulb temperature is that measured by an ordinary thermometer, the wet-bulb temperature is that achieved by a thermometer with a wetted wick over which the airflows, and the dew-point temperature is that temperature to which the existing air would have to be cooled before condensation would occur. Outside of the laboratory, direct methods of determining dew-point temperature are woefully inaccurate and the wet-bulb psychrometer remains the practical instrument in widest use.

2.3.3 The Psychrometric Chart

The psychrometric chart is designed to allow graphical analysis of air systems, eliminating the need to perform tedious calculations to compute the conditions of air at any point within the system. Figure 2-3 shows the psychrometric chart for normal temperatures and pressures in both IP and SI units. Charts are published by American Society of Heating,

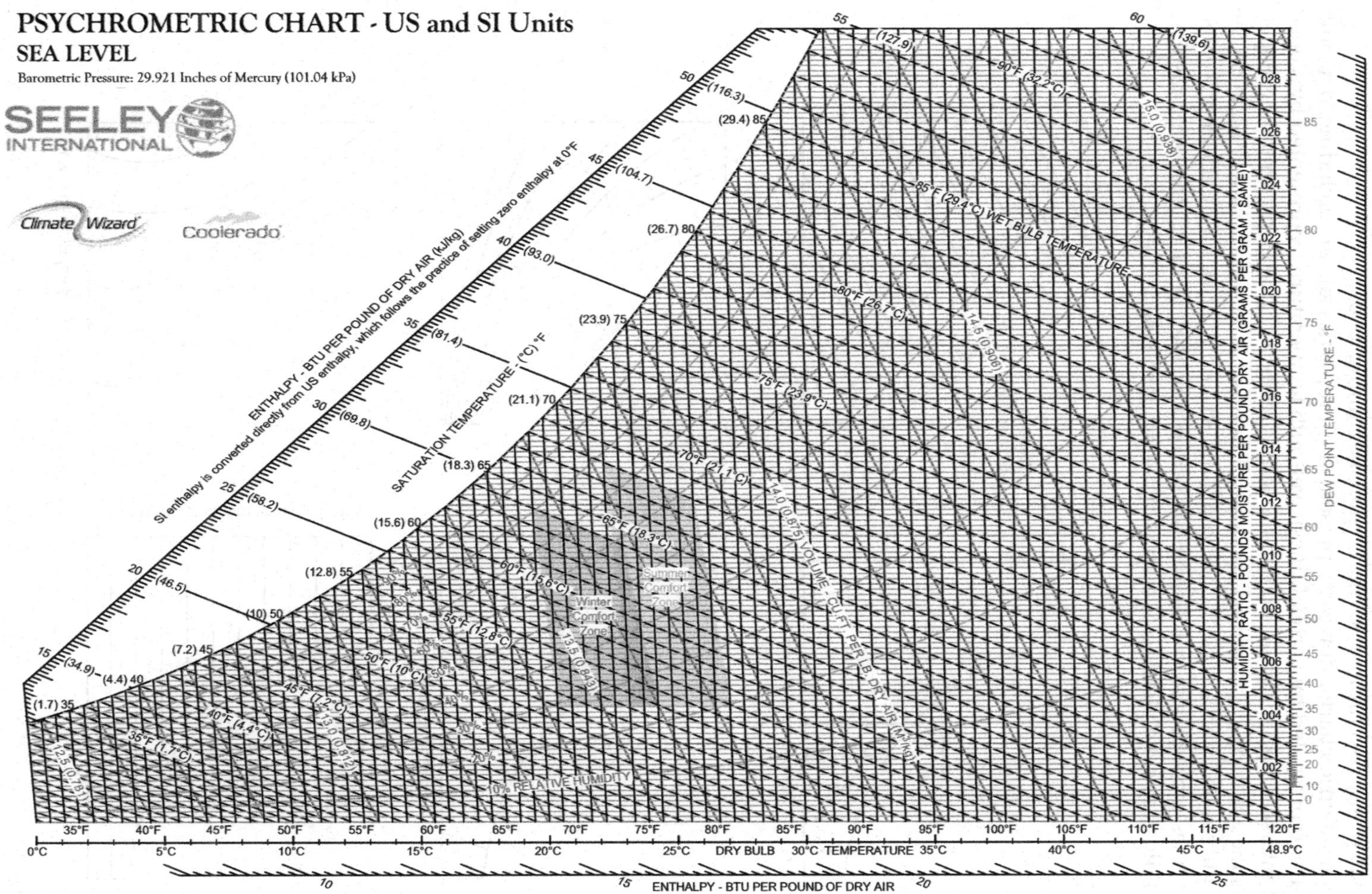

FIGURE 2-3
Psychrometric chart for normal temperatures, IP and SI units. (Coolerado, 4430 Glencoe St., Denver, CO 80216 http://www.coolerado.com/pdfs/Psychrmtrcs/0000Psych11x17US_SI.pdf).

Refrigerating, and Air-Conditioning Engineers and numerous other organizations, some of which have slightly different organization and configuration, so it is well to study the construction of any psychrometric chart before beginning its use.

All psychrometric charts, however, present the properties of air, including dry-bulb temperature (represented by vertical lines), wet-bulb temperature (represented by sloped lines that intersect the saturation line on the left), specific volume (represented by sloped lines), specific enthalpy (represented by sloped lines that intersect the linear scale on the left), humidity ratio (represented by horizontal lines), and relative humidity (represented in curved lines more or less paralleling the saturation line), in much the same fashion. Figure 2-4 illustrates the plotting of each of these properties separately.

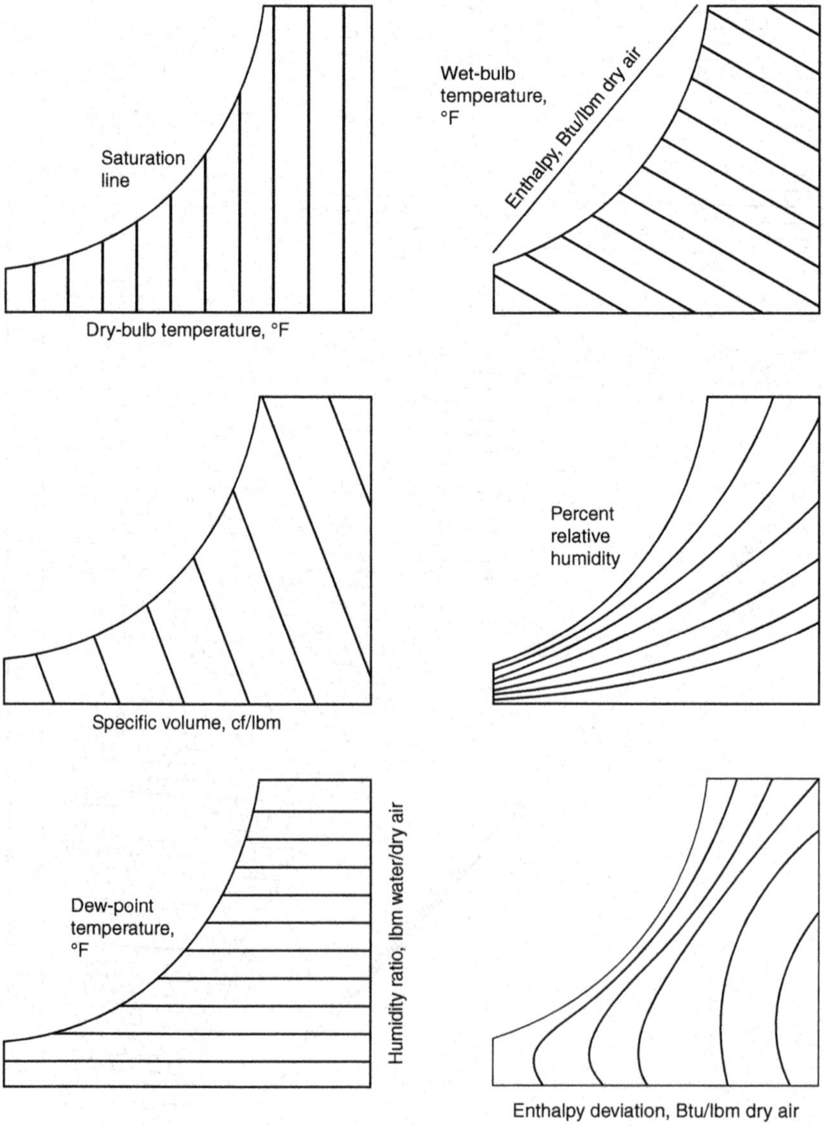

FIGURE 2-4
Primary moist air parameters on psychrometric chart.

2.4 Typical Air-Conditioning Processes

The use of the psychrometric chart is best explained by examples illustrating typical air-conditioning processes. Unless otherwise noted, the atmospheric pressure is assumed to be standard, which is accurate enough for normal applications up to 2,000 ft [600 m] (see Section 9.3.6 for required corrections for higher altitudes). The typical processes that can be easily evaluated via the psychrometric chart are sensible heating and cooling, latent cooling, dehumidification/humidification, evaporative cooling, and air mixing.

2.4.1 Sensible Heating or Cooling

Heating is the term used to describe the process of increasing the temperature of air without changing its moisture content. Therefore, this process is represented as a horizontal line on the psychrometric chart, as illustrated in Figure 2-5, and $W_1 = W_2$ as noted on the schematic drawing for the heating apparatus shown by Figure 2-6. In this schematic, heat

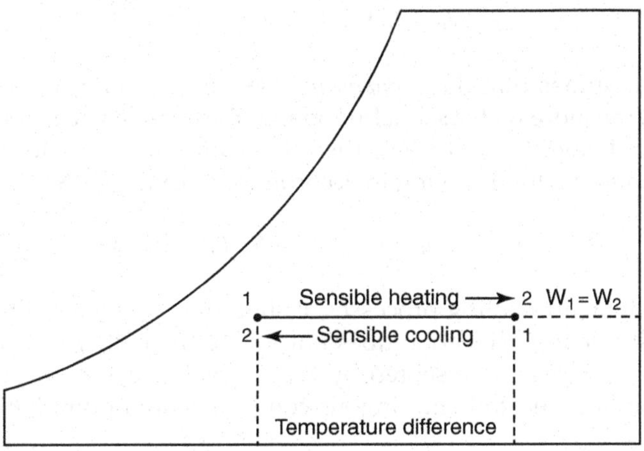

FIGURE 2-5
Skeleton psychrometric chart illustrating sensible cooling and heating.

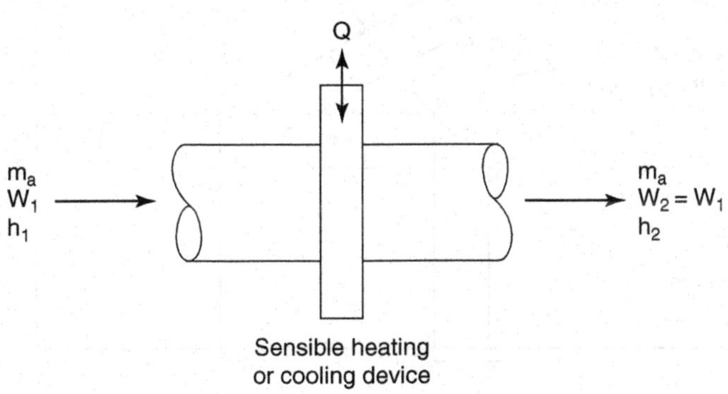

FIGURE 2-6
Schematic of heating or cooling devices.

is added by means of a heating device, such as a hot water or steam coil heating coil, direct resistance electric heating coil, and fuel-fired furnace.

"Sensible" cooling (i.e., cooling without humidity ratio change) is exactly the reverse of heating illustrated in Figure 2-5, so that the movement on the psychrometric chart is now from right to left. The schematic shown in Figure 2-6 would, for cooling, involve a chilled water coil or direct-expansion refrigeration coil.

2.4.2 Combined Sensible and Latent Cooling (Dehumidification)

If moist air is cooled below its dew point, the condensation of vapor from the air must be taken into account. This process is illustrated in Figure 2-7 as the air changes conditions from point 1 to point 3. Note that the total enthalpy change is now composed of two parts, a "latent" cooling enthalpy change, representing the energy released upon condensation of some of the vapor in the air from W_1 to W_3, and the sensible cooling enthalpy change, representing the energy removed in cooling the air from temperature T_1 to T_3.

Therefore, the energy removed in the cooling process, q, is the sum of the latent cooling from W_1 to W_3, plus the sensible cooling from T_1 to T_3, as defined by Eq. (2-17).

$$q = (h_1 - h_4) + (h_4 - h_3) = h_1 - h_3 \tag{2-17}$$

This formulation assumes that the *condensate* (condensed liquid) is drained away at the temperature T_2. In the more realistic likelihood that the condensate is cooled to the final air temperature before being drained away, then the energy removed to *subcool* the condensate must be added to the total cooling in accordance with Eq. (2-18)

$$q = h_1 - h_3 + C_{p,w}(T_2 - T_3)(W_1 - W_3) \tag{2-18}$$

As a final comment on air-cooling processes, it should be pointed out that the actual process of cooling with dehumidification does not follow the path 1 to 2 to 3, but it actually follows a path more or less represented by the dashed line 1–3. In practice, the important aspects are point 1 (the coil entering air conditions) and point 3 (the coil leaving air

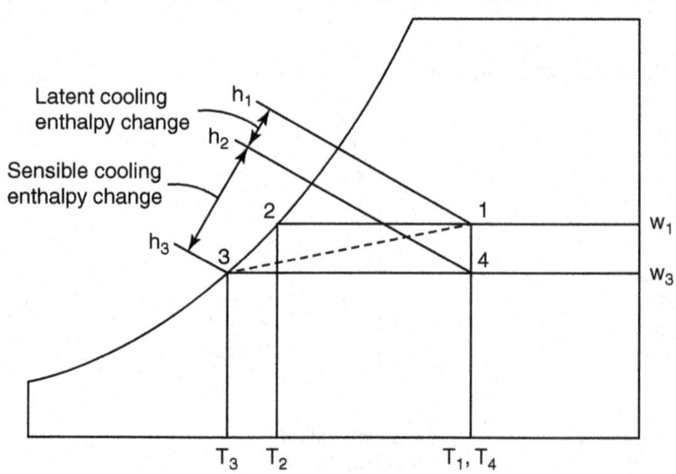

FIGURE 2-7
Skeleton psychrometric chart illustrating cooling with dehumidification.

conditions). The actual cooling line will vary based on the coil depth, fin spacing, water and air velocities, and so on, and its exact shape and form is immaterial to the designer.

2.4.3 Heating and Humidification

The process of heating and humidification is typical of that encountered in HVAC systems in colder climates since dry air in the winter time may allow the indoor air humidity condition to fall below acceptable levels (see Chapter 5).

This process is illustrated in Figure 2-8. Moisture may be added to the air after it is heated by a jet of steam (process from point 1 to point 2) or it could be added by a spraying water at the wet-bulb temperature. In the latter case, the process from points 3 to 2 is adiabatic (i.e., no heating occurs), but the dry-bulb temperature decreases while the moisture evaporates. The apparatus is shown schematically in Figure 2-9. Of course, many other paths on the psychrometric chart are possible to achieve the same results.

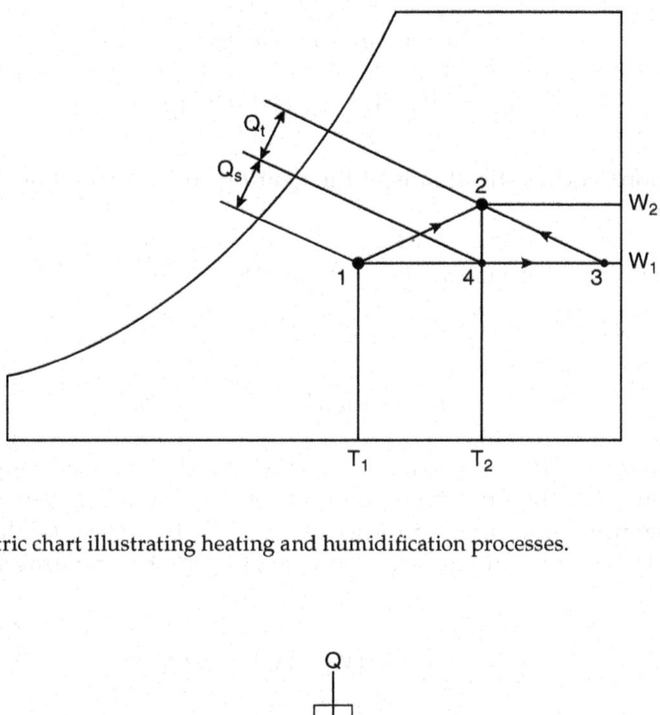

FIGURE 2-8
Skeleton psychrometric chart illustrating heating and humidification processes.

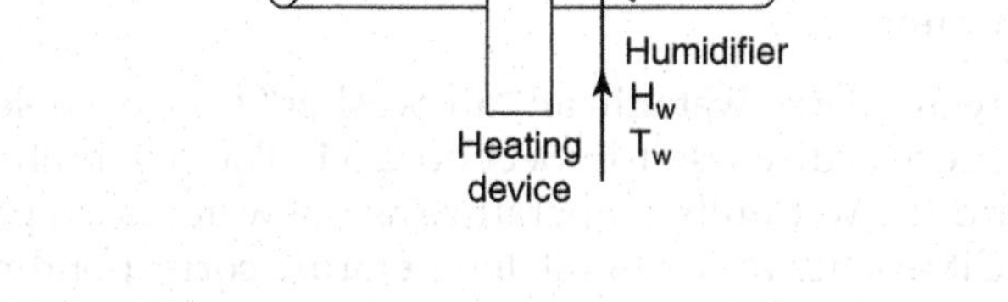

FIGURE 2-9
Schematic of heating and humidifying apparatus.

The final state of the moist air at point 2 in Figure 2-8 depends on (1) the heat added by the heating coil, (2) the moisture addition $W_2 - W_1$, and (3) the specific enthalpy of the injected water (or steam) h_w. An energy balance yields the following:

$$m_a h_1 + m_w h_w + q = m_a h_2 \tag{2-19}$$

and a mass flow balance gives

$$m_a W_1 + m_w = m_a W_2 \tag{2-20}$$

or

$$W_2 - W_1 = m_w / m_a \tag{2-21}$$

Therefore, the heating required is

$$q = m_a (h_2 - h_1) - m_w h_w \tag{2-22}$$

$$q = m_a (h_2 - h_1) - m_a (W_2 - W_1) h_w \tag{2-23}$$

If the humidification is adiabatic, that is, if the heating coil is turned off (or if steam injection is used), then

$$q = 0 - m_a (h_2 - h_1) - m_a (W_2 - W_1) h_w \tag{2-24}$$

or

$$h_2 = h_1 + h_w (W_2 - W_1) \tag{2-25}$$

Equation (2-24) shows that the final state of the air depends on the enthalpy of the injected water (or steam) and that the final state point lies on a straight line through h_1 with slope determined by the initial and final humidity ratios as well as the enthalpy of the injected water (or steam). This slope can be shown conveniently by rearranging Eq. (2-25) to form Eq. (2-26).

$$h_w = (h_2 - h_1)/(W_2 - W_1) = \Delta h / \Delta W \tag{2-26}$$

While the variable $\Delta h/\Delta W$ is plotted on many psychrometric charts as a "protractor," the slope of the line may be determined by using a "sensible heat ratio," (SHR) which has a one-to-one correspondence with the $\Delta h/\Delta W$ ratio.

2.4.4 Evaporative Cooling

The process of using recirculated water in an "air-washer" type of cooler approximates the adiabatic saturation process discussed in Section 2.3.1. The dry-bulb temperature of the air is decreased toward the wet-bulb temperature as the water is evaporated. The air may leave the cooler before it reaches the dry-bulb temperature corresponding to saturation, but in any case the process occurs at a constant wet-bulb temperature (except when makeup water to the cooler enters at a temperature different from the wet-bulb temperature of the air, in which case the deviation is usually insignificant). Air washer-type coolers or other evaporative coolers are used where hot, dry conditions lend themselves to the process and where the increase in humidity is tolerable.

2.4.5 Air Mixing

The adiabatic mixing of two streams of air is a common occurrence in HVAC systems. This process is illustrated schematically in Figure 2-10.

Application of the first law of thermodynamics to the adiabatic mixing process gives

$$m_1 h_1 + m_2 h_2 = m_3 h_3 \tag{2-27}$$

Application of the continuity equation to the airflow gives

$$m_1 + m_2 = m_3 \tag{2-28}$$

and to the moisture content of the air gives

$$m_1 W_1 + m_2 W_2 = m_3 W_3 \tag{2-29}$$

Now, if the variable m_3 is eliminated, we obtain

$$m_1 (h_1 - h_3) = m_2 (h_3 - h_2) \tag{2-30}$$

and

$$m_1 (W_1 - W_3) = m_2 (W_3 - W_2) \tag{2-31}$$

These equations may be expressed as

$$(m_1/m_2) = (h_3 - h_2)/(h_1 - h_3) = (W_3 - W_2)/(W_1 - W_3) \tag{2-32}$$

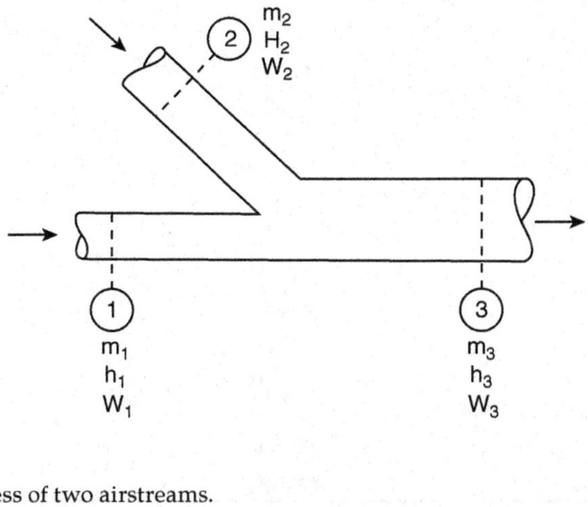

FIGURE 2-10
Schematic of mixing process of two airstreams.

An examination of this result shows that the mixed air condition (point 3) must lie on a straight line on the psychrometric chart connecting the two inlet conditions (points 1 and 2) and that point 3 divides the line into segments, with the same ratio to each other as the mass flow ratio of the dry air in the two inlet streams. Of course, Eq. (2-32) may be solved for h_3 and W_3:

$$h_3 = (m_1 h_1 + m_2 h_2)/(m_1 + m_2) \tag{2-33}$$

$$W_3 = (m_1 W_1 + m_2 W_2)/(m_1 + m_2) \tag{2-34}$$

2.5 Psychrometric Analysis of Complete Systems

2.5.1 Space Heating and Cooling Loads

The maximum probable net rate of heat loss from a conditioned space (room) which would have to be made up by addition of heat from the heating system to maintain some desired temperature and humidity conditions in the space is called the *heating load*. Similarly, the maximum probable net rate of heat gain in the space which would have to be extracted by the space cooling system to maintain some desired conditions of temperature and humidity is called the *cooling load*.

The results of the heating and cooling load calculations for a small, prototypical office building are shown in Table 2-1. We are concerned at this point with the cooling load, which is composed of two parts: a sensible cooling load component of 246,760 Btu/h and a latent cooling load component of 72,356 Btu/h.

TABLE 2-1

Results of Heating/Cooling Load Analysis of a Small, Prototypical Office Building (IP Units)

Building Element	Heating Load (Btu/h)	Cooling Load (Btu/h)
Sensible Heat Loss/Gain		
Roof	33,174	13,311
Walls	142,322	56,997
Floor	26,922	10,580
Glass (conduction)	63,482	25,256
Glass (radiation)	0	40,615
Infiltration	43,686	17,406
Appliances/equipment	0	25,598
People	0	56,997
Net sensible load	309,586	246,760
Latent Heat Loss/Gain		
People	0	56,997
Infiltration	0	15,539
Net latent load	309,586	72,356
Total net load	309,586	319,116

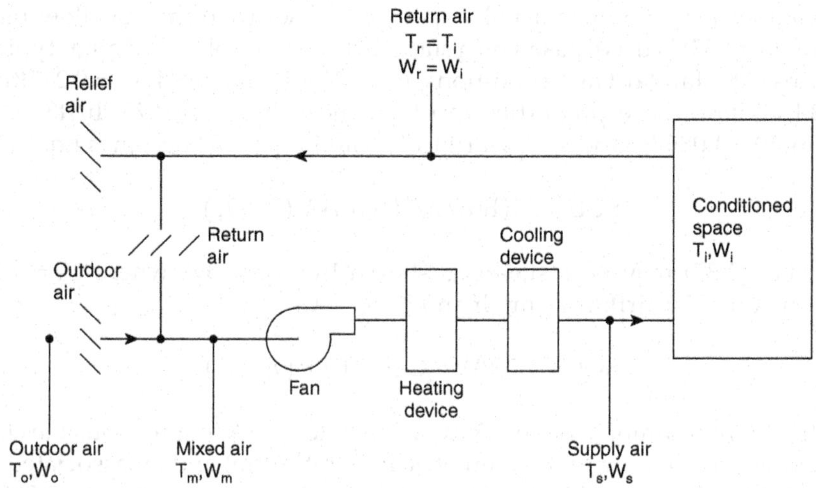

FIGURE 2-11
Schematic of air-conditioning system.

To maintain the occupied space of the building at desired conditions, the HVAC system must remove energy from the building at this same rate during peak-load conditions. This is accomplished by supplying air to the space at precisely the conditions that would cause the supply air to gain energy at exactly the rate at which it is entering the space from the sources of cooling load. If the supply air gains too much energy as it circulates through the space, the space is overcooled, and if it gains too little energy, the space becomes progressively warmer and more humid.

This process is shown in the schematic diagram for the system in Figure 2-11.

Applying the first law of thermodynamics to the space, we know that

$$\text{energy rate in} - \text{energy rate out} = \text{rate of change of space energy}$$

Since we want no change in the space conditions, the energy rate in must equal the energy rate out, or

$$\text{SCL} + \text{LCL} = m_a(h_r - h_s) \tag{2-35}$$

where SCL and LCL are the sensible and latent cooling loads for the space, respectively, and h_r and h_s are the enthalpies of the air within the room and of the supply air, respectively.

2.5.2 Sensible Cooling Load

It is customary practice to size the airflow within an HVAC system on the basis of the sensible cooling load, since the latent cooling load is typically much smaller. In this case, the sensible cooling load is related to the supply airflow by

$$\text{SCL} = m_a(\Delta h_{\text{air,sensible}}) = V_{\text{air}}/v_{\text{air}}(h_r - h_s) \tag{2-36}$$

or

$$\text{SCL} = m_{\text{air}}C_{p,\text{air}}(T_r - T_s) = V_{\text{air}}/v_{\text{air}}(T_r - T_s) \tag{2-37}$$

Since the volumetric flow rate is usually specified rather than the mas flow rate, a shortened form of Eq. (2-37) is used, based on a standard specific volume of air at typical cooling coil conditions. At "standard air" conditions, $P = 14.7$ psia, $T_{db} = 68°F$, $C_p = 0.24$ Btu/(lbm-°F), and v = 13.1 cf/lbm, these properties result in the value 1.1 [0.075 lb/ft3 × 0.241 Btu/lb°F ×60 min/h) = 1.085 (rounded upward to 1.1)] and Eq. (2-37) becomes Eq. (2-38).

$$\text{SCL} = 1.1(\text{Btu/h/°F-cfm})\ V(T_i - T_s) \tag{2-38}$$

In SI units, Eq. (2-37) may be restated as shown by Eq. (2-39), where C_p = 1.0048 kg K, $\rho = 1.202\,\text{kg/m}^3$, and V = airflow as m^3/h and T_i, T_s = °C.

$$\text{SCL} = 1.2(\text{kW/°C-m}^3\text{/h})\ V(T_i - T_s) \tag{2-39}$$

Of course, the more common case in HVAC system design is that the sensible cooling load is known and we are looking for a volumetric flow rate (cfm) and an associated supply air temperature that will satisfy the sensible cooling load. For this case, Eqs. (2-38) and (2-39) can be rearranged to yield the flow rate.

$$V(\text{cfm}) = \text{SCL}/\left[1.1(\text{Btu/h/°F-cfm})(T_i - T_s)\right] \tag{2-40}$$

and

$$V\left(\text{m}^3\text{/h}\right) = \text{SCL}/\left[1.2\left(\text{kW/°C-m}^3\text{/s}\right)(T_i - T_s)\right] \tag{2-41}$$

2.5.3 Latent Cooling Loads

An equation similar to Eq. (2-36) may be derived for the latent cooling loads. The change in enthalpy of the circulating air due to changes in moisture content must equal the latent cooling load, or

$$\text{LCL} = m_a(\Delta h_{\text{air,latent}}) \tag{2-42}$$

or

$$\text{LCL} = mw(\Delta h_{\text{water}}) = m_a\,(W_i - W_s)\ h_{fg,\text{water}} \tag{2-43}$$

or

$$\text{LCL} = V_{\text{air}}/v_{\text{air}}\,(W_i - W_s)\,h_{fg,\text{water}} \tag{2-44}$$

where h_{fg} is the latent heat of vaporization of water at the prevailing conditions. Using standard air conditions as defined in Section 2.5.2, and 1,056 Btu/lbm as h_{fg}, we may develop the following approximate relation:

$$\text{LCL} = 4{,}840(\text{Btu/h})/\left[(\text{cfm})(\text{lbm water/lbm air})\right]V(W_i - W_s) \tag{2-45}$$

In SI units, Eq. (2-45) may be stated as Eq. (2-46).

$$\text{LCL} = 0.83\ \left(\text{kW/°C-m}^3\text{/hr}\right)V\left(W_i - W_s\right) \tag{2-46}$$

2.5.4 Psychrometric Chart Representation of Space Conditions

The choice of supply airflow rate and temperature must be such that both the sensible and latent cooling loads are satisfied simultaneously. This process of supplying conditioned air to satisfy (or balance) the space loads may be represented on the psychrometric chart. To have both the latent and sensible loads satisfied simultaneously, the latent change in enthalpy per pound of supply air must always be in the same ratio to the sensible change in enthalpy per pound of supply air, no matter what supply airflow rate is chosen. If that were not the case, the product of the latent and sensible enthalpy changes and the airflow rate would not equal the latent and sensible loads or Eqs. (2-38) and (2-45) would not be satisfied. On the psychrometric chart, this means that the supply air condition must always lie on a line with the same slope as the ratio of latent to total cooling load. This line is called the SHR line.

Thus, the SHR line, representing the satisfaction of space loads, is drawn through the point representing the design space conditions and at a slope equal to the ratio of the sensible to the total space loads. The line is shown in the skeleton psychrometric chart of Figure 2-12. The guideline is drawn in first, then the SHR line is drawn parallel to it and through the space conditions point. It is important to note that any combination of supply air temperature and flow rate that satisfies the given latent and sensible loads will lie on the SHR line.

The question naturally arises as to what happens if the supply air is set at conditions that do not lie on the SHR line. Such would be the case if the loads were different than calculated (i.e., if the ratio of sensible to latent energy were different than design values) or if the air supply flow rate or temperature was different than anticipated due to improper sizing or changes in temperature settings. If the supply air conditions do not fall on the SHR line describing the actual loads, for any reason, the system cannot provide the desired occupied space conditions. Remedial action would involve a change in the supply air temperature, humidity, and/or flow rate. It is not generally possible to correct such a problem by changing the thermostat setting, since that action does not generally change the condition of the supply air but only the flow rate, so it would only be by happenstance that the new SHR line would intersect the supply air conditions point. More probably, such a change would simply result in a room temperature that is too cold in order to try to provide the proper humidity conditions.

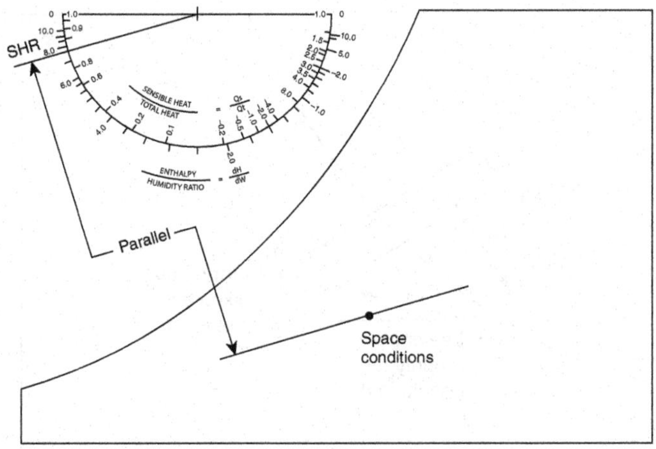

FIGURE 2-12
Skeleton psychrometric chart illustrating method of locating SHR line.

2.5.5 The Coil Line

Although it is not generally possible to draw a line on the psychrometric chart showing the actual process of cooling (or heating) of air within a coil, it is possible to plot points representing the average conditions of air entering and leaving the coil. The line connecting these two points is referred to as the coil line or the coil process line, even though the actual process between the points may not be a straight line.

An example of the coil line is given in Figure 2-13. The method of determining the sensible and latent cooling enthalpy changes is discussed in Section 2.4.3. Neglecting any subcooling of the condensate, the latent cooling heat exchange is $(h_m - h_p)$ and the sensible cooling heat exchange is $(h_p - h_s)$, where the subscripts m and s refer to the mixed (outside and return) air and supply air, respectively, while p refers to the point denoting the split between latent and sensible enthalpy components, as shown in Figure 2-13.

Thus, if the coil airflow rate is given at coil inlet conditions as V, the sensible load on the coil is

$$Q_{\text{sens}} = m(h_p - h_s) = (V_m / v_m)(h_p - h_s) \tag{2-47}$$

Of course, the approximate relation from Eq. (2-38) may be used if desired (and if the standard conditions exist):

$$Q_{\text{sens}} = 1.1\ V_m(T_{m,db} - T_{s,db}) \tag{2-48}$$

The latent coil load is given by

$$Q_{\text{lat}} = m(h_m - h_p) = (V_m / v_m)(h_m - h_p) \tag{2-49}$$

Again, the approximate relation from Eq. (2-45) could be used, as restated in Eq. (2-50)

$$Q_{\text{lat}} = 4{,}840\ V_m(W_m - W_p) \tag{2-50}$$

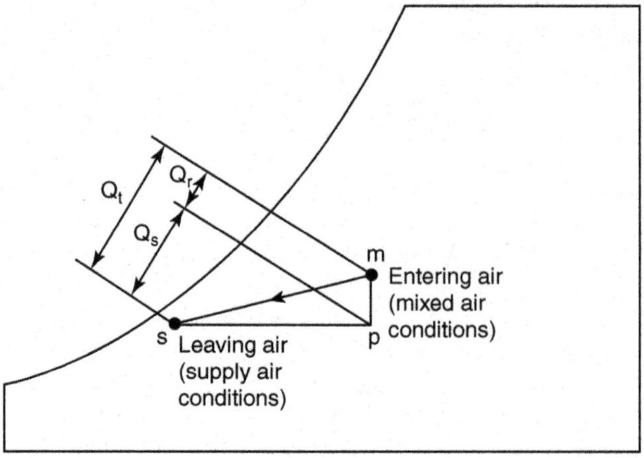

FIGURE 2-13
Skeleton psychrometric chart illustrating coil process line for cooling coil.

or in SI units, Eq. (2-46) may be used

$$Q_{lat} = 0.83\, V_m \left(W_m - W_p\right) \tag{2-51}$$

2.5.6 Coil Contact and Bypass Factors

Not all of the air that passes through a cooling coil is cooled equally. Some of the air passes nearer the coil surface than other portions of the flow and is thus cooled to a greater degree. The contact factor, C, is defined as that fraction of the air that is assumed to be cooled to the surface temperature of the coil. Conversely, the bypass factor, B, is the fraction of the air that does not "touch" the coil surface nor cooled at all. Then, by definition, $C+B=1$.

Of course, these two fractions of air are mixed, resulting in a leaving air temperature somewhere between the entering air temperature and the coil surface temperature. Also, the surface temperature of the coil is not the same at all points. However, it is possible to define an *effective surface temperature*, T_{es}, which is illustrated in Figure 2-14. This temperature, which is located on the psychrometric chart at the intersection of the saturated air curve and the extension of the coil process line, is also called the *apparatus dew point temperature*, T_{adp}.

It may be seen from Figure 2-14 that $C=b/(a+b)$ and $B=a/(a+b)$. If $C=1$, then $T_{es}=T_s$, or the leaving coil air temperature equals the effective surface temperature. Manufacturers of coils sometimes describe their coils as having certain cooling efficiencies, which is another way of defining the contact factor (for a certain flow rate and apparatus dew-point temperature).

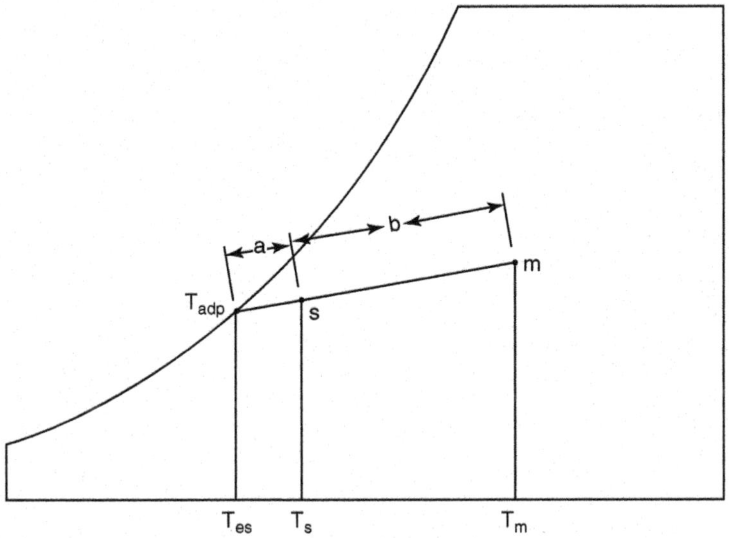

FIGURE 2-14
Skeleton psychrometric chart illustrating apparatus dew point temperature (effective surface temperature) for a cooling coil.

2.5.7 Psychrometric Analysis of Complete Systems

The typical procedure used for "designing" the system is to base the analysis on (1) the sensible and latent cooling loads of the space, (2) outside and inside air conditions, and (3) ventilation (outside air) requirements. Then, either the volumetric flow rate of return air into the space or the dry-bulb temperature of the air is selected, and the other is then determined. The range of typical values of the difference between room air and supply air temperature is usually 15°F–30°F [5°C–17°C] (with 20°F [11°C] most commonly selected), but good practice dictates consideration of such factors as the type and location of air supply outlets in sections of the supply temperature. The supply airflow rate must be within a range of acceptable values as well, it must be sufficient to provide "throw" to all parts of the room, and it must not be so great as to create discomfort due to drafts. In Chapter 6, we present details of the selection of air distribution components.

Bibliography

ASHRAE Handbook-Fundamentals, American Society of Heating, Refrigerating, and Air-Conditioning Engineers, Inc. (Atlanta, GA, 2017).

Fundamentals of Psychrometrics, James H. Carpenter, Carrier Air-Conditioning Company (Syracuse, NY, 1962).

3

HVAC Systems Concepts

3.1 Introduction

One of the primary objectives for the heating, ventilating, and air-conditioning (HVAC) systems designer, after having determined the net heating and cooling requirements, is to determine how these requirements are to be satisfied. Based on a number of criteria (see Chapter 4), the designer will select one or more systems for the building. Thus, it is necessary to have a thorough understanding of HVAC systems concepts, together with an understanding of each system's specific advantages and disadvantages.

Generally, HVAC systems are divided into two basic categories:

1. *Primary systems*: Central apparatus that provides the primary heating or cooling media, usually boilers and refrigeration machines, which consume energy directly to produce steam, hot water, chilled water, and so on
2. *Secondary systems*: Terminal systems that satisfy the individual space heating and/or cooling requirements, including air-handling units (AHUs), duct systems, heating and cooling coils, fan coil units (FCUs), and radiators

In some applications, the two systems are combined, such as in packaged units where compressors, heating equipment, and air handlers are combined in one unit. But in most cases, it is easier to separate the two system concepts and evaluate each more or less independently. In this chapter, then, we address the various types of *secondary systems* and the ways in which they can be applied.

Secondary HVAC systems must be designed with sufficient capacity to offset the peak heating and cooling loads imposed on them. However, peak-load conditions occur only for short periods throughout the year, and it is necessary that the secondary system be capable of responding to changes in these loads in the areas they serve. The "part-load" performance of the system is equally important (if not more so) with the "peak-load" performance of the system.

3.2 All-Air Systems

With these systems, air is circulated to the individual spaces by an AHU, either factory-fabricated or field-assembled (called "built-up" in the industry). Typically, the AHU consists of a supply air fan, heating and/or cooling coils, filters, dampers, controls, and,

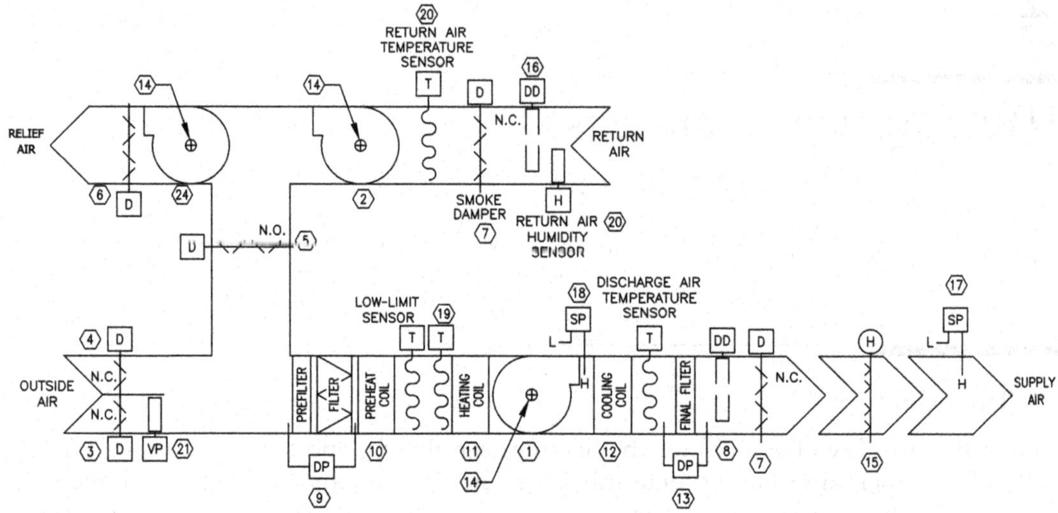

FIGURE 3-1
General air-handling unit schematic. (SW Standard Detail GC03.)

commonly, a return air fan. A general schematic of a larger capacity commercial air handler is shown diagrammatically in Figure 3-1. This schematic is designed to illustrate the many variations possible in air handler configuration, dictated by application and/or code requirements. Each of the elements shown is discussed in the following chapters.

The components of the AHU illustrated in Figure 3-1 are as follows:

Num.	Component	Num.	Component
1	Supply air fan(s).	12	Cooling coil.
2	Return air fan(s) option when an economizer cycle is used.	13	Final filter, generally required for health care and some laboratory applications. Arrange AHU for "draw-through" configuration.
2A	Relief air fan(s) option when an economizer cycle is used.	14	Fan inlet airflow monitoring station. Option for constant volume AHU, required when system is variable air volume (VAV).
3	Minimum outside air damper. Modulating damper when system is VAV and/or when an economizer cycle is used.	15	Primary humidifier, generally required for health care and some laboratory applications.
4	Modulating maximum outside air damper(s) required when economizer cycle is used.	16	Smoke detector. Required at return air connection to each floor or smoke compartment in the area served by the AHU.
5	Return air damper(s). May be two positions for constant volume systems, modulating when system is VAV and/or when an economizer cycle is used.	17	Supply duct static pressure sensor for supply air fan volume control.
6	Relief air damper(s). Not required unless system is VAV and/or when an economizer cycle is used.	18	Supply duct static pressure sensor to monitor fan high-limit discharge pressure.
7	Smoke detector required for systems rated for 15,000 cfm [7,000 L/s] or more.	19	Mixed air temperature sensor required when system uses a preheat coil and/or an economizer cycle.

(Continued)

Num.	Component	Num.	Component
8	Smoke detector required for systems rated for 2,000 cfm [950 L/s] or more.	20	Return air temperature humidity sensor (optional).
9	Main filter. Provide prefilter if minimum efficiency reporting value (MERV) rating of main filter exceeds 12.	21	Minimum outdoor airflow sensor required when demand control ventilation is used.
10	Preheat coil required if a minimum outdoor airflow results in mixed air temper ≤40°F [4°C] in winter and/or system is VAV.	22	Not used.
11	Heating coil required only for single-zone system.	23	Not used.

AHUs are typically defined as *blow-through,* as illustrated in Figure 3-2, when the supply fan is located upstream of the cooling coil and *draw-through* when the supply fan is located downstream of the cooling coil, as illustrated in Figure 3-3.

The *blow-through* AHU configuration is recommended for most applications for the following reasons:

1. Even though the draw-through configuration offers advantage in both AHU size and duct connections, it has the disadvantage that fan heat is added downstream of the cooling coil, meaning the actual coil temperature must be reduced to compensate. For chilled water cooling coils, this may mean that the chilled water supply temperature may also have to be reduced, increasing energy costs. Thus, a blow-through configuration is more energy efficient.
2. The coil pressure drop is imposed on the suction side of the fan in a draw-through configuration, which may result in fan performance problems.

But, in some health care applications, a *final filter* is required to be installed downstream of the fan(s) and cooling coil of the AHU, which is most easily achieved in a blow-through configuration. However, in this configuration, the supply air leaving the cooling coil is at or near saturation, and the pressure drop through the high MERV filter (see Chapter 10) can result in enough reduction in the partial pressure of water vapor to result in condensation within the filter. *Thus, the filter gets wet and stays that way.* To avoid this problem, while

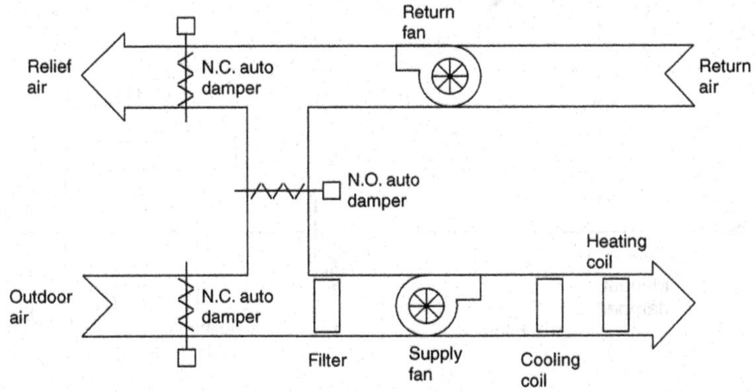

FIGURE 3-2
Blow-through air-handling unit configuration.

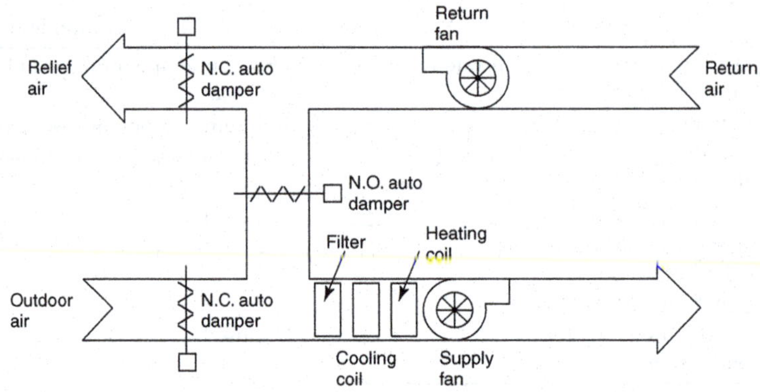

FIGURE 3-3
Draw-through air-handling unit configuration.

providing even airflow across the entering side of the final filter bank, the use of a supply air fan array (see Chapter 9) in a draw-through AHU configuration is preferred. The fan heat added downstream of the cooling coil provides just enough reheat to "move" the supply air condition away from the saturation line far enough to avoid the condensation problem within the final filter.

3.2.1 Single-Zone Systems

The *single-zone system* is perhaps the simplest and most common type of all-air system, as illustrated in Figure 3-4.

This system consists of an AHU with heating and cooling coils controlled in sequence in response to the room conditions as indicated by the room thermostat. In cooler, dryer climate zones, the heating coil is typically installed upstream of the cooling coil and the single-zone system operates as a *changeover system*, providing heating or cooling, but not both simultaneously. However, in hot, humid climate zones, the heating coil must be installed downstream of the cooling coil and controlled to allow the use of reheat (simultaneous heating and cooling) when humidity control is needed.

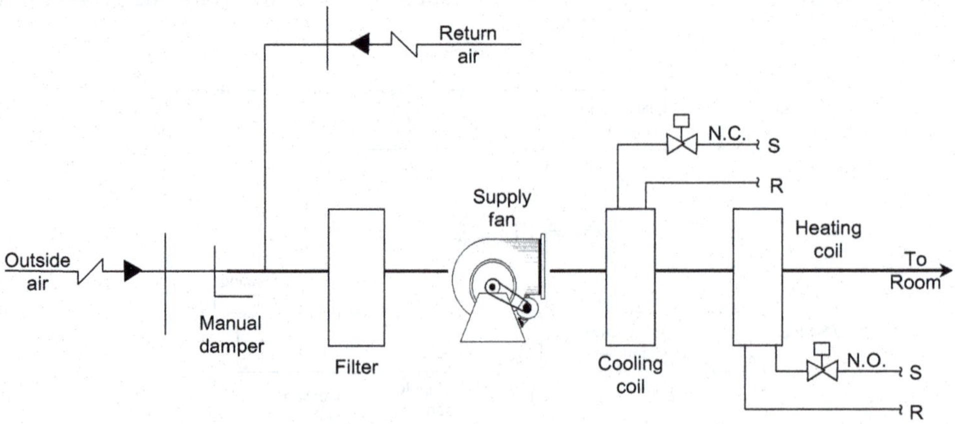

FIGURE 3-4
Single-zone system schematic.

This type of system is typically configured for *constant air volume* (CAV) in smaller capacities but is required by most building codes to operate in a VAV mode when the required fan capacity exceeds a specific motor size.

The maximum *sensible cooling load* (SCL) or maximum *heating load* (HL), as shown in Chapter 2, dictates the design system airflow as the greater airflow computed by Eq. (3-1) or (3-2) in IP units.

$$V_c = \text{SCL} / \left[1.1 \times (T_r - T_c)\right] \tag{3-1}$$

where

SCL = sensible cooling load, Btu/h
T_r = room temperature, °F
T_c = maximum cooling coil discharge temperature, °F
V_c = cooling air quantity, cfm

and

$$V_h = \text{HL} / \left[1.1 \times (T_h - T_r)\right] \tag{3-2}$$

where HL is the sensible heating load in Btu/h and T_h is the maximum heating coil discharge temperature in °F.

In SI units, V_c and V_h are computed in accordance with Eq. (3-1a) and (3-2a).

$$V_c = \text{SCL} / \left[1.2 \times (T_r - T_c)\right] \tag{3-1a}$$

where

SCL = sensible cooling load, kW
T_r = room temperature, °C
T_c = maximum cooling coil discharge temperature, °C
V_c = cooling air quantity, m^3/s

and

$$V_h = \text{HL} / \left[1.2 \times (T_h - T_r)\right] \tag{3-2a}$$

where HL is the sensible heating load in kW and T_h is the maximum heating coil discharge temperature in °C.

The final airflow for the unit is selected as the larger of V_c or V_h. During the part-load conditions, the supply airflow remains constant and T_c or T_h is varied to maintain the desired room temperature T_r.

Obviously, the application of this system is best in a space that is large enough to warrant its own single AHU, such as a computer room and a single large area, but it is often applied to serve a number of small areas with *very similar* load conditions. The difficulty in applying this type of system in a multiroom environment is that these rooms, typically, will have dissimilar load conditions at different times and, consequently, one space or the other cannot be satisfied if there is a single controlling thermostat. Spaces with different orientations and/or internal load conditions will require different responses by the air system in order to maintain conditions. This type of system, with only a single point of control, cannot respond properly to multiple (different) conditions. To mitigate this issue, there are control modes that include monitoring of each space and then adjusting the single-zone unit performance in attempt to "minimize discomfort" (see Chapter 13).

The advantage of this type of system is that it is simple and relatively inexpensive to install. The limitation, again, is that it can be satisfactorily applied only to a single space, or a number of spaces with very similar load conditions. At the part-load conditions, this system reacts well, supplying only the heating or cooling energy needed, unless there is a need to reduce humidity in the space(s) and the system is configured to provide reheat.

For large single-zone systems (such as used in gymnasiums and auditoriums), the International Energy Conservation Code requires the use of VAV to reduce fan energy consumption if the supply fan motor exceeds 5 hp [3.75 kW]. This may be accomplished by incorporating a variable frequency drive (VFD) or other types of motor control so as to limit the fan motor demand to not more than 30% of the design value at 50% of design airflow. The control sequence for these systems becomes more complex and requires integrating fan speed with the basic space temperature control (see Chapter 13).

3.2.2 VAV Systems

Since the late 1970s, the *VAV system* has become the most widely applied type of multiple-zone system. Basically, the VAV AHU delivers *cooling only* and varies the supply air volume to deliver more or less air to a particular space based on space requirements. (This is similar to the air-conditioning unit in your car, where you turn the fan speed up or down to satisfy your need for cooling.)

The simple VAV system central apparatus consists of a supply fan, the cooling coil, filters, and dampers as shown in Figure 3-5. In some climates, a heating/preheating coil may also be required to provide freeze protection under reduced airflow conditions.

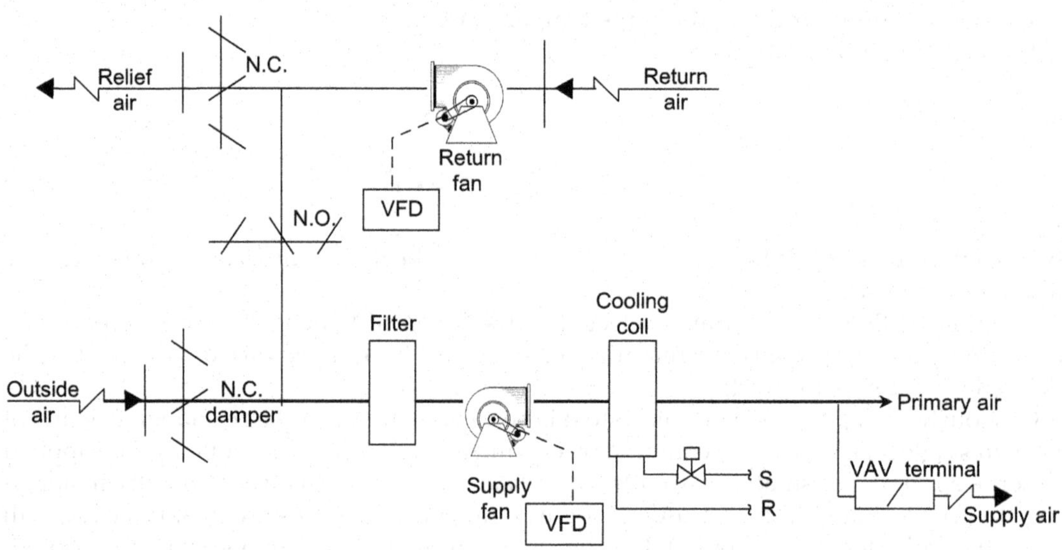

FIGURE 3-5
Variable air volume system schematic.

VAV terminal units are controlled from room temperature sensors to vary the volume of the air to that particular space (see Chapter 6). Fan static pressure controls are provided so that, as the volume of air in each space is reduced (based on load conditions), the fan will respond to the resulting increase in static pressure in the duct system by slowing the fan speed via a VFD. This type of system works very well in core areas of office buildings that usually require cooling only.

Since this system provides only cooling, the system airflow, V, is dictated by the maximum SCL, as shown in Eq. (3-3) and Eq. (3-3a).

$$\text{In IP units,} \quad V = V_c = \text{SCL} / \left[1.1 \times (T_r - T_c)\right] \tag{3-3}$$

$$\text{In SI units,} \quad V = V_c = \text{SCL} / \left[1.2 \times (T_r - T_c)\right] \tag{3-3a}$$

During the part-load conditions, T_c remains constant and V is varied to maintain the desired room temperature, *Tr*.

3.2.3 Dual–Duct Systems

Until about 1970, the *CAV dual–duct system* was considered *the* best of the all-air systems for buildings that had a large number of spaces that needed individual heating and cooling. For example, many large multi-tenant office buildings were conditioned by dual–duct systems.

Basically, the dual–duct system central apparatus consists of a common supply fan delivering air that is split between a heating coil ("hot deck") and a cooling coil ("cold deck") arranged in parallel. Dual–duct terminal units, one for each control zone, blend the two airstreams (one hot and one cold) to yield a blended supply air temperature that satisfies the room, either heating or cooling, as indicated by the room load. Effectively, each terminal unit serves as a single-zone unit.

The maximum SCL normally dictates the system airflow. Since all terminal units are wide open to the cold deck under these conditions, the flow may be dictated by Eq. (3-3) or Eq. (3-3a), as applicable.

At any part-load condition, the fan volume flow obviously remains constant and is equal to the sum of the hot and cold deck flows:

$$V = V_h + V_c \tag{3-4}$$

To determine these individual flows, it is necessary to consider the blending that is occurring and the loads imposed on the cold deck by the hot deck and vice versa.

At any time other than all spaces being at peak cooling load, there is "blending" at the mixing boxes in a double-duct system. This blending amounts to reheat and represents an increased heating and cooling load imposed on the respective coils.

With two airstreams at two different temperatures, we can evaluate the system performance using the following definitions:

V total airflow through the system, cfm [L/s]

V_h hot deck airflow, cfm [L/s]

V_c cold deck airflow, cfm [L/s]

V_r return airflow, cfm [L/s]

V_o outdoor (ventilation) airflow, cfm [L/s]
Q_h heating coil (hot deck) load, Btu/h [kW]
Q_c cooling coil (cold deck) load, Btu/h [kW]
T_h hot deck supply air temperature, °F [°C]
T_c cold deck supply air temperature, °F [°C]
T_r room (and return) air temperature, °F [°C]
T_o outdoor air temperature, °F [°C]
T_m mixed air temperature, °F [°C]
Q_{hr} sum of all room heating loads, Btu/h [kW]
Q_{cr} sum of all room cooling loads, Btu/h [kW]

Equations (3-5)/(3-5a) and (3-6)/(3-6a) can be used to express the load conditions at both the heating coil and the cooling coil, as follows:

$$\text{In IP units,} \quad Q_c = 1.1\ V_c(T_m - T_c) \tag{3-5}$$

$$\text{In SI units,} \quad Q_c = 1.2\ V_c(T_m - T_c) \tag{3-5a}$$

and

$$\text{In IP units,} \quad Q_h = 1.1\ V_h(T_h - T_m) \tag{3-6}$$

$$\text{In SI units,} \quad Q_h = 1.2\ V_h(T_h - T_m) \tag{3-6a}$$

Since, from Eq. (3-5), the total airflow is equal to the sum of the hot and cold deck airflows, the following relationships are valid:

$$V_c = V - V_h \tag{3-7}$$

$$V_h = V - V_c \tag{3-8}$$

By applying a first-law balance to both the hot and the cold decks, we obtain

$$Q_c = Q_{cr} = 1.1\ V_c(T_m - T_c) \tag{3-9}$$

$$Q_h = Q_{hr} = 1.1\ V_h(T_h - T_m) \tag{3-10}$$

Concentrating for the moment on the cold deck, Eqs. (3-6) and (3-9) can be substituted in Eq. (3-10) to yield the Eq. (3-11), in IP units:

$$1.1\ V_c(T_m - T_c) = Q_{cr} + 1.1\ (V - V_c)\ (T_h - T_m) \tag{3-11}$$

Therefore, simplifying this relationship yields

$$V_c = \left[Q_{cr} +\ 1.1\ V(T_h - T_m)\right] / \left[1.1(T_h - T_c)\right] \tag{3-12}$$

Similarly, for the hot deck,

$$V_h = \left[Q_{hr} + 1.1\ V(T_m - T_c)\right] / \left[1.1(T_h - T_c)\right] \tag{3-13}$$

where

$$T_m = (V_r T_r + V_o T_o)/V \tag{3-14}$$

There are basically two types of CAV dual–duct systems: the *traditional dual–duct* system described earlier and the *reheat dual–duct* system, in which the single hot deck heating coil is replaced by reheat coils at each terminal unit. *Neither of these systems are used today because they are "energy hogs" that consume excessive fan energy and energy for reheating.* But designers may encounter during this type of system during renovation projects.

The *VAV dual–duct* system, typically referred to as the *sequenced-air dual–duct* system or the *two-fan dual–duct* system, shown in Figure 3-6, has been developed to eliminate the energy burdens imposed by the older CAV system configurations.

The central apparatus of the sequenced-air dual–duct system consists of a cooling coil and "cold deck" fan, a heating coil and a "hot deck" fan, and an optional return fan. The system is arranged so that the hot deck operates only with return air. The cold deck, however, is capable of using 100% outdoor air for cooling under "enthalpy control." Traditional dual–duct systems simply mix cold air with hot air to deliver air to the room at a temperature to maintain desired conditions. With the sequenced-air concept, the two airstreams are delivered to the room separately: cold air in varying quantities to satisfy the cooling requirements and hot air in varying quantities to satisfy the heating needs. Blending of the two airstreams is done only if the total airflow tends to fall below the minimum required for the space. Due to the varying volumes of hot and cold air required throughout the system, both the hot deck and the cold deck must have 100% VAV capacity and control.

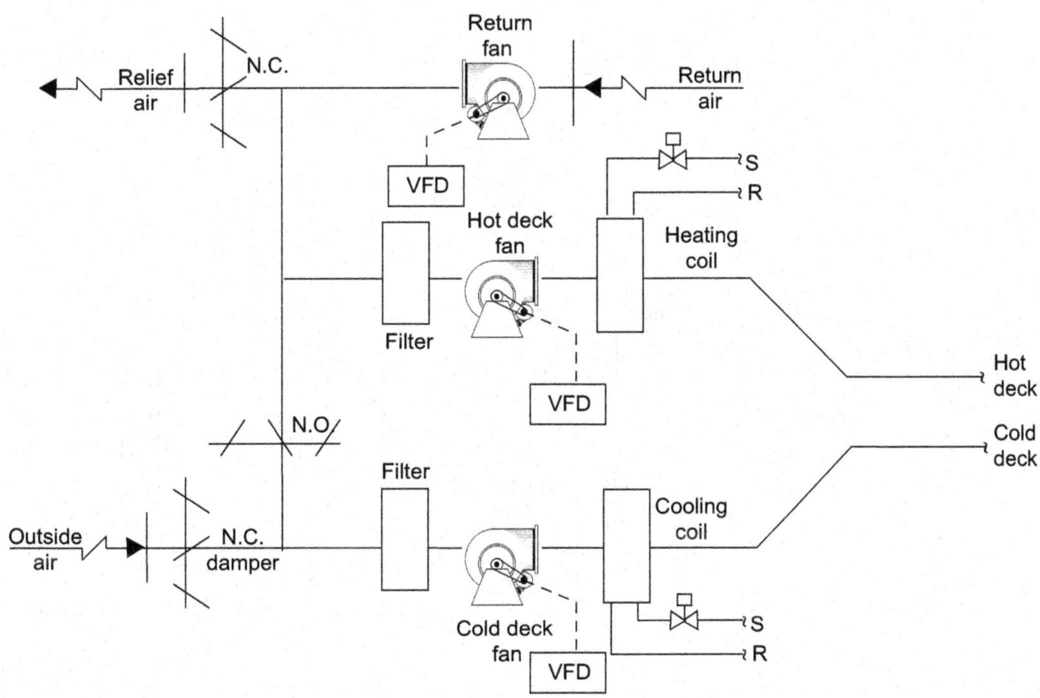

FIGURE 3-6
Sequenced-air dual–duct system schematic.

3.2.4 Multizone Systems

The CAV *conventional multizone* system is thermodynamically identical to the CAV dual–duct system. The primary difference between the two is that the blending of the two airstreams in the multizone system is done at the AHU by a series of zone dampers, responding to zone (room) conditions, as shown in Figure 3-7. *The multizone system suffers from the same energy inefficiencies as the CAV dual–duct system and is almost never used today except for certain specialized applications, such as museums.* They may be encountered during renovation projects.

To improve the energy performance of the multizone system, engineers have developed the *bypass multizone system* shown in Figure 3-8.

The traditional multizone system has heating and cooling coils in parallel, with a single-supply fan. Mixing dampers, in response to zone requirements, blend hot and cold air as necessary to provide the required supply air temperature. However, the bypass multizone eliminates the unit heating coil, creating a *bypass hot deck*, and uses heating coils in each individual zone. Thus, the mixing dampers and heating coils can be operated in sequence, closing off the cold deck before heating is energized. The cold deck temperature is set on the basis of the zone requiring the coldest air. Also, this system normally operates with minimum outdoor air to reduce the excess HLs imposed on the heating side of the system.

3.2.5 VAV/Variable Temperature Systems

Initially applied in residential applications, this system was conceived as a low first cost approach to providing some level of multiple zone control by a CAV single-zone AHU or

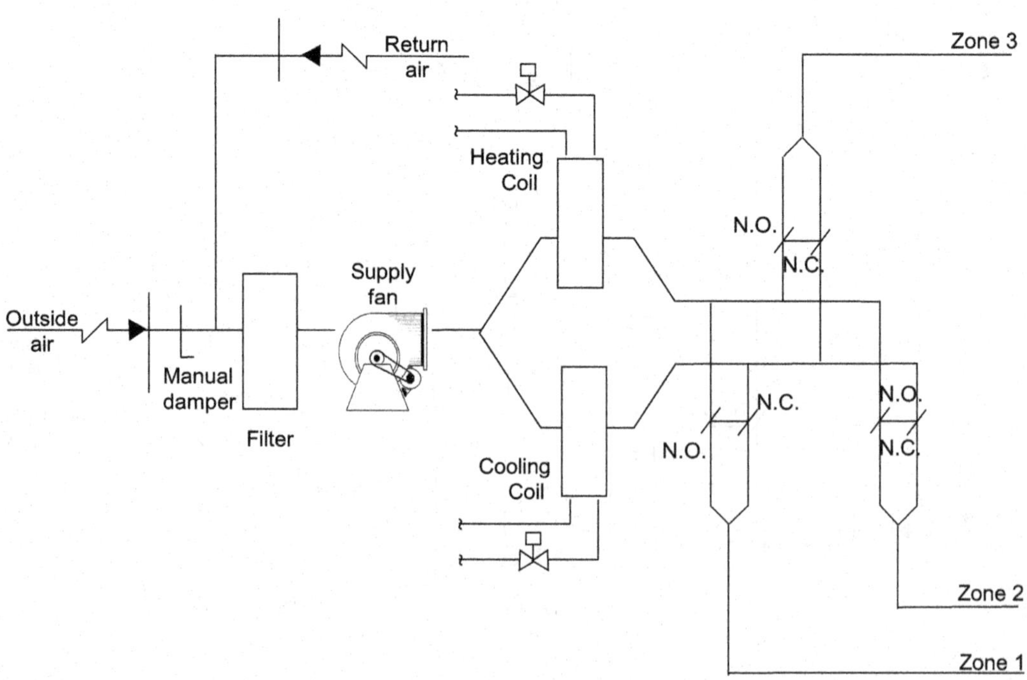

FIGURE 3-7
Conventional multizone system schematic.

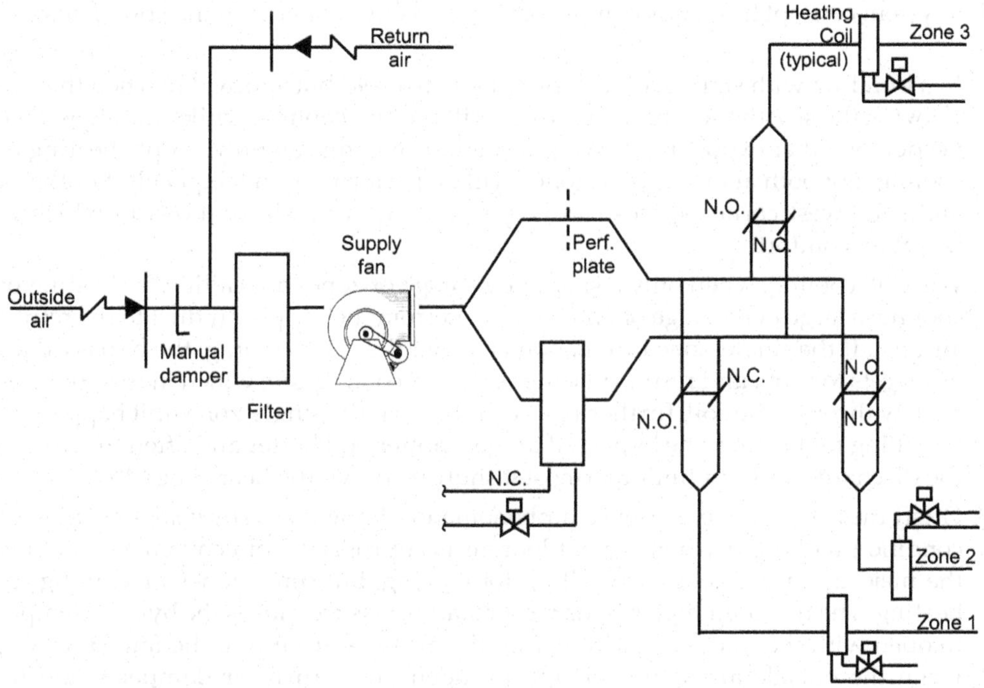

FIGURE 3-8
Bypass multizone system schematic.

packaged unit using a direct expansion (DX) refrigeration for cooling. Later, manufacturers expanded the concept to address smaller commercial buildings.

Each control zone is provided with a modulating damper to control zone airflow in response to the zone heating or cooling load. A bypass damper between the supply and return ducts is modulated to relieve excess duct static pressure as the zones reduce airflow in response to load. Thus, the system concept combines a CAV single-zone air handler with individual VAV zone control.

Each zone is provided with a temperature sensor and controller that controls its respective zone damper and communicates with a master unit that polls all zones for their heating or cooling requests. Based on the greater need, either heating or cooling is provided by the single-zone unit.

If all zones require cooling or all zones require heating, the system operates more or less as a conventional VAV system, providing enough supply air to each zone to satisfy its corresponding load. As the zone loads reduce and the zone dampers throttle the supply airflow, the supply duct static pressure at the single-zone unit increases and is sensed by a duct static pressure controller that, in turn, modulates the bypass damper to relieve the pressure and maintain the duct static pressure at set point.

Since the single-zone unit can provide only heating or cooling at any given moment, the system attempts to satisfy the "calling" zones while closing the zone dampers serving those zones needing the opposite of the current unit-operating mode. After all of the "calling" zones are satisfied, the unit switches over to the opposite mode in an attempt to satisfy the zones that had been closed. The previously open dampers will close, the closed dampers will open, and the bypass damper will continue to modulate to maintain the duct static pressure at set point.

This system concept has a number of comfort and energy performance problems:

1. In a building with significant exterior glass on a cold but sunny day when the sun is low in the sky, the southern exposure will require cooling, while heating will be needed on other exposures. Since the single-zone unit can only supply heating or cooling, not both at once, some zones will be performing inadequately. Southeast and southwest corner spaces can be particularly difficult locations in which to maintain comfort.
2. With DX cooling, when only a small percentage of zones are calling for heating or cooling, a large percentage of air is bypassed from the supply air duct to the return air duct. If the single-zone unit happens to be calling for cooling, the bypassed air often gets too cold and results in icing of the DX cooling coil, which in turn reduces airflow through the coil, furthering the problem. If the single-zone unit happens to be calling for heating, the bypassed air gets hotter and hotter and often times trips the discharge air high-limit thermostat shutting down the heat altogether.
3. Often these systems experience a phenomenon known as *return air spilling*. This condition will occur when the outdoor air is appropriate for economizer cooling; the majority of the zones are calling for cooling, but some zones are calling for heating. The resultant high discharge air static pressure causes the bypass damper to open with the intent of recirculating the air back around to the fan. However, when the outside air damper is mostly open, the return air damper is mostly closed, eliminating much, if not all, of the return path to the fan. In this case, cold economizer bypass air has nowhere to go except back down the return air duct to the nearest return air grill and spill back out into the space.
4. Since the decision for the single-zone unit to provide heating or cooling is predicated on polling the zones, an undesirable situation can occur when a particular zone does not satisfy its zone controller, swings the "vote," and keeps the unit in either heating or cooling for a prolonged amount of time. This often occurs when one zone is driving the cooling load, such as an IT room, or if a zone is undersized. During this condition, a significant number of the zones will not be getting their desired request.
5. Because some of the zone dampers will be closed at times, adequate ventilation is not delivered to these zones, whereas open zones are overventilated.
6. These systems are not energy efficient. Systems with DX cooling that run with iced coils consume electricity with little or no beneficial work being done. As the single-zone unit switches back and forth from heating to cooling (often several times per hour), there are "swinging" space temperatures. Spaces are often overventilated, using more energy than necessary. Finally, since supply fan static pressure control is via a bypass damper and not a VFD, there is no fan energy savings.
7. This system type can be difficult and expensive to maintain. The systems often switch frequently from heating to cooling and back again, causing many starts and stops that shorten the equipment life and result in frequent failures. And, since these systems have a high comfort complaint factor, more frequent service calls and maintenance expense result.

Use of this system is not recommended (even for residential applications): Often contractors will offer this system concept for use in smaller buildings since the cost of a conventional

VAV system would be "too expensive." Naive owners may accept the proposal as "value engineering," but the system provides only marginal ability to satisfy the definition of *air conditioning* given in Chapter 1 and, thus, has little value.

3.3 Air–Water Secondary Systems

3.3.1 Active and Passive Chilled Beam Systems

Passive chilled beams are used to offset space SCLs only since condensation cannot be allowed. Thus, they cannot be used for latent cooling/dehumidification.

A passive chilled beam consists of a fin-and-tube heat exchanger, contained in a housing (or casing), that is suspended below the ceiling. Chilled water passes through the tubes. As warm air from the space rises toward the ceiling, air surrounding the chilled beam is cooled, causing it to sink toward the floor, thus creating convective air current to cool the space. Since this system has no air supply, a separate air system for delivery of the required minimum ventilation airflow is also required.

Passive chilled beams have two advantages over all-air systems:

1. Their application "decouples" space sensible and latent cooling loads, allowing better humidity control by the additional required air systems.
2. Their application reduces supply airflow rates and, thus, fan energy requirements. Separate air system air flow rate(s) need only be the greater of the air flow required to offset the SCL not met by chilled beams, the air flow required to meet ventilation requirements, or the air flow required to offset space latent cooling loads.

However, there are disadvantages:

1. Since the beams have cold surfaces within each occupied space, to prevent surface condensation, beam surface temperatures can never be lower than the space dew-point temperature. Typically, the surface temperature is maintained at about 60°F–65°F [16°C–18°C], which is about 5°F–10°F [3°C–6°C] above the anticipated maximum space dew-point temperature.

   To ensure that surface condensation on the chilled beams does not occur, the chilled water system must be designed so that the chilled water supply temperature delivered to the chilled beams is guaranteed never to be less than 57°F [14°C]. Configure the chilled beam piping so that the beams will be supplied with return chilled water from AHU cooling coils, at about 57°F [14°C], and leave each beam at 61°F [16°C]. This will eliminate any water blending or heat transfer losses, saving energy and significantly simplifying the chilled water temperature control for the chilled beams. *It will also make it virtually impossible for human error or control failure to result in supply water delivered to the chilled beams in the future from being too cold and creating condensation problems.*
2. The beams have limited sensible cooling capacity (about 600 Btu/h ft [270 W/m] maximum), which means that a significant percentage of the space SCL may still have to be met by air systems.

3. The response to space sensible load changes is relatively slow since heat is transferred via convective air currents in each space. In larger, open areas, this is typically not a problem. However, if applied in smaller spaces, over- or undercooling can easily occur if loads fluctuate relatively quickly.

An active chilled beam also consists of a fin-and-tube heat exchanger contained in a housing recessed within the ceiling. Unlike the passive chilled bean, active chilled beams contain an integral *primary air supply* provided by an air-handling system. This primary air passes through nozzles, inducing air from the space up through the cooling coil and allowing an active chilled beam to provide more cooling capacity than a passive chilled beam of the same size, ranging from 815 Btu/h ft [800 W/m] to 1,425 Btu/h ft [1,400 W/m].

Typical induction ratios for these units range from about 1:3 to 1:5. The primary air system must deliver the required amount of ventilation air to each space, must be dehumidified enough to offset space latent loads, and must be at high enough pressure to induce room airflow sufficient to offset the space SCL.

In general, chilled beam systems can be installed wherever the ratio of sensible cooling to latent cooling is not lower than 0.7. In other words, at least 70% of the cooling load is sensible cooling. If the aforementioned ratio is lower than 0.7, additional primary air is required to dehumidify the space, which in turn increases the size of the air handler and ductwork, requires more energy to move the air, and begins to compromise the benefits a chilled beam system delivers.

The best applications for chilled beams are spaces that exhibit a significant difference between the SCL and the airflow required for ventilation. A good example is a laboratory where processes or equipment generates a considerable amount of dry heat. A conventional all-air HVAC system may require 12–18 air changes/h to handle the high internal gain cooling load. With a chilled beam system, the number of air changes could be as low as six, reducing airflow by up to 300% and resulting in considerable energy savings.

Active chilled beams are a variation in the *induction system* concept invented by Dr. Willis Carrier in the 1930s and widely applied to serve the perimeter zones in high-rise buildings in the 1940s–1960s. These units suffered from three major problems that remain with chilled beams:

1. Great care and careful attention to operations by the building staff is required to ensure that chilled supply temperatures remain high enough to avoid latent cooling and the creation of condensate, whereas it remains low enough to provide the design sensible cooling capacity.
2. Noise is due to the high airflow velocity through the induction nozzles.
3. Dust and dirt is accumulated on the perforated face of the beam, since the induced air is unfiltered, and the need for routine cleaning is required to avoid adversely affecting the induction ratio.

Overall, the application of active chilled beam systems must be carefully evaluated by the HVAC designer. Numerous published case studies, including a 2013 study by Jeff Stein, P.E., and Steven T. Taylor, P.E., Taylor Engineering in Alameda, California, in 2013 found that in three separate buildings, designed by three different engineering firms that "… VAV reheat system(s) had the lowest first costs and the lowest energy costs of the three systems. The analysis showed that many of the supposed advantages of active chilled beams with a dedicated outdoor air system … such as improved indoor air quality and a lower floor/floor height, also turned out to be largely overstated."

Chilled beams, passive or active, may not be suitable for the following locations and/or conditions, so designers must take extra precautions when considering their application:

1. Vestibules or atriums in which the latent cooling load is difficult to predict and control
2. Spaces with high latent cooling requirements, including gymnasiums, theaters, and other high occupancy spaces
3. Hospitals, laboratories, and so on where the recirculation of room air is not permitted or high airflow rates (air changes/hour) are required
4. Spaces with ceiling heights exceeding 14 ft [4.3 m]

3.3.2 FCUs and Unit Ventilators

FCUs are simply small CAV single-zone units used as room terminal units and typically served with hot water and chilled water to maintain room conditions. These systems are generally classified as follows:

1. *Two-pipe system*: Either hot or chilled water is piped throughout the building to a number of FCUs. This system may provide heating or cooling, but cannot provide both simultaneously since only hot or cold water is circulated. Often called a *changeover* system, care must be taken when switching the system from heating to cooling and then back again. One of the drawbacks of this system is that the changeover period, particularly in moderate climates, can take hours and even days.
2. *Three-pipe system*: Two supply pipes, one carrying hot water and the other chilled water, make both heating and cooling available at any time needed. One common return pipe serves all FCUs. This proprietary system is no longer used due to its poor energy efficiency and noncompliance with energy codes, but may be encountered during renovation projects.
3. *Four-pipe system*: Two separate piping circuits are used: one for hot water (or steam) and the other for chilled water. The best-performing fan coil system, it is capable to providing heating or cooling to spaces upon demand.

FCUs can be configured as (a) horizontal units, typically located at or above the ceiling (common in hotels and motels), and (b) vertical units, typically located against an exterior wall (typical for schools and hospital patient rooms), or chase-enclosed units designed to be installed in the corner of rooms in multistory residential applications.

FCUs, with or without cabinets, are available in capacities from about 200 cfm [100 L/s] through 2,000 cfm [950 L/s]. These units include a chassis (frame), coils, removable fan and motor assembly, drain pan, filter, insulation, hydronic piping and control valve package, and other accessories required for the project. A cabinet (enclosure) with internal insulation is required for units exposed to view.

Individual heating and cooling coils are always required, even though manufacturers offer a coil configuration. In hot, humid climates, the heating coil must be located downstream of the cooling coil so that reheating for humidity control can be provided when required.

Due to limited factory-installed insulation (typically 1/2″ [13 mm] or less), units installed above ceilings are at risk of surface condensation ("sweating") if there is any outdoor air infiltration into the ceiling plenum. Therefore, the use of *blower coil units* (BCUs), which are available with 1″ [25 mm] of insulation, is recommended for these applications. *Above ceiling, ducted FCUs should be used only as a last resort.*

Finally, any FCU or BCU installed above a ceiling must be equipped with an auxiliary drain pan and a drain pan-level sensor to shut down the unit in the event that the condensate level rises to an overflow condition (usually caused by a clogged drain line, a common problem when units are poorly maintained).

A special configuration of FCUs is called a *unit ventilator.* The unit ventilator is a larger capacity vertical FCU typically used in classrooms requiring high ventilation air. The unit is arranged to use up to 100% outdoor air to provide cooling when the outdoor air temperature is low enough to offset the imposed cooling loads. There are four standard control cycles for two-pipe unit ventilators, which are as follows:

Cycle I. 100% outdoor air is admitted at all times (except during the warm-up period).

Cycle II. The minimum required amount of outdoor air is introduced when the unit is providing heating, increasing up to 100% when needed for cooling.

Cycle III. Except during the warm-up stage, a variable amount of outdoor air is introduced to maintain a fixed temperature (typically 55°F [13°C]) of air entering/leaving the coil.

Cycle IV. When mechanical cooling is provided, the unit outdoor airflow rate is again controlled at the required minimum and the unit provides cooling via a coil using either chilled water or integral DX refrigeration system.

While classroom unit ventilators have been used for years, they are generally no longer applied due to their complexity, poor energy performance, and noise impact in the classroom, and are not recommended.

3.4 Unitary Systems

There are a large number of types of unitary systems and combinations of system, which are generally classed as *unitary*. These systems are the factory-assembled units that include DX refrigeration for cooling (see Chapter 11). Heating may be provided by separate hot water or electric heating coils, or by reverse-cycle refrigeration (see Section 3.5).

3.4.1 Incremental Units

Incremental units are basically packaged full heating, cooling, and air-handling systems that are mounted in a through-the-wall or under-the-window type of unit (i.e., the typical motel unit). It consists of a heating coil, a cooling coil, an air-cooled refrigeration circuit with an air-cooled condenser to provide cooling, and a fan that supplies the air to the space. This unit is a small version of a CAV single-zone system in which the primary and secondary system components are packaged as one device. These units are called *packaged terminal air conditioners* or *packaged terminal heat pumps,* since their application is limited to a single space in a building.

Generally, these units are available to provide cooling capacity from 6,000 Btu/h [2 kW] to about 18,000 Btu/h [5 kW] and typically include an electric resistance coil to provide heating. However, there are also available in an air-source heat pump configuration, and some manufacturers offer these units with compact, gas-fired (natural gas or LPG) furnaces for heating. This last type is often used in nursing homes, assisted living facilities, and so on that are required by licensing agencies to have a backup system to provide heating in the event of a power outage.

3.4.2 Packaged Units

Packaged units are larger capacity units that are designed for outdoor installation and include a heating coil, a cooling coil, an air-cooled refrigeration circuit with an air-cooled condenser to provide cooling, and a fan that supplies the air to the building area served. Again, these units can be version of a CAV single-zone system or a VAV multizone system in which the primary and secondary system components are packaged in one enclosure. Generally, these units are available to provide cooling capacity from 24,000 Btu/h [7 kW] to about 300,000 Btu/h [88 kW].

Heating may be provided by an integral gas-fired furnace (LPG) or by air-source heat pump operation.

These units may be installed at ground level but are more commonly installed as *rooftop units* (see Chapter 1 for their support requirements).

3.4.3 Split Systems

Split systems are available when the designer wants to separate the air-handling component, installed indoors, from the refrigeration component, located outdoors. Conventional split systems consist of an indoor BCU and an outdoor *condensing unit*, connected via refrigerant lines. The condensing unit contains the refrigeration compressor and the air-cooled condensing coil. These units are available to provide cooling capacity from 24,000 Btu/h [7 kW] to about 300,000 Btu/h [88 kW].

In lieu of the typical incremental unit or small, ducted split system, *ductless, mini-split system air conditioners* ("mini-splits") may be used. These units have two main components: an outdoor variable speed compressor/condenser and an indoor VAV AHU. A conduit, which houses the power cable, refrigerant tubing, suction tubing, and a condensate drain, links the outdoor and indoor units.

The main advantages of "mini-splits" are their small size and flexibility for zoning by providing heating and cooling to individual spaces. Many models allow as many as four indoor AHUs (for four separate control zones) to be connected to one outdoor unit. And, since mini-splits have no ducts, they avoid the energy losses associated with the ductwork of conventional split systems.

However, the designer must correctly size each indoor unit and judge the best location for its installation. Oversized or incorrectly located air handlers often result in short cycling, which wastes energy and does not provide proper temperature or humidity control.

The primary disadvantage of mini-splits is their inability to integrate ventilation air supply required by building codes (see Chapter 10). Thus, a separate ventilation air system is normally required. Also, since these units have poor sensible heat ratio capabilities, careful attention to the unit sensible cooling capacity is required by the designer when using the units in a high sensible heat gain applications.

These small units are available to provide cooling capacity from 6,000 Btu/h [2 kW] to about 24,000 Btu/h [5 kW].

3.4.4 Variable Refrigerant Flow Systems

Variable refrigerant flow (VRF) systems were introduced in Japan by Daikin in 1982 and use refrigerant as the cooling and heating medium. Liquid refrigerant is provided by a single outdoor unit, typically rated at 6–24 tons [21–84 kW], and is circulated within the building to multiple DX indoor units, equivalent to the mini-splits discussed earlier, with capacities that range from 0.5 to 2.5 tons [1.8–8.8 kW]. These systems use DC inverter-driven, variable speed compressors in order to support VRF over the full range of the compressor turndown (typically 10%–100%).

These systems come in two configurations: two-pipe and three-pipe systems. In a two-pipe (liquid and suction piping) system, all of the zones must be either all calling for cooling or all calling for heating (which generally defeats the purpose of zoning). A three-pipe (liquid, suction, and hot gas piping) system has the ability to simultaneously heat some zones while cooling others (technically, as a heat pump) and is more commonly applied.

To control refrigerant flows, automatic diverter valves are incorporated in the piping system to divert refrigerant flows and to ensure proper oil return.

There are two basic types of VRF systems: *heat pump* and *heat recovery*. Heat pumps can operate in heating or cooling mode. A heat recovery system, by managing the refrigerant through a gas flow device, can simultaneously heat and cool, some indoor FCUs providing heating while some provide cooling, depending on the requirements of each building zone. The majority of VRF systems are equipped with variable speed compressors. Often called inverter-driven compressors, this component responds to indoor temperature changes, varying the speed to operate only at the levels necessary to maintain a constant and comfortable indoor environment. Due to this flexibility, VRF systems that include inverter compressors are inherently energy efficient. Heat recovery systems increase VRF efficiency because, when operating in simultaneous heating and cooling, energy from one zone can be transferred to meet the needs of another.

These systems, necessarily, come with fully integrated original equipment manufacturer controls. But experience has shown that these controls often provide poor humidity control and are very difficult to integrate into typical BACnet-based building control systems (see Chapter 13).

Given the narrow capacity range for these systems, they are generally applied in small office buildings that need multiple zoning or each floor of larger, multistory offices. VRF systems are not suitable for all applications. *Their applications in clinics, schools, and other facilities with significant ventilation air requirements have been less successful and should be avoided.* The key challenges include the following:

1. *Refrigerant piping*: A VRF system piping configuration is restricted by distance criteria between the condensing unit and the most remote evaporator. The maximum length of refrigerant path for a VRF system is determined by the compressor's ability to overcome the pressure drop and for the system to maintain proper oil return.

   Each manufacturer specifies both the sizes for refrigerant piping required for their system and the maximum permissible vertical and total refrigerant piping developed length. A word of caution: although manufacturers' literature may state that refrigerant lines can be as long as 500 ft [150 m], the fine print may stipulate that, after the first "tee" from the condensing unit, there is a limit of about 135 ft [40 m] to the furthest indoor unit.

2. *Compliance with American National Standards Institute (ANSI)/(American Society of Heating, Refrigerating, and Air-Conditioning Engineers (ASHRAE) Standard 15*: VRF systems must comply with ASHRAE *Standard 15, Safety Standard for Refrigeration Systems*, which limits the type and amount of refrigerant allowed in an occupied space. VRF systems raise the specter of refrigerant leaks, which can be difficult to find and repair, particularly in inaccessible spaces. The refrigerant leak, especially if the system serves small rooms, can cause oxygen depletion. So designers must limit each system size based on the smallest room area served. Compliance with ASHRAE *Standard 15* is sometimes difficult, especially where a long length of refrigerant piping is involved. The total refrigerant charge in the refrigerant loop must be within the limits prescribed by ASHRAE *Standard 15*.

   VRF system manufacturers have developed products and protocols in an attempt to address concerns over refrigerant leakage. Typically, all joints must be brazed, *no flared fittings allowed*. Headers and splitters within the piping system are specifically designed not to require flared connections or to change piping wall thicknesses. But, leaks happen and the location and detection of refrigerant leaks that do occur can be very difficult within VRF systems. Some factors contributing to this are as follows:

   a. The piping systems tend to be large, so it can take weeks or months for a slow refrigerant leak to become apparent.
   b. Physical leak detection is difficult as all refrigerant pipes are insulated, and is even harder where they are routed through inaccessible or difficult-to-access ceiling cavities. Also, leaks on internal parts of equipment, such as indoor units, can be difficult to locate without disassembly.
   c. If a leak does occur, it is almost impossible to determine how much refrigerant has been lost. Where a loss of refrigerant has been diagnosed, the only sure way to recharge the system is to decant all remaining refrigerant from the system and replace it with a known (weighed) quantity.

3. *Oil management*: Long refrigerant piping loops also raise concerns about oil return. Typically, each compressor has its own oil separator, which is optimized for the VRF system. Periodically, the VRF goes into "oil retrieval mode" during which time the thermostatic expansion valve opens and the compressor cycles at high pressure to flush oil out of any location where it has accumulated. But this approach may not be 100% reliable, and lack of oil return can trigger early compressor failure.

4. *Ventilation air requirements*: VRF systems do not provide ventilation of their own, so a separate ventilation system is necessary. Therefore, a separate *dedicated outdoor air system* is typically required, and the VRF system function is restricted to recirculation of air within each occupied space.

5. *Particulate matter removal*: Particulate filters (air cleaners) are required to have an MERV of not less than 6, rated in accordance with ANSI/ASHRAE *Standard 52.2*. The standard filter provided with most VFD indoor units has a rating equivalent to only MERV 1 or 2.

   Higher MERV rating filters may be an available option for selected indoor units, but higher rating filters are not available for all types. Designers must be careful when selecting indoor units to insure that the filters meet guide specification requirements.

6. *Lack of economizer*: No VRF indoor unit is configured to accommodate an airside economizer, and because the system is refrigerant based (and not water based), there is no option for a waterside economizer. This represents a potential conflict with most energy conservation codes based on ASHRAE *Standard 90.1*.
7. *VRF systems are proprietary*: All VRF system components, other than the refrigerant piping itself, are fully proprietary, including their controls. Designers do not have the flexibility to use "anybody's" building control and automation system to run these systems. A gateway is required to connect a building control system to the VRF system controls, but the building control system can only monitor the VRF controller input/output (I/O) points, not control operation of the VRF system itself.
8. *Reliability and maintenance*: Although suppliers claim that VRF systems are very reliable, some contractors and engineers believe that a VRF system with many compressors (e.g., as many as 20–25 compressors are required to produce 100 tons [350 kW] of cooling) is inherently less reliable than a water chiller (especially if water cooled) that has a smaller number of compressors (e.g., 4–6 compressors for 100 tons capacity [350 kW]).
9. *Performance guarantee*: Prior to the release of ASHRAE *Standard 90.1–2010*, there were no efficiency requirements for VRF systems. The certification standard referenced in ASHRAE Standard *90.1–2010* is ANSI/AHRI *Standard 1230*, also issued for the first time that year. *Designers must review these requirements carefully and specify VRF system performance requirements and the documentation necessary to demonstrate compliance.*

In early 2017, the U.S. Army Corps of Engineers updated the *United Facilities Criteria* to ban the use of VRF systems in U.S. Air Force Facilities and to "strongly discourage" the use of VRF systems in U.S. Army Facilities. In the U.S. Navy, VRF systems may be used only with the explicit approved of the Naval Facilities Engineering Command. Other owners who have experienced problems with this type of system have taken similar actions. Thus, *the use of VRF systems is not recommended for any application other than a small commercial building (≤ 4,000 gsf [375 m²]) with minimal zoning requirements, not located in a hot, humid climate zone.*

3.5 Heat Pumps

3.5.1 The Heat Pump Concept

Heat pumps are "reverse-cycle" building heating/cooling units or systems that can extract heat from a building and reject that heat to the environment, providing cooling for a building, and can switch from providing cooling to providing heating by extracting heat from the environment and rejecting that heat into a building. This heat transfer cycle can be accomplished using a vapor compression refrigeration cycle or an absorption refrigeration cycle, though by far vapor compression systems are more widely used.

All refrigeration cycles hinge on one common physical characteristic: if a chemical compound (which we can call a refrigerant) changes phase from a liquid to a gas, a process called evaporation, the compound must absorb heat to do so. Likewise, if refrigerant changes phase back from a gas to a liquid, the process of condensation, the absorbed heat must be rejected. Thus, all refrigeration cycles depend on circulating a refrigerant between a heat "source" (from which heat is removed, thus producing cooling) and heat "sink" (somewhere to which the collected heat can be rejected).

For any heat pump using the vapor compression refrigeration cycle (see Chapter 11), there is a fifth element required: a reversing valve. This allows the condenser and evaporator heat exchangers to switch roles as the system switches from providing cooling to providing heating and vice versa by reversing the direction of refrigerant flow between these two components.

The amount of energy consumed by the compressor is dictated by the "lift" or pressure increase that the compressor must provide to raise the pressure/temperature of the low pressure, low temperature vapor that enters to the high pressure, and high temperature vapor that leaves. On the low pressure, low temperature side of the system, refrigerant conditions depend on the temperature needed to provide required sensible cooling and dehumidification (typically about 45°F [7°C]), whereas the high pressure/hot side refrigerant must be at conditions that allow it to condense at ambient temperatures of 105°F–125°F [51°C–65°C]. If condensing temperatures are lowered to the heat sink's temperature condition, then the required lift is decreased and required energy input to the compressor also decreases.

The "type" of heat pump is typically defined by the environmental component that is used as a heat source, from which heat is extracted when providing heating, and heat sink, to which heat is rejected when providing cooling. The heat source/sink may be atmospheric air, directly or indirectly; the ground; or large water sources, such as lakes, rivers, or oceans. Subtypes of heat pumps are also sometimes defined on basis of heat transfer media used, air, refrigerant, or water.

Vapor compression cycle heat pumps have been available since the 1950s, and factory-built heat pump units are widely used in residential and light commercial applications in mild climate regions. However, since the early 2000s, the energy efficiency of these units, both air source and water source, has improved dramatically by the following design changes:

1. *Variable air volume supply air*: Variable indoor airflow provided by the use of electronic commutation motors. By varying the indoor airflow rate in response to the imposed cooling and HLs, indoor fan energy consumption is reduced significantly since no heat pump system is required to operate at peak capacity more than a few hours each year. Additionally, when providing cooling, variable airflow results in much better humidity control than provided by older, single speed systems, a real boon for installations in the south.
2. *Variable speed compressor(s)*: Variable compressor speed/load control provides a better match between the imposed load and the compressor capacity. Older, single-speed compressors cycled on and off as the imposed cooling load/HL was reduced, resulting in thermal losses at the beginning and end of each run cycle. The next level of improvement was to provide two-speed compressors and, more recently, fully variable speed compressors. Some very high-efficiency systems even use multiple variable speed compressors configurations in which one compressor is used during normal "part-load" operation and the second is used only during "peak-load" conditions.
3. *Variable heat sink/source medium flow*: For both air-source and water-source heat pumps, variable sink/source flow is being applied to reduce energy consumption required for air or water transport.

4. *Demand defrost control*: For air-source heat pumps, demand defrost control is incorporated to minimize energy losses associated with the build-up of frost or ice on the outdoor heat exchanger when the unit is providing heating.

Vapor compression cycle heat pumps now typically use R-410A or R-134a refrigerant, significantly reducing the potential for atmospheric ozone depletion and warming in the event of a refrigerant leak. The refrigerant change has resulted in the heat transfer area of both indoor and outdoor heat exchanges being increased, improving both cooling/heating performance and energy consumption even further.

3.5.2 Air-Source Heat Pumps

Air-source heat pumps use atmospheric (outdoor) air as their ultimate heat source/sink, either extracting heat from outdoor air to provide heating or rejecting excess heat into it when cooling. Thus, the performance and seasonal efficiency of these heat pumps are significantly impacted by outdoor air temperature.

ASHRAE *Standard 90.1* defines efficiency on the basis of specific required minimum performance measurements:

1. For cooling by units rated with a cooling output of less than 65,000 Btu/h [19 kW], *seasonal energy efficiency ratio* (SEER) is defined as the total cooling output of the heat pump during its normal cooling usage period (in Btu) divided by the total electric energy input during that same period (in Watt-hours).
2. For cooling by units rated with a cooling output of 65,000 Btu/h [19 kW] or greater, *energy efficiency ratio* (EER) is defined as the ratio of net cooling capacity (in Btu/h) to the total rate of electrical input (in watts) under designated operating conditions.
3. For heating by units rated with a cooling output of less than 65,000 Btu/h [19 kW], *heating seasonal performance factor* (HSPF) is defined as the total heating output of the heat pump during its normal heating usage period (in Btu) divided by the total electric energy input during that same period (in Watt-hours).
4. For heating by units rated with a cooling output of 65,000 Btu/h [19 kW] or greater, heating efficiency is defined in terms of coefficient of performance (COP). COP is a dimensionless value defined as the net heating capacity of the heat pump (in Btu/h [kW]) divided by the energy input (in Btu/h [kW]) under designated operating conditions.

ASHRAE has established current minimum performance requirements for packaged and split system (separate indoor and outdoor components) air-source heat pumps. But most manufacturers offer "high"-efficiency units that have SEER ratings and HSPF ratings that are much higher and comply generally with the EPA's Energy Star requirements.

For larger capacity air-source heat pump units, Table 3-1 summarizes ASHRAE's current minimum requirements:

An additional important part of achieving high-efficiency air-source heat pump energy performance is to select each heat pump capacity as close as possible to the anticipated peak load. Since most heat pumps are installed in Climate Zones 1–3, where cooling is more of a consideration than heating, units are typically selected on the basis of the peak-imposed cooling load in order to minimize oversizing and loss of cooling efficiency. For example, a residence or commercial area may require 40,800 Btu/h [12 kW] or 3.4 tons of cooling.

TABLE 3-1

Minimum Energy Efficiency for Commercial Heat Pumps

System Cooling Capacity (Btu/h)	Minimum Cooling Performance (EER)	Minimum Heating Performance (COP)
65,000–134,999 [19–39 kW]	11.0	3.3 at 47°F [9°C]/2.25 at 17°F [–8°C]
135,000–239,999 [40–70 kW]	10.6	3.2 at 47°F [9°C]/2.05 at 17°F [–8°C]
≥240,000 [≥70 kW]	9.5	

Because heat pump units are manufactured in specific capacity ranges, typically 2–6 tons [7–21 kW] (in 1 ton [3.5 kW] increments), 7-1/2 tons [26 kW], and 10 tons [35 kW], contractors would normally recommend installation of a 4.0 ton [14 kW] heat pump for this example.

But the more efficient and less costly approach would be to install a 3.0 ton [11 kW] heat pump. Doing so would result in better match between imposed load and available capacity during part-load operating periods, reducing energy consumption. The shortfall in capacity at peak load may result in indoor temperatures rising 1°F–2°F [1°C] above cooling set point temperature, but since the peak cooling period occurs for only a few hours each year, the negative comfort impact is minimal.

For applications located where heating requirements are more significant, sizing the heat pump unit on the basis of heating needs rather than cooling needs may result in improved heat pump performance and improved energy efficiency. While this selection may result in some oversizing relative to cooling and a small loss in cooling efficiency, the larger system will be able to provide compressor-based heating for longer periods, reducing the amount of low-efficiency supplement heating required, saving energy, and reducing operating costs.

Careful review of imposed heating and cooling loads is required to select an air-source heat pump unit and maximize energy efficiency. It is important that consumers and designers evaluate alternative unit selections and make the final decision on the basis of the lowest life-cycle cost, considering both first cost and anticipated energy costs of the anticipated life of the unit.

For all air-source heat pumps, and based on system thermodynamics, as the outdoor air temperature falls, the heat pump is able to extract less and less heat from the environment while requiring more and more energy input to the compressor. Ultimately, the heat pump will be unable to extract the amount of heat needed to offset building losses, and additional supplemental heating will be required. Typically, the supplemental heating capacity is provided by electric resistance heating coils that, by themselves, have a COP of 0.98. When the outdoor air is too cold for the refrigerant to extract any heat from it, there is no need to operate the compressor at all and all heating is done by the supplemental heater, at low efficiency and high expense.

However, since the 1990s, research to develop an air-source heat pump that would work in colder climates has been underway. One manufacturer now offers a cold climate heat pump that features a two-speed, two-cylinder compressor for efficient operation, a backup booster compressor that allows the system to operate efficiently down to 15°F [–9°C], and a plate heat exchanger called an "economizer" that further extends the performance of the heat pump to well below 0°F [–18°C].

An alternative to the use of electric resistance supplemental heating, the dual fuel heat pump has been in use since about 1980. This type of heat pump uses a high-efficiency condensing furnace firing LPG (propane) to provide heating when fossil fuel firing is more cost-effective than operating the unit compressor(s) and consuming electricity.

While the COP of a condensing furnace is essentially the same as for electric resistance heating, the much lower cost of LPG relative to cost of electricity can result in much more economical heating. Typically, the "tipping" point for switching from the heat pump's compressor-based heating cycle to fossil fuel-fired heating is when the outdoor temperature falls below about 30°F [–1°C].

3.5.3 Water-Source Heat Pumps

Air-source heat pump units cannot easily be applied to larger, particularly multistoried, buildings that require a large number of independently controlled thermal zones. Larger apartment buildings, hotels, schools, office buildings, and so on simply will not have the outdoor space available for a large number of independent single-zone air-source heat pumps.

An alternative, then, is to use water-source heat pumps. Typically, a water-source heat pump system consists of one or more individual single-zone, water-cooled packaged units with a supply fan, indoor air coil, compressor, and water-to-refrigerant heat exchanger, connected to a common water supply system that acts as the heat source/sink for each individual heat pump unit.

3.5.3.1 Closed Circuit Heat Pump Systems

The most common, and very efficient, approach to using water-source heat pumps is to use a closed circuit water heat exchanger as illustrated in Figure 3-9.

In this configuration, each packaged water-source heat pump refrigerant system reverses the direction of refrigerant flow from summer to winter with the following results: during summer, heat is removed from the space via the air coil (evaporator) and rejected to the central circulating water heat exchanger via the water-to-refrigerant heat exchanger (condenser). Excess heat rejected to this heat exchanger is, in turn, rejected to the outdoors through a closed circuit evaporative cooler in order to maintain the heat exchanger temperature at or below 85°F [29°C].

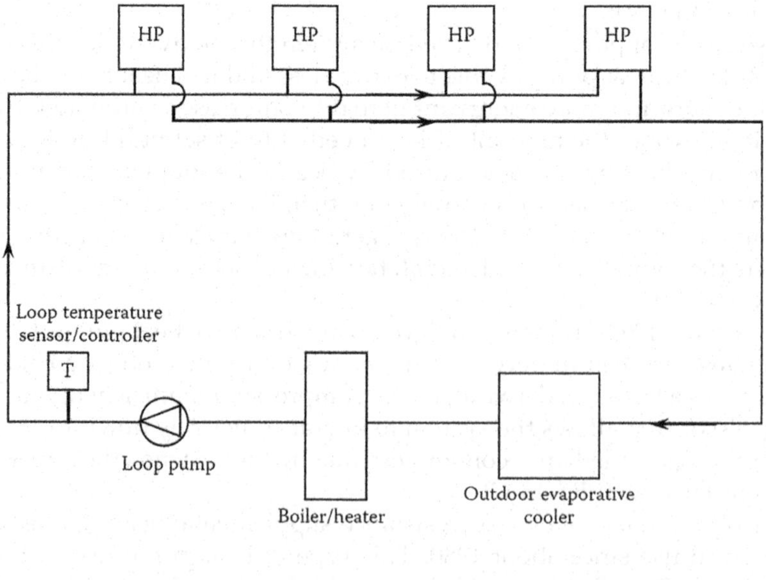

FIGURE 3-9
Schematic of basic closed circuit water-source heat pump system. (Fig. 9.2, *Energy Management and Conservation Handbook*, Frank Kreith and D. Yogi Goswami, Eds. (CRC Press/Taylor & Francis Group, 2017).)

During winter, heat is removed from the circulating water heat exchanger via the water-to-refrigerant heat exchanger (evaporator) and rejected to the space via the air coil (condenser). If enough heat is removed from the heat exchanger to lower its temperature to below 65°F [18°C], an auxiliary heater (typically a high-efficiency condensing hot water boiler) adds supplemental heat to the heat exchanger to maintain the required temperature.

ASHRAE stipulates minimum energy performance requirements only for closed heat exchanger water-source heat pump units of less than 135,000 Btu/h [40 kW] cooling capacity.

3.5.3.2 Closed Circuit Geothermal Heat Pump Systems

Closed circuit geothermal heat pumps are a variation to the basic closed circuit heat pump system, as discussed in the Section 3.5.3.1, and they are designed to take advantage of the earth (ground) or a large body of water as heat source/sink. The supplemental heater(s) and evaporative cooler(s) that represent the heat source and sink in the basic closed circuit heat pump system are augmented by a ground-coupled heat exchanger or water-coupled heat exchanger that is used as the primary system of heat source and sink. The supplemental heater(s) and evaporative coolers(s) are maintained to augment the ground-coupled heat exchanger and/or to operate under emergency conditions.

Figure 3-10 illustrates the common ground-coupled geothermal heat pump system configuration.

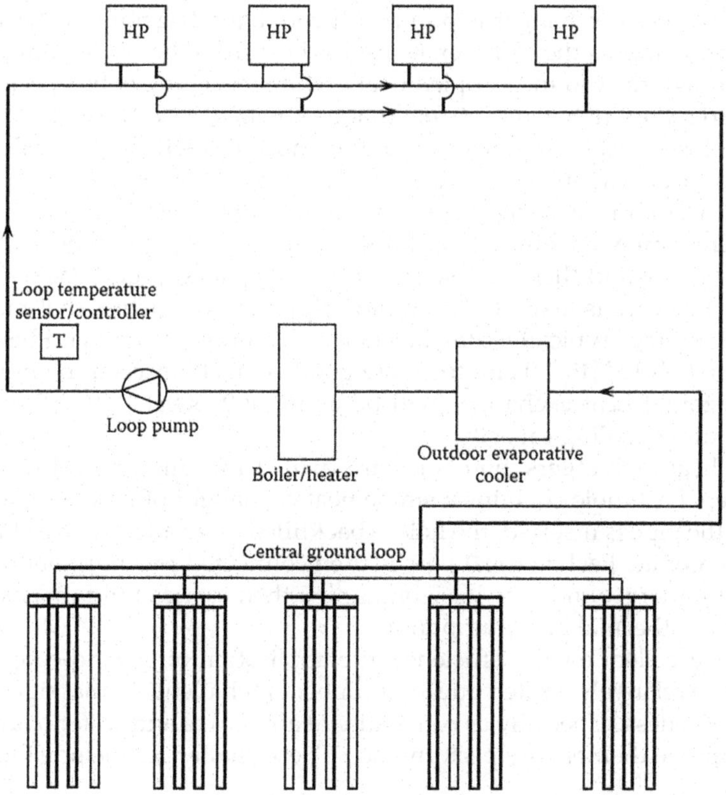

FIGURE 3-10
Schematic of basic closed circuit geothermal heat pump system. (Fig. 9.3, *Energy Management and Conservation Handbook*, Frank Kreith and D. Yogi Goswami, Eds. (CRC Press/Taylor & Francis Group, 2017).)

The ground-coupled heat exchanger may be installed either horizontally directly in the ground or vertically through one or more wells, typically referred to as "boreholes." The fluid could be either water or a refrigerant.

Geothermal heat pumps take advantage of the natural constant temperature of the ground 5–6 ft [1.5–1.8 m] below grade where the ground mean annual temperature remains between 45°F and 65°F [7°C–18°C], depending on location and water table. The ground temperature is warmer than the air above it in the winter and cooler than the air in the summer, resulting in the need for less compressor lift and more efficient heating and cooling.

The geothermal heat pump system has three major parts: the ground coupled heat exchanger, the heat pump unit, and the air delivery system (ductwork and air outlets/inlets). The heat exchanger is a system of pipes that is buried in the shallow ground near the building. A fluid (usually water or a mixture of water and antifreeze) circulates through the pipes to absorb or deposit heat within the ground.

In the winter, the heat pump removes heat from the heat exchanger and pumps it into the indoor air delivery system. In the summer, the process is reversed, and the heat pump moves heat from the indoor air into the heat exchanger. The heat removed from the indoor air during the summer can also be used to heat water.

The ground-coupled heat exchanger provides the means of transferring heat to the earth in summer and extracting heat from the earth in winter. Physically, the heat exchanger consists of several lengths of plastic pipe typically installed either in horizontal trenches or in vertical boreholes that are backfilled to provide a close contact with the earth.

Fluid inside the heat exchanger is pumped through a refrigerant heat exchanger in the geothermal heat pump. In the summer, it absorbs heat from the refrigerant and carries it to the ground through the ground-coupled heat exchanger piping to be rejected. In winter, it absorbs heat from the earth and transfers that heat to be extracted by the refrigerant.

Horizontal closed heat exchangers are the most cost-effective configuration when (1) adequate land is available, (2) soil conditions are such that trenches are easy to dig (e.g., no rock), and (3) trench depths are not defined by frost lines exceeding 36″ [1 m] below grade. Trenching machines or backhoes are used to dig the trenches normally 5–6 ft below finished grade and then lay a series of parallel plastic pipes. Backfilling the trench requires that great care is taken to eliminate sharp rocks or debris that may damage the heat exchanger piping. Typically, compacted sand or a mix of sand and bentonite is backfilled for the first 6″–12″ [15–30 cm] to create good contact between the piping and earth. A typical horizontal heat exchanger will be about 400–600 ft [120–180 m] long per ton [3.5 kW] of required cooling capacity.

For vertical heat exchangers, boreholes are drilled into the ground from 150 to 450 ft [45–140 m] deep. Each hole contains a single heat exchanger of pipe with a U-bend at the bottom. After the pipe is inserted, the hole is backfilled or grouted with a slurry of mixture of sand and bentonite. Each vertical pipe is then connected to a horizontal pipe, which is also concealed underground. The horizontal pipe then carries fluid in a closed system to and from each geothermal heat pump unit.

Despite their excellent energy efficiency, the application of ground-coupled heat pump systems has been relatively limited due to a number of "challenges" that significantly reduce their cost-effectiveness, especially as conventional air-source and water-source heat pumps continue to improve their energy performance. These challenges include the following:

1. *Land type and space for a ground-coupled heat exchanger*: The soil in which a horizontal ground-coupled heat exchanger is installed must be reasonably finely textured, no rock and soil that contains little clay. The soil in which boreholes and vertical

heat exchangers are installed is less critical since the borehole is backfilled with a mix of sand and bentonite after the piping is installed to insure good heat transfer with the surrounding undisturbed earth.

Horizontal trench heat exchangers installed with two pipes per trench typically require 2,000–4,000 ft^2 [600–1,200 m^2] of ground surface area per ton [3.5 kW] of cooling capacity required. Obviously, that requirement limits the applicability of this type of heat exchanger. Vertical heat exchangers require only about 200–400 ft^2 [60–120 m^2] of ground surface area per ton [3.5 kW] of cooling capacity required.

2. *Costs of ground-coupled heat exchanger*: Horizontal trenching, assuming no rock is present near the ground surface, is relatively inexpensive since it is accomplished by conventional backhoes or trenching machines fairly quickly. Drilling boreholes for vertical heat exchangers is more difficult and expensive than trenching. And, even with vertical heat exchangers, some horizontal trenching is required to connect each borehole piping heat exchanger to a common piping systems.

   Another cost element associated with vertical heat exchangers is the need to drill a test bore prior to designing the system in order to determine actual subsurface conditions. The soil thermal conductivity can vary as much as 400% depending on the type of soil and rock encountered, groundwater hydrology, and so on. This means that the number and depth of geothermal boreholes may vary significantly from site to site for the same system capacity.
3. *Circuit pumping energy consumption*: Geothermal heat pumps require a flow ranging from 2.5 to 3.5 gpm [9.4–13.2 l/min] for each ton [3.5 kW] of cooling load. In older systems, the pumps to provide this flow are operated at all times at a constant speed, and thus, a significant pumping energy was consumed. Today's systems used variable flow pumping, and the costs of pumping have been reduced.
4. *System complexity*: Geothermal heat pump systems have more components, more complex controls, and higher maintenance costs than the conventional systems.

A more efficient alternative to the ground-coupled heat exchanger is a *water-coupled heat exchanger*. Large ponds, lakes, rivers, and even the ocean represent excellent heat sources/sinks for use with heat pumps. While water temperatures vary in these bodies of water summer to winter, the water temperature swing is always more moderate than air temperature, rarely falling below about 32°F [0°C] or rising above 85°F [29°C]. Even water that freezes on the surface is above freezing below the ice.

Water-to-water heat exchange is 20%–50% more efficient than water-to-ground heat exchange, significantly improving the efficiency of heat exchange of water-coupled systems over ground-coupled systems. These systems typically have a cooling EER of 18–20 and a heating COP of 3–4.

Water-coupled heat exchangers may be configured with horizontal coiled piping submerged in the body of water. The alternative is to install a high-efficiency plate-and-frame heat exchanger with the closed circuit flow to the heat pump unit(s) on one side and flow from the body of water on the other. This second configuration, despite having somewhat higher pumping energy consumption, is preferred to avoid the problems with silting, drift, and mechanical damage that can occur to submerged piping loops.

While the applicability of water-coupled heat pump systems is limited simply because availability of usable water sources/sinks is more limited, the systems themselves are both less expensive and more energy efficient than ground-coupled systems and represent a very cost-effective approach when applicable.

3.5.3.3 Open Circuit Geothermal Heat Pump Systems

Open circuit geothermal heat pump systems, often called *groundwater heat pump systems,* use two or more groundwater wells for their heat source/sink, one or more wells from which groundwater is withdrawn, and one or more wells into which the water is "injected" back into the ground after passing through the heat pump units. These systems are as efficient, sometimes even more so than, as water-coupled heat pump systems.

However, these systems have a significant negative environmental impact. Since the water flow rate required for heat pump operation is relatively high, the amount of water withdrawn from an available aquifer amounts for a significant demand on the aquifer. Then, since the water reinjection point is typically far outside the "recharge zone" for the aquifer, simply pumping the water back into the ground really does not benefit the aquifer from which it was removed.

Since most aquifer levels in the United States have already fallen significantly over the past 50 years and continue to fall, the use of this heat pump system configuration and further stress on aquifers is not recommended.

3.5.3.4 Gas-Fired Engine-Driven Heat Pumps

To many, the use of electricity as a primary energy source is not desirable. The electrical power industry routinely releases huge quantities of CO_2 into the atmosphere, along with large quantities of SO_2 and NO_x. On a comparative basis, these poor greenhouse gas emissions make grid electricity a poor sustainable choice for any building heating and cooling system.

In the early 1990s, one U.S. manufacturer introduced a small natural gas-fired engine-driven packaged air-source heat pump for the residential and light commercial markets. While the engine itself is no more efficient than grid electricity power production, its use of clean-burning LPG (propane) resulted in significantly reduced greenhouse gas emissions and a lower operating cost due to the low cost of gas fuels compared to electricity.

But that heat pump was a dismal failure! Of more than 4,000 units sold, 100% of the units failed within the first 3 years of their operation and the manufacturer was forced to replace almost all of these systems with conventional units. The problem with this concept was not with the heat pump, but with the engine. In this case, the manufacturer attempted to use a modified engine designed for lawn tractors and small engine generators originally designed for 150–500 h annual use in an application that required the engine to operate 4,000–5,000 h/year.

Theoretically, an engine-driven heat pump, coupled with recovered heat from the engine jacket and exhaust, can achieve a heating COP of 1.4–2.0. And, today, manufacturers have now teamed with automotive engine manufacturers to offer gas engine-driven packaged air-source heat pumps in the 8–20 tons cooling capacity range, primarily for commercial applications. These units have an advertised SEER of 18+ and heating COP of 1.5.

3.5.3.5 Heat Recovery Chiller/Heat Pump System

Large buildings are often cooled by chilled water systems that circulate water at 40°F–45°F [4°C–7°C] to cooling coils located in AHUs located throughout the building. Whereas small residential or light commercial buildings may require only cooling or heating at any one time, large buildings may require simultaneous cooling and heating year-around. Plus, there is almost always a need to generate service hot water in large buildings.

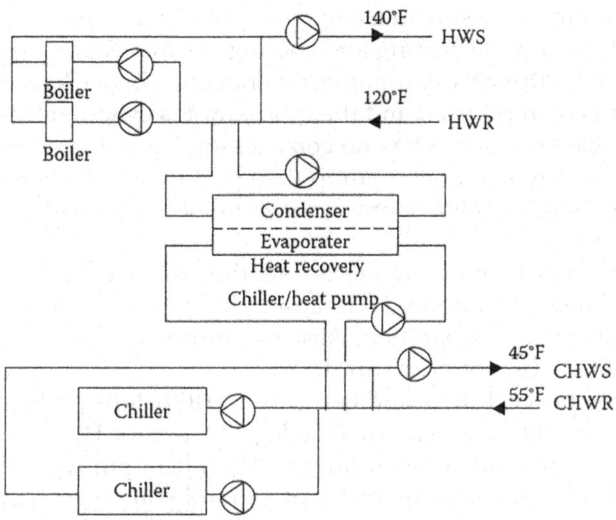

FIGURE 3-11
Schematic of heat recovery chiller/heat pump system. (Fig. 9.9, *Energy Management and Conservation Handbook*, Frank Kreith and D. Yogi Goswami, Eds. (CRC Press/Taylor & Francis Group, 2017).)

When providing cooling, the chilled water system rejects the excess collected heat to the outdoor air via cooling towers via condenser water at 85°F–95°F [29°C–35°C] temperature. These temperatures are typically too low to allow condenser water to be used for heating, so the dedicated heat recovery chiller (HCR) heat pump system is sometimes applied as illustrated in Figure 3-11.

Though not strictly a "reverse-cycle" system, this system does represent a heat pump configuration. A "lead" HRC is configured with a primary/secondary piping loop on the chilled water return piping upstream of the remaining cooling-only chillers. The HRC capacity is selected based on the need to provide heating during the summer. Thus, the HRC removes heat from the return chilled water, lowering its temperature, and rejects that heat to the hot water system at 130°F–140°F [54°C–60°C]. The effect is to reduce the imposed load on the cooling-only chillers by "precooling" the return chilled water while simultaneously producing hot water at sufficiently high temperature to be useful.

For this concept to be cost-effective, there must be a need for heating during the summer cooling period. This heat may be needed for reheat associated with space humidity control, service hot water heating, pool heating, and so on. To maximize the energy efficiency of this concept, though, there must be a need for chilled water cooling as long as feasible. This means that the use of airside economizer cycles, required by most energy conservation building codes, need to be carefully evaluated to determine when it is more efficient to provide mechanical cooling and capture the rejected heat versus using the outdoor air to provide "free" cooling and operating boilers to provide heat.

3.5.4 Advanced Technology Heat Pumps

Absorption cycle heat pumps are essentially air-source heat pumps driven not by electricity, but by a heat source such as natural gas, propane, solar-heated water, or geothermal-heated water. Because natural gas is the most common heat source for absorption heat pumps, they are also referred to as "gas-fired heat pumps."

Absorption heat pumps for residential and light commercial application use an *ammonia–water absorption cycle* to provide heating and cooling. As in a vapor compression cycle heat pump, the refrigerant (in this case, ammonia) is condensed in one heat exchanger to release its heat; its pressure is then reduced and the refrigerant is evaporated to absorb heat.

The absorption cycle heat pump has no compressor. Evaporated ammonia is absorbed into water, and a relatively low-power pump then pumps the ammonia–water solution to a heat exchanger at a slightly higher pressure where heat is added to essentially boil the ammonia out of the water.

The generator–absorber heat exchanger boosts the efficiency of the unit by recovering the heat released when the ammonia is absorbed into the water. Other innovations include high-efficiency vapor separation, variable ammonia flow rates, and low-emission, variable-capacity combustion of the natural gas.

Large gas–fire absorption chillers and heat pumps (100+ tons [350+ kW] capacity), using a *water/lithium bromide solution*, are also available (see Chapter 11).

Solar-assisted (sometimes called "solar-augmented") heat pumps use rooftop solar heat collectors configured to augment air-source or ground-source geothermal heat exchangers. These systems are generally series configured with solar panels that are arranged to act as an additional heat source for the heat pump.

For air-source or water-source heat pumps, solar panels can provide higher temperature heat during the winter, reducing the need for supplemental heating by electric coils or boilers. This configuration can also reduce surface land requirements for ground-coupled heat pumps, expanding their range of applicability.

Bibliography

HVAC Water Chillers and Cooling Towers: Fundamentals, Application, and Operation (2nd Edition), Herbert W. Stanford III, PE (CRC Press, Taylor & Francis Group, Boca Raton, FL, 2010).

Commercial/Institutional Ground-Source Heat Pump Engineering Manual, American Society of Heating, Refrigerating, and Air-Conditioning Engineers (Atlanta, GA, 1995).

"A Review of Domestic Heat Pumps," Iain Staffell, Dan Brett, Nigel Brandon, and Adam Hawkes (*Energy Environmental Science*, 2012, Vol. 5, Pgs. 9291–9306).

Chapter 9, authored by Herbert W. Stanford III, PE, *Energy Management and Conservation Handbook*, 2nd Edition, Edited by Frank Kreith and D. Yogi Goswami (CRC Press, Taylor & Francis Group, 2017).

"Blow-Through vs. Draw-Through," Stephen W. Duda, PE (*ASHRAE Journal*, January 2018, Pgs. 48–52).

Section II

The Design Method

4

HVAC Systems Design

4.1 Introduction

The design of an effective air-conditioning system must proceed in a systematic sequence involving the following basic steps:

1. *Obtain a detailed understanding of the owner's criteria and performance requirements for the project*: In addition, the designer must determine whether there are design constraints, such as a limited construction budget or limited ability of the owner to understand and/or maintain sophisticated heating, ventilating, and air-conditioning (HVAC) systems or equipment. The specific functional requirements for each area of the building must be identified (proposed use, number of people, hours of use, owner equipment to be provided, etc.). Limitations imposed by building codes and/or regulatory agencies must also be considered. Determine whether the project must meet specific sustainability design requirements, such as Leadership in Energy and Environmental Design (LEED).
2. *Evaluate alternative systems concepts to satisfy project requirements*: Most building types can be provided with more than one type of air-conditioning system that will meet the design criteria defined in Step 1. Each alternative system concept must be identified and evaluated, formally or informally, by the designer to determine the best system for the particular project. The use of Building Information Modeling (BIM), discussed in Chapter 16, has made the comparison and evaluation of the life-cycle costs of alternative systems significantly simpler.
3. *Define control zones and compute heating/cooling loads*: Typically, a "zone" is defined as a space or a group of spaces with similar requirements that is "controlled" from a single temperature sensing point. To determine whether a group of spaces can be combined, three criteria must be examined for each space:
   a. *Heating/cooling criteria*: inside design conditions, envelop load variations, internal heat gains, required degree of control, and so on
   b. *Operational criteria*: daily and/or weekly use of spaces (e.g., hours per day of use, changes in occupancy, and partial use)
   c. *Ventilation criteria*: ventilation rates and space relationships required for odor/contaminant control

Only if spaces have similar criteria in all of these categories can they be combined into a single control zone.

4. *Locate major mechanical equipment elements, and define building space requirements for the air-conditioning system*: Space for air-conditioning systems must be provided in any building in four locations:
   a. Space for primary equipment such as boilers, chillers, and pumps. Typically, one central mechanical equipment room is required to house these systems. The designer must provide adequate space for the primary equipment, space for the additional support equipment, and space to allow for proper maintenance of all types of the equipment.
   b. Space for air-handling equipment. Whereas smaller building may use outdoor rooftop air-handling units, larger buildings will require "fan rooms" to house this equipment.
   c. Sufficient ceiling space for air distribution (ductwork); HVAC, plumbing, and fire protection piping; and electrical raceway is required. These space requirements are in addition to the minimum space required for the ceiling itself and recessed light fixtures.
   d. Outdoor space for condensing units, cooling towers, exhaust fans, and so on. Some of this space may be located on the ground, whereas roof space is required for certain equipment.
   e. When locating and arranging HVAC system elements, take into account required access and clearances for effective maintenance, all of which is easier when using BIM (See Chapter 16) rather than manual or computer-aided drafting methods.
5. *Design air distribution systems and effective air diffusion*: Once the air diffusion devices are located and the required airflows to each device determined on the basis of the heating and cooling load calculations, the duct systems for distributing supply air and returning air to an air-handling unit can be designed. Exhaust systems and special duct systems for fume hoods, etc., can then be designed.
6. *Design piping systems*: Piping design starts with determining configuration and routing of the different piping systems, moves on to pipe sizing, and finishes with defining the details of piping installation and equipment connections.
7. *Define equipment performance criteria*: The designer should define all specific equipment requirements but should not select a specific equipment item (brand, model, etc.) to meet these requirements only for "basis of design." By defining performance requirements (capacity, size, weight, electrical characteristics, etc.), the designer places the burden of performance compliance on the contractor and equipment vendor. Then, if the equipment item does not perform satisfactorily, the designer is in the position of requiring that the contractor and equipment vendor be responsible for any remediation that is required.
8. *Design system operating controls*: Even though this item is listed lastly, it is, in fact, *the most important part of any HVAC design* and must be ongoing throughout the design sequence.

Each of these design steps is described in more detail in the following chapters of this book.

4.2 Criteria for HVAC Design

Over many years of HVAC design experience, the authors have learned that some things either simply do not work or, even if they do at least work, they may be more trouble than they are worth for both the owner and the designer. *The following systems and design practices are not acceptable unless their application has been discussed with the owner and the designer is satisfied that the project requires their use*:

1. Rooftop HVAC systems or equipment on buildings more than one story in height and/or for which reasonable roof access for maintenance and repair has not been provided. Without adequate access, rooftop HVAC equipment tends to be ignored and maintenance suffers.
2. Rooftop systems or equipment on any building in "snowy" climates
3. Air-cooled chiller systems larger than 150–200 tons, *unless required by the owner*, air-cooled chillers consume 40%–50% more energy than water-cooled chillers.
4. Direct resistance electric heat, except for terminal unit heating/reheating coils *when dictated by owner requirements*
5. Two-pipe, changeover heating/cooling systems of any type, except in Climate Zones 6 and 7
6. *Unit ventilators*: These devices are relatively complex as terminal units, provide poor humidity control, and are noisy.
7. Window air-conditioning units, packaged terminal air conditioners or packaged terminal heat pumps, except for residential occupancies
8. Variable air volume (VAV) systems using direct expansion (DX) cooling, unless a manufacturer's packaged system with integral original equipment manufacturer controls is used
9. Parallel-configured fan-powered air terminal units *unless dictated by the owner or for special applications*
10. Fiber glass duct liner or fiberboard duct. Historically, this material has proved to be a breeding ground for mold.
11. Direct buried underground ductwork
12. Use of glycol for chilled water systems, except for process chilled water or computer room air-conditioning systems or for chilled water systems that use ice-type thermal storage
13. Use of chase or shaft wall construction or unducted underfloor air distribution for air ducting or plenums, *air must be ducted*. Architectural materials can never be constructed to have low enough air leakage to be satisfactory as air ducts.
14. Air conditioning of buildings and/or spaces, other than residential occupancy, that have operable windows
15. Air conditioning of commercial kitchens and food service areas (except, perhaps, spot cooling systems)
16. In cold climates, locating supply, return, and/or transfer ductwork in unheated spaces

17. In cold climates, locating HVAC heating and/or cooling equipment in unheated spaces
18. Lithium bromide absorption water chillers or combustion engine-driven water chillers
19. Passive chilled beam cooling *unless dictated by the owner. Never use active chilled beams.*
20. Wireless networked direct digital control systems or components. This is an open invitation for the building controls to be "hacked."

4.3 Compliance with Building Codes and Standards

All projects must comply with the applicable building codes. In addition, special code requirements, including special inspections by a third-party expert, may apply for seismic design, wind load design, and so on. By reference, building codes may include standards issued by different technical associations, particularly the National Fire Protection Association (NFPA) and the American Society of Heating, Refrigerating, and Air-Conditioning Engineers (ASHRAE) The designer must conduct a thorough code analysis to determine all applicable codes and standards that apply and document their impact on the design as part of the schematic design phase of each project.

4.3.1 International Mechanical Code

The *International Mechanical Code* (IMC) is published by the International Code Council through governmental consensus process and is updated on a 3-year cycle to include the most current technology and safest HVAC practices. The IMC is designed to protect public health and safety for all building heating, cooling, and ventilation systems design, installation, and inspection by providing minimum required standards to be met.

The IMC is the most widely used mechanical code in the United States and is also used as the basis for the mechanical code of several other countries.

4.3.2 International Energy Conservation Code

Internationally, code officials recognize the need for an energy conservation code addressing the design of energy-efficient building envelopes and installation of energy-efficient mechanical, lighting, and power systems through requirements emphasizing performance. The *International Energy Conservation Code* (IECC) is designed to meet these needs through model code regulations that will result in the optimal use of fossil-fuel and nondepletable resources. The IECC is based on the requirements of ASHRAE *Standard 90.1* and *Standard 90.2* and establishes regulations for energy-efficient buildings using prescriptive and performance-related provisions.

This code contains separate provisions for commercial buildings and for low-rise residential buildings (three stories or less in height above grade). Each set of provisions is separately applied to buildings within their respective scopes.

The IECC, unlike the IMC, is a design document and heavily impacts design decisions made by the HVAC designer by the following significant requirements:

1. Heating/cooling load calculation software must be "approved" by the building official.
2. Minimum equipment efficiency requirements should be incorporated into equipment schedules on the drawings.
3. Each individual outdoor air connection to an AHU must have an automatic shutoff damper that closes when the AHU is off. Motor-operated dampers are generally required, but gravity dampers are acceptable when the AHU's outdoor airflow is ≤300 cfm [140 L/s].
4. Incorporate *demand control ventilation* for any space 500 sf [46 m^2] or larger that has a design occupancy load, based on the IMC, that is, ≥40 occupants/1,000 sf [92 m^2].
5. Exhaust air heat recovery is required for any AHU rated at ≥5,000 cfm [2300 L/s] supply air, with a minimum outdoor airflow ≥70% of the total airflow.
6. Fan horsepower limits apply.
7. Fan motors of 5 hp [3.7 kW] or greater or DX units with 110,000 Btu/h [32.2 kW] cooling capacity are required to be VAV.
8. Airside economizer cycles are required for all AHUs with ≥65 Mbh [18 kW] cooling capacity in all U.S. Climate Zones except 1A, 1B, 2A, 7, and 8 (see Chapter 1).
9. Designers must make sure that specified control sequences include the following:
   a. Auto changeover thermostats must have 5°F [2.8°C] deadband within which no heating or cooling energy is expended.
   b. Air source heat pumps must have controls that prevent supplemental heat operation "when the heat pump can meet the heating load."
   c. Control systems, including stand-alone thermostats, must include "setback" control to ≤55°F [13°C] for heating and ≥85°F [29°C] for cooling. Setback humidity set point, typically 60% relative humidity, may override temperature set points as needed.
   d. For VAV multiple-zone systems, provide reset of supply air pressure set point based on maintaining *at least* one VAV unit damper 100% open.
   e. Reheat-type air systems serving multiple control zones are required to have supply air temperature reset of at least 25% of the temperature difference between the design supply air temperature and the design zone air temperature. Note that humidity conditions can be used to limit temperature reset.

In addition, the IECC requirements address numerous other aspects of HVAC elements (duct air leakage, duct insulation, pipe insulation, etc.)

Finally, the IECC allows designers to adopt a *performance-based* compliance method. Code compliance by this method requires that a proposed building and its systems have an annual energy cost less than or equal to the same building if it did comply with the prescriptive and performance-related requirements of the code. However, comparison is made on the basis of annual energy costs, so the energy sources and costs selected become a part of the analysis.

4.3.3 NFPA Standards and Guidelines

HVAC design, as applicable, must comply with the referenced standards published by the NFPA, as follows:

NFPA 30	Flammable and Combustible Liquids Code
NFPA 31	Standard for the Installation of Oil-Burning Equipment
NFPA 54	National Fuel Gas Code
NFPA 58	Liquefied Petroleum Gas Code
NFPA 70E	Standard for Electrical Safety in the Workplace
NFPA 90A	Standard for the Installation of Air-Conditioning and Ventilating Systems
NFPA 90B	Standard for the Installation of Warm Air Heating and Air-Conditioning Systems
NFPA 91	Standard for Exhaust Systems for Air Conveying of Vapors, Gases, Mists, and Particulate Solids
NFPA 92	Standard for Smoke Management Systems
NFPA 92A	Standard for Smoke-Control Systems Using Barriers and Pressure Differences
NFPA 92B	Standard for Smoke Management Systems in Malls, Atria, and Large Spaces
NFPA 96	Standard for Ventilation Control and Fire Protection of Commercial Cooking Operations
NFPA 99	Health Care Facilities Code
NFPA 101	Life Safety Code
NFPA 241	Standard for Safeguarding Construction, Alteration, and Demolition Operations

4.3.4 ASHRAE Standards and Guidelines

The following ASHRAE guidelines and standards are typically applicable to HVAC systems design:

Guideline 0	The Commissioning Process
Guideline 1.5	The Commissioning Process for Smoke Control Systems
Guideline 4	Preparation of Operating and Maintenance Documentation for Building Systems
Guideline 12	Minimizing the Risk of Legionellosis Associated with Building Water Systems
Guideline 13	Specifying Building Automation Systems
Guideline 16	Selecting Outdoor, Return, and Relief Dampers for Air-Side Economizer Systems
Standard 15	Safety Standard for Refrigeration Systems (Supersedes ANSI B9.1, Safety Code for Mechanical Refrigeration) and Addenda
Standard 52.2	Method of Testing General Ventilation Air-Cleaning Devices for Removal Efficiency by Particle Size
Standard 55	Thermal Environmental Conditions for Human Occupancy
Standard 62.1	Ventilation for Acceptable Indoor Air Quality
Standard 62.2	Ventilation and Acceptable Indoor Air Quality in Low-Rise Residential Buildings
Standard 90.1	Energy-Efficient Design of New Buildings Except Low-Rise Residential Buildings
Standard 90.2	Energy-Efficient Design of Low-Rise Residential Buildings
Standard 110	Method of Testing Performance of Laboratory Fume Hoods
Standard 111	Measurement, Testing, Adjusting and Balancing of Building HVAC Systems
Standard 135	BACnet—A Data Communication Protocol for Building Automation and Control Networks
Standard 135.1	Method of Test for Conformance to BACnet

(Continued)

Standard 170	Ventilation of Health Care Facilities
Standard 188	Prevention of Legionellosis Associated with Building Water Systems
Standard 189.1	Standard for the Design of High-Performance, Green Buildings Except Low-Rise Residential Buildings
Standard 202	Commissioning Process for Buildings and Systems

ASHRAE standards and guidelines are often reference standards developed by specialized associations, including, but certainly not limited to, the following:

AABC	Associated Air Balance Council
ACGIH	American Conference of Governmental Industrial Hygienists
ADC	Air Diffusion Council
AMCA	Air Movement and Control Association International
ANSI	American National Standards Institute
AHRI	Air-Conditioning, Heating, and Refrigeration Institute
ASME	American Society of Mechanical Engineers
ASTM	American Society for Testing and Materials
AWS	American Welding Society
CTI	Cooling Technology Institute
EN	European Standards
HI	Hydraulic Institute
IEC	International Electrotechnical Commission
IEEE	Institute of Electrical and Electronics Engineers
ISO	International Organization for Standardization
MSS	Manufacturers Standardization Society of Valve and Fittings Industry
NAIMA	National American Insulation Manufacturers Association
NEMA	National Electrical Manufacturers Association
SMACNA	Sheet Metal and Air Conditioning Contractors Association
UL	Underwriters Laboratories

4.4 Designing for Maintainability

"Maintainability is a characteristic of design, whereas maintenance is the result of design…." the definition provided in Ref. [1]. Therefore, designers must focus on ensuring that the HVAC systems they design and specify can be maintained in order to achieve their design service life.

Two aspects of building design have major impacts on the maintainability of HVAC systems: where HVAC equipment is located and the access provided for its maintenance and, ultimately, replacement and the indoor and outdoor space provided for HVAC equipment maintenance.

Accessible is defined as not requiring a contortionist for routine maintenance chores. Examples of poor accessibility include equipment installed above inaccessible ceilings; equipment access panels on the wrong side of equipment, that is, where there was no space for access; and air-handling units with no coil access.

4.4.1 HVAC Equipment Location and Access

HVAC system components are typically located on the roof outdoors, on the ground outdoors, in mechanical equipment rooms indoors, and within ceiling cavities indoors. There are specific requirements relative to maintainability of HVAC equipment at each of these locations.

Outdoor equipment is easier to install, easier to service, and much easier to replace in the future if located on the ground rather than on the roof. If a rooftop location is selected, building codes may require safety railings to be installed to protect service personnel if the equipment is installed within 6 ft or less of the roof edge. If the roof slopes, level service platforms may also be required. And, ultimately all rooftop equipment must be replaceable, so staging areas for cranes, and so on, must be provided.

The following guidelines should be followed to ensure proper outdoor equipment operation and accessibility for maintenance:

1. Air-cooled condensing equipment rejects heat to the outdoor air, and therefore, these units cannot be enclosed or placed in a position where the rejected heat recirculates into equipment intakes. Provide space for the units, adequate separation between multiple units, and space around the units for maintenance and air movement in accordance with the manufacturer's requirements.
2. The outdoor area requirement for cooling towers, evaporative condensers, and so on is discussed in Chapter 11.
3. Packaged hot water boilers designed for outdoor installation are available in capacities up to about 2,000,000 Btu/h [600 kW]. Typically, these boilers are about 3–6 ft square [0.9–3.6 m^2] and 4–6 ft [1.2–2 m] high. A single boiler requires 75–100 sf [7.5–10 m^2] of outdoor area to allow for service, whereas two boilers require 120–140 sf [12–14 m^2] of area, with each additional boiler adding 50–60 sf [5–6 m^2] to this requirement. Since the hot surfaces of these boilers represent a potential public safety hazard, access to this equipment must be controlled via secure screening or fencing with lockable access openings.

Outdoor ground-level equipment access for routine maintenance is rarely a problem. However, when time comes to replace the equipment at the end of its service life, it is necessary to locate the equipment in an area where future truck and/or crane access is possible. Design any equipment screening or sound barriers in removable sections. Finally, allow adequate space for the potential that future equipment may be larger and/or configured differently.

Adequate access to outdoor rooftop equipment is a routine problem. A roof hatch with a vertical ladder (often located in a small janitor's closet or on the back wall of an electrical closet) is totally inadequate for even routine maintenance, *a stair to the roof is required*. In a multistory building, it is recommended that at least one elevator serve the roof level.

Indoor equipment must be located within mechanical equipment rooms sized to accommodate the equipment and to provide adequate clearances around the equipment for maintenance purposes. Mechanical equipment room access must include at least one double door, 5–6 ft [1.7 m] wide, to allow the replacement of equipment components such as motors, fan shafts and wheels, and pumps. If the room houses a chiller, removable wall panels or even roll-up doors to the outside should be provided so chiller tubes can be cleaned and replaced and the chiller itself can be replaced in the future.

Ceiling cavity space should not be used to house any HVAC equipment except for terminal units or small fan coil units. Ceiling cavity access to HVAC equipment can be provided by using a removable panel (layin) ceiling. If the ceiling itself is not an accessible type, access panels or doors are required to provide for routine service of terminal units, fire and smoke dampers, duct-mounted coils and humidifiers, controls, and so on. If the equipment contains a fan and/or a refrigeration compressor, provisions for removal and replacement in the future must also be provided.

4.4.2 Mechanical Equipment Rooms for Air-Handling Equipment

Whereas smaller buildings may use outdoor rooftop air-handling units, larger buildings will require "fan rooms" to house this equipment. The fan room dimensions are based on the size of the air-handling unit and its required maintenance space. The length (L) and width (W) of a typical air-handling unit required to serve a given building area can be determined from manufacturer data and, then, the minimum room dimensions established as follows:

Room length = $1.25 \times L$

Room width = $2.5 \times W$

Room height = 12–14 ft [4 m] (minimum to underside of structure)

On construction documents, designers should show coil pull, filter pull, and motor access space requirements to ensure that piping, raceway, or other equipment does not intrude into these spaces.

Designers must ensure that flood control is provided in all fan rooms, as a single, uncontrolled water leak can do hundreds of thousands of dollars of damage to a building. All floor openings should be protected with 2–4 in. [50–100 mm] high curbs or pipe sleeves to "dam" any loose water on the floor. Adequately sized floor drains must be provided, with floors sloped to these drains.

4.4.3 Mechanical Equipment Rooms for Boilers and Chillers

A boiler room, if required by the application, will house one or more boilers and required support equipment. For a hot water boiler system, space for pumps must be provided. For a steam boiler system, the space requirements are significantly greater, requiring support space for the feedwater and/or deaerators and water treatment systems.

Chiller rooms must be separated from boiler rooms and require space for the chiller(s), chilled water pump(s), condenser water pumps, and water treatment systems. ASHRAE *Standard 15,* adopted by many code authorities, classifies a chiller room as a "refrigeration machinery room" if the amount of refrigerant in the largest chiller exceeds certain specified levels. *If so classified, the room must be constructed to higher standards, including refrigerant leak monitoring and alarm, and a detailed code review for these spaces is critical.*

The space required for a boiler room and chiller room can vary significantly due to the number and arrangement of boilers and chillers. A reliable primary system will include at least two boilers and/or chillers to provide standby capacity in the event of equipment failure. Hospitals, research facilities, and so on must have at least three boilers and chillers (each sized for 50% of the load) so that the failure of one unit will not impact the ability to meet peak loads.

As a general guide, the following are the *minimum* space allocations for primary HVAC equipment rooms:

Hot water boiler room	0.7% of the gross building area
Steam boiler room	1.0% of the gross building area
Chiller room	1.5%–1.7% of the gross building area
Room height	14–18 ft [4.3–5.5m] clear below the bottom of the structure supporting the roof or floor above

On construction drawings, show tube pull/clean space requirements so that piping, raceway, or other equipment does not intrude into these spaces.

The designer must provide flood control measures in all primary HVAC equipment rooms. Floor drains must be provided adjacent to pumps, chillers, boilers, cooling towers, and so on, with floors sloped to each drain. Floor openings (except floor drains, of course) must be protected with 2–4 in. [50–100 mm] high curbs to dam loose water on the floor.

Spill control is also required around areas containing water treatment chemicals. The most cost-effective approach is for the HVAC engineer to specify the use of double-walled containers, with spill lips, for all chemicals. If this is not possible, a 6 in. [150 mm] high curb is required to contain the volume of the largest storage container. A 6–12 in. [150–300 mm] deep "pump-out" recess is also required to facilitate cleanup of a chemical spill.

4.4.4 Ceiling Cavity Space

Sufficient ceiling space for HVAC air distribution ductwork; HVAC, plumbing, and fire protection piping; and electrical raceway is required. These space requirements are in addition to the minimum space required for the ceiling itself and recessed lighting fixtures (including an allowance for their removal).

Table 4-1 reflects typical minimum ceiling space requirements for commercial and institutional building when air secondary systems are used.

Thus, the minimum floor-to-floor height in a multistory building is 44" [1100 mm] + floor-to-ceiling height + structure depth.

When water-type secondary systems are used, the ductwork space requirement can be eliminated in single-story buildings and reduced to only 12"–18" [300–450 mm] in multistory buildings.

The space requirements listed earlier must be increased for special occupancies such as laboratories, hospitals, and kitchens because of the much higher "density" of distribution systems in these types of facilities (see Chapter 6 for a discussion on the location and access requirements for aforementioned ceiling terminal units).

TABLE 4-1

Ceiling Space Requirements for Maintainability

Lighting	8" [200 mm] (6" [150 mm] for fixture plus 2" [50 mm] to lift and move fixture above the ceiling.)
Ductwork	24" [600 mm] (Ducts with high aspect ratio width to height increase installation costs.)
Piping and electrical raceway	12" [300 mm] (may increase to incorporate fire protection and/or roof drainage systems)
Total	44" [1,100 mm]

Never use fire-rated roof/ceiling or floor/ceiling assemblies in mechanical equipment spaces as it complicates the installation of hangers for piping, ductwork, equipment, and so on. Solid ceilings severely restrict access for maintenance, and the clip retainers for layin ceiling tiles are lost every time the ceiling is accessed, ultimately destroying the fire rating of the assembly.

4.5 Designing for Energy Conservation

Design of HVAC systems to minimize energy burdens imposed by them is an important design criterion, generally dictated by compliance with energy conservation building codes and standards listed as follows.

4.5.1 Introduction to ASHRAE Standard 90.1

The United States does not have a national energy code or standard, even though the federal government supports the development of energy codes and standards. Because there is no national energy code, energy codes and standards are adopted at the state and/or local jurisdiction levels. The path that energy codes and standards take for adoption varies by locality. The *IECC*, and the vast majority of state and local energy conservations codes that use it as a model, is based on ASHRAE *Standard 90.1, Energy Standard for Buildings Except Low-Rise Residential Buildings*, which applies to commercial and institutional buildings and to residential buildings exceeding three stories in height. The IECC also addresses low-rise residential buildings on the basis of ASHRAE *Standard 90.2*, but this type of construction is outside the scope of this book. Therefore, to understand most energy conservation codes, it is necessary to understand ASHRAE *Standard 90.1*.

ASHRAE *Standard 90.1* has been around since 1975, when the first Arab oil embargo emphasized the worldwide dependence on fossil fuels, particularly with respect to buildings. Since 1975, the standard has proceeded through four "generations" of updates (where major changes took place) and is currently routinely updated every 3 years.

With each new edition of ASHRAE *Standard 90.1*, the Department of Energy (DOE) issues a "determination" on whether the new edition will improve energy efficiency in commercial buildings over the existing edition. The publication of a positive review by DOE sets in motion statuary requirements directing states and local authorities to certify that they have reviewed their building codes and that all necessary updates are completed to meet or exceed the current edition of ASHRAE *Standard 90.1*.

Thus, for HVAC designers, it may be better to simply ignore current or local code requirements and design to comply with the most recent edition of ASHRAE Standard 90.1.

At time of this writing, ASHRAE *Standard 90.1* is comprised of 12 sections and 7 appendices that provide additional information but are not part of the requirements promulgated by the standard. All buildings must comply with the *mandatory* requirements within each section of the standard. From there, there are two additional alternative means, or *paths* for building compliance:

1. *Prescriptive path*: All components of the building meet the minimum standards specified by the standard, which, for the HVAC designer, includes all or part of the following sections:

   Building envelope (Section 5) defines minimum wall insulation, minimum roof insulation, roof reflectance, and minimum glazing performance.

HVAC (Section 6) defines minimum equipment efficiency, minimum system features, limitations on reheat, limitations on fan power, requirements for ventilation control, economizer cycles, and VAV requirements.

Lighting (Section 9): maximum indoor *lighting power density*, expressed in Watts/sf, minimum lighting controls, exterior lighting, parking garage lighting, and daylighting application and control.

Other equipment (Section 10): electric motors, potable water booster pumps, elevators, and escalators.

2. *Performance path*: A proposed building design is demonstrated, via building energy modeling (BEM), to use less energy than a "baseline" building.

   In the performance approach, a baseline *energy cost budget (ECB) is established, based on the building size and program. This baseline* ECB is established using building energy simulation to model a building with the same size and program as the project building, but designed in full compliance with the prescriptive requirements of Sections 5–10. Oddly, the ECB is expressed in units of dollars, which means that energy sources and costs are a factor in the analysis.

   The performance approach may also be used to demonstrate design energy efficiency, often expressed as percent better than a previous edition of ASHRAE *Standard 90.1*.

Outside the United States, India's *Energy Conservation Building Code* has a similar form and scope to ASHRAE *Standard 90.1*. Other countries have different forms, such as Ireland's *Building Energy Rating* and Hong Kong's *Buildings Energy Efficiency Ordinance*.

4.5.2 Simplified Energy Analysis

In the United States, prior to the widespread use of computers for building energy analysis and modeling, two simplified manual methods were used by designers. And, for preliminary analysis of design alternatives, these methods may still have use.

1. *Degree-day method*: The most common single-measure methods are the *degree-day concept* for heating and the *equivalent full-load hours* concept for cooling. For residences and other simple structures and systems, the heating degree-day concept assumes that on a long-term average solar and internal gains will offset heat loss when the mean daily outdoor temperature is 65°F and that fuel consumption will be proportional to the difference between the mean daily temperature and 65°F. This basic concept can be represented in the following equation, stating that energy consumption is directly proportional to the number of degree-days:

$$E = C_D \times (\text{HL} \times \text{HDD} \times 24)/(\text{TD} \times k \times \text{HHV}) \tag{4-1}$$

   where

   E = fuel or energy consumption for the estimate period

   HL = design heat loss, including infiltration and ventilation, Btu/h

   HDD = number of 65°F degree-days for the estimate period

   TD = design temperature difference, °F

k = correction factor that includes the effect of rated full-load efficiency, part-load performance, oversizing, and energy conservation devices:

$k = 1$ for electric resistance heating

$k = 0.55$ for conventional fossil-fuel heating systems

$k = 0.65$ for high-efficiency condensing fossil-fuel heating systems

HHV = higher heating value of fuel, units consistent with HL and E

C_D = empirical correction factor for heating effect versus 65°F degree-days, as shown in Table 4-2

Single-measure methods for estimating cooling energy are not widely applied because of the more complex factors that influence cooling energy requirements.

2. *Modified temperature bin method*: While no longer widely used, this methodology was an early attempt to address the need to consider "multiple measures" in evaluating building energy consumption. This method allocates monthly weather data into discrete 5°F temperature "bins" and summarizes the monthly hours of occurrence within each temperature bin, dividing the hours into "occupied" and "unoccupied" periods. Then, peak heat transmission gains/losses, solar radiation heat gain, internal heat gains, and so on are prorated for each bin. Next, each bin load condition is applied to compute the energy consumed by a specified type of HVAC system. Finally, these bin energy consumptions are multiplied by the hours of occurrence and totaled on a monthly and annual basis.

4.5.3 Building Energy Modeling

BEM is physics-based software simulation of building energy use. A BEM program takes as input a description of a building including geometry, construction materials, and lighting, HVAC, refrigeration, water heating, and renewable generation system configurations, component efficiencies, and control strategies. It also takes descriptions of the building's use and operation including schedules for occupancy, lighting, plug loads, and thermostat settings. A BEM computer program combines these inputs with information about local weather and uses physics equations to calculate thermal loads, system response to those loads, and resulting energy use, along with related metrics such as occupant comfort and

TABLE 4-2

Factor C_D for Heating Degree-Days

Degree-Days	C_D
0	0.86
1,000	0.80
2,000	0.75
3,000	0.70
4,000	0.66
5,000	0.60
6,000	0.62
7,000	0.65
8,000+	0.67

energy costs. BEM programs perform a full year of calculations on, typically, an hour-by-hour basis.

BEM applications may answer questions that cannot be easily answered by other means. Major use cases include the following:

1. *Architectural design*: Architects use BEM to design energy-efficient buildings, specifically to inform quantitative trade-offs between up-front construction costs and operational energy costs
2. *HVAC design and operation*: Commercial building HVAC systems can be large and complex. BEM helps mechanical engineers design HVAC systems that meet building thermal loads efficiently. It also helps design and test control strategies for these systems.
3. *Electrical and lighting design and operation*: Everything from the performance of different lighting systems and source (including daylighting), cogeneration, onsite renewable energy sources, and so on, including their impact on HVAC system energy consumption
4. *Building performance rating*: BEM can be used to assess the inherent performance of a building compliance with ASHRAE *Standard 90.1* and *Standard 189.1*.

Analysis software for BEM must meet the requirements of ANSI/ASHRAE *Standard 140, Standard Method of Test for the Evaluation of Building Energy Analysis Computer Programs*. Current software that performs BEM and meets this standard includes the following:

1. *EnergyPlus* is a state-of-the-art BEM "engine" developed by USDOE that is capable of modeling low-energy buildings and HVAC systems, in addition to more conventional buildings. Replacing the DOE 2 engine, EnergyPlus provides more accuracy and fewer work-arounds, and enables the analysis of more innovative and complex mechanical system and building designs.
2. *OpenStudio* is a software development kit (SDK) that reduces the effort of EnergyPlus-based application development. It also includes a graphical application that demonstrates use of the SDK.
3. *TRACE 700*: "Trane Air Conditioning Economics," or TRACE, was developed by the Trane Inc., a subsidiary of Ingersoll Rand, is a design and analysis tool developed to help professionals optimize the design of a building's heating, ventilating, and air-conditioning systems based on energy use and life-cycle cost.
4. *Carrier HAP*: Like TRACE, the HAP software was developed by Carrier Corporation, a brand of the UTC Climate, Controls and Security division of United Technologies Corporation, UTC Building & Industrial Systems, Inc., as a design and analysis tool developed to help professionals optimize the design of a building's heating, ventilating, and air-conditioning systems based on energy use and life-cycle cost.
5. *DesignBuilder*: Developed in the United Kingdom by DesignBuilder Software, Ltd. This software provides advanced modeling tools in an easy-to-use interface to EnergyPlus. This enables the whole design team to use the same software to develop comfortable and energy-efficient building designs from concept through

to completion. DesignBuilder v5 includes significant productivity improvements for LEED and ASHRAE *Standard 90.1* compliance and a new set of scripting tools to allow EnergyPlus simulations to be customized. Other highlights include combined heat and moisture simulation and a HVAC fluid-to-fluid heat exchanger component performance modeling.

6. *eQUEST* is the DOE 2 engine with wizards and graphics built on top of it. eQUEST is the most popular energy modeling program in existence. It is used by energy modelers and engineers all over the world. One big contributing factor to its popularity is its cost, FREE, and the fact that it is built on the DOE 2 simulation engine. The other benefit of eQUEST is that it can be used at every stage of building development, from the early designs to final stages.
7. *TRNSYS* is a simulation program primarily used in the fields of renewable energy engineering and building simulation for passive as well as active solar design. TRNSYS is a commercial software package developed at the University of Wisconsin.
8. *Lesosai* is a user-friendly commercial software helping in energy analysis including life-cycle cost analysis. It is based on internationally standardized procedures. Languages are English, French, German, and Italian. Lesosai allows the calculation of environmental impacts of the energy consumption, taking into account the energy used in the building, but also imbedded in the building's construction materials. This calculation is based on a building life-cycle approach and uses the list of impacts maintained by KBOB (the Swiss coordinator of federal, cantonal, and communal building and real estate authorities), and the methodology conforming to Swiss *Standard SIA 2032*. To simulate buildings anywhere in the world, Lesosai includes the meteorological data generator, Meteonorm, and includes many official meteorological datasets.

4.5.4 Exhaust Air Heat Recovery

Air-to-air energy recovery heat exchangers can significantly reduce the energy needed to cool and heat ventilation makeup air and are of particular import in high occupancy buildings, hospitals, laboratories, and so on. The technology can be used effectively in any building that is reasonably tightly constructed, with the return/exhaust air duct(s) located close to the fresh makeup air intake(s).

Any project that requires 30% of more outdoor intake to satisfy general ventilation requirements defined by ASHRAE Standard 62.1 is a candidate for exhaust heat recovery.

There are two basic types of exhaust air heat recovery:

1. *Sensible heat recovery*: The four most common exhaust air-sensible heat recovery systems are air-to-air plate heat exchangers, thermal wheels, heat pipe coils, and *run-around* systems.
    a. Air-to-air plate heat exchangers use thin metal plates arranged so that the exhaust and outdoor airstreams are in a crossflow configuration, as shown in Figure 4-1.
    b. Thermal wheels use a rotating wheel of light metal or ceramic material that, while in one airstream, is heated and, as the heated medium enters the adjacent colder airstream, gives up that heat, as illustrated in Figure 4-2.

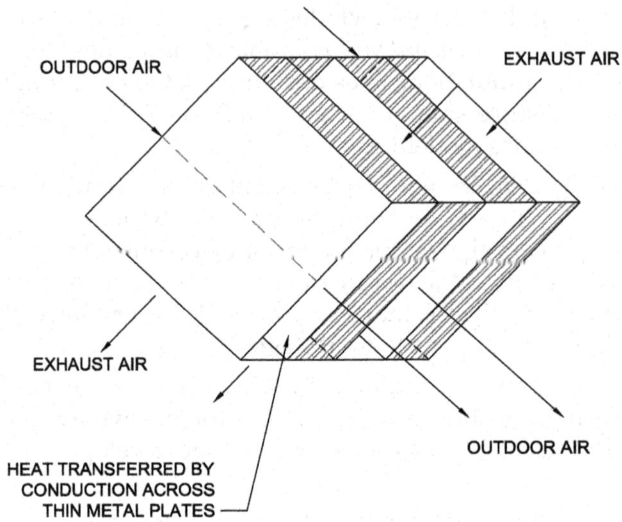

FIGURE 4-1
Crossflow air-to-air heat exchanger.

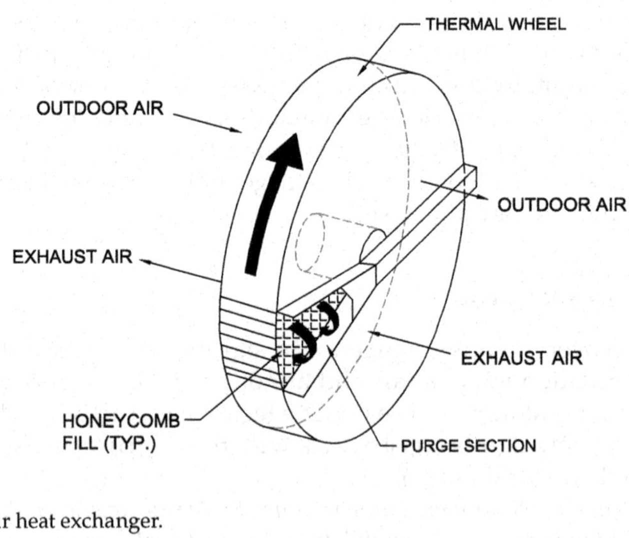

FIGURE 4-2
Rotary wheel air-to-air heat exchanger.

c. The heat pipe coil, a by-product of the space program, contains a refrigerant that, in the hot airstream, absorbs heat and is evaporated. A capillary action forces the liquid down the center "wick" and causes the gas to move to the cold airstream. Here, it gives up heat (is cooled) and condenses.

d. The run-around system, as illustrated in Figure 4-3, consists of two water coils and a circulation system of water/polyethylene glycol. Heat is absorbed in one coil and pumped to the other, where it is rejected.

Using any of these systems, air-to-air heat recovery can be analyzed as follows:

a. Determine the airflow rates for both the exhaust air and the intake air to be heated and cooled.

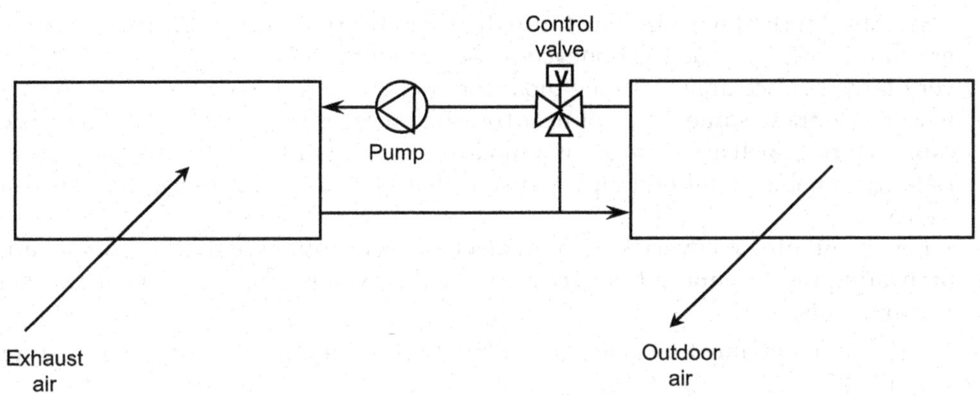

FIGURE 4-3
Run-around system air-to-air heat exchanger.

b. From the manufacturer's data, determine the heat recovery factor, R_x, for both the exhaust side and the intake side. Note that the total heat recovery factor is dictated by the *smaller* airflow rate. The value of R_x is typically 0.55–0.65.
c. Determine the average exhaust air temperature, T_e (°F) [°C]. (Since latent heat transfer is not possible except in dirty heat wheels or heat wheels using a desiccant media, it can be ignored.)
d. At any outdoor air temperature, T_o (°F) [°C], the total heat transfer, Q (Btu/h) [kW], can be computed using Eq. (4-2) or (4-2a).

$$\text{For IP units,} \quad Q = 1.1 \times V \times (T_e - T_o) \times R_x \tag{4-2}$$

where V is the smaller of the exhaust or intake airflows (cfm).

$$\text{For SI units,} \quad Q = 1.3 \times V \times (T_e - T_o) \times R_x \tag{4-2a}$$

where V is the smaller of the exhaust or intake airflows (m^3/s).

For most exhaust air/ventilation air configurations, there is no need for heating or cooling when the outdoor temperature is between the average space set point temperature and the cooling coil discharge air temperature, so no heat recovery is required. But the heat content of the exhaust air still has value, and at least with the run-around system, this heat can be used to preheat service water with the addition of a water-to-water heat exchanger to the run-around system piping loop.

2. *Enthalpy or total heat recovery*: Enthalpy heat recovery systems, typically thermal wheels, recover both sensible heat and latent heat from the exhaust airstream, and thus, their heat recovery factor improves from a maximum of about 65% to a maximum of 80%.

For an enthalpy recovery system to meet its maximum potential precooling and preheating performance, the exhaust flow rate must be at least 75% of the outdoor airflow rate.

Sensible thermal wheels, which use disks made up of thin metal, plastic, paper, or ceramic surfaces, such as honeycomb or a random woven screen mesh, to create very large surface areas, are the most common system in use. Enthalpy thermal wheels use these same types of heat transfer surfaces but also incorporate desiccant material, such as silica gel or a molecular sieve (adhered to the matrix material), that enables total enthalpy transfer, that is, both mass (moisture) and heat transfer.

For an enthalpy recovery system to meet its maximum potential precooling and preheating performance, it is necessary for the exhaust airflow to meet two key requirements:

1. The flow rate must be a significant fraction of the makeup airflow rate (typically 75+%).
2. The temperature and humidity of the exhaust air must be close to those of the conditioned space (i.e., heat loss or gain in the return or exhaust ductwork must be small).

4.5.5 Internal Source Heat Recovery

All buildings produce heat. This heat may be from internal heat gains from lights, people, and appliances or may be more concentrated heat sources such as computer rooms, telephony rooms, and process equipment. This heat must be removed by cooling in order to maintain the space temperature set point required for proper HVAC system operation.

Heat absorbed by cooling systems is typically rejected to the outdoors via condensing units or cooling towers. However, if there is a simultaneous need for that heat in other areas of the building, it may be cost-effective to use that heat to satisfy this need. The most common way of applying this concept is to recover the heat from condenser water.

Condenser water typically leaves a chiller at 85°F–95°F [27°C–35°C]. If the heating water system is designed for about 100°F–105°F [38°C–40°C] supply temperature and a 20°F–25°F [11°C–14°C] temperature range, then condenser water is sufficiently high enough to preheat the return heating hot water.

Since heating is normally required when outdoor temperatures are low enough for HVAC systems to provide "free" cooling via an airside or waterside economizer system, it is necessary to determine which operating mode is more cost-effective:

1. Satisfy the building internal heat gains via mechanical cooling and use the rejected condensing heat to help offset building heat losses.
2. Satisfy the building internal heat gains via an economizer cycle and use a boiler system to offset building heat losses.

To implement this concept, a water-to-water heat exchanger, either plate-and-frame or shell-and-tube type, designed for a 2°F–4°F [1°C–2°C] approach is installed in the condenser water return piping and piped so that return heating water passes through the heat exchanger as it is routed back to the boiler. The heat exchanger, then, becomes the first stage of heat rejection for the refrigeration condenser and the first stage of heating for the boiler system.

Adding to the overall efficiency of this concept is the fact that the heat rejected to the building heating water does not have to be rejected by the cooling tower and the tower fans do not have to operate as much, reducing tower energy use.

4.5.6 Thermal Energy Storage

Cooling thermal energy storage (TES) systems should only be considered for facilities with large swings in cooling loads and electrical demand and where electrical rate demand charges are high. Due to the complexity and economic risks associated with TES, only knowledgeable, sophisticated owners should be considered for TES installations.

There are three general thermal storage system technologies: water, ice, and phase change eutectic salts, though the last one has little application for HVAC applications. These technologies can vary significantly in terms of size, efficiency, reliability, and cost, as summarized in Table 4-3.

The goal for thermal storage in a chilled water system is to reduce the *demand* for purchased cooling energy during on-peak periods by shifting cooling energy consumption to off-peak periods. This reduces electrical charges, particularly demand charges, on a year-around basis.

Aside from shifting cooling energy consumption from the on-peak to the off-peak period, *rarely is there any reduction in overall cooling energy consumption with TES,* simply because every thermal energy system will have losses. Water and ice storage systems have heat gains that increase the cooling energy consumption. Water systems have thermal mixing problems that require more cooling to be stored than can actually be used. Ice systems, due to the reduced temperatures, require greater compressor energy than an equivalent water storage system. All of these losses may or may not be offset by improved chiller coefficient of performance (COP) due to reduced condensing temperatures during off-peak periods when lower outdoor temperatures prevail.

Control of the thermal storage system is critical to ensure that enough cooling energy is stored during the off-peak period to offset the shortfall in chiller capacity during the on-peak period. Conversely, it is a waste of energy to store more cooling than can be used during the following on-peak period.

TABLE 4-3

Chilled Water TES Advantages/Disadvantages

Storage Medium	Volume (cf/ton-h) [L/kWh	Storage Temp (°F) [°C]	Discharge Temp (°F) [°C]	Advantages	Disadvantages
Chilled water	12–15 [96–120]	36–44 [2–7]	40–46 [4–8]	Can use existing chillers.	Large storage volume.
				Stored water can be used for fire protection.	Higher heat gains. Difficult to maintain supply temperature set point.
Ice	2.4–3.3 [19.3–26.6]	32 [0]	34–36 [1–2]	Low storage volume	Poor chiller COP due to low temperature requirements.
				High discharge rates.	Expansion problem.
				Easy to maintain supply temperature set point	Poor thermal conductivity of ice.
Eutectic salts	6 [48]	47	48–50	Can use existing chillers.	Relatively large storage volume. High chilled water supply temperature.

The analysis of thermal storage economics requires that an accurate peak 24-h cooling load profile be developed. Basically, this consists of developing, for the cooling design day, the outdoor temperature and imposed cooling load at each hour. Then, the day must be divided into the on-peak and off-peak periods defined by the applicable electric rate schedule.

The total daily cooling requirement must still be met by the chiller(s), but using thermal storage, the chiller(s) runs 24h, not 12–14h, each day. The chiller(s), then, can be sized on the basis of the "split" between the energy required for *charging* during the off-peak period versus the energy consumed in *discharging* during the on-peak period.

Charging represents the off-peak period production of cooling energy by the chiller that is stored, whereas discharging is the use of that stored energy during the on-peak period. The amount of cooling required during the on-peak discharge, divided by the amount of cooling energy stored during the off-peak charging, is defined as the *storage efficiency*, typically 0.95 or greater for ice systems and 0.90–0.95 for water systems.

Selecting a chiller and the optimum amount of storage requires an evaluation of all of the cost factors under different chiller capacity/storage ratios, and selecting the combination of chiller and storage yields the best economic performance. There are three general thermal storage system technologies: chilled water, ice, and eutectic salts. These technologies can vary significantly in terms of size, efficiency, reliability, and cost, as summarized in Table 4-3.

The use of thermal storage will be economically attractive only if the following conditions apply:

1. The peak cooling load is much higher than the average cooling load over the design day; that is, peak loads are relatively short duration (more typical of colder climate zones).
2. The electric utility rate includes high demand charges, ratchet charges, and/or high differential between on-peak and off-peak rates.

The following steps are required to determine the feasibility of a thermal storage system:

1. Consider the owner's ability to operate a chilled water system with thermal storage. These systems require ongoing monitoring and adjustment to work properly and efficiently. And, *system failure can have significant negative financial impacts* if cooling demand must be met by mechanical cooling if the TES fails.
2. Determine the cooling load profile, hour-by-hour during the peak cooling period, in order to determine required chiller and storage capacities. Then, using a detailed annual cooling load profile for existing facilities or hour-by-hour computer modeling for new facilities, compute the annual operating cost for the system. Compute electrical costs using the exact rate schedule(s) to be applied to this system.
3. Select the thermal storage technology to be used. Ensure that all operating modes are fully defined by the design and incorporated into the system controls.
4. Determine the capital requirements and anticipated economic performance of the thermal storage system. While simple payback methods can be applied, more sophisticated methods such a rate of return or life-cycle cost analysis should be applied.

5. Negotiate a long-term contract with the electric utility. This contract should establish the specific rates to use and provide for exemptions from demand peaks caused by occasional system failure. *The contract period must be longer than the estimated payback period for the system.*
6. While the thermal storage system is being installed, throughout the commissioning period, and through the first year of operation, the operating staff must be fully involved, trained, and monitored to ensure successful system performance.

4.6 Designing for Sustainability

4.6.1 High-Performance Building Elements

One of the first and most oft-cited definitions of *sustainability*, and almost certainly the one that will survive for posterity, is the one created by the Brundtland Commission. The Commission defined sustainable development as development that "meets the needs of the present without compromising the ability of future generations to meet their own needs." The Brundtland definition thus implicitly argues for the rights of future generations to raw materials and vital ecosystem services to be taken into account in decision-making.

The following definitions of sustainability are more specific:

> Sustainable means using methods, systems and materials that won't deplete resources or harm natural cycles.
>
> *(Rosenbaum, 1993)*

> Sustainability identifies a concept and attitude in development that looks at a site's natural land, water, and energy resources as integral aspects of the development.
>
> *(Vieira, 1993)*

> Sustainability integrates natural systems with human patterns and celebrates continuity, uniqueness and place-making.
>
> *(Early, 1993)*

In review of these definitions, environmental context is an important variable to most working definitions of sustainability. This emphasis is expressed in the following composite definition: "Sustainable developments are those which fulfill present and future needs while only using and not harming renewable resources and unique human-environmental systems of a site: air, water, land, energy, and human ecology and/or those of other off-site sustainable systems."

Thus, sustainability boils down to two environmental elements:

1. *Resource depletion*: This element can be simply defined as "not exceeding the capacity of our planet to support our life styles." This includes not only physical resources such as fossil fuels and materials (minerals, metals, timber, etc.) but also human resources, ensuring a good quality of life for individuals and communities alike.

2. *Pollution*: Basically, this element means "not fouling our nest" and includes the need to reduce waste, air and water pollution, and ozone depleting and greenhouse gas emissions.

But, while the definition of what constitutes sustainability continues to evolve, there are six fundamental principles that nearly everyone agrees on that defines *sustainable building design*, which are as follows:

1. *Sustainable building planning*: Creating sustainable buildings starts with giving priority to adaptive reuse or rehabilitation of existing buildings. In 2040, 40%–50% of the buildings expected to be in use then are in existence today. *Therefore, to make a significant impact on the sustainability of the built environment, we need to focus 50%–60% of our design efforts on those buildings that are already built.*

   If new construction is required, the location, orientation, and landscaping of the new building will affect local ecosystems, transportation methods, and energy consumption, and this impact must be thoroughly evaluated to mitigate it. Siting for physical security has become a critical issue in optimizing site design, the location of access roads, parking, vehicle barriers, and perimeter lighting must be integrated into the design along with sustainable site considerations.

   Architectural design elements such as interior space planning, envelope design, and the use of natural daylighting and/or ventilation can have major impacts, both directly and indirectly, on the energy consumed by the building and the resulting greenhouse emissions.
2. *Energy efficiency*: Given the impact of greenhouse gases (measured by equivalent CO_2 levels) on the world's climate, it is essential to find ways to reduce heating/cooling loads, increase efficiency, and use renewable energy resources in buildings.

   Energy efficiency, aside from reducing the rate of depletion of limited natural resources and reducing the building operating cost, also reduces the buildings "carbon footprint," a media term that reflects the direct or indirect release of CO_2 into the atmosphere, contributing to global warming.

   Direct release of CO_2 may result from a building's direct combustion processes for heating, service water heating, and process heating. Atmospheric CO_2 is also a by-product of engine-driven systems in a building, such as oil or natural gas fuel electrical power generators, and combined heat and power systems. In Chapters 4 and 5, improving building system's combustion efficiency and reducing CO_2 emissions are discussed.

   The indirect release of CO_2 results from the building's purchased electrical energy consumption. In the United States, about 70% of all electrical power is produced by coal-, oil-, and/or natural gas-fueled generating methods. These boiler and engine combustion processes release CO_2 directly into the atmosphere, unlike nuclear, hydroelectric, geothermal, solar, and wind electrical power generation systems. *Today, each kWh of purchased power, on average, results in the release of 1.760 lb [0.8 kg] of CO_2 into the atmosphere.*

   In addition to reducing the building's energy requirements, sustainable design must include evaluation of the sources for the energy that is used. Electrical power can be purchased on the basis of "green" generation by nuclear, hydroelectric, geothermal, solar, and/or wind. Onsite power can be generated using solar energy through photovoltaic arrays.

3. *Protect and conserve water*: In many parts of the country, clean water is an increasingly scarce resource. A sustainable building should reduce, control, or treat site runoff, use water efficiently, and reuse or recycle water for on-site use when feasible.
4. *Environmentally preferable products*: A sustainable building should be constructed of materials that minimize life-cycle environmental impacts such as global warming, resource depletion, and human toxicity. These environmentally preferable materials are products that have a lesser or reduced effect on human health and the environment when compared with competing products or services that serve the same purpose. They have lower net energy content, low pollutant emissions, have acceptable recycled material and bio-based content, and are not ozone depleting.
5. *Indoor air quality*: It, within a building, has a significant impact on occupant health, comfort, and productivity. Among other attributes, a sustainable building should maximize daylighting, have appropriate ventilation and moisture control, and limit noise and vibration imposed on the occupants.
6. *Operational and maintenance practices*: Incorporating operating and maintenance considerations into the design of a facility will greatly contribute to improved working environments, higher productivity, and reduced energy and resource costs. Designers are encouraged to specify materials and systems that simplify and reduce maintenance requirements; require less water, energy, and toxic chemicals and cleaners to maintain; and are cost-effective; and reduce life-cycle costs. Owners, in turn, must provide adequate preventative and renewal maintenance to ensure that the building and its systems remain functional and efficient. Building operation and maintenance is discussed in Chapter 9.

Therefore, there is a need to establish specific, *science-based* performance goals for new buildings and renovations to existing buildings in order to produce high-performance buildings:

1. *Planning*: A new planning concept is required, including justification for new construction, emphasis must be placed on using existing facilities, not building new ones. Each building's energy and water consumption and cost must be computed and included as part of the life-cycle cost analysis.
2. *Integrated design principles*: A design team, consisting of the owner(s), architects, engineers, and, perhaps, energy and environmental consultants, must be assembled to plan each building. To ensure that the performance goals of this order are met, these team members must be experienced in the design of high-performance sustainable buildings, perform extensive evaluation of design alternatives using life-cycle methods, and, at no later than the end of the schematic design phase, have the design and the design decisions reviewed.
3. *Sustainable design criteria and goals*: Specific design criteria and goals for energy efficiency, water protection and conservation, indoor air quality, daylighting, and the environmental impact of materials must be met. In meeting the goals that are established, owners and designers must use life-cycle cost analysis to evaluate the economic performance of design alternatives.
4. *Measurement and verification*: No process can be managed without the measurement and evaluation of performance and results. Therefore, this method establishes

requirements for installing utility metering, tracking building utility use over the first 2–5 years of life of each building, and implementing any remedial actions necessary to correct problems with buildings that do not meet established performance goals.

4.6.2 Introduction to ASHRAE Standard 189.1

The primary standard applying to sustainable design is ASHRAE *Standard 189.1, Standard for the Design of High-Performance Green Buildings Except Low-Rise Residential Buildings*. This is a consensus standard with scientifically proven basis and demonstrated performance, unlike the LEED "Green Building Rating System" developed by the U.S. Green Building Council.

In recent years, there has been increasing concern that dependence on the LEED rating system is not supportable and, in fact, its use increases professional liability exposure for designers. *It is far better to base sustainable design on the ASHRAE Standard 189.1 requirements.* ASHRAE Standard 189.1 defines high-performance building requirements using enforceable, code-friendly language to ease the process of adoption by local jurisdictions.

Although the site sustainability features of the standard seem distant from HVAC designs, site shading and solar reflectance index do impact HVAC loads and the energy consumption of the building. Designers must pay extra attention when evaluating these opportunities to maximize the total benefit. In the water section, cooling coil condensate collection must be incorporated for air-conditioning units above 5.4 tons [19 kW] in the climate locations specified by the standard. Designers must also follow the requirements in reducing cooling tower water use.

In the energy section, the mandatory provisions refer to ASHRAE *Standard 90.1* equipment efficiencies. Both prescriptive-option and performance-option provisions include renewable energy systems. For prescriptive-option provisions, the Standard Renewables Approach uses ASHRAE *Standard 90.1* equipment efficiencies with requirements for more renewable energy, whereas the Alternate Renewables Approach uses ASHRAE *Standard 189.1* equipment efficiencies without requiring as much renewable energy.

For performance-option provisions, performance option A uses ASHRAE *Standard 90.1* Appendix G (as modified by Section 7.5.2 of ASHRAE *Standard 189.1*), whereas performance option B uses ASHRAE *Standard 189.1* Appendix C.

Most HVAC-related requirements for the indoor environmental quality are mandatory provisions whereby designers must comply with specific sections in both ASHRAE *Standard 62.1* and *Standard 55*. Minimum ventilation rates, outdoor air delivery monitoring, filtration and air cleaning, thermal comfort, acoustical control, and moisture control are all part of the mandatory requirements.

In addition to the mandatory requirement of no CFCs, multiple-attribute product declaration is required by the Materials and Resources section. EPD-certified products can be found from various Program Operators' websites (e.g., UL *Sustainable Product Guide*), including a handful of HVAC-related products.

Finally, the whole construction process and plans for operation section of the standard is mandatory. This section is to ensure that the high-performance green building has been constructed, commissioned, operated, maintained, and sustained throughout its lifetime.

4.6.3 Net Zero/Sum Zero-Energy Buildings

The "zero energy," "net zero energy," and "sum zero" are concepts that are relatively new, and there are, as yet, no definitive, widely accepted zero-energy metrics to guide designers.

In the United States, ASHRAE, the DOE, and the National Renewable Energy Laboratory (NREL) have spearheaded much of the work on net zero-energy buildings (NZEB) to date. NREL presents several definitions for *net zero energy*, and they encourage building designers, owners, and operators to select the metric that best fits their project. NREL suggests four ways in which net zero energy may be defined:

1. *Net zero site energy*: Site energy refers to the energy consumed and generated at a site (e.g., a building), regardless of where or how that energy originated. In a net zero site energy building, for every unit of energy the building consumes over a year, it must generate a unit of energy, preferably from renewable sources.
2. *Net zero source energy*: Source energy refers to primary energy needed to extract and deliver energy to a site, including the energy that may be lost or wasted in the process of generation, transmission, and distribution. ASHRAE has postulated that, on an annual basis, a NZEB can produce or consume as much energy from renewable sources as it uses, while maintaining an acceptable level of service and functionality. *NZEBs can exchange energy with the power grid as long as the net energy balance is zero on an annual basis.*
3. *Net zero-energy costs*: Net zero-energy cost is, perhaps, the simplest metric to use: it means that the building has an energy utility bill of zero over the course of a year. In some cases, building owners or operators may take advantage of selling renewable energy credits (RECs) from on-site renewable generation.
4. *Net zero-energy emissions*: Net zero-energy emissions buildings either use no energy that results in greenhouse gas emissions or offset all emissions by exporting emissions-free energy (typically from on-site renewable energy systems).

In Europe, the push for NZEBs (referred to as "nearly zero-energy buildings" in the EU) is much stronger than in the U.S. *Directive 2010/31/EU*, generally referred to as the "Energy Performance of Buildings Directive" (EPBD), requires that *new public buildings* constructed in and after 2018 be NZEBs and *all new buildings* be NZEB by 2020. Further, in December 2017, EU countries agreed to require that *all existing buildings* be renovated to meet NZEB criteria by 2050 and included specific progress goals for 2030 and 2040. The exact definition of an NZEB is, however, left to each EU member country to promulgate, within specific criteria defined by the EPBD.

In the United Kingdom, the government now requires that all new residential construction be nearly zero-energy buildings. Many other countries (Canada, India, Japan, etc.) have zero-energy building demonstration projects under construction or completed, and the concept is being closely evaluated.

Regardless of the definition or metric used for a NZEB, minimizing energy consumption through efficient building design is the fundamental design criterion and the highest priority. Energy efficiency is generally the most cost-effective strategy with the highest return on investment, and maximizing efficiency opportunities before developing renewable energy plans will minimize the cost of the renewable energy projects needed. Using advanced energy analysis tools, design teams can optimize efficient designs and technologies.

Energy efficiency measures include design strategies and features that reduce the demand-side loads such as high-performance envelopes, air barrier systems, daylighting, sun control and shading devices, careful selection of windows and glazing, and passive solar techniques.

Once building loads are reduced, these loads should be met with efficient equipment and systems. This may include energy-efficient lighting, electric lighting/daylighting controls, and high-performance HVAC systems and equipment.

Once efficiency measures have been incorporated, the remaining energy needs can be met using renewable energy technologies. Common on-site electricity generation strategies include photovoltaics, solar water heating, and wind turbines. Priority should be given to renewable approaches that are readily available, replicable, and most cost-effective.

Depending on the NZEB metric and guidelines used, buildings may be permitted to use energy generated off-site from renewable sources to offset energy used in a building. If space is limited, a facility owner may install dedicated wind turbines, solar collectors, and so on at a separate location or take advantage of renewable energy supplies offered by third parties.

Credit for off-site renewable generation can be gained by purchasing RECs. RECs are available from many renewable energy technologies. Large, utility-scale wind farms, solar plants, geothermal plants, and hydropower facilities generate electricity without using fossil fuels or primary energy. The costs of constructing and operating these generation facilities are often paid for by selling the "credit" for generating energy renewably (as well as selling the energy itself). The structure and market for RECs is evolving, and it varies regionally.

Ultimately, the only way to measure if a building is a NZEB is to look at the energy crossing the site boundary. Other definitions, including source, emissions, and cost, are based on this measured information and include weighing factors and algorithms to get to the metric of interest.

While this concept is just beginning to gain traction in United States, it is well entrenched in Europe and requires that HVAC designers take a very active role in any NZEB project. *But even if NZEB is not an immediate design goal for a new building or major renovation of an existing building, designers should always strive to create buildings that are as near to NZEB as economically feasible* (see Chapter 15).

4.7 Construction Documents Quality Control

No design is perfect and no designer is expected to be perfect, *but he or she is expected to not be negligent.* A major part of meeting the *professional standard of care* and avoiding negligence is to implement and follow specific procedures to ensure the quality of the design and the design documents.

The first step in this process is, obviously, to eliminate (or least reduce to a minimum) having the design fail to incorporate an owner requirement or expectation. This is, basically, an impossible goal, but every designer must strive to come as close as possible to attaining it, the more the project satisfies the owner, the less the potential for dispute and claims.

The *schematic design phase* defines the basic project scope and elementary space plan, systems definition, and level of finish for the project. During the *design development phase,* the details of the design must be developed. Too often, though, critical design decisions do not get made until construction drawings are underway simply because the designer fails to properly execute the design development stage (or even the schematic design phase) of the project. *At the end of this phase,* ***every*** *major and most minor design decisions must be made,* *the space plan must be developed in significant detail, and all support systems (structural, HVAC, plumbing, electrical, etc.) must be well defined so that no significant changes to these elements are required during the production of construction drawings.*

To meet this goal, the owner's needs, wants, and wishes must be well understood (and documented) by the design team. To ensure this, especially with the "inexperienced" owner, designers must make sure that the owner understands the project design in these early stages well enough to actually define his or her needs, wants, and wishes. Even if the client says he/she understands, document each review and owner decision. If the project gets to the construction documents phase, or worse yet to actual construction, before the client says "I didn't understand that it was like that…that won't work for me," then the project is on the road to trouble.

At the end of the design development phase, the owner should review and approve the design elements, in writing. Since nothing is perfect, any modifications to design during the construction drawings phase must be recorded and approved by the owner.

The construction drawings should not include any new, major decisions about the project. Construction drawings are produced to define all of the design decisions that have been made during the schematics and design development phases in sufficient detail to allow the project to be built.

Today, there are several goals the designer must meet in producing construction drawings:

1. They must be complete.
2. They must be accurate.
3. They must not be ambiguous.

Failure to meet any of these goals, even in part, may bring charges of negligence against the designers. While no set of construction documents is perfect, the aforementioned goals must be met to the maximum extent possible. *Since other designers are doing this, every designer can be held to that standard of care.*

To meet this goal for completeness, standardization of sheets and sheet layouts, the use of standard symbols and abbreviations, standard construction details, and, most importantly, office guidelines and checklists are critical steps to ensuring that everything required is included in the documents. Schedules must be standardized within the office. *The goal of standardization is to require the designer to provide all needed information, every time.* This has the added benefit of creating an "office" product, rather than a product that varies from designer to designer (probably with varying levels of quality, also).

To meet the goal for accuracy, a *quality control* system must be used. This system must address three factors:

1. *Information accuracy*: Does the information on the drawings match the "definition file" requirements and is the information correct? For example, one type of inaccurate information may be the interior finish for a space being defined differently than required by the owner. A more basic type of inaccurate information is simple math errors, such as three subdimensions that do not add up to the overall dimension shown.
2. *Missing information*: Missing information can range from the very simple to the very complex, like how did this project get on the street for bids with no wall fire ratings shown?

3. *Conflicting information*: An HVAC equipment schedule may call for a pump to have a 20 hp [15 kW] motor, but the electrical panel schedule indicates that the pump has a 10 hp [7.5 kW] motor. This type of conflict, based on inaccurate information transfer between disciplines or simply due to typographic errors, is deadly!

The worse enemy of providing accurate and adequate construction drawings is time. Many designers will become so focused on their production schedule that they will fail to adequately review and check the drawings before they are issued. Some will try to use the bid period to make corrections, issuing new drawings right before bids are due, with the effect of irritating of all the bidders and increasing bid prices. Others will even wait until after bidding to make corrections, hoping to negotiate "trade-offs" with the successful bidder. And, a few will not make required corrections until problems begin to show up during construction.

Design time on any project must be considered secondary to the quality of the project! A client may be unhappy if the project is delayed a week or even a month while accurate and adequate construction drawings are produced. But he or she will probably be unhappier if poor documents result in cost increases and time delays during the construction period.

Bibliography

Maintainability: A Key to Effective Serviceability and Maintenance Management, Benjamin S. Blanchard, Dinesh C. Verma, and Elmer L. Peterson (Wiley-Interscience, 1995).

Effective Building Maintenance, Herbert W. Stanford III, PE (Taylor & Francis, Ltd. and The Fairmont Press, 2010).

Directive 2010/31/EU of the European Parliament and of the Council of 19 May 2010 on the Energy Performance of Buildings.

Towards Nearly Zero-Energy Buildings, Definition of Common Principles Under the EPBD, Final Report, Ecofys, Politecnico di Milano/eERG, and University of Wuppertal, 14 February 2013, Project Number BESDE10788.

5

Heating/Cooling Load Calculation

5.1 Introduction

5.1.1 Building Thermal Balance

As illustrated in Figure 1-1, buildings may be viewed as thermodynamic *systems.* The advantage of such an approach is that it allows an overall view of the processes of heat gain and loss which is operating continually to change the environmental conditions felt by the occupants. The system "boundary" consists of the building construction elements, walls, roofs, floors, windows, and so on, that separate the *inside* from the *outside.*

Heating, ventilating, and air-conditioning (HVAC) designers are concerned with maintaining some desired conditions of temperature and humidity within the boundary, and to accomplish that, we must design HVAC systems that are capable of offsetting the energy lost or gained across the boundary with equal energy added or removed from the space by that system.

Heat loss is defined as the transfer of energy from the inside to the outside of the boundary, while *heat gain* is defined as the transfer of energy from the outside to the inside of the boundary or the addition of heat within the boundary. The mechanisms can occur simultaneously depending on the driving functions at any particular time.

5.1.2 Purpose of Load Calculations

Peak heat losses and gains, or the maximum anticipated, of each building space represent the *heating and cooling loads* and must be determined in order to establish the *capacity* of the HVAC system being designed. At the individual room or space level, the designer must determine the amount of supply air required to offset that room's peak heat gain or loss (whichever is larger). At the zone or system level, the sum of the individual room air flows defines the zone or system capacity required.

5.2 Mechanics of Building Heat Loss and Gain

Sensible heat gain is added to a space by *conduction, convection,* and/or *radiation,* driven by temperature differences across the building boundary, direct solar radiation, and internal heat sources. Heat added by conduction or convection immediately becomes cooling load. However, heat gain by radiation is not immediately converted to cooling load. Radiant energy must first be absorbed by the surfaces that enclose the space and the objects within the space (furniture, and so on). As soon as these objects become warmer than the space air, some of the heat is transferred to the air by convection, becoming cooling load. This mechanism takes time and, therefore, it may be some hours after a load is turned on in a space before 100% of the heat released appears as heat gain. Likewise, even after a load is turned off, the heat absorbed earlier will still be released into the space.

The composite *heat storage capacity* of the space surfaces and objects determines the rate at which their respective surface temperatures increase for a given radiant input and thus governs the relations between the radiant portion of heat gain and its corresponding part of the space cooling load. The thermal storage effect is critically important in differentiating between instantaneous heat gain for a given space and its cooling load for that moment. Predicting the nature and magnitude of this elusive phenomenon in order to estimate a realistic cooling load has been the basis of much research and, since about 1960, development of increasingly complex calculation methodologies. The same basic mechanisms govern all of these various load components and are described in the following subsections.

Latent heat gain occurs when occupants, equipment, and/or infiltration add moisture to a space. Latent heat gain is considered to be instantaneous.

Figure 1-2 graphically illustrates the complex components that contribute to heat loss and gain in a building.

5.2.1 Design Conditions

Outdoor design conditions are selected on the basis of the project location and how closely indoor conditions must be maintained. The American Society of Heating, Refrigerating, and Air-Conditioning Engineers (ASHRAE) *Handbook-Fundamentals* lists outdoor design conditions for selected locations throughout the world.

In the ASHRAE data, heating design temperatures are provided for two different criteria: "99.6%" and "99%." Each criterion indicates the percentage of annual hours at which the temperature is at or above the listed temperature. For critical applications (laboratories, hospitals, and so on), the 99.6% criteria are used. For all other applications ("comfort" air-conditioning), the 99% criteria are used.

For cooling design, three different criteria are used: "0.4%," "1%," and "2%." Again, these criteria represent the percentage of annual hours at which the temperature is at or above the listed temperature. For critical applications such as hospitals, and research facilities, the 0.4% criteria is used, while the 1% criteria is used for most commercial and institutional applications. The use of the 2% criteria should be limited to residential applications only.

A number of factors enter into the determination of human comfort, including dry-bulb temperature, humidity, air movement, and adjacent radiant/absorbent surfaces (such as a cold glass window wall in the winter). Generally, the term *effective temperature* is the measure utilized to define the effects of the first three of these factors on the human body. ASHRAE Standard 55 defines a *comfort zone* within which a lightly clothed person is generally *comfortable* which is summarized in Table 5-1.

The center of this range is 75°F [24°C] and 50%RH, the typical comfort air-conditioning indoor design conditions. However, energy codes often call for a "deadband" between cooling and heating, so the range of 72°F [22°C] to 77°F [25°C] is often used for indoor temperature set points.)

The inside design conditions for a certain types of buildings, or a specific area within a building, (e.g., laboratories and hospitals) may be dictated by factors other than comfort and the owner must define its specific needs.

TABLE 5-1

Indoor Design Temperature and Relative Humidity

Temperature	Relative Humidity
(°F) [°C]	(% RH)
70 [21]	50
71 [22]	50
72 [22]	50
73 [23]	50
74 [23]	50
75 [24]	50
76 [24]	50
77 [25]	50
78 [26]	50
79 [26]	45
80 [27]	40

5.2.2 Heat Transfer through Walls and Roofs

The two driving functions behind heat transfer through walls and roofs are the *temperature difference* between the indoors and outdoors and *solar radiation* incident on outdoor surfaces. The effect of solar radiation, some of which is *reflected* and some of which is *absorbed* by the wall or roof, is to raise the outside surface temperature. Thus, the outside surface temperature will typically be higher than the ambient air temperature.

The amount of solar radiation on a surface perpendicular to the sun's rays varies during the year due to the elliptical nature of the earth's sun orbit, with the average value being 433 Btu/h-sf [1.37 kW/m^2]. Under normal *clear sky* conditions, only about 85% of the incident radiation makes its way through the atmosphere as *direct radiation*, averaging 368 Btu/h-sf [1.16 kW/m^2], with the remainder being reflected back to space by the atmosphere.

Upon entering the atmosphere, some solar radiation is "scattered" by the dust, ozone, water, and so on, in the earth's atmosphere and reaches the earth's surface in the form of *diffuse* radiation. Finally, some radiation is *reflected* from surrounding surfaces. Thus, the total radiation incident on a surface is given by Eq. (5-1):

$$E_t = (E_D \times \cos\theta_v) + E_d + E_r \tag{5-1}$$

where

E_t = total shortwave (SW) solar radiation reaching a surface, Btu/h-sf [W/m^2]

E_D = direct solar radiation, Btu/h-sf [W/m^2]

θ_v = incidence angle between the incoming direct solar radiation and a line perpendicular to the surface

E_d = diffuse solar radiation, Btu/h-sf [W/m^2]

E_r = reflected solar radiation, Btu/h-sf [W/m^2]

There are several mathematical models for computing all of these solar factors for both cloudless conditions and a variety of atmospheric conditions that the designer can define (amount of clouds, air pollution clearness levels, and so on).

The impact of the total SW radiation reaching the wall or roof is a function of many variables including mass, reflectance, the location, and amount of insulation, and so on. Each heating/cooling load calculation model addresses these factors with varying degrees of sophistication, as discussed in Section 5.3.

The conductive heat transfer through walls and roofs at any instant can be computed utilizing Eq. (5-2):

$$Q = A \times \mathrm{TD}/R_t = A \times \mathrm{TD} \times U \tag{5-2}$$

where

$$U = 1/R_t \tag{5-3}$$

and

Q = heat flow, Btu/h [W]
R_t = sum of individual thermal resistances, (°F-sf)/Btu/h [°C-m²/W]
A = surface area of boundary element, sf [m²]
TD = design temperature difference between indoors and outdoors, °F [°C]
U = overall heat transfer coefficient, Btu/h(°F-sf) [W/°C-m²]

The resulting heat gain through walls is considered to be 63% by radiation and 37% by convection. Heat gain through roofs is considered to be 87% by radiation and 16% by convection.

For most building walls and roofs, the construction is made up of various components assembled to create series or parallel heat flow resistances. For example, a wall made up of *layers* of different materials will have series heat flow resistances, while a wall made up of different materials, one beside the other such as insulation between studs, will have parallel heat flow resistances.

To compute the *total resistance to heat transfer* R_t, Eqs. (5-4) and (5-5) can be used:

$$R_t = R_1 + R_2 + R_3 + \cdots + R_i \tag{5-4}$$

for series components and

$$1/R_t = 1/(R_1 + R_2 + R_3 + \cdots + R_i) \tag{5-5}$$

for parallel components, where $R_1 \ldots R_i$ are the individual component thermal resistances.

In the ASHRAE *Handbook-Fundamentals*, the thermal properties of most common building materials are provided. These data are provided in one of two formats: for materials of a particular thickness, the thermal resistance is given as the resistance (R) for that thickness, while materials that vary in thickness, the *specific conductance, k*, is listed in terms of unit thickness. The relationship between these two formats is given as follows:

$$R = L/k \tag{5-6}$$

where

k = specific conductance, Btu/h-°F-sf-in. [W/(m² °K)]
L = length of conduction path (thickness), in. [m]

To demonstrate U computation, Figures 5-1 and 5-2 can be examined as follows.

In Figure 5-1, the paths through the cavity and through the studs are parallel, while all other components are in series. Therefore, the *average U* for this wall is computed as shown in Table 5-2 in terms of IP units.

Stud framing is assumed to occur at 16-in. centers with nominal 2 × 4 studs forming 11% of the wall area. Therefore, the average wall resistance is computed as

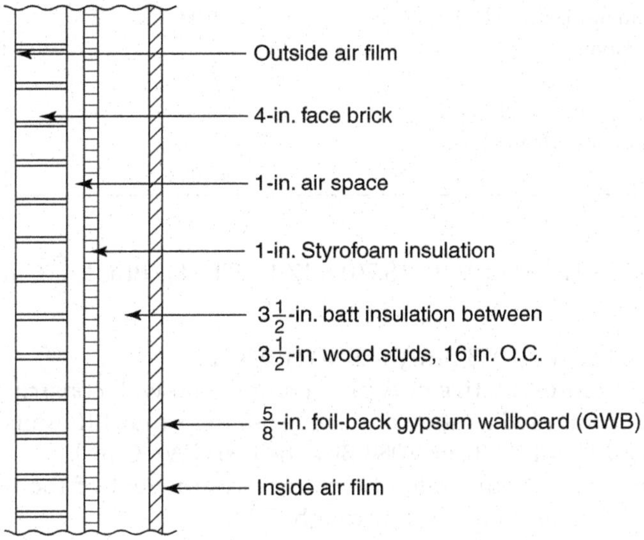

FIGURE 5-1
Cavity wall with brick veneer.

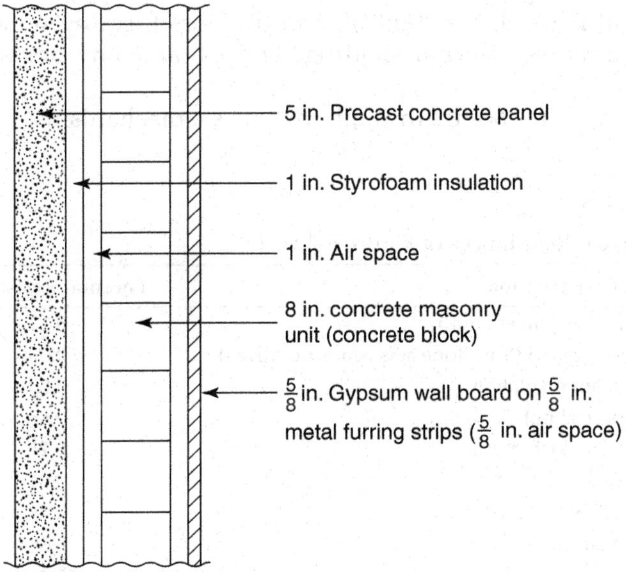

FIGURE 5-2
Concrete panel wall energy.

TABLE 5-2

Thermal Resistances of Figure 5-1 Wall

	Thermal Resistance (*R*)	
Wall Component	**Stud Path**	**Cavity Path**
Outside air film (15 mph)	0.17	0.17
Face brick (4 in.)	0.44	0.44
Air space (1 in.)	0.85	0.85
Styrofoam insulation (1 in. [25.4 mm]	5.00	5.00
Batt insulation	0	11.00
Wood studs	4.38	0
Foil-back gypsum wall board	0.56	0.56
Inside air film (still air)	0.68	0.68
R_t	**12.08**	**18.70**

$$R_t = (0.11 \times 12.08) + (0.89 \times 18.70) = 17.13\ (°\text{F-sf})/\text{Btu/h}\ \left[°\text{C-m}^2/\text{W}\right]$$

and the wall *U* is 1/17.13 or 0.058 Btu/(h-°F-sf) [W/°C-m^2].

Figure 5-2 is more representative of typical commercial wall construction in the United States [and the *U* is computed as shown in Table 5-3 in terms of IP units.

Therefore, the wall *U* is 1/11.31 or 0.088 Btu/(h-°F-sf) [W/°C-m^2].

It is necessary to compute the value *U* for each component of the building boundary using the ASHRAE data and Eqs. (5-4) through (5-6).

5.2.3 Heat Transfer through Fenestration

Fenestration refers to any aperture in a building envelope, including the glazing material (glass) that is used, the framing and dividers forming window construction, external shading devices, internal shading devices, and any between-glass shading devices.

Fenestration heat transfer occurs through three basic mechanisms:

TABLE 5-3

Thermal Resistances of Figure 5-2 Wall

Wall Construction	Thermal Resistance
Outside air film (15 mph)	0.17
Concrete panel (5 in. stone aggregate, not dried)	2.00
Styrofoam insulation	5.00
Air space (1 in.)	0.85
Concrete block (8 in.)	1.11
Air space	0.94
GWB, 5/8″	0.56
Inside air film (still air)	0.68
R_t	**11.31**

1. Conductive and convective heat gain or loss driven by the temperature difference between the indoor and outdoor air.
2. Solar heat gain due to the solar radiation.
3. Infiltration heat gain or loss due to air leakage.

At any given instant, the heat flow due to conductance and convection is defined by Eq. (5-2), the same as for walls and roofs. Of this heat gain, 63% is considered to be via radiation, while 37% is via convection. Fenestration U values, however, cannot be calculated but must be provided by the glass manufacturer or obtained from other sources, such as Table 5-4. This table is by no means comprehensive and the designer must make sure that the correct U for the fenestration products utilized in any building is obtained for load calculations. If the manufacturer data are unavailable, the ASHRAE *Handbook-Fundamentals* has more extensive information.

The same solar radiation incident on walls is incident on the fenestration incorporated in those walls. Figure 5-3 depicts the processes involved in solar radiation through glass areas. Some of the incident radiation is reflected back outside before penetrating the glass. The amount of reflectance is a complex function of the angle of incidence, the makeup of the glass, surface coatings, and colors, and so on.

To account for all of these conditions, the term *solar heat gain coefficient* (SHGC) has been developed to provide for computation of the instantaneous fenestration solar heat gain as follows:

$$Q = \text{SHGC} \times A \times E_t \tag{5-7}$$

where

Q = fenestration solar heat gain, Btu/h [W]
SHGC = solar heat gain coefficient
A = fenestration glass area, sf [m^2]
E_t = total SW solar radiation reaching a surface, Btu/h-sf [W/m^2]

The SHGC is the fraction of total incident solar radiation on the fenestration that enters the glazing and becomes heat gain. It includes both the directly transmitted portion and the absorbed and reemitted portion and, again, represents the data that must be provided by the fenestration product manufacturer.

Exterior, interior, and between-glass shading devices (blinds, drapes, and so on) effectively reduce the solar heat gain. However, the mechanisms differ for each shading method.

External shading devices typically consist of fins, overhangs, and reveals that reduce the direct sunlighted area of the fenestration as a function of time of day and the geometry of the fenestration. Figure 5-4 illustrates that these various components and the sunlighted area, A_s, are computed as follows:

$$Z = W - (D \times \tan B) \tag{5-8a}$$

$$Y = H - \left[(D \times \tan A)/\cos B\right] \tag{5-8b}$$

TABLE 5-4

U-Factors for Typical Fenestration (Btu/h-sf-°F) [W/°C-m²]

Glazing Type	Glass Only: Center	Glass Only: Edge	Vertical Fixed Installation[a]: Alum. w/o Thermal Break	Vertical Fixed Installation[a]: Alum. with Thermal Break	Vertical Fixed Installation[a]: Wood/Vinyl	Vertical Fixed Installation[a]: Insulated Fiberglass/Vinyl
Single Glazing						
1/8″–1/4″ [3–6 mm] glass	1.04 [5.90]	1.04 [5.90]	1.13 [6.42]	1.07 [6.08]	0.98 [5.56]	0.94 [5.34]
Double Glazing						
¼″ [6 mm] airspace	0.55 [3.12]	0.64 [3.63]	0.69 [3.92]	0.63 [3.58]	0.56 [3.18]	0.53 [3.01]
½″ [12 mm] airspace	0.48 [2.73]	0.59 [3.35]	0.64 [3.63]	0.57 [3.24]	0.50 [2.84]	0.48 [2.73]

Glazing Type	Glass Only: Center	Glass Only: Edge	Vertical Curtain Wall Installation: Alum. w/o Thermal Break	Vertical Curtain Wall Installation: Alum. with Thermal Break	Vertical Curtain Wall Installation: Structural Glazing
Single Glazing					
1/8″–1/4″ [3–6 mm] glass	1.04 [5.90]	1.04 [5.90]	1.22 [6.93]	1.11 [6.30]	1.11 [6.30]
Double Glazing					
¼″ [6 mm] airspace	0.55 [3.12]	0.64 [3.63]	0.79 [4.49]	0.68 [3.86]	0.63 [3.58]
½″ [12 mm] airspace	0.48 [2.73]	0.59 [3.35]	0.73 [4.15]	0.62 [3.52]	0.57 [3.24]

Glazing Type	Glass Only (Skylights): Center	Glass Only (Skylights): Edge	Manufactured Skylight (Sloped Installation): Alum. w/o Thermal Break	Manufactured Skylight (Sloped Installation): Alum. with Thermal Break	Manufactured Skylight (Sloped Installation): Reinforced Vinyl/Alum. Clad Wood	Manufactured Skylight (Sloped Installation): Wood/Vinyl
Single Glazing						
1/8″–1/4″ [3–6 mm] Glass	1.19 [6.76]	1.19 [6.76]	1.98 [11.24]	1.89 [10.73]	1.75 [9.94]	1.47 [8.35]
Double Glazing						
¼″ [6 mm] airspace	0.58 [3.29]	0.66 [3.75]	1.31 [7.44]	1.11 [6.30]	1.05 [5.96]	0.84 [4.77]
½″ [12 mm] airspace	0.57 [3.23]	0.65 [3.69]	1.30 [7.38]	1.10 [6.25]	1.04 [5.91]	0.84 [4.77]

[a] Multiply by 1.12 for operable fenestration.

$$A_s = Z \times Y \qquad (5\text{-}8c)$$

The shaded area of fenestration is subject only to the diffuse and reflected solar radiation components, which are significantly smaller than the direct radiation value.

Fenestration products, when fully shaded from the outside, have solar heat gains as much as 80% lower. One way to accomplish this is to use *louvers* or *sunscreens* to cover the entire

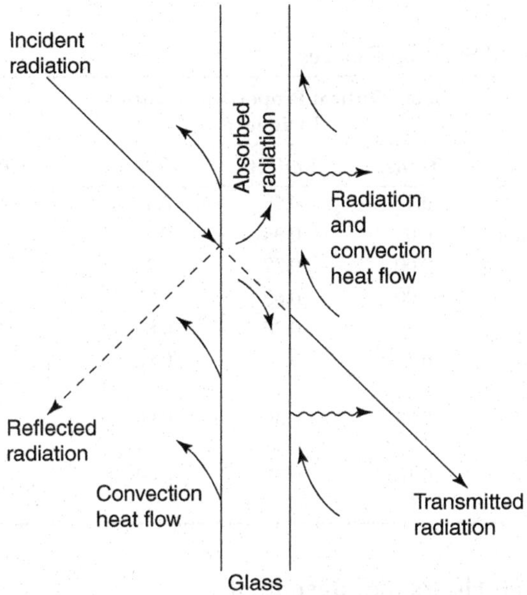

FIGURE 5-3
Solar radiation transmission, reflection, and absorption in glass.

fenestration at all times rather to relay on the partial shading provided by fins and overhangs. Manufacturers of these devices can provide the detailed solar performance data required for analysis.

For *inside or between-glass shading devices*, which can include blinds, drapes, shades, and so on, the SHGC must be modified to take into account the shading device. For this computation, the properties of the shading device must be defined by the designer, similarly to the data provided in Table 5-5.

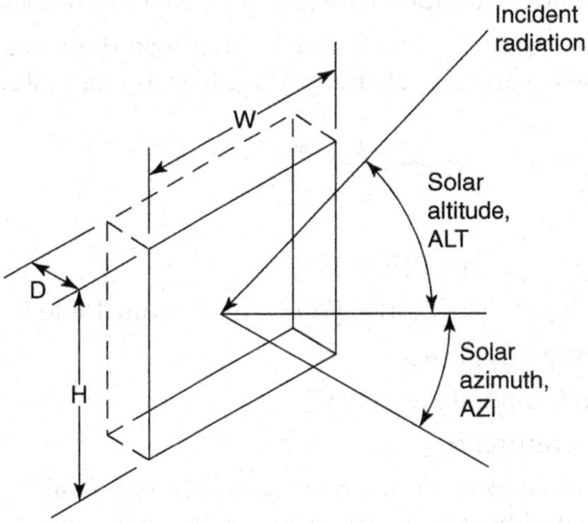

FIGURE 5-4
External shading by fins and overhangs.

TABLE 5-5
Properties of Indoor Typical Shading Devices

Indoor Shading Device	Solar-Optical Properties at Normal Incidence			Total Shading Coefficient with Clear Double Glazing (1/2″ [12 mm] Airspace)
	Trans.	Reflect.	Absorp.	
Venetian blinds, light color	0.05	0.55	0.40	0.58
Venetian blinds, medium color	0.05	0.35	0.60	0.62
Vertical blinds, white	0.00	0.77	0.23	0.60
Roller shades, white	0.00	0.65	0.35	0.35
Roller shades, dark color	0.00	0.20	0.80	0.71
Draperies, open weave, medium color	0.45	0.25	0.00	0.58
Draperies, semiopen weave, medium color	0.30	0.30	0.00	0.54
Draperies, closed weave, medium color	0.10	0.35	0.00	0.48

5.2.4 Heat Loss through Floors and Basements

There are three special cases for computing floor heat flow: floors above grade (including over crawl spaces), floors on grade, and floors below grade. For floors on grade and floors below grade, there is no heat gain since the ground temperatures are normally below the indoor design temperature.

1. *Floors above grade* include floors over crawl spaces, over unheated basements, or exposed to the outdoors (overhangs and cantilevers). If the floor is over a crawl space, the heat gain or loss depends on the amount of ventilation and other factors, such as perimeter insulation, HVAC ducts (and their losses), and the HVAC plant. If the space is well ventilated, the temperature may be taken to be close to the outdoor temperature and the heat gain or loss is defined by Eq. (5-2). If the space is well sealed and the floor of the crawl space is not insulated, the crawl-space temperature will approach the indoor temperature and the heat gain or loss will approach zero.
2. *Floors on grade* are typically concrete slabs that lose their heat primarily through the perimeter. The magnitude of the heating load is calculated as follows:

$$Q = F_2 \times P \times (T_r - T_o) \tag{5-9}$$

where
Q = heat loss from floor slab, Btu/h [W]
F_2 = heat loss coefficient, Btu/h-ft-F [W/°C-m²/] from Table 5-6
P = slab perimeter length, ft [m]
T_r = indoor (room) temperature, °F [°C]
T_o = outdoor temperature, °F [°C]

3. *Floors below grade* (basement floors) have a much lower heat loss than the basement walls. The anticipated heat loss can be computed as follows:

$$Q = F_{bf} \times A \times \text{TD} \tag{5-10}$$

TABLE 5-6

Approximate Heat Loss Coefficient F_2 for Floor Slab Edge Heat loss

Insulation	0–2,950 Degree Days	3,000–5,350 Degree Days	5,400–7,450 Degree Days
Uninsulated edge	0.80	0.84	0.93
R5.4 insulation from edge to footer	0.47	0.49	0.54

where

Q = heat loss from basement floor, Btu/h [W]

F_{bf} = heat loss coefficient, Btu/h-sf-F from Table 5-7

A = basement floor area, sf [m^2]

TD = $T_r - (T_a - S)$ where T_a is the average annual outdoor temperature and S is the annual amplitude of the ground temperature. S has a value of 18°F–22°F for the coastal, southern, and western states and 22°F–25°F in the midwestern and upper plains states.

4. *Basement walls* represent a special case. The heat flow paths from the basement wall to the ground surface are nonparallel, partial concentric circles. To calculate the heat loss by the wall, it must be divided into sufficient small sections, each representing a potential heat flow path and then summing the losses for each path to yield a total. Equation (5-10) is then applied in the same manner as for basement floors except for the heat loss coefficient F_{bw} (Table 5-8).

TABLE 5-7

Heat Loss Factor F_{bf} for Basement Floors

Depth (ft) [m] Below Grade	Heat Loss (Btu/h-sf-°F) [W/m²-°C]
5 [1.52]	0.026 [0.148]
6 [1.83]	0.025 [0.142]
7+ [2.13+]	0.023 [0.131]

TABLE 5-8

Approximate Heat Loss Factor F_{bw} through Basement Walls

	Heat Loss (Btu/h-sf-°F) [W/m²-°C]			
Depth (ft) [m] Below Grade	No Insulation	Insulation R = 4.17	Insulation R = 8.34	Insulation R = 12.5
1 [0.30]	0.410 [2.33]	0.152 [0.86]	0.093 [0.54]	0.067 [0.38]
2 [0.61]	0.222 [1.26]	0.116 [0.66]	0.079 [0.45]	0.059 [0.34]
3 [0.91]	0.155 [0.88]	0.094 [0.53]	0.068 [0.39]	0.053 [0.31]
4 [1.22]	0.119 [0.68]	0.079 [0.45]	0.060 [0.34]	0.046 [0.26]
5 [1.52]	0.096 [0.55]	0.069 [0.39]	0.053 [0.31]	0.044 [0.25]
6 [1.83]	0.079 [0.45]	0.060 [0.34]	0.049 [0.28]	0.040 [0.23]
7+ [2.13+]	0.069 [0.39]	0.054 [0.31]	0.044 [0.25]	0.037 [0.21]

5.2.5 Internal Heat Gains

Lighting, people, equipment, and other heat-producing appliances within the air-conditioned spaces of a building are defined as sources of *internal heat gain.*

The heat gain from people has two components: *sensible* and *latent*. The total and relative amounts of sensible and latent heat vary depending on the level of activity, and in general, the level of latent heat increases with the level of activity. Table 5-9 summarizes typical occupant heat gain for a number of normal activities. The latent heat gain is assumed to instantly become cooling load, while the sensible load is partially delayed depending on the occupancy profile and the nature of the conditioned space. Fifty to sixty percent of the heat release from a person is via radiation, while the remaining 40%–50% is via convection.

The primary source of heat from lighting comes from the lamps, although significant additional heat may be generated from associated components such as the ballast in a fluorescent or metal halide light. Each watt of lighting load (lamp plus ballast) produces 3.1413 Btu/h.

Where lighting is installed below ceilings or in ceilings which are not part of return air plenums, all of the light heat is allocated as space heat gain. Where return air ceiling plenums are utilized, there are two potential conditions that must be evaluated by the designer:

1. With *unventilated lights*, 15%–20% of the light heat will be absorbed by the return air passing over the light and will not contribute to the space heat gain. However, most designers do not take this split into account and assign all of the light heat to the space.
2. For *ventilated lights*, 40%–60% of the light heat will be absorbed by the return air passing through the light and will not contribute to the space heat gain. The light manufacturer must provide the test data indicating the return air heat absorption rate to be expected.

In IP units, motors produce 2,545 Btu/h for each brake horsepower of motor load, divided by the motor efficiency:

$$Q = 2,545 \times \text{BHP}/\text{EFF} \tag{5-11}$$

TABLE 5-9

Occupant Heat Release Rates (Btu/h) [W]

Activity	Typical Application	Sensible Heat	Latent Heat
Seated at rest	Theater	210 [62]	140 [41]
Seated, very light work, writing	Hotels, residential	230 [67]	190 [56]
Seated, eating	Restaurant (dining)	255 [75]	325 [95]
Seated, light work, typing	Offices	255 [75]	255 [75]
Standing, light work, or walking slowly	Retail store, bank	315 [92]	325 [95]
Light bench work	Factory (assembly, testing)	345 [102]	435 [128]
Walking, light machine work	Factory (light fabrication)	345 [102]	695 [204]
Moderate dancing	Dance club	405 [119]	875 [256]
Heavy work, lifting	Factory (heavy fabrication, warehouse)	565 [166]	1,035 [303]
Heavy work, athletics	Gymnasium	635 [186]	1,165 [342]

where

Q = motor heat, Btu/h

BHP = motor brake horsepower

EFF = motor efficiency, decimal, 0–1.0

In SI units, the motor load [W] represents the motor heat release.

However, the amount of this heat that actually contributes to the heat gain of a space is dependent on the motor and load arrangements, as summarized by Table 5-10.

For other heat-releasing equipment and appliances within a space, the manufacturer will typically provide the heat release rate for the equipment and the radiation/convection ratio. However, if no data are available, the conservative approach is to utilize the factors defined for lights and motors. The maximum hourly heat gain can be estimated as 50% of the *nameplate* rating because of the diversity of typical appliance utilization. The radiant and convection components, if unknown, can be estimated to be 70% and 30%, respectively.

5.2.6 Infiltration Loads

Infiltration is the uncontrolled flow of outdoor air into a building through cracks and other openings in the building boundary and through the normal use of exterior doors and, perhaps, windows. *Exfiltration* is the leakage of indoor air out of the building. In any building, the amount of infiltration is exactly offset by the amount of exfiltration. Figure 5-5 illustrates the typical distribution of infiltration and exfiltration on "highrise" buildings. For "lowrise" buildings, infiltration generally occurs on the windward side(s) of the building, while exfiltration occurs on the leeward side(s).

TABLE 5-10

Typical Motor Heat Release

Motor/Load Configuration	Heat to Space (%)
Both motor and load within the space	100
Motor outside space but load within	33
Motor within space but load outside	67

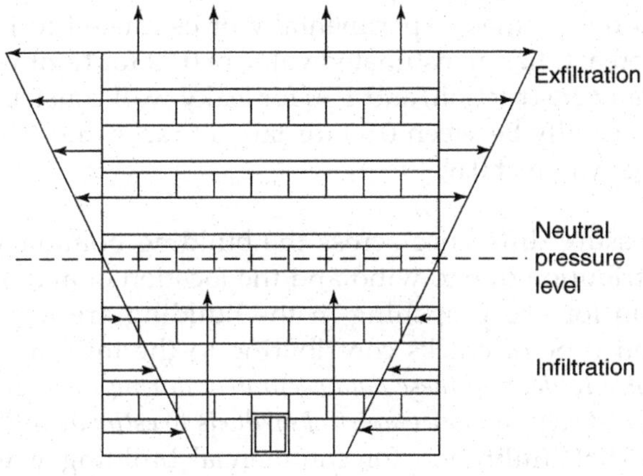

FIGURE 5-5
Air flow due to stack effect in a multistory building with no wind.

For the HVAC designer, infiltration represents a heating or cooling load, depending on the temperature and humidity condition of the outdoor air relative to the indoor air. The sensible and latent load values are computed using Eqs. (2-48) and (2-52), respectively, where the temperature and specific humidity differences are those between the indoor and outdoor air design conditions.

In both of these equations, the critical value is V, the infiltration air flow rate (cfm)[m³/s], which must be determined by the designer.

Infiltration is driven by (1) pressure differences across the building boundary caused by wind; (2) air density differences due to temperature differences between indoor and outdoor air (buoyancy or *stack effect*); and (3) the impact of combustion air appliances, air leakage from duct and air distribution devices, and mechanical ventilation systems, all of which act to change the building interior pressure. The effect of all of these mechanisms can be summarized in one equation that defines the pressure total pressure difference across the building boundary:

$$\mathrm{PD} = \mathrm{PR} + \mathrm{PW} - \mathrm{PI} + \mathrm{PS} \tag{5-12}$$

where

PD = pressure difference between outdoors and indoors across building boundary, in. wg [Pa]
PR = static pressure at reference height of building, in. wg [Pa]
PW = wind pressure, in. wg [Pa]
PI = interior reference pressure, in. wg [Pa]
PS = pressure difference due to temperature difference between outdoors and indoors, in. wg [Pa]

Once the pressure difference PD is determined, the infiltration airflow rate through the walls can be computed utilizing Eq. (5-13):

$$V = C(\mathrm{PD})^n \tag{5-13}$$

where

V = infiltration air flow rate, cfm [L/s]
C = flow coefficient determined experimentally or estimated to include overall cracks and opening area. C has an estimated value of 0.22 for tight wall construction, 0.66 for average wall construction, and 1.30 for leaky wall construction.
n = flow exponent, usually between 0.50 for larger cracks and 0.85 for hairline cracks, with an average value of 0.65.

To determine the pressure difference across the building boundary, the data about the exterior pressure distribution due to wind and the location of and air flow rate/pressure difference relationship for every opening in the building are needed. Additional data about the amount and type of cracks contributing to the infiltration are also required. *It is essentially impossible to develop these data accurately during the design of the building and, therefore, the designer must rely on more empirical methods to estimate infiltration.*

Research indicates that infiltration for commercial building walls can be estimated using of Table 5-11, based on 0.30 in. wg [75 Pa] of pressure difference, equivalent to 25 mph [40 km/h] wind speed).

TABLE 5-11

Estimate Wall Infiltration Rate

Quality of Wall Construction	Infiltration Rate (cfm/sf) [L/s-m²]
Tight	0.10 [5]
Average	0.30 [15]
Leaky	0.60 [30]

In applying these data, it must be remembered that only half of the building wall area, that is, the *windward* side, should considered as infiltrating area, while the other half of the building is the *exfiltrating* area.

Cracks and openings at windows and doors must be considered in addition to the overall wall infiltration. Windows in commercial and institutional buildings are rarely operable, and their contribution to the infiltration airflow can be assumed to be included in the estimates for the wall. (Note, though, that a building with a lot of small windows may have a wall that is more *leaky* than average.)

The literature provides detailed calculation methods for estimating the air leakage for doors, first, due to the cracks around the door opening and, second, for the door opening and closing due to occupant traffic. Most designers ignore the second component since there is seldom any way for him or her to determine occupant traffic patterns for a "scientific guess." Even when trying to estimate crack length and type for detailed calculation of the closed door condition, the lack of information and the resulting poor quality of any assumptions rarely justifies the attempt to apply rigorous methods. Lacking more detailed information, the designer can use the infiltration rates listed in Table 5-12, based on a 25 mph [40 km/h] wind pressure:

Experienced designers may also *estimate* infiltration by the *air change method.* This method simply requires an assumption of the number of air changes/hour that a space will experience based on an appraisal of the building type, construction type and quality, and building use. The infiltration air flow rate can be computed by this method utilizing Eq. (5-14):

$$V = (\mathrm{AC/h} \times \mathrm{Vol/AC}) \tag{5-14}$$

where

V = infiltration air flow rate, cf/h [m³/h]

AC/h = estimated air changes/hour

Vol/AC = space volume per air change, cf [m³]

Research indicates that infiltration air changes/hour for commercial buildings are significantly lower than for residential buildings and values of 0.1–0.3 AC/h are representative.

TABLE 5-12

Infiltration Rate, cfm/ft [L/s m]

Door Type	Tight	Average	Leaky
Swinging	0.10 [0.15]	0.22 [0.34]	0.66 [1.02]
Revolving	-	11.1 [17.2]	-

Ventilation air introduced by the HVAC systems tends to pressurize the building and reduce the pressure difference across the building boundary, while mechanical *exhaust systems* tend to increase the pressure differential. If the net *positive* air low is sufficiently high, the interior pressure will be *equal to or higher* than the outside pressure, eliminating the pressure difference and any resulting infiltration. *Typically, if the ventilation air flow, less than the exhaust air flow, exceeds about 10% of total air flow, infiltration may be assumed to be zero for load calculation purposes.*

5.3 Load Computation Methodologies

5.3.1 Heat Balance and Radiant Time Series

Cooling load calculation using the *heat balance* methodology involves calculating a surface-by-surface conductive, convective, and radiative heat balance for each surface and a convective heat balance for the room air. Sometimes called the *exact solution,* these principles form the foundation of all cooling load calculation methods.

The heat balance generally is modeled with four coupled heat transfer components:

1. Conduction through the building boundary elements (walls, roofs, fenestration, and so on)
2. Convection (from the surface) of the building boundary elements to the inside air.
3. SW radiant absorption and reflection for the building boundary surfaces.
4. Longwave (LW) radiant interchange between all indoor surfaces.

The SW radiation is from the solar radiation entering through windows and radiation emitted from internal heat sources. The LW radiant exchange includes the absorption and emittance from low temperature sources such as wall, roof/ceilings, equipment, and people.

A twelve (12) surface model, as summarized in Table 5-13, is generally used to define each thermal zone (any space controlled by a single temperature sensor and maintained to be at uniform temperature):

The basic heat balance for each surface in the zone is defined by Eq. (5-15):

$$q_{LWX} + q_{SW} + q_{LWS} + q_{ki} + q_{\text{sol}} + q_{\text{conv}} = 0 \tag{5-15}$$

where

q_{LWX} = net LW radiant exchange flux between zone surfaces
q_{SW} = net SW radiation flux to surface from lights
q_{LWS} = LW radiation flux from equipment in zone
q_{ki} = conductive flux through the surface
q_{sol} = transmitted solar radiation flux absorbed at face
q_{conv} = convective heat flux to zone air

TABLE 5-13
Twelve Surfaces

Surface	Description
1	Wall 1
2	Wall 2
3	Wall 3
4	Wall 4
5	Roof/ceiling
6	Floor
7	Wall 1 window
8	Wall 2 window
9	Wall 3 window
10	Wall 4 window
11	Roof/ceiling skylight
12	Zone thermal mass

The heat balance processes are formulated for a 24-h *design day* defined by the designer. The primary variables in the heat balance are the 12 outside surface temperatures and 12 inside surface temperatures for each of the 24h in the design day ($T_{so,I,j}$ and $T_{si,I,j}$ for each surface I at each hour j). For each $T_{si,I,j}$, q_{conv} can be computed as $A_i \times h_{c,i} \times (T_{si,I,j} - T_a)$, where T_a is the indoor design air temperature.

The zone cooling load is computed as follows:

$$Q = q_{ce} + q_I + q_{\text{conv}} \tag{5-16}$$

where

q_{ce} = convective part of internal loads
q_I = sensible load due to infiltration

The iterative heat balance calculation procedure for a thermal zone is simple (but tedious), consisting of a series of initial calculations that proceed sequentially, followed by a double iteration loop for each surface, each hour, as follows:

1. Initialize areas, properties, and face temperatures for all surfaces for 24h.
2. Calculate incident and transmitted solar flux for all surfaces for 24h.
3. Distribute transmitted solar energy to all inside faces for 24h.
4. Calculate internal load quantities for all 24- h based on defined load conditions and operating profiles.
5. Distribute LW, SW, and convective energy from internal loads to all surfaces for 24h.
6. Calculate infiltration loads for 24h.

7. Iterate the heat balance according to the following:
   a. For each hour $j = 1–24$
   b. For each surface face $k = 1$ to maximum number of faces
   c. For each surface $i = 1–12$
   d. Compute $T_{so, I, j}$ and $T_{si, I, j}$
   e. Iterate i
   f. Test for convergence
   g. Iterate k
   h. Compute zone heat gain
   i. Iterate j
8. Display results.

Convergence typically occurs after about four surface (k) iterations.

In an attempt to retain some means of performing heat gain and heat loss calculations manually, ASHRAE has introduced the *radiant time series* (RTS) method, a simplified method derived from the heat balance method. ASHRAE states that "... it effectively replaces all of the other simplified methods...." The RTS calculation uses radiant time factors and conduction time factors to distribute hourly heat gains over a 24-h period instead of iteratively calculating the time delay effects. A table of time factors published in the ASHRAE *Handbook-Fundamentals* determines how long heat is held by a wall or other surface before creating a cooling load in the space. However, even a cursory review of the example manual calculation presented in the *Handbook-Fundamentals* shows that the method is far too tedious and time-consuming to be used by the practicing HVAC designer.

Interestingly, ASHRAE goes on to state that the older methods of cooling and heating load calculation, described in Sections 5.3.2 and 5.3.3, as follows "... those methods are not invalidated or discredited" and goes on to state the following:

> The accuracy of cooling load calculations in practice depends primarily on the availability of accurate information and the design engineer's judgement in the assumptions made in interpreting the available data. Those factors have a much greater influence on a project's success than does the choice of a particular cooling load calculation method.

5.3.2 Total Equivalent Temperature Difference with Time Averaging

The *Total Equivalent Temperature Difference with Time Averaging (TETD/TA)* method of load calculation was introduced in 1967 to provide a method for computing cooling loads which was more rigorous and accurate than the steady-state methods used until that time.

The basic TETD concept is very simple: In Eq. (5-2), replace the simple TD with a TETD that accounts for the mass of the building walls and roofs by introducing *decrement factors* and *time lags* for each of these components.

Instantaneous heat gains computed using the TETD are added to the solar and internal heat gains in the thermal zone to yield a total instantaneous gain. This gain is then converted to cooling load by applying a *time averaging* technique of averaging the radiant portions of the heat gain for the current hour with related values from an appropriate period of preceding hours.

The fundamental weakness of the TETD/TA method is that the simple averaging technique to account for mass was a poor approximation of the actual physics involved. Also, the choosing of an appropriate averaging period was subjective and often amounted to no more than an engineering "guess."

While originally presented as a manual calculation method, the actual TETD/TA computation process was so tedious that this quickly one of the first cooling load calculation methods transferred to computers (primarily by the software developed by Terry Y. Sun, PE, for Automated Procedures for Engineering Consultants, Inc. in 1967–1968).

5.3.3 Transfer Function

Although similar in principle to TETD/TA, the *transfer function (TF)* method applies a series of weighting factors or conduction TF coefficients to the walls and roofs and the temperature differences across these components to determine heat gain with appropriate reflection of the thermal inertia (mass). Solar heat gain through glass and internal heat gain is calculated directly for the load hour. Then, this method applies a second series of weighting factors or room TFs to the radiant heat gain components to account for the thermal storage effect in converting instantaneous heat gain into cooling load.

Both series of weighting factors consider data for several previous hours preceding the current hour, as well as the current hour.

While the TF method is deemed by ASHRAE as "scientifically appropriate and technically sound for a specific cooling load analysis," the calculation procedures require computer support to be used. When introduced in 1972, very few HVAC designers had access to computers, so a simplification of the TF method had to be made so that calculations could be done manually. This simplification is called the *cooling load temperature difference/solar cooling load/cooling load factors (CLTD/SCL/CLF)* method.

For a representative group of walls and roofs, detailed TF calculations were performed, and from this data, the CLTD for each wall and roof was generated and tabulated. The CLTD could be used in Eq. (5-2) in lieu of TD for a one-step computation that produced load results equivalent to the more complicated TF computation.

CLF for similar one-step calculations of solar load through glass and for loads from internal heat sources were developed. Later, in 1992, the CLF for glass solar gain was replaced with the improved SCL factor, which allowed additional accuracy.

CLTDs, SCLs, and CLFs all included the effect of time lag of conductive heat gain through opaque surfaces and the time delay by thermal storage in converting radiant heat gain to cooling load. However, as time went by, it was determined that modern construction materials and techniques were not adequately represented by the earlier data and, thus, application of the methodology to modern buildings became more and more difficult. Also, with the expansion of the availability and use of computers in the 1990s, the rationale for a simplified manual load calculation method became questionable.

5.3.4 Computer-Based Calculation Tools

Today, heating/cooling calculations for commercial, institutional, and industrial buildings are routinely performed using computer-based methods. There are, literally, dozens of computer software packages available to the designer for use in computing heating/cooling loads. The methodologies used may be one of those described previously, but each software developer may have made modifications to expand the software's capability, improve its speed of calculation, and so on. *Providing accurate calculations, which will result in good HVAC systems design, is the primary goal that must be met by any computer-assisted calculation tool.*

In selecting a software package for his or her everyday use, the designer must do the following:

1. Investigate the various methodologies and algorithms used and be satisfied that these meet the criterion established in the latest ASHRAE publications. Currently, the most scientifically derived method for estimating the heating and cooling loads for a defined building space is the *heat balance* method. However, other methods described previously may be acceptable based on the designer's experience and confidence.
2. Determine if the software is "user-friendly," allowing the designer to describe the load components in ways that reduce (or even eliminate) the potential for ambiguity or error. Detailed software error checking and trapping is absolutely necessary.
3. Insure that the software will support multilevel load calculation to at least the room/space, zone/system, and building levels.
4. Determine that the output clearly shows the results of each component calculation *and* provide basic *check figures* (cfm/sf, Btu/sf, sf/ton, and so on) to flag to the designer where potential load anomalies may exist.

5.5 Benchmark Heating/Cooling Loads

The one advantage to performing heating/cooling load calculations by hand was that the younger, less experienced designer developed a familiarity with the details of the calculation elements and developed a "feel" for the quality of the computation results. This advantage is lost with computer-based computation methods, and consequently, there is a real danger that "garbage in-garbage out" load calculations can occur when the less experienced designer simply accepts the printed output without evaluating the results for *reasonableness.*

For heating loads, the steady-state computation methods are simple enough that the designer can easily do a quick calculation to test any software output that looks unreasonable. For example, a zone in an office building will rarely have a heat loss exceeding 30 Btu/sf [95 W/m^2], but a high infiltration load in a lobby or atrium can drive this value to 60 Btu/sf [189 W/m^2] or higher. The designer then must determine if the infiltration computation is correct.

Cooling load calculations are much more complex, and quick test calculations are difficult. Therefore, as a general guide, Table 5-14 is provided to demonstrate the normal range of cooling load anticipated for different types of facilities located in the United States. *The values in Table 5-14 are not recommendations but rather "check figures" to allow the designer to evaluate his load calculations. Wide variations from these values can and do happen, but the designer should validate any calculation when they do.*

TABLE 5-14

Benchmark Cooling Loads

Building Type	**Lighting (W/sf) [W/m²]**	**sf/Person [m²/Person]**	**Cooling Load (sf/ton) [m²/kW]**
Residential			
Single family housing	0.8 [8.6]	400 [37.2]	600 [196]
Condominiums and apartments (high rise)	1.0 [10.8]	175 [16.3]	450 [147]
Retail Facilities			
Strip shopping centers	1.9 [20.4]	75 [7.0]	350 [114]
Regional shopping centers	1.9 [20.4]	75 [7.0]	330 [108]
Discount and outlet stores	1.9 [20.4]	55 [5.1]	350 [114]
Supermarkets	1.9 [20.4]	100 [9.3]	350 [114]
Department stores	1.9 [20.4]	50 [4.6]	350 [114]
Multiple use complexes	1.9 [20.4]	75 [7.0]	350 [114]
Commercial and Public Buildings			
Correctional facilities	1.2 [12.9]	90 [8.4]	350 [114]
Court houses	1.4 [15.1]	60 [5.6]	300 [98]
Office buildings	1.3 [14.0]	110 [10.2]	360 [118]
Dining and entertainment facilities	1.8 [19.4]	15 [1.4]	135 [44]
Libraries and museums	1.6 [17.2]	60 [5.6]	340 [111]
Assembly Buildings			
Houses of worship	2.2 [23.7]	11 [1.0]	250 [82]
Auditoriums	1.5 [16.1]	11 [1.0]	400 [131]
Arenas and stadiums	1.5 [12.9]	10 [0.9]	400 [131]
Convention and exhibition centers	1.4 [15.1]	15 [1.4]	350 [114]
Community centers	1.4 [15.1]	25 [2.3]	400 [131]
Motels, Hotels, and Dormitories			
Hotels and motels	1.7 [18.3]	150 [13.9]	350 [114]
Dormitories	1.5 [12.9]	150 [13.9]	340 [111]
Nursing homes	1.6 [17.2]	150 [13.9]	400 [131]
Education Facilities			
Preschools and elementary schools	1.5 [12.9]	25 [2.3]	240 [78]
Middle schools	1.5 [12.9]	25 [2.3]	240 [78]
High schools	1.5 [12.9]	25 [2.3]	240 [78]
Colleges and universities (classrooms, admin)	1.5 [12.9]	25 [2.3]	240 [78]
(Laboratories)	1.8 [19.4]	50 [4.6]	185 [60]
Gymnasiums	1.7 [18.3]	40 [3.7]	250 [82]
Health Care Facilities			
Hospitals	1.6 [17.2]	60 [5.6]	275 [123]
Outpatient facilities	1.4 [15.1]	80 [7.4]	325 [106]
Research and development laboratories	1.8 [19.4]	100 [9.3]	150 [49]

Bibliography

ASHRAE Handbook-Fundamentals, American Society of Heating, Refrigerating, and Air-Conditioning Engineers, Inc. (Atlanta, GA, 2017).

6

Air Tempering and Distribution

6.1 Heating and Cooling Coils

Air is heated and cooled by passing it over a heat transfer device called a "coil," which is a tube-and-plate heat exchanger with air on the outside and the primary heating/cooling medium (water, steam, electricity, refrigerant, and so on) on the inside. Coils are generally categorized as follows:

Heating: hot water, steam, electric resistance
Cooling: chilled water, refrigerant

Except for electric resistance heating coils, all coils are constructed with fins, usually aluminum or copper, bonded mechanically to copper tubes. The primary heating/cooling medium passes through the tubes and heat is transferred by conduction to the plates. Here, heat is transferred to or from the air primarily by convection. The use of fins bonded to the tubes simply increases the contact area for heat transfer with the air and thus the overall heat transfer efficiency.

Water coils are arranged for *counterflow configuration*. As shown in Figure 6-1, the final temperature of the cooler fluid may exceed the outlet temperature of the hotter fluid.

This is particularly important for chilled water systems that have small temperature-difference requirements. Another advantage of the counterflow configuration is that less surface area is required for a given rate of heat transfer.

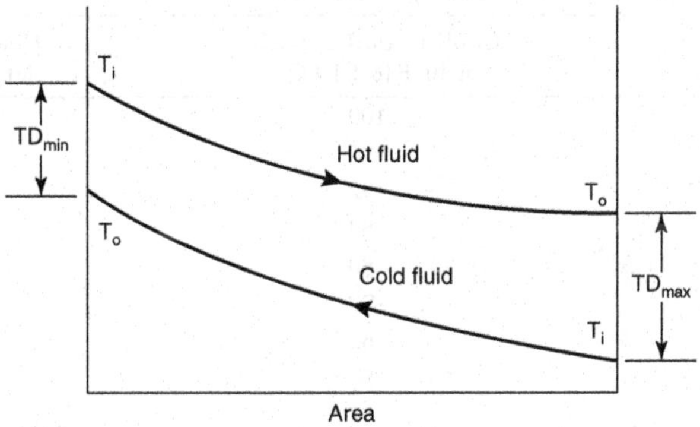

FIGURE 6-1
Temperature distribution in a counterflow water coil.

The overall heat transfer by the coil is determined by the mass flow rates of the two media. The surface area, the temperature gradients, and the overall coefficient of heat transfer, U, as defined by Eq. (6-1).

$$Q = -m_h C_{p,h} dT_h = m_c C_{p,c} dT_c = U\,dA(T_h - T_c) \tag{6-1}$$

where m is the mass flow rate and T is the temperature of the fluids, hot (h) and cold (c), respectively. dA is the differential area of the coil under study.

Integrating Eq. (6-1) over the total area of the coil yields

$$Q = UA(\mathrm{TD}_i - \mathrm{TD}_o)/\ln(\mathrm{TD}_i/\mathrm{TD}_o) \tag{6-2}$$

where $\mathrm{TD} = T_h - T_c$ at the inlet (i) and outlet (o) conditions, respectively. The temperature term in Eq. (6-2) is called the *logarithmic mean overall temperature difference* (LMTD).

Since U is defined by the design of the coil and the materials used in construction, which is fairly consistent between manufacturers, the basic performance of any coil is defined by the coil surface area and the temperatures required or LMTD. If large temperature differences are available, as for most heating applications, the required coil area is small. For small temperature differences, as for chilled water systems, the coil area is large.

Water coil surface area is defined by the number of rows of tubes and the spacing, or density, of the fins (in terms of fins/inch of coil width). Thus, a heating coil may have one or two rows and 8–10 fins/in. [0.8–1.0 fins/mm], a cooling coil will often have six to eight rows and up to 16 fins/in. [1.6 fins/mm].

The capacity of any water coil is controlled by modulating the flow of liquid through the coil (see Chapter 13). However, as the flow is reduced, the capacity of the coil is not reduced linearly, as shown in Table 6-1.

This nonlinearity establishes the need for a method of controlling flow through coil to make the resulting coil performance as linear as possible, as shown in Figure 6-2 and discussed further in Chapter 13.

Steam and refrigerant coils are *phase-change heat exchangers,* since the primary medium enters as one state and leaves as another. For these coils, heat extracted or released as the primary medium changes phase produces the temperature change in the air. For all

TABLE 6-1

Typical Water Coil Flow versus Capacity Relationship

Water Flow Rate, %	Cooling Coil Capacity (at 10°F [6°C] TD) %	Heating Coil Capacity (at 20°F [11°C] TD) %
100	100	100
90	95	98
80	92	97
70	88	96
60	80	95
50	70	90
40	62	85
30	50	75
20	38	70
10	25	50
0	0	0

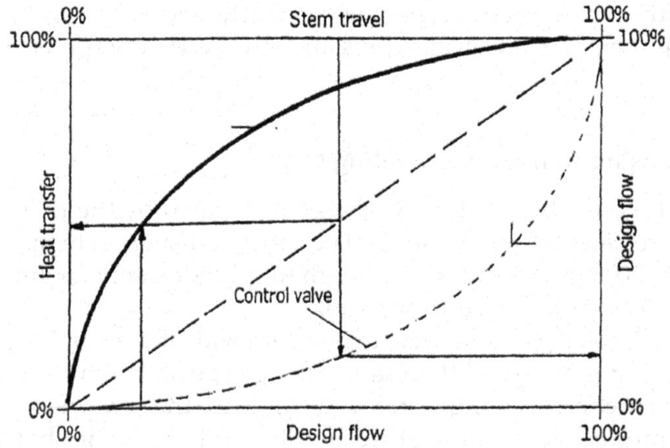

FIGURE 6-2
Water coil capacity versus flow relationship.

practical purposes, the temperature of the primary medium remains constant and, unlike water coils, the coil surface temperature remains constant, not varying with coil depth.

Selection of a heating or cooling coil requires that the designer define the heat transfer rate required, an entering temperature for water or an entering condition for steam or refrigerant, and the required leaving air temperature. From manufacturer's data, then, it is only necessary to select one or more coils that will satisfy these requirements and conditions. Final selection will be based on (1) the space available in which to install the coil (height vs. width), (2) air-pressure-drop limitations, and, (3) for water coils, water-pressure-drop limitations.

All coils are manufactured in standard sizes from 12 in. [300 mm] high, in increments of 4 in. [100 mm], and 24 [600 mm] to 144 in. [3.6 m] wide, in increments of 6 in. [150 mm] up to 72 in. [1.8 m] width and increments of 12 in. [300 mm] above that. Thus, for a large airflow, a large "coil bank" must be assembled from standard coil modules within a structural frame.

Cooling coils represent a more difficult problem for the designer than does any other type of coil. First, performance is more critical, since the LMTD available is small. This, in turn, produces the need for a large surface area, which results in relatively high airside and waterside pressure losses. Finally, dehumidification requires that the designer incorporate drain pans and drain piping into the installation to carry away the condensate.

Cooling coil, both water and refrigerant type, moisture carryover must be avoided. In hot, humid climates with high ventilation air rates, the latent load imposed on the cooling coil may be 25%–50% of the total load. The resulting condensate forms on the coil fins, tends to "pile up" at the bottom of the coil, and water droplets can *carryover* with the supply air if the "face" velocity it too high. The maximum cooling coil airflow velocity should be selected on the basis of Table 6-2.

TABLE 6-2

Coil Face Velocity to Prevent Carryover

Percent of Outdoor Air	Maximum Coil Face Velocity (FPM) [m/s]
≤20%	400–500 [2.0–2.5]
20%–60%	350 [1.8]
61%–80%	300 [1.5]
81% +	200–250 [1.0–1.3]

Heating coils, other than electric types, are typically selected at up to 750 fpm [3.8 m/s] face velocity since there are no serious pressure drop considerations and no condensate forms.

6.1.1 Direct Expansion Refrigerant Cooling Coils

Direct expansion (DX) cooling coil performance is dictated by the rate of refrigerant flow and the saturated refrigerant temperature. The saturated suction temperature is typically selected to be 40°F–50°F [5°C–10°C], low enough to provide latent cooling but high enough to minimize compressor energy consumption.

While DX cooling coils perform well in the removal of sensible heat, their ability to perform latent cooling is often less than satisfactory, especially when used in a single zone system with "on/off" control configuration. At full load, the compressor may run for long periods and will produce low temperature supply air that satisfies both the sensible and latent cooling loads. However, as the sensible load is reduced, a constant speed DX compressor will "cycle" on for short periods to meet the sensible load but will not remain on long enough to satisfy the latent cooling load so as to "dehumidify," and thus, the room humidity level increases. This problem is illustrated by the data developed during American Society of Heating, Refrigerating, and Air-Conditioning Engineers (ASHRAE) Research Project RP-1254, summarized in Table 6-3.

The data show that, as the run-time fraction decreases with the decrease in sensible in cooling load, the ability of these systems to dehumidify decreases as the effective SHR approaches 1.00 (i.e., no latent cooling capability). *By the time space cooling load reduces to 50% or less, the ability for the coil to do latent cooling essentially disappears.*

This problem is much more severe with constant supply fan operation, usually required in all but residential occupancies, in order to maintain the required minimum ventilation airflow rate. Since the greatest source of moisture introduced in most building is by the ventilation airflow, indoor humidity rapidly increases as the sensible load decreases.

If DX systems *must* be used (due to project size and loads, zoning requirements, and/or owner's inability to maintain more complex chilled water systems), the designer must consider the following to improve latent cooling ability:

1. "Judicious" *undersizing* of single compressor equipment will tend to reduce compressor cycling. Since comfort cooling systems will operate for far more hours at part load than at peak load, it may be better to have short periods when the design temperature condition cannot be maintained than have long periods when the design humidity condition cannot be maintained.

TABLE 6-3

DX System Latent Cooling as Indicated by the Coil Effective

Fraction of Compressor Run Time	Effective Evaporator Coil Sensible Heat Ratio
1.0	0.76
0.8	0.82
0.6	0.92
0.4	1.00
0.2	1.00
0.0	1.00

SHR at a Given Fraction of Compressor Run Time.

2. Design for higher temperature rise, which has the effect of lowering the discharge air dew-point temperature and reducing airflow velocity across the cooling coil, both of which improve the coil's latent cooling capacity.
3. Utilize variable air volume (VAV) supply airflow control. Packaged DX VAV systems with vendor-installed controls typically work well. Note that with both of these options, special attention to providing required ventilation airflow is required by the designer.
4. Use multiple compressors with multiple refrigerant circuits and a common evaporator coil with any system 5 tons or larger. Even, better, make use of variable speed compressors to more closely match capacity to imposed load. To maximize latent cooling capacity, use *fully intertwined* coil configurations. *Split face* coils offer enhanced part load dehumidification control with return air only applications, but the inactive circuits of a split face coil can bypass excessive humidity when applied in high outside airflow applications.
5. Use heat pipe or "run-around" heat recovery cycle coils to "wrap" the DX cooling coil as shown in Figure 6-3.
   This has the effect of using DX discharge air to "precool" the air entering the DX coil, effectively shifting DX capacity from sensible cooling to latent cooling. The wraparound heat pipe is a heat pipe wrapped around a cooling coil. It consists of two sections, the *precool* (evaporator) section placed before the cooling coil and the *reheat* (condenser) section placed after the reheat coil.
   The precool section is located in the incoming airstream before the cooling coil. When warm air passes over the first section, the liquid refrigerant vaporizes, moving heat to the reheat section (downstream from the cooling coil). Since heat has been removed from the air before the cooling coil, air passing through the cooling coil drops to a lower temperature, resulting in more condensate

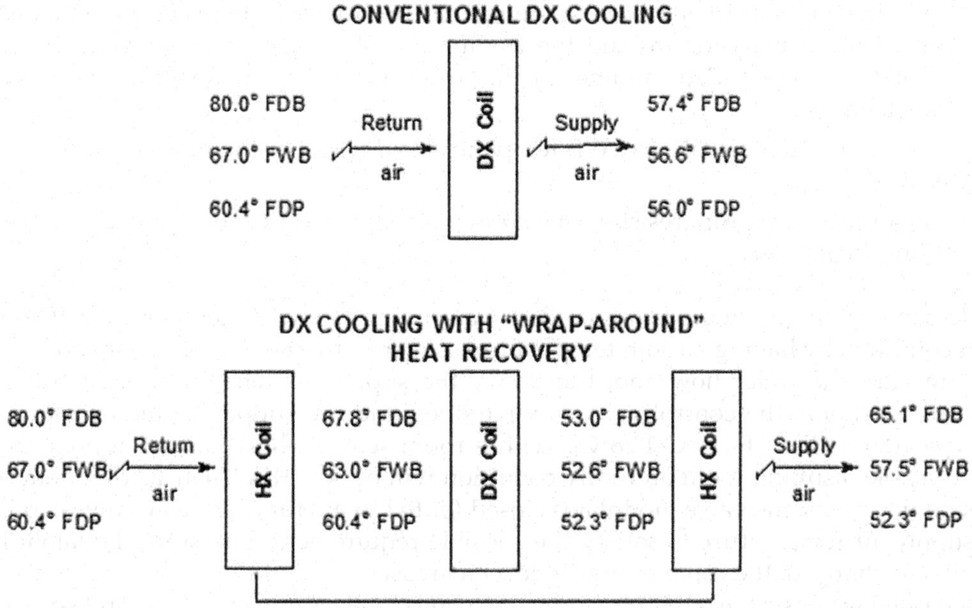

FIGURE 6-3
DX cooling coil run-around heat recovery cycle.

removal. The overcooled air is then reheated to a comfortable temperature and a lower relative humidity (RH) by the reheat section, using the same heat originally absorbed by the first section.

This approach, however, requires increased airflow and higher fan static pressure that must be taken into account.

6. Use pulse-*width modulation* (PWM) *expansion* valves in lieu of thermal expansion valves. PWM expansion valves, with the proper electronic controller and proportional sensors, can provide more precise control and less need for compressor cycling
7. Utilize *hot gas reheat*. To reduce operating costs associated with reheat, manufacturers are now offering packaged units with integral hot gas reheat and separate liquid line control so that dehumidification and cooling can be controlled independently.
8. Utilize outdoor air preconditioning or a dedicated outdoor air system and, thus, eliminate the need for the DX system to deal with any latent cooling loads at all.

6.1.2 Chilled Water Cooling Coils

In general, chilled water cooling coil performance varies as follows at standard coil rating conditions of 44°F [7°C] entering chilled water temperature, with entering air conditions of 80°F dry bulb (DB)/68°F wet bulb (WB) [26.7°C DB/20°C WB]:

1. Both the leaving air DB and WB temperatures *increase* as the water flow range (TD) *increases*.
2. Both the leaving air DB and WB temperatures *decrease* as the number of coil rows *increase*. Also, the difference between leaving DB and leaving WB temperatures gets smaller as the fins/in increase and the airside effectiveness (or efficiency) of the coil improves with the increased fin area.
3. Coil air temperature *approach*, the temperature difference between the entering chilled water temperature and the leaving air DB temperature, *decreases* as the number or rows and/or number of fins/in. *increase* (i.e., the result of more heat transfer area).
4. Both the leaving air DB and WB temperatures *decrease* as the number of fins/in. *increase*.
5. Coil air pressure drop (resistance) *increases* as the number or rows and/or number of fins/in. *increase*.

Single-zone cooling systems using chilled water maintain space temperature at the set point condition by having a room temperature sensor or thermostat modulate the control valve to vary the water flow and, thus, vary the supply air temperature. At full cooling load condition, the control valve is wide open and the supply air temperature will be (typically) 50°F–55°F [10°C–12.8°C]. With a room sensible heat ratio of approximately 0.85–0.90, the resulting room humidity condition will be 50% RH. Then, as the room cooling load decreases, the valve modulates closed (or to bypass for three-way valves), raising the supply air temperature to satisfy the sensible requirement. But, since the latent load remains unchanged, the room humidity level increases.

For chilled water systems with two-position control valves (fan coil units, and so on), the operating results and poor humidity control are exactly the same as for DX systems with on–off control, since the "open–close" operation of the chilled water control valve mimics the "on–off" control of a compressor in a DX system. *Thus, two-position control valves should never be used for control of a chilled water coil.*

6.1.3 Hot Water Heating Coils

Hot water heating coils are much simpler to select than cooling coils. First, since these coils are always dry, carryover is not an issue and a coil face velocity of 700–750 fpm [3.6–3.8 m/s] is typical. Then, because of the higher waterside differential temperatures, the coils are almost never more than two rows deep, meaning that the air pressure losses through these coils are much lower than chilled water cooling coils.

Select hot water heating/reheating coils for a maximum leaving air temperature of 90°F–95°F [32°C–35°C]. The ASHRAE *Handbook—Fundamentals* provides specific guidance that the maximum supply air temperature should not be greater than 15°F [8.3°C], higher than the space temperature set point. Thus, if the space temperature is 75°F [24°C], the maximum discharge temperature should be 90°F [32°C]. *Due to its buoyancy, supply air that is warmer than 95°F [35°C] will not drop enough for fully mixed air distribution, producing stratification and significant vertical temperature variation within the space.*

6.1.4 Steam Heating Coils

Steam heating coils are available in two configurations: *conventional finned tube* configuration and *non-freeze* coils. The proper description of "non-freeze" coils is "steam distributing," since they use a tube-in-a-tube and header-in-header construction. These coils have an inner tube down the entire length of an outer tube and were originally designed to evenly distribute the steam and condensate throughout the coil so there were no "dead spots" or "cold spots." However, it was also found that this type of coil was much more difficult to freeze than the conventional steam coil.

For HVAC applications, perforated inner tubes are typically used. Steam is distributed from the inner supply header down the inner tubes evenly and is released to the outer condensing tubes through holes spaced about 9″–12″ [230–300 mm] apart. The outer tubes are "capped" with no return bends, so that the condensate generated in the outer condensing tubes and any non-condensable gases have no place to go but back down the outer tube toward the outer condensate header and out the condensate connection.

See Section 6.2.6 for how these coils must be piped and controlled to avoid *steam stall* and the collection of condensate in the lower tubes that may ultimately freeze when these coils are used in preheating applications.

6.1.5 Electric Resistance Heating Coils

Electric resistance heating coils are basically large *resistors* that dissipate heat. The surface temperatures of the coils are very hot, and little actual surface area is required for a given heat transfer requirement. Two aspects of the selection of these coils are important: (1) provision must be made to de-energize the coils if airflow is interrupted to prevent their overheating, and (2) capacity variation can be done only in "steps," unless expensive SCR (*silicon-controlled rectifier*) controllers are utilized.

Electric resistance heating coils are manufactured in two configurations: *open coil* and *enclosed coil*. The open coil has its electrical elements fully exposed to airflow within a frame, while the enclosed coil has finned sheathing covering the electrical elements. For most HVAC applications, the open coil is utilized.

Select electric resistance heating coils on the basis of heating requirement imposed on it in terms of kW [3,413 Btu/h]. To determine electrical load (amps) imposed by a heating coil, Eqs. (1-3) and (1-4) apply.

To prevent hot spots, airflow must be uniformly distributed across the coil face. The coil's UL Listing requires that it not be installed closer than 48" [1,200 mm] downstream or upstream from a fan outlet, abrupt transition, or other obstructions. Elbows or other offset duct fittings must be located at least 48" [1,200 mm] from inlet of the heater and 24" [600 mm] from outlet of the heater.

Sufficient minimum airflow must be provided to prevent overheating and nuisance tripping of the thermal cutouts. The minimum required velocity is determined from Figure 6-4 on the basis of entering air temperature and KW per square foot of cross-sectional duct area. *The maximum air inlet temperature for open coil heaters is 100°F [38°C] and 80°F [27°C] for finned tubular heaters.*

Determining the number of stages of electric heat requires consideration of the total TD of the heating coil and the degree of control required. For "good" space temperature control, the TD for each stage should be generally no more than 5°F [3°C], though 10°F [5°C] per stage may be "adequate" for noncritical comfort control. For typical overhead supply air, the air temperature leaving the heating coil should not exceed 95°F [35°C] in order to avoid air buoyancy problems. Based on a room (heating) temperature set point of 70°F [21°C], the total TD would be 25°F [14°C]. Thus, a three-stage heating coil would provide *adequate* control," while a five-stage coil would provide *good* control.

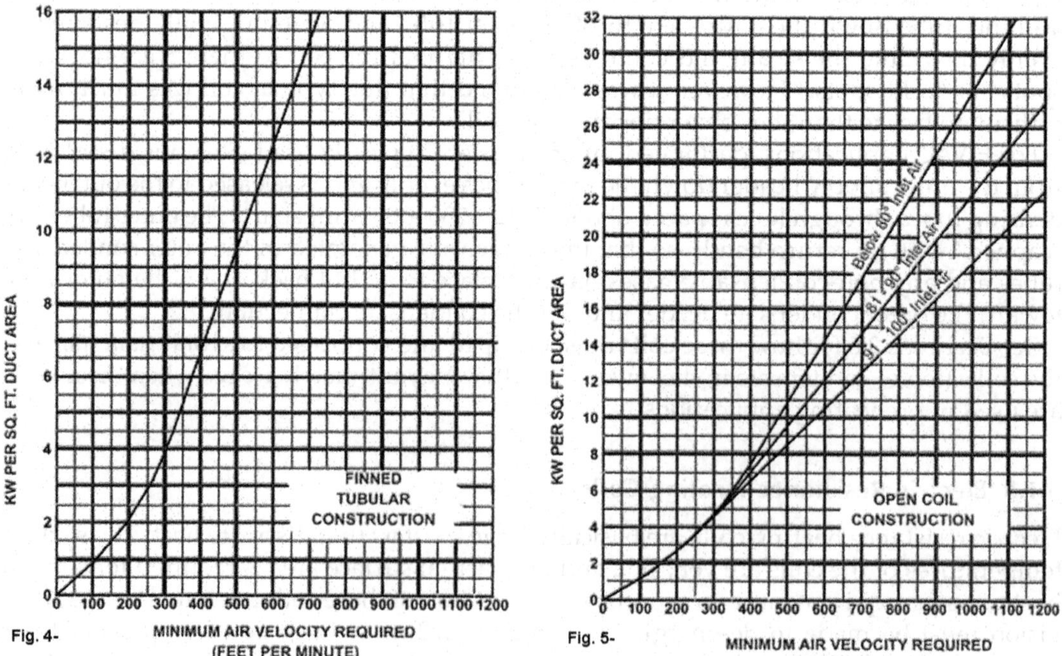

FIGURE 6-4

Minimum air velocity required for electric heating coil. (Fig. 4 and Fig. 5, "Installation, operating and maintenance instructions for Indeeco electric duct heaters," Indeeco, 425 Hanley Industrial Court, St. Louis, Missouri 63144.)

Control of a multistage electric heating coil requires some method of switching to turn power on/off to each stage, and three methods are commonly applied, as follows:

1. The most common method for doing this is *via* use of a *mechanical relay or contactor* to control power delivery to the electrical circuit of each coil stage, resulting in "on/off" control of each stage.
2. *Mercury displacement relays* (MDRs) can cycle faster than mechanical relays. However, installation of MDRs can be an issue, since they have to be installed perfectly vertical, and overheating of the relay, due to excessively fast cycling or overloading, can cause it to explode, a situation which creates a hazardous materials problem. In addition, shipping and disposing of MDRs is becoming increasingly more difficult due to more stringent federal environmental regulations. Thus, *the use of MDRs is not recommended.*
3. The *solid-state relay* (SSR) is a popular alternative to mechanical power control. A common characteristic of all solid-state devices (including SSRs) is that they generate the heat that needs to be removed. SSRs generate more heat than SCRs. Almost all SSRs are rated for maximum output at a temperature of 25°C [77°F]. However, in real-world operating conditions where temperatures inside electrical enclosures exceed 40°C [104°F], a SSR will fail if used at full output. Most SSR manufacturers have a derating chart for their product to compensate for this discrepancy. Unfortunately, many vendors will use the advertised maximum rating when selecting a SSR, resulting in early failure. Another drawback to SSRs is cost, since they are significantly more expensive than mechanical relays.

When more precise temperature control is required (e.g., hospitals and laboratories), an SCR controller can be used to provide fully proportional control of heating output. An SCR controller is a time proportioned controller that modulates the heater to supply the exact amount of heat required to satisfy the temperature requirements by modulating the time the electric heater is powered on, not the kW of the heater. An SCR can cycle as fast as 0.08 s and, with proper selection and use, can cycle on and off 1,000,000,000 times without any operating problems. Unlike a mechanical relay, an SCR controller has no mechanical parts to wear out and it will not arc and is not affected by dirty contacts. SCR controllers are typically rated for ambient operating temperatures of around 50°C [122°F] at full power, significantly surpassing SSRs. But, as for all solid-state controllers, protection against overtemperature, voltage spikes, and short circuits must be taken into account at the outset of a system design.

6.1.6 Preheating and Precooling Coils

When the ventilation airflow is high enough so that the mixed air temperature is 40°F or less [≤4°C] at the heating design outdoor temperature, a preheat coil is required to provide freeze protection for a chilled water cooling coil and, for single zone units, to provide heating to raise the AHU airflow temperature up to the normal unit discharge temperature set point. Almost always, a VAV air system will require a preheat coil since the amount of return air will decrease as the supply airflow decreases, but the minimum ventilation airflow, incompliance with ASHRAE *Standard 62.1*, must remain constant. This will result in the mixed air temperature falling to 40°F or less [≤4°C] on a design winter day.

The required preheat coil capacity is computed using Eqs. (6-3) and (6-3a), as follows:

$$\text{For IP units, } Q_c = 1.1 \times V_c \times (T_L - T_E) \quad (6\text{-}3)$$

$$\text{For SI Units, } Q_c = 1.3 \times V_c \times (T_L - T_E) \quad (6\text{-}3a)$$

where

Q_c = Required heating capacity, Btu/h [kW]
V_c = Coil airflow rate, cfm [m³/s]
T_L = Coil leaving air temperature, °F [°C]
T_E = Coil entering air temperature, °F [°C]. For 100% outdoor air units, T_E equals the design winter outdoor ambient temperature. For units with return air, T_E equals the mixed air temperature defined by the ratio of outdoor airflow to total airflow. For VAV systems, the lowest T_E occurs when the total airflow is reduced to its design minimum.

Steam is the recommended heating media for preheat coils, if available. Today, however, HVAC contractors and control vendors have little understanding of steam systems and even less understanding of how to control steam coils. So, the designer must take care to ensure that steam can be used effectively for a specific project.

For preheating, "non-freeze" coils with either (a) two control valves (a two-position first-stage valve for freeze protection and a second-stage modulating valve) or (b) modulating integral face and bypass dampers are required. These control methods are required to prevent *steam stall.*

Steam stall is a condition which occurs when the necessary pressure differential across the coil's trap becomes negative, causing condensate to no longer be discharged from the drainage device and instead to accumulate inside the coil, typically resulting in freezing and a ruptured coil, water hammer, and/or uneven heating temperatures across the coil.

Steam traps do not have the ability to draw out condensate on their own. Instead, condensate is discharged due to the pressure difference between the trap inlet (primary) pressure and outlet (secondary) pressure. The trap inlet pressure must therefore be higher than the trap outlet pressure to allow for proper condensate flow.

While steam systems are designed to discharge condensate through this pressure differential, various factors can interfere with this mechanism. For example, the use of single modulating steam control valve can lead to a reversal in the trap operating differential pressure and cause stall under light load conditions.

These coils require a minimum steam pressure *at the coil* of 10 psig [70 kPa], otherwise back pressure in the condensate system can result in "steam stall," causing flooding and freeze-up. *For a minimum pressure of 10 psig [70 kPa] at the coil, a minimum system supply pressure of at least 20 psig [140 kPa] is required, since the control valve is sized for a full-open pressure drop of 50% of the available pressure.*

One option for control of a steam preheat coil requires two control valves piped in parallel: a first stage two-position valve to ensure flow through the coil and that pressure is maintained (0%–20% capacity) and a second stage modulating valve (20%–100% capacity). Size the first stage valve for a pressure drop of 50% of supply steam pressure. Size the second stage valve for a pressure drop of 25% of the supply steam pressure. Set the first stage valve to open at 2.5°F [1.4°C] below the design coil leaving air temperature and to close at 2.5°F [1.4°C] above the design coil leaving air temperature.

A better alternative is using a single modulating control valve and integral face and bypass dampers. When the outdoor air temperature is above freezing, the control valve can be modulated to maintain discharge air temperature set point. But, as the outdoor air temperature drops to freezing, the control valve can be opened to provide full flow through the coil to avoid the potential of steam stall and freezing and the face and bypass dampers are modulated to maintain discharge air temperature set point.

Hot water coils used for preheat duty require continuous hot water circulation when in operation *via* an individual pump to provide freeze protection, as shown by Figure 6-5.

The required preheat pump flow rate, V, is computed in accordance with Eqs. (6-4) and (6-4a).

$$\text{For IP units, } V = Q_c/\left[500\times(T_S - T_R)\right] \tag{6-4}$$

$$\text{For SI units, } V = Q_c/\left[4{,}200\times(T_S - T_R)\right] \tag{6-4a}$$

where

V = Water flow rate, gpm [m³/s]

Q_c = Heating load, Btu/h [kW]

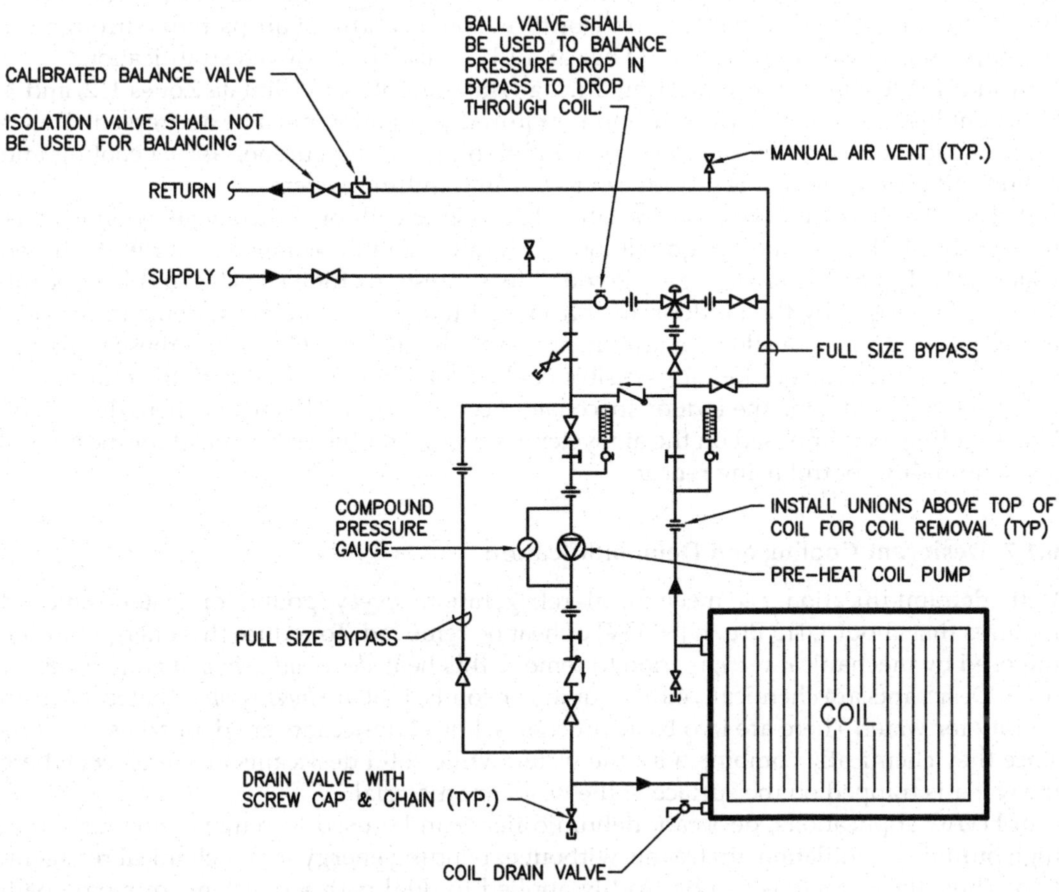

FIGURE 6-5
Hot water preheat coil piping configuration. (SW Standard Detail CL03B.)

T_S = Design hot water supply temperature, °F [°C]
T_R = Design hot water return temperature, °F [°C]

The hot water preheat coil must be selected for higher tube velocities than typical heating or cooling water coils in order to prevent localized freeze-up when the entering air temperature falls below 32°F [0°C], a minimum of 3 fps [1 m/s]. When preheating is required, the pump must operate continuously to maintain this flow velocity throughout the coil.

An alternative design for hot water preheat control is to incorporate a modulating two-way control valve and integral face and bypass dampers. When the outdoor air temperature is above freezing, the control valve can be modulated to maintain discharge air temperature set point. But, as the outdoor air temperature drops to freezing, the control valve can be opened to provide full flow through the coil as a "freeze protection" measure and the face and bypass dampers can be modulated to maintain discharge air temperature set point.

When either steam or hot water preheat coils are applied in AHUs that utilize return air, care must be taken to arrange the return air and outdoor ventilation airstreams so that thorough mixing of the two occurs upstream of the preheat coil. Otherwise, temperature stratification can occur, resulting in poor coil leaving air temperature control and a greatly increased chance of spot freeze-up of the coil. Arrangement of the duct connections of the two airstreams, use of air baffling, and/or directional control of air patterns through the use of parallel blade dampers should be carefully considered to avoid stratification.

In addition to a need for preheating during winter months, in Climate Zones 1, 2, and 3, the latent load imposed by the minimum required ventilation must be addressed. This is typically accomplished by separate "preconditioning" of the outdoor air *via* cooling and dehumidification before it is introduced to the air handling system.

Precooling can be accomplished *via* either DX cooling coils or chilled water cooling coils, though the ability to handle high outdoor air humidity ratios is much better with chilled water coils. Typically precooling is provided to address the high sensible and latent cooling loads imposed by the outdoor air introduced to one or more air systems in order to minimize ventilation airflow requirements (see Chapter 10). Generally, these coils are selected to satisfy the outdoor air sensible cooling load but also to satisfy the outdoor air latent cooling load, plus the indoor space latent cooling load. Effectively, then, there is no latent cooling load imposed on the air systems served, and there is no need for individual space humidity control using reheat.

6.1.7 Desiccant Cooling and Dehumidification

With dehumidification *via* mechanical refrigeration, every pound of water removed requires that about 1,040 Btu/h [300 W] of heat be removed. To reduce this energy burden imposed by mechanical refrigeration to remove this heat, *desiccant dehumidification* can be used. Desiccants are chemicals, either liquids or solids, that are *hygroscopic*, that is, have an affinity for water. There are two basic processes: liquid desiccants are defined as *absorbers* since they chemically combine with the water, while solid desiccants are *adsorbers* where the water is trapped on the surface of the desiccant material.

In HVAC applications, desiccant dehumidifiers can be used to remove moisture from high humidity ventilation airstreams without expending energy for mechanical refrigeration. Thus, these systems can be readily applied to dual path air systems or single path systems with ventilation air preconditioning. However, the moisture that is removed must be eliminated from the desiccant so it can be recycled for continued use within a closed cycle system. These systems are defined as *regenerative,* since the water removal process "regenerates" or "reactivates" the desiccant, making it usable again to remove moisture.

Typical materials used in regenerative systems include silica or alumina gel, activated alumina, lithium chloride salt or salt solution, or glycol solution.

It is generally not economical to install a desiccant system in facilities where the requirement for indoor air dew point is higher than 50°F or where the latent to total cooling load ratio is less than 25%. Another way to express this is that desiccant systems are most effective in facilities where the indoor air needs to be dry, such as a library, or where a lot of internal moisture is generated, such as a health club or a marine laboratory.

Desiccants can reduce cooling loads and peak demand by as much as 50% in some applications by reducing the latent cooling loads imposed on mechanical refrigeration systems. Estimating energy savings in specific applications is difficult because of the complexity in modeling annual performance and humidity. A sophisticated hourly building analysis program should be used. Regeneration requires heat input that is roughly 80%–90% of the latent heat of the moisture removed. If regeneration is done with waste heat, this process has little cost; otherwise, heating fuel costs can be high.

6.2 Air Distribution Design

Virtually all heating and cooling systems for commercial buildings rely on the distribution and circulation of air as the medium for heat transfer to, within, or from the occupied space. In the large majority of commercial buildings, however, the air distribution system serves the dual functions of accomplishing heat transfer between the central systems and rooms as well as providing the medium for load pickup within each room.

The designer is concerned with specifying the collection of fans, plenums, ducts, and air distribution devices that will serve the needs of each space in a cost-effective fashion. To do so requires consideration of many related factors, such as the heating and cooling loads to be satisfied in each space, the ventilation requirements for the particular room activity, the acceptable noise level for the room, the distribution of heating and cooling loads within the room, the initial costs of system components and installation, and the energy and maintenance costs associated with the air distribution system. Although it is not generally possible to determine a true optimum design based on all these factors, the design engineer can take into account the most important for any particular application and apply sound judgment based on experience to complete the design.

6.2.1 Room Air Diffusion

Room air diffusion is the composite of space air needs that will determine the size and characteristics of the duct and fan system required. First, however, some definitions must be introduced:

Air distribution assembly (terminal unit [TU]): a damper and actuator mechanism housed in a sound-attenuating enclosure and controlled by a thermostat to regulate the flow and/or temperature of air into a control zone. Depending on the system type, the unit may contain a heating coil and/or a secondary supply fan.

Damper: air volume control device that operates by varying the cross-sectional area open to flow.

Diffuser: supply air outlet that distributes air into the room in some desired pattern.

Drop (rise): vertical difference between the height of a horizontal airstream at the outlet and at the end of its throw, called *rise* if the airstream rises and *drop* if it falls.

Grille: any covering for an opening through which air passes.

Radius of diffusion: horizontally measured distance through which a horizontally spreading airstream travels before diminishing to some specified velocity.

Terminal velocity: maximum velocity of an airstream at the end of its throw.

Throw: distance an airstream travels in a horizontal or vertical direction before the maximum airstream velocity diminishes to a specified level.

6.2.2 Typical Air Distribution Patterns

Occupant comfort limits defined by ASHRAE *Standard 55* and ISO *Standard 7730* have been established as follows:

1. Air temperature and humidity levels maintained in accordance with the data provided in Section 4.3.2.
2. Maximum air motion in the occupied zone (6′6″ [2 m] above floor, with 2′0″ [600 mm] of walls) to be less than 30 fpm [0.15 m/s] for heating and 50 fpm [0.25 m/s] for cooling.

Before proceeding to an examination of the criteria that have been developed for evaluating the effectiveness of room air distribution, Figures 6-6 through 6-10 show typical patterns of room air distribution that occur with various types and placement of supply outlets.

In Figure 6-6, for a non-spreading upward projection outlet, we may note the relatively constant temperature from floor to ceiling during cooling. But there is a relatively large stagnant zone on the opposite wall near the floor during heating. In Figure 6-7, for a spreading upward-projection outlet, there is a relatively large stagnant zone in the upper portion of the room during cooling, resulting in large variations in temperature in that area. And again, there is a significant stagnant zone, with relatively large temperature variations during heating. In Figure 6-8, for horizontal projection from a wall outlet, the performance during cooling usually involves some stagnant areas near the ceiling and significant stagnant zones and temperature gradients during heating. These stagnant zones may be greatly reduced if the projection (throw) is sufficiently long to cause the airflow to follow the opposite wall down to the floor. In Figure 6-9, for a non-spreading downward projection outlet, such as a ceiling diffuser, the temperature variations are relatively small during cooling, but the velocities could be excessive for occupant comfort, and there is, again, a stagnant cold air zone near the floor during heating. In Figure 6-10, for the radially distributed pattern from a ceiling-mounted circular diffuser, effective and uniform spreading of the air results in a quite uniform space temperature during cooling, but natural convection currents act to produce a large stagnant zone and wide temperature variations during heating.

The location of a *return air intake* has little effect on the room air distribution patterns except in the immediate vicinity of the intake. However, if the return intake is located

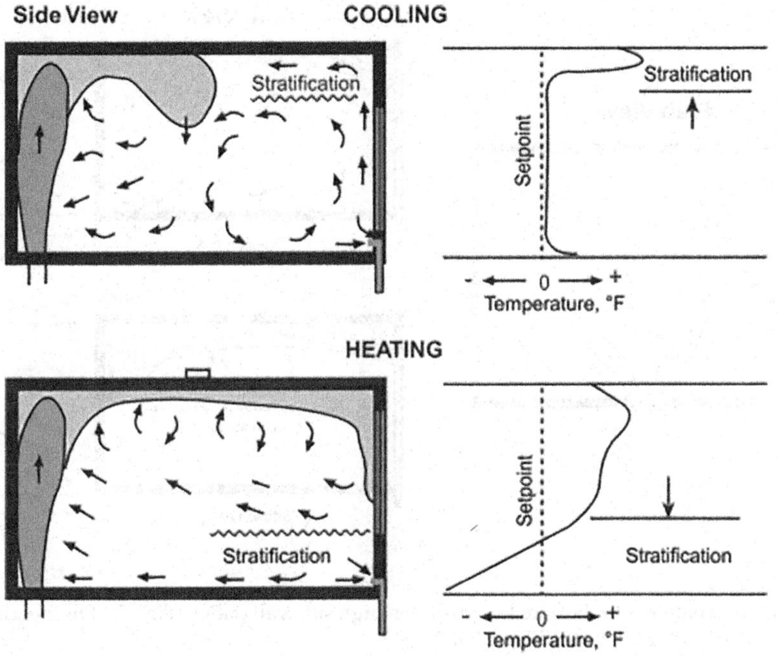

FIGURE 6-6
Room air distribution patterns for floor-mounted spreading upward-projection outlet. (Fig. 31, "Engineering Guidelines," Section B, Titus HVAC, 605 Shiloh Rd., Plano, TX 75074.)

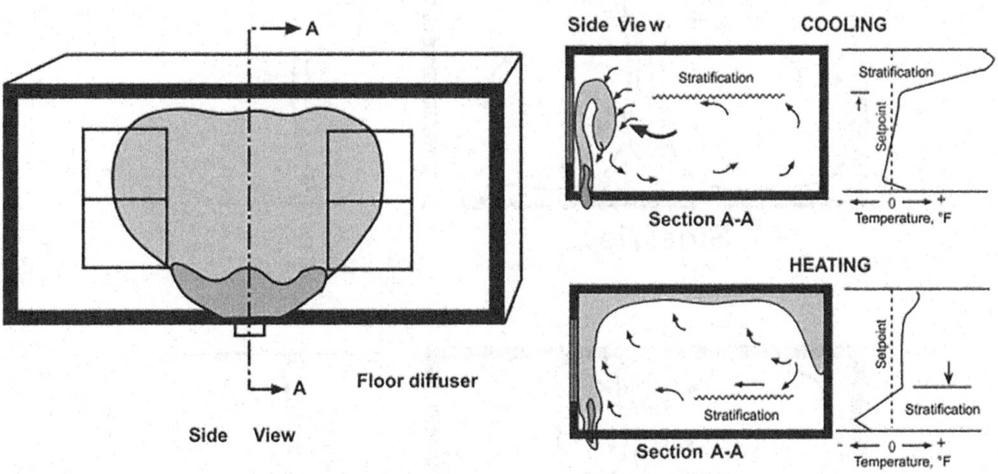

FIGURE 6-7
Room air distribution patterns for floor-mounted non-spreading upward-projection outlet. (Fig. 29, "Engineering Guidelines," Section B, Titus HVAC, 605 Shiloh Rd., Plano, TX 75074.)

within the stagnant zones, it will provide some local relief, and the warmest air will be returned to the air-handling unit during cooling and the coolest air will be returned during heating. Care should always be exercised that the return is not placed in such a position as to cause *short-circuiting* of the flow from an outlet directly to the return rather than being distributed through the room.

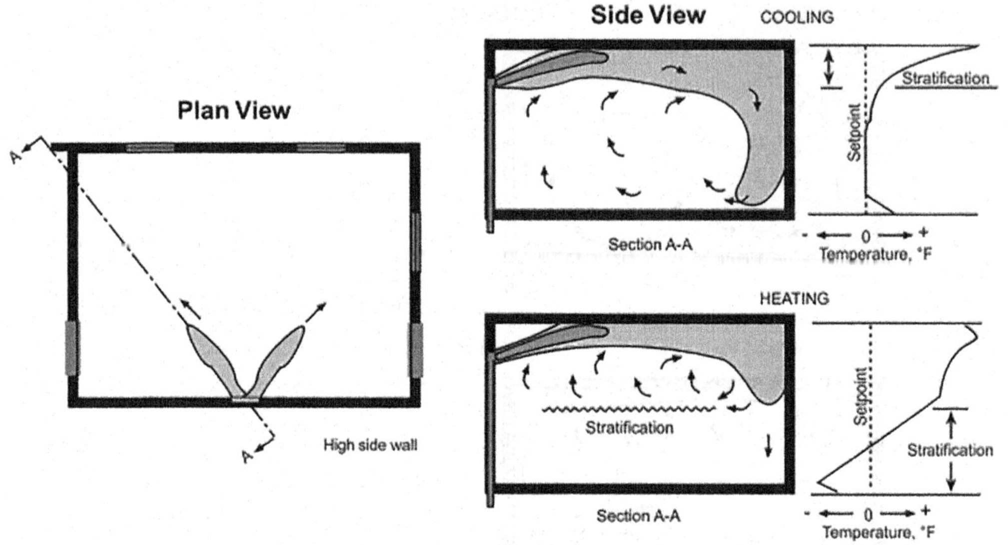

FIGURE 6-8
Room air distribution patterns for horizontal projection high sidewall outlet. (Fig. 17, "Engineering Guidelines," Section B, Titus HVAC, 605 Shiloh Rd., Plano, TX 75074.)

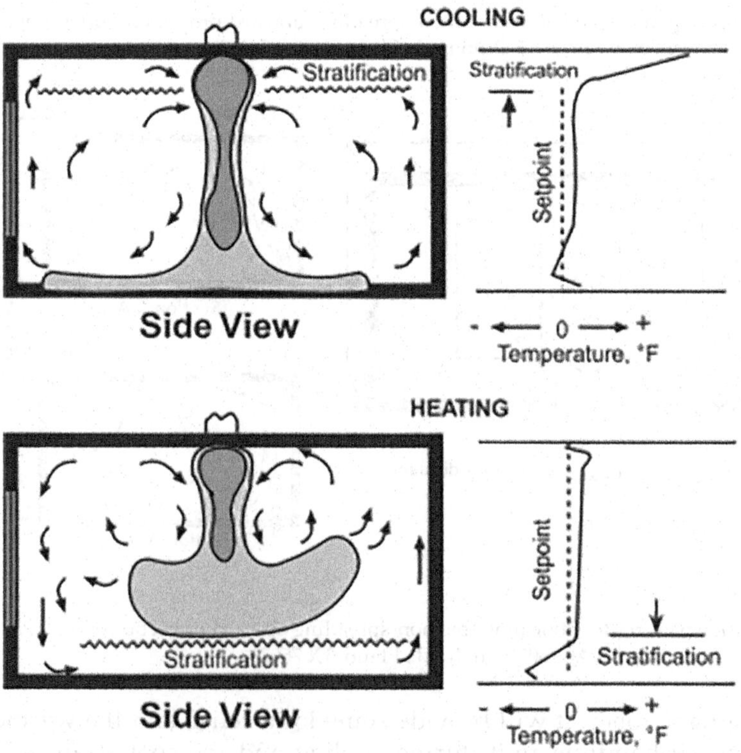

FIGURE 6-9
Room air distribution patterns for ceiling-mounted non-spreading downward-projection outlet. (Fig. 21, "Engineering Guidelines," Section B, Titus HVAC, 605 Shiloh Rd., Plano, TX 75074.)

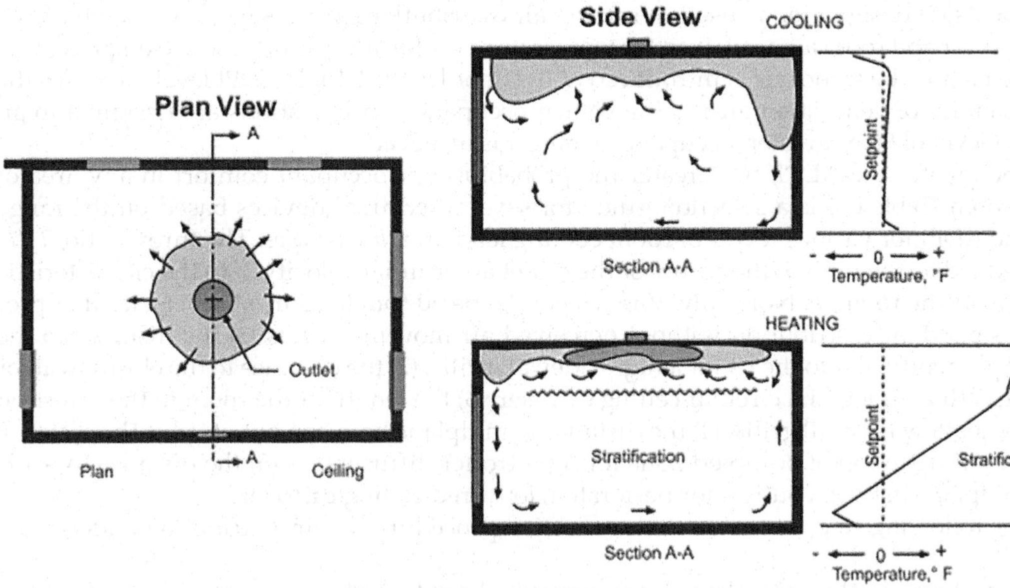

FIGURE 6-10
Room air distribution patterns for ceiling-mounted radial diffuser. (Fig. 18, "Engineering Guidelines," Section B, Titus HVAC, 605 Shiloh Rd., Plano, TX 75074.)

Even if the average room air DB temperature, humidity, and mean radiant temperatures are within the limits for comfortable design conditions, the occupants may feel discomfort if the air movement is excessive or if the local temperature in various areas of the room differs greatly from the average room temperature. The effect of these considerations is represented in a measure called the *effective draft temperature* (T_{ed}):

$$T_{ed} = (T_x - T_c) - A(V_x - B) \tag{6-5}$$

where

T_x = Local temperature, °F [°C]
T_c = Average room temperature, °F [°C]
V_x = Local air velocity, ft/min [m/s]
A = 0.07°F/(ft/min) [7.65°C/(m/s)], a constant
B = 30 ft/min [0.15 m/s], a constant

A high proportion of people engaged in sedentary office work would be comfortable in rooms where the effective draft temperature is between –3°F [–1.7°C] and +2°F [+1.1°C], where the air velocity is less than 70 ft/min [0.36 m/s].

To provide an overall measure of the comfort conditions with respect to room air diffusion, an *air diffusion performance index* (ADPI) has been defined by ASHRAE *Standard 113* as the percentage of points measured in a room that are within both the ASHRAE temperature and velocity ranges for comfort when the supply air temperature is providing cooling. ADPI is based on the acceptance and recognition that it is not possible to achieve a comfort level of 100%, but 80% acceptance is achievable and measurable.

The ADPI designation is used in mixing air distribution systems (systems in which the supply air and room air are mixed for temperature uniformity) as opposed to displacement systems (in which cool air is introduced at the floor level). A high ADPI level increases the probability of ventilation air mixing. When occupancy specifications are designed to an ADPI level of 80 or greater, occupancy comfort is achieved.

The greater the ADPI, the greater the probability for occupant comfort in any area of the room. Table 6-4 is a selection guide for several terminal devices based on the range of the ADPI for various levels of room cooling load and *throw ratios*. The throw ratio, T_v/L, is defined as the ratio of the throw of the outlet at terminal velocity V to the characteristic length of the room, L. Normally, V is selected to be 50 fpm [0.25 m/s], a velocity that produces good air distribution without perceived air movement. L is (1) the distance to the wall perpendicular to the jet for a high-sidewall grille, (2) the distance to the closest wall or intersecting air jet for a circular ceiling diffuser, (3) the length of the room in the direction of the jet flow for a sill grille, (4) the distance to midplane between outlets plus the distance from ceiling to top of occupied zone for light troffer diffusers, or (5) the distance to walls or midplane between outlets for perforated, louvered ceiling diffusers.

The following steps summarize the selection procedure for air distribution outlets:

1. Determine the air volume requirements, based on the room cooling load and the temperature difference between the supply air temperature selected and the design room air temperature. These airflow rates must be compared to the rates determined from other design constraints, such as the minimum flow required to accommodate latent loads, minimum outdoor air for various activities, minimum circulation for odor control, and so on, and the largest required flow selected.

TABLE 6-4

Air Outlet Section Guide Based on ADPI

Air Outlet Type	Room Load, Btu/h sf [w/m²]	T50/L for Max. ADPI	Maximum ADPI
High sidewall grille	80 [25]	1.8	68
	60 [19]	1.8	72
	40 [13]	1.6	78
	20 [6]	1.5	85
Circular ceiling diffuser	80 [25]	0.8	76
	60 [19]		83
	40 [13]		88
	20 [6]		93
Sill/floor grille, straight vanes	80 [25]	1.7	61
	60 [19]	1.7	72
	40 [13]	1.3	86
	20 [6]	0.9	95
Sill/floor grille, spread vanes	20–80 [6–25]	0.7	94
Ceiling slot diffuser	80 [25]	0.3	85
	60 [19]		88
	40 [13]		91
	20 [6]		92
Light troffer diffuser	60 [19]	2.5	86
	40 [13]	1.0	92
	20 [6]	1.0	95
Perforated face or louvered ceiling diffuser	11–51 [4–16]	2.0	96

2. The tentative outlet type and location are selected based primarily on architectural or aesthetic considerations.
3. The room's characteristic length, L, is determined.
4. The throw length is calculated by multiplying the throw ratio, T_{50}/L, from Table 6-4, by the characteristic length, L.
5. An outlet is selected from manufacturers' data which has an appropriate throw length and which meets noise, static pressure drop, and any other special criteria imposed.

The ADPI methodology was developed primarily for cooling loads only. As illustrated in Figures 6-6 through 6-10, using the same ceiling diffuser for both heating and cooling may result in poor performance when delivering warm air, which has a tendency to stay at the ceiling due to its low density. This condition may sometimes lead to selection of separate perimeter heating systems, such as perimeter radiation. If air is to be used for both heating and cooling, a separate perimeter air system with upward vertical discharge may be ideal, since in the winter, the cold surfaces at the windows will be warmed, reducing the cold radiant effects, and in the summer, the cool air will reduce the warm radiant effect from the same surfaces.

Heating *and* cooling from the same overhead diffusers are more typical, and several design precautions are in order as follows:

1. Supply outlets should be located in the ceiling a few feet from the outside wall and the discharge should be directed toward the wall.
2. Outlets must be selected on the basis of both maximum and minimum airflows required.
3. Outlet throws should be selected to reach the wall on heating but not so great as to cause excessive drafts on cooling.
4. Several outlets, or spaced linear diffusers, should be used to allow even distribution of the warm air over the wall. Keep the amount of air delivered per outlet as small as economically feasible to keep throw distances short.
5. The heating supply air temperature should be held to ≤95°F [35°C] to reduce the buoyant force of the heated air, which otherwise can overcome the momentum of the supply flow and reverse its direction back toward the ceiling.
6. Generally, the amount of warm airflow is lower than the amount of cold airflow to a space simply because different design temperature differences are used. Select a diffuser so that it may be slightly smaller than ideal for the cooling airflow, but only slightly larger than ideal for heating airflow.

As discussed in Chapter 7, the airflow to each outlet is dictated by the pressure loss in each airflow delivery path in supply to the duct system. To compensate for the normal pressure loss variations in systems, *balancing dampers* are normally required to adjust outlet airflows to their design value. These dampers should never be installed as part of the air outlet device but should be installed as far as possible upstream of the outlet in order to (1) be located in a higher air velocity and, thus, provide better control, and (2) not create air noise in the space.

There are literally dozens of types of ceiling diffusers, sidewall grilles, and floor grilles available from manufacturers. Selection is based on location (ceiling, floor, or sidewall), duty (supply, return, exhaust, and so on), performance, need to be protected from mechanical damage, and esthetics. While most air outlets and inlets are available of steel or aluminum construction, aluminum should always be used in hot, humid climates. In wet locations (kitchens and so on), or where aseptic conditions are maintained (hospital operating rooms, pharmaceuticals, manufacturing, and so on), stainless steel construction is required.

6.2.3 Large Space Air Distribution

The distribution of air in large spaces is much more dependent on air distribution device location and the methods using ADPI and throw versus spacing cannot be used. Rather, *jet projections* must be analyzed and the air distribution devices selected accordingly.

There are two basic principles in analyzing air distribution in large spaces. First, based on their densities, hot air rises and cold air falls. Second, air that is discharged as a free jet, as opposed to air that is constrained along a surface, acts independently of the supply device outlet pattern, following only the rules of jet theory.

Air in large spaces is most commonly discharged horizontally, either at the perimeter of the space or in regular patterns throughout the space, *via* sidewall grilles. Here, the type of grille, discharge or blade angle, and adjacent surfaces all impact the discharge or *air jet*.

Table 6-5 summarizes the performance of a typical airfoil blade double-deflection sidewall diffuser under two installation conditions.

For 0° deflection, the drop at each jet velocity essentially doubles. At a 45° spread, throw is reduced by about 50% for the same terminal velocity.

Sidewall grilles must be selected so that the throw and drop combination does not result in the supply jet entering the occupied zone at above 50 fpm [0.25 m/s].

Stratification can also be applied to large spaces so that air conditioning is provided only within the occupied zone. To achieve stratification, the grilles must be located so that the air jet does not break up the desired stratification layer. Therefore, they should be located just above this boundary and selected so that air velocities have reached terminal speed

TABLE 6-5

Performance of Typical Airfoil Blade Sidewall Grille

		Vertical Drop (ft) [m]	
"Jet" or Discharge Velocity (fpm) [m/s]	Horizontal Throw (ft) [m]	Grille Just Below Ceiling	No Ceiling
300 [1.5]	5 [1.5]	1 [0.3]	1 [0.3]
	10 [3.0]	4 [1.2]	4 [1.2]
	15 [9.6]	8 [2.4]	8 [2.4]
	20 [6.1]	13 [4.0]	13 [4.0]
500 [2.5]	20 [6.1]	2 [0.6]	3 [0.9]
	30 [9.1]	5 [1.5]	7 [2.1]
	40 [12.2]	10 [3.0]	13 [4.0]
700 [3.6]	30 [9.1]	3 [0.9]	3 [0.9]
	40 [12.2]	5 [1.5]	7 [2.1]
	50 [15.2]	8 [2.4]	10 [3.0]
1000 [5.1]	30 [9.1]	1 [0.3]	2 [0.6]
	40 [12.2]	3 [0.9]	3 [0.9]
	50 [15.2]	6 [1.8]	4 [1.2]

Cooling application, 20° upward deflection, 0° spread, 50 fpm terminal velocity.

(50 fpm) before entering the occupied zone. To enhance stratification, returns should be located within the stratification zone, not the occupied zone.

In gymnasiums, assembly spaces, industrial areas, and so on, a *fabric air diffuser* (more accurately called a *fabric air dispersion device*) that is fabricated from flexible, permeable, polyethylene fabric and classified as an air distribution device by UL is often applied for high volume airflows and to minimize the potential for mechanical damage. These diffusers are furnished with suspension mounting hardware systems arranged to allow the fabric to be freely suspended and fully inflate at no more than 0.25″ wg [62 Pa] air pressure.

6.2.4 Displacement Ventilation Air Distribution

The vast majority of air distribution designs utilize overhead supply diffusers to deliver air to the space served, including the ventilation air component. Even when dedicated outdoor air systems (see Chapter 10) are applied, air is typically supplied overhead. As defined in ASHRAE *Standard 62.1*, overhead distribution of cool ventilation air has a 1.00 effectiveness factor. But, ventilation effectiveness falls to 0.80 during the distribution of warm ventilation air.

In Scandinavia and other northern, cold climate countries in Europe, the concept of *displacement ventilation* has been used since the late 1970s with some success to offset the reduced ventilation effectiveness when warm air is supplied overhead. This concept uses a room air distribution strategy by which conditioned outdoor air is supplied at low velocity from air supply diffusers located in a floor plenum and extracted above the space occupied zone at the ceiling. This method improves ventilation effectiveness factor to 1.00.

The use of low velocity air supply from floor grilles takes advantage of the natural *thermal plums*, air that rises due to the heat release from occupants, appliances, and equipment in the conditioned space.

Expanding on this concept, some buildings in the United States and United Kingdom have used this type of system to provide floor level supply air sufficient for ventilation and to offset space heat gains and losses. In this case, air is supplied *via* an unducted *underfloor air distribution* (UFAD) system that uses a raised floor plenum concept. *But, performance by these systems have been mixed, for several different reasons.*

The U.S. General Services Administration (GSA) allowed the use of unducted UFAD in buildings prior to 2007. Several buildings were constructed between 2001 and 2005 for the GSA incorporating unducted UFAD, and numerous problems with the design and operation of UFAD soon became evident. In 2006, six buildings with unducted UFAD were selected for detailed study by the National Institutes of Building Sciences. This study identified several significant failures in design and construction, resulting in poor performance by unducted UFAD.

Consequently, in 2007, unducted UFAD was banned by the GSA, and the GSA 2010 *Design Guidelines for Public Buildings* states the following:

> A *fully ducted* underfloor air distribution (UFAD) system or displacement ventilation system is permitted except in courtrooms, restrooms, cafeterias, kitchens, laboratories, loading docks, mail rooms, U.S. Marshal's areas, and detention areas. The insulated supply air ductwork from the AHU must be connected directly to each floor diffuser or sidewall grille. Unducted or partially ducted UFAD systems or displacement ventilation systems with pressurized plenums are not permitted. Unducted supply is allowed only in computer rooms.
>
> Equipment such as air-handling units, VAV boxes, or other equipment that requires maintenance, is not permitted below a raised-access floor.

The No. 1 problem with unducted UFAD is plenum air leakage. While plenum air leakage is an architectural design and general construction issue, it is a mechanical engineering concern to the extent that the capacity of the air-handling unit must be adequately sized to compensate for the leakage at design load, and controls must be designed to compensate for the air leakage at partial loads. To minimize these design issues, a tight plenum is required.

Plenum air leakage is defined in two categories:

1. General construction leaks from the plenum into other building cavities. An example would be leakage around and within annular spaces in conduit, openings around electrical device boxes, and so on.
2. Leakage through the raised-access flooring into conditioned spaces, such as leaks of conditioned air from the plenum through floor-panel seams and edge closures, in-floor electric-power connections and outlet service units, as well as around air diffusers that are not fitted well and/or do not close tightly.

Plenum air leakage tests at design static pressures conducted in the six federal buildings by the GSA ranged from 30% to 200% of the design airflow rates; compared to the design criteria of maximum 5% air leakage.

Construction of an airtight plenum requires strict coordination of 10–12 trades. Special construction techniques to guarantee plenum tightness have not yet been developed for concrete, masonry, drywall, millwork, sealant and joint specialists, raised-access floor installers, carpenters, plumbers, electricians, communications, and IT installers.

Some of the consequences of plenum air leakage include the following:

1. Occupant complaints of being cold around the feet and legs from air leakage during partial-load conditions or when the thermostatic set point was satisfied.
2. Energy wasted because AHUs were required to run at higher than expected airflow rates and/or static pressures and for longer periods to compensate for the air leakage.
3. Indoor air quality issues and housekeeping difficulties caused by accumulation of inert and biological matter and pest infestation within the raised-floor plenum.

The second common problem with unducted UFAD is that the thermal mass of the floor slab and plenum walls is a significant issue for energy management and control. Observations during the GSA study revealed that longer than expected operational periods by the HVAC systems were required to maintain plenum temperatures.

Heat and moisture transmission and condensation in the plenum also are issues because gradients across the plenums can result in nonuniform temperatures in the occupied spaces, and surfaces within the plenums were more likely to support condensation.

Even with ducted UFAD, one aspect must be clearly understood: there will be temperature variations (gradient) between the floor and the "stratified height" (typically about 7 ft above the floor). A temperature gradient of about 10°F can be expected, with 60%–70% of that gradient occurring within the first 3 ft above the floor. This may result in some "cold floor" of "cold draft" complaints. After the building is occupied, some relocation of floor outlets may be required to address these complaints.

Other problem areas with UFAD include the following:

1. Separate perimeter and core zone heating and cooling systems are required, along with segregated UFAD, and when that is not done, the concept tends to provide poor comfort control.

2. Occupancy, over time, results in space rearrangement... moving desks and other office furniture, piling books and papers on the floor, bringing that nice throw rung from home to dress up a drab cubical, other things that will negatively impact the design air distribution. Moving floor vents and electrical connections is not easy once the system is under pressure and working. Once you lift a tile to do some work you are losing static pressure for the whole system and if the floor is open for any length of time, temperature control goes out of the window. And, even after the floor is re-closed, are the new diffuser locations going to provide satisfactory air distribution and even zone temperature?
3. Access for maintenance to the connections and equipment within the floor plenum, once furniture and other fixtures are in place, can create real headaches, and this is the reason for GSA's prohibition of equipment being located underfloor.
4. Every change within the plenum will introduce the need to *reseal* leakage pathways because of new or changed electrical, communications, and/or plumbing penetrations. And, finally, there will be a need to routinely check and reseal the plenum cavity as sealants and caulks age and deteriorate, creating new air leakage pathways.

6.3 Terminal Units

Constant- and variable-flow TUs consist of a damper and actuator (two, if a dual duct system) in an enclosure and a unit controller designed to operate in a *pressure independent* mode. The unit controller monitors the unit's inlet velocity pressure and modulates the damper/actuator assembly to maintain set point airflow as the supply pressures change. For VAV TUs, a room temperature sensor provides an input signal to establish the required airflow to offset the load condition.

Installation of VAV TUs, both single duct and single-duct fan-powered types, should be in accordance with Figure 6-11.

6.3.1 VAV Single-Duct TUs

For VAV applications, single-duct TUs must be sized for *both* maximum and minimum required airflows. Typically, TUs are rated for a maximum airflow at 2,000 fpm [10 m/s] inlet velocity. But, TU velocity sensors are only accurate down to about 600 fpm [3 m/s]. Therefore, TUs should be selected for the controllable range of airflows as summarized in Table 6-6.

The *total pressure loss* through a TU is the true indicator of the fan energy required to deliver the design airflow through the TU. Unfortunately, manufacturers typically list only the static pressure drop, which is always lower than the total pressure drop since the velocity at the TU inlet is much higher than the outlet velocity, resulting in static pressure regain. For all except very noise-sensitive applications, VAV TUs should be selected for a total pressure (static pressure plus velocity pressure) loss of ≈0.5 in. wg [125 Pa] or less. For most applications, this provides the optimum energy balance.

To size TUs when total pressure loss is not cataloged by the manufacturer, the total pressure loss across the TU may be computed using Eq. (6-6) or (6-6a).

$$\text{In IP units, TP} = \text{SP} + \left[(V_{in}/4{,}005)^2 - (V_{out}/4{,}005)^2 \right] \quad (6\text{-}6)$$

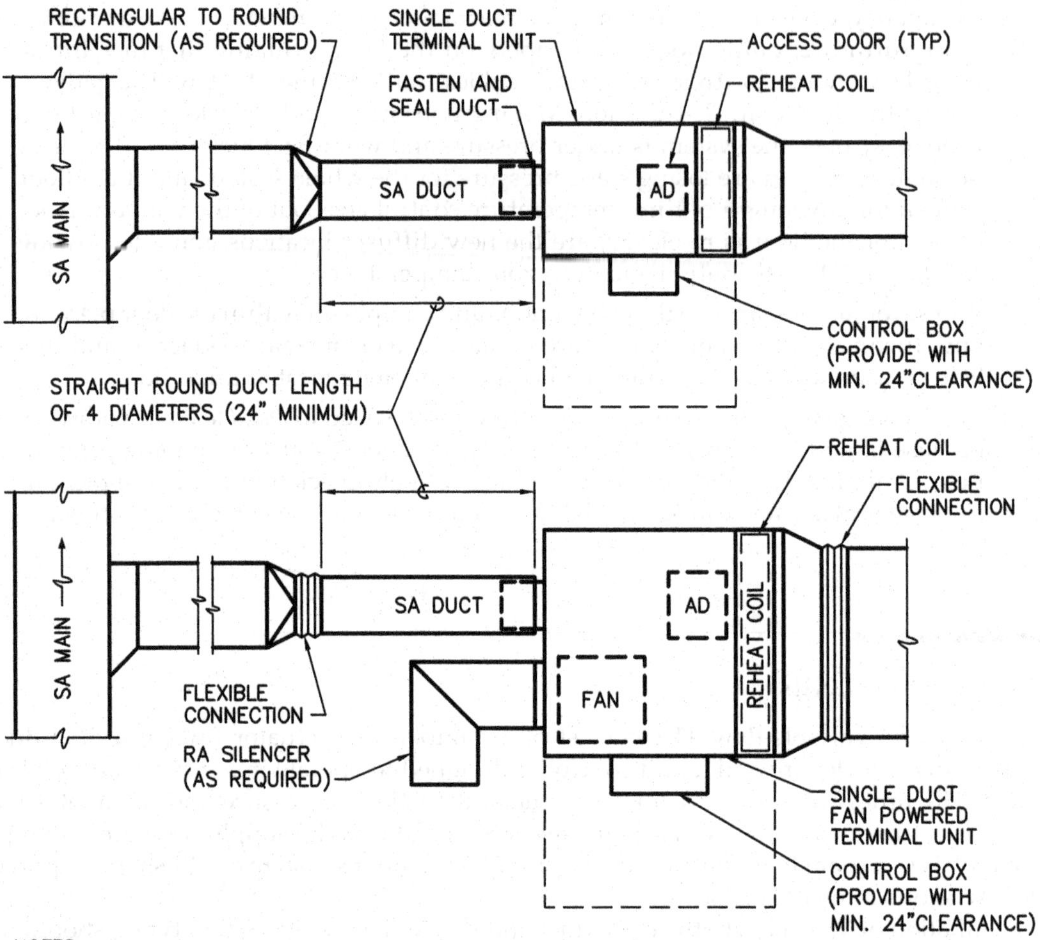

NOTES:

1. WHEN MINIMUM UPSTREAM STRAIGHT DUCT CONNECTION TO TERMINALS AS INDICATED ABOVE CANNOT BE MAINTAINED, PROVIDE ORIFICE PLATE, STRAIGHTENING VANES OR OTHER DEVICE AS RECOMMENDED BY TERMINAL UNIT MANUFACTURER AND SUBMIT TO ENGINEER FOR REVIEW PRIOR TO INSTALLATION.
2. MANUFACTURER OF TERMINAL UNIT SHALL PROVIDE CONTROLS ON LEFT OR RIGHT SIDE AS REQUIRED BY FIELD CONDITIONS (VERIFY PRIOR TO ORDERING).
3. ARRANGE ACCESS TO FACILITATE EASY FIELD BALANCE AND MAINTENANCE OF TERMINAL UNIT.
4. MAINTAIN CLEARANCE REQUIRED BY NEC FOR UNITS WITH ELECTRIC COILS AND DISCONNECTS.

FIGURE 6-11
VAV single duct TU installation details. (SW Standard Detail AT01.)

$$\text{In SI units, TP} = \text{SP} + \left[0.602 \times \left(V_{in}^{\ 2} - V_{out}^{\ 2}\right)\right] \tag{6-6a}$$

where

TP = TU total pressure loss, in. wg [Pa]
SP = Cataloged static pressure loss, in. wg [Pa]

TABLE 6-6

TU Sizing Criteria

Inlet Connection Diameter (in.) [mm]	Airflow Range (Min–Max) (CFM) [L/s]	Inlet Velocity Range (FPM) [m/s]
4 [100]	55–200 [26–94]	
6 [150]	120–400 [57–189]	
8 [200]	205–700 [98–330]	
10 [250]	310–1,100 [146–519]	600–2,000 [3–10]
12 [300]	475–1,550 [224–732]	
14 [350]	640–2,100 [302–991]	
16 [400]	840–2,800 [396–1,320]	
24 × 16 [600 × 400]	1,700–5,000 [802–2,360]	

V_{in}, V_{out} = velocity (fpm) [m/s] at the TU inlet and outlet, calculated by dividing the TU airflow rate (cfm) [m^3/s] by the inlet and outlet area (ft^2) [m^2].

TU manufacturers typically provide two types of sound data for their equipment: *discharge* and *radiated*. Discharge noise is rarely an issue if the TU has hard duct on the inlet and outlet, with flex duct between the supply duct and air outlet(s). As a general rule, VAV boxes located above acoustical ceilings should have radiated *Noise Criteria* (NC) levels (see Chapter 14) no more than ≈5 NC above the desired room NC rating.

Finally, when it is necessary to install TUs at a high elevation above finished floors, especially above a ceiling, service and maintenance accessibility must be carefully analyzed and mounting heights in excess of 12 ft [3.5 m] above finished floors will normally require special access accommodations.

The simplicity of the VAV system is only one step removed from the single-zone system and the cost is only somewhat higher. The obvious disadvantage of the simple VAV system is that it provides only cooling, and where heating is required, the system must be configured for VAV with *terminal reheat*. As room cooling loads are reduced from their peak, the room temperature sensor responds by first reducing the primary airflow to its minimum set point(s) adding heat to offset further loss in cooling load and imposed heating load.

However, when providing overhead heating, the discharge air at a temperature more than 15°F [8.3°C] above the room temperature set point will likely result in significant temperature-based air stratification and uneven temperatures in the space. The typical VAV terminal reheat unit is selected for a single maximum airflow set point that occurs when the unit is providing cooling. As the cooling load is reduced, the unit damper "throttles" the airflow to a preset minimum rate that is used when providing heat. However, one or both of the following conditions may occur that make the space difficult to heat:

1. The heating load may be such that the supply air temperature must be much greater than the recommended 15°F [8.3°C] above the room temperature set point with minimum airflow. This creates stratification and failure to provide adequate heating, as described previously.
2. The diffusers serving the space may not provide adequate air dispersion and mixing when delivering minimum airflow.

Both of these problems will be exacerbated with a ceiling height greater than 9 ft [3 m] or so.

These problems have led to the development of the VAV *fan-powered TU* described in the following section. But, these units have certain disadvantages and another way of avoiding poor air distribution/heating is to utilize single-duct VAV terminal reheat units with "dual maximum airflow set point" control as illustrated in Figure 6-11. This allows for a very low minimum airflow rate during no- and low-load periods but also provides for sufficient airflow to meet heating requirements.

6.3.2 VAV Single-Duct Fan-Powered TUs

A VAV fan-powered TU consists of a primary air volume control damper arranged in series with a low-pressure supply fan. Fan-powered TUs may be configured as "series" or "parallel" and in both cases. The volume damper delivers a varying amount of cold supply air from the air handler to provide cooling and ventilation, down to the minimum airflow required by codes. With the series configuration, the TU fan runs at all times to maintain a constant airflow to the space and good overall air distribution. With the parallel configuration, the TU fan runs only when heating/reheating is needed.

When comparing energy consumption between series- and parallel-configured fan-powered TUs, numerous studies have found that air leakage by the backdraft damper required to prevent airflow back through when the fan is off is the primary cause of poorer energy performance by parallel-configured TUs.

Fan-powered TUs should be used for any space that requires a high minimum flow rate in order to maintain good air mixing, to prevent "dumping" by air outlets, and/or to meet the heating load at a reasonable supply air temperature (e.g., ≤95°F [35°C]). For fan-powered TUs, select only series-configured units and make sure that the minimum required downstream static pressure loss (typically 0.2″–0.3″ wg [50–75 Pa]) is present. Otherwise, the TU's forward curved blade fan may overload and the fan motor fail.

While fan-powered TUs are used to maintain better air distribution in a space and to deal with the problem of providing adequate heating, their energy consumption, especially for series configured units, can be reduced significantly by using a second level of VAV control *via* applying speed control to the TU fans by using electronically commutated motors. Based on studies by ASHRAE and AHRI in the early 2000s, it was found that the use of the concept did work and resulted in energy savings. Basically, this control concept includes varying the TU fan speed, while cooling in direct proportion to the cooling load but running the TU fan a full speed when heating is required. This approach reduces both fan energy and the use of reheat.

These industry studies, as reported in the ASHRAE Journal in October and November 2017, clearly indicate that the series TU configuration is superior in performance, initial cost, and energy consumption.

6.3.3 VAV Dual Duct TUs

The VAV dual duct TUs consist of dual air dampers and operators, one normally closed for cold airflow control and one normally open for hot airflow control, each operating in a pressure independent mode. These TUs are selected on the same basis defined in Chapter 4.

As the space cooling load is reduced, the cold air damper modulates toward a closed position. At a defined point, before the cold airflow goes to zero, the hot air damper begins to modulate toward the open position. Thus, the TU operation may be shown graphically by Figure 6-11.

When both dampers are partially open, the hot airflow serves to provide reheat of the cold air. So, the *crossover point* must be selected on the basis of providing the required minimum ventilation airflow on the basis of partial ventilation by both airstreams.

Bibliography

ASHRAE Handbook-Fundamentals, American Society of Heating, Refrigerating, and Air-Conditioning Engineers, Inc. (Atlanta, GA, 2017).

Standard 410, Forced-Circulation Air-Cooling and Air-Heating Coils, Air-Conditioning, Heating, and Refrigeration Institute (AHRI) (Arlington, VA, 2001).

Engineer's HVAC Handbook, Price Industries Ltd. (Winnipeg, MB, Canada, 2012).

"Engineering Guidelines," published online by Titus HVAC, 605 Shiloh Rd., Plano, TX 75074.

7

Duct Design

7.1 Introduction

Once air distribution *outlets* have been selected and located based on the heating and cooling loads in each space, the duct system that will supply the proper amount of air to each space may be designed. The overall task is to route and size each duct section from the supply fan to each space or terminal unit, including the ducted return, as applicable, within the constraints of available space allocated for this purpose and acceptable noise levels, while minimizing total life-cycle costs. This is generally an iterative problem, in that the optimum system design cannot usually be predetermined from the available system loads and room air diffusion characteristics but must be approximated from successive trial designs and modifications.

7.1.1 Steps in Duct System Design

The duct system design procedure may be summarized in the following steps:

1. A tentative, single line duct system is sketched, connecting the supply outlets and return intakes with the central system air-handling units or fans.
2. The layout is checked against structural, piping, and electrical drawings on each floor to avoid interference with obstructions and equipment, and appropriate corrections are made. (See Chapter 16 for building information modeling [BIM] requirements to address this aspect.)
3. The sizes of all main and branch ducts are calculated using one of the accepted design methods. When using BIM software (see Chapter 16), the designer must select the design method to be used and establish duct sizing limits defined in this chapter.
4. The total pressure requirements of all duct sections, both supply and return, are calculated, based on the flow rates, velocities, and loss coefficients of each.
5. For constant air volume (CAV) systems, the results of step 4 are examined to determine the amount of balancing the system requires (i.e., the amount of flow restriction, or *dampering,* necessary in duct branches to avoid excessively high static pressures in those branches, while obtaining sufficient static pressure to obtain design flows in the branches with greatest pressure drop). If necessary, revisions are made in the duct system components to reduce the imbalance between branches.

6. The duct system design is analyzed to determine the expected noise level and to be revised as necessary to reduce noise transmission (see Chapter 14). At this stage, the system should also be evaluated for construction, operating, maintenance, and energy costs to provide the data for making valid life-cycle comparisons, then the design process may be repeated (at least steps 3, 4, and 5) with different duct sizes to obtain a practical optimum design.

7.1.2 Duct Air Balancing Rationale

Variable air volume (VAV) systems are *self-balancing*, since the airflow through each section of ductwork is controlled by the VAV damper in each terminal unit served. However, for CAV systems, the implication of the balancing step, which is invariably the bane of any duct system design and installation, is that the typical methods of duct design do not produce a system that provides the proper amount of air to each branch by default.

Another way of looking at this process is to see the system as automatically self-regulating as to pressure (i.e., the pressure loss in each complete duct run from the fan through to the terminal unit or air outlet is, to the closest extent possible, always the same). Any pressure "imbalances" in the design procedure will be resolved by flow changes away from the design specifications.

The practical solution to this problem most commonly used is to restrict the flow via balancing dampers to those branches receiving too much flow. If some branches have too little flow at full open damper, it may be possible to increase the fan speed, thereby increasing the static pressure, and restrict the flow to all branches except those requiring more flow. Thus, even a poor duct design can usually be made workable if the fan can be operated to provide sufficiently high static pressure.

7.2 Energy and Pressure Relationships

The first law of thermodynamics may be written for the incompressible flow of air in ducts between any two points 1 and 2 at steady state as

$$\text{rate of energy input} = \text{rate of energy output}$$

or

$$E_1(\text{kinetic}) + E_1(\text{potential}) + E_1(\text{internal}) + \text{flow} + e(\text{fan}) + Q$$
$$= E_2(\text{kinetic}) + E_2(\text{potential}) + E_2(\text{internal}) + \text{flow}$$

or

$$1/2mV_1^2 + mgz_1 + mh_1 + e(\text{fan}) + Q = 1/2mV_2^2 + mgz_2 + mh_2 \quad (7\text{-}1)$$

where the enthalpy at either point, $h = u + P_v$, is actually the sum of internal energy, u, and "flow energy," P_v. We now invoke this definition and also divide Eq. (7-1) by the mass flow rate, m, to obtain an expression in terms of unit mass, which could be called *specific energy*, as shown by Eq. (7-2).

$$1/2V_1^2 + gz_1 + P_1v_1 + q + e(\text{fan}) = 1/2V_2^2 + gz_2 + P_2v_2 + (u_2 - u_1) \tag{7-2}$$

The total pressure change between any two points 1 and 2 may be found by solving Eq. (7-2). If we approximate the specific volume as the average value between the two points, $v = (v_1 + v_2)/2$ and install the conversion factors g_c, F_1, and F_2, to yield Eq. (7-3).

$$P_2 - P_1 = \frac{V_1^2 - V_2^2}{2g_c v} + \frac{g(z_1 - z_2)}{g_c v} + \frac{e(\text{fan})}{F_2 v} + \frac{q}{F_1 v} - \frac{u_2 - u_1}{F_1 v} \tag{7-3}$$

or

total pressure difference	=	velocity pressure difference	+	elevation pressure difference	+	fan energy pressure difference	+	pressure difference due to heating or cooling	+	pressure difference due to internal energy changes

where

V = average duct velocity, ft/s [m/s]
g = gravitational acceleration, 32.2 ft/s² [9.8 m/s²]
g_c = conversion factor, 32.2 lbm-ft/(lbf-s²) [9.8 m/s²]
z = elevation, ft [m]
P = total pressure, lbf/sf [Pa/m²]
v = specific volume, cf/lbm [m³/kg]
e = fan energy, W (lbm/s) [W]
q = heat addition, Btu/lbm [kJ/kg]
F_1 = conversion factor, 1/778 Btu/(ft-lbf) [1/334 kJ/kg]
F_2 = conversion factor, 1.356 W-s/(ft-lbf) [W-s/J]
u = internal energy, Btu/lbm [kJ/kg]

Equation (7-3), and the associated verbal equation below it, can be seen as expressions of the sources of pressure changes in air ducts. The difference between the total pressure and the velocity pressure at any point in the duct is typically called the *static pressure.* Static pressure is the pressure measured inside the duct with a probe oriented so as not to be influenced by the velocity of the air. The total pressure is that measured by a probe oriented directly toward the airflow to measure the impact pressure of the air. The difference between the two is typically measured with a pitot tube as an indication of the velocity of the air, as defined by Eq. (7-4).

$$P = (V_2/2g_c v) + P_s \tag{7-4}$$

where P is the total pressure and P_s is the static pressure.

The last two terms in Eq. (7-3) represent the heat added to the flowing mass of air and its increase in internal energy, respectively. If we lump the heating of the air due to the fan (i.e., losses due to fan inefficiency and the inefficiency of the motor, if the motor is located in the airstream) with any intentional heating or cooling of the air, these last two terms represent all of the "thermal" energy changes in the airstream. The difference in the magnitudes of these two terms represents the lost pressure due to friction which is defined by Eq. (7-5).

$$P_f = \left[(u_2 - u_1)/F_1 v\right] - (q/F_1 v) \tag{7-5}$$

Of course, this relationship does not tell us how to calculate the magnitude of the friction head but merely shows what happens to the energy. If the friction head is zero, for ideal frictionless, incompressible flow, the difference in internal energy must equal the heat added. Thus the internal energy of the air can typically be increased in two ways, through friction or through direct heating. The result of either is seen as an increase in the temperature of the air.

7.3 Friction Losses

7.3.1 Calculation of Friction Losses in Ducts

Although Eq. (7-5) shows that the difference between the heat added to flowing air and the observable rise in internal energy is equal to the loss in total pressure due to friction, we must go to empirical methods to estimate the losses in advance. Typically, this is done by selecting values of friction pressure loss per unit length of straight duct for a specific duct size and flow velocity conditions from published tables and charts for various types of duct construction. Figures 7-1 and 7-2 are representative charts for sizing smooth typical sheet metal ducts.

Figures 7-1 and 7-2 can be used to select a duct size for a given airflow rate and friction loss. For example, a duct to carry 5,000 cfm [2,400 L/s] at a pressure loss of 0.1 in. wg [25 Pa] per 100 ft [30 m] of duct length would be 26 in. [660 mm] in diameter, with a resulting flow velocity of 1,400 fpm [7 m/s]. Knowing a duct size and required airflow quantity, the anticipated friction loss can be determined from Figures 7-1 and 7-2.

In certain cases, the loss due to friction is calculated by an empirical loss coefficient, expressed as a function of the velocity pressure by Eq. 7-6.

$$P_f = f \times (L/D)\left(V^2/2vg_c\right) \tag{7-6}$$

where

f = friction factor depending on the Reynolds number (Re) and the roughness of the duct (usually read from a plot of/versus R, referred to as a Moody diagram)

L = duct length, ft [m]

D = duct diameter, in. [mm]

For a noncircular duct, D is taken to be the hydraulic diameter, which is four times the cross-sectional area divided by the perimeter of the duct.

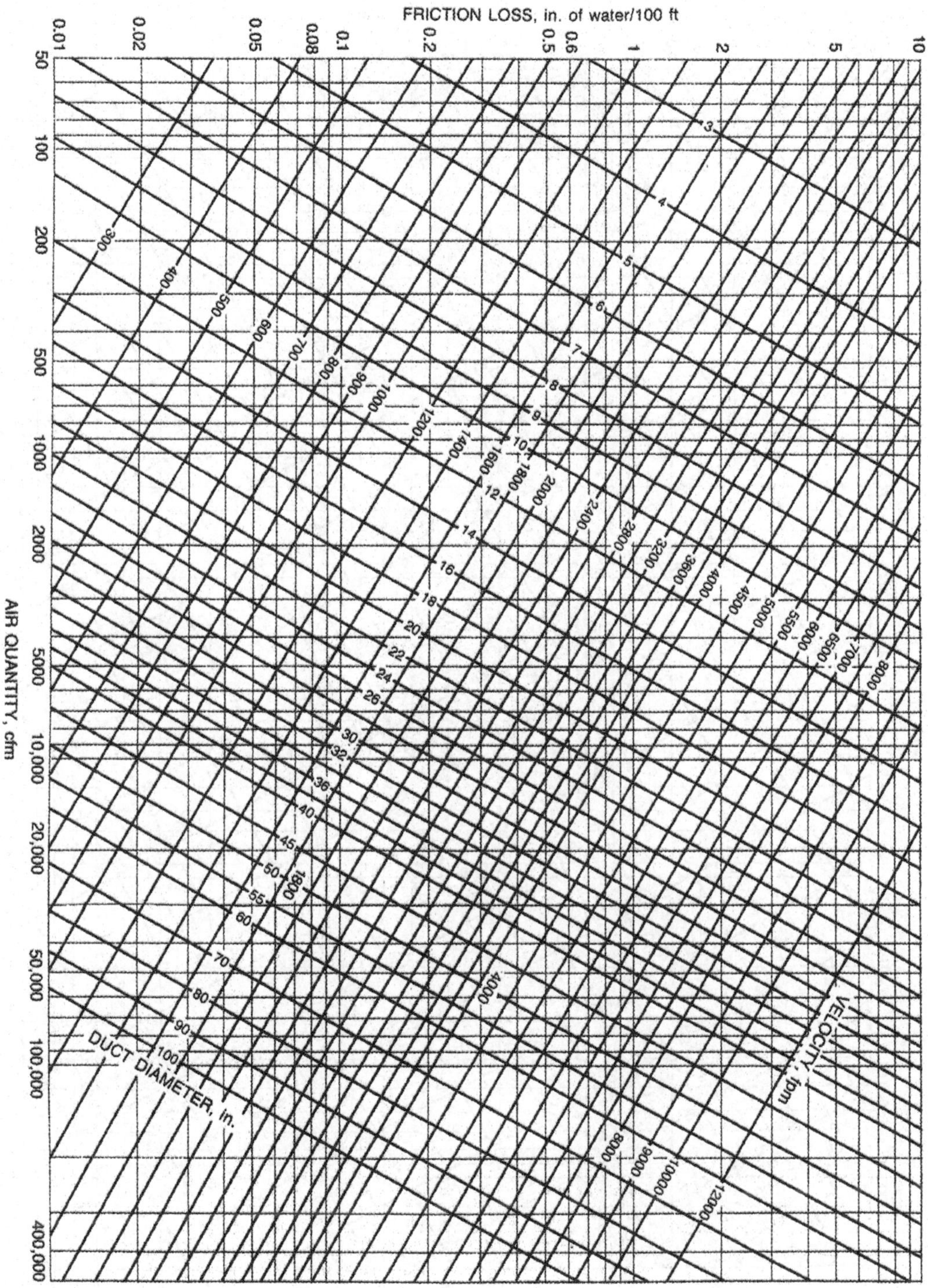

FIGURE 7-1
Friction loss versus airflow rate for various duct diameters (IP units). (Fig. 10, Page 21.9, 2017 *ASHRAE Handbook-Fundamentals*, Inch-Pound Edition (ASHRAE, 1791 Tullie Circle, N.E., Atlanta, GA, 30329.))

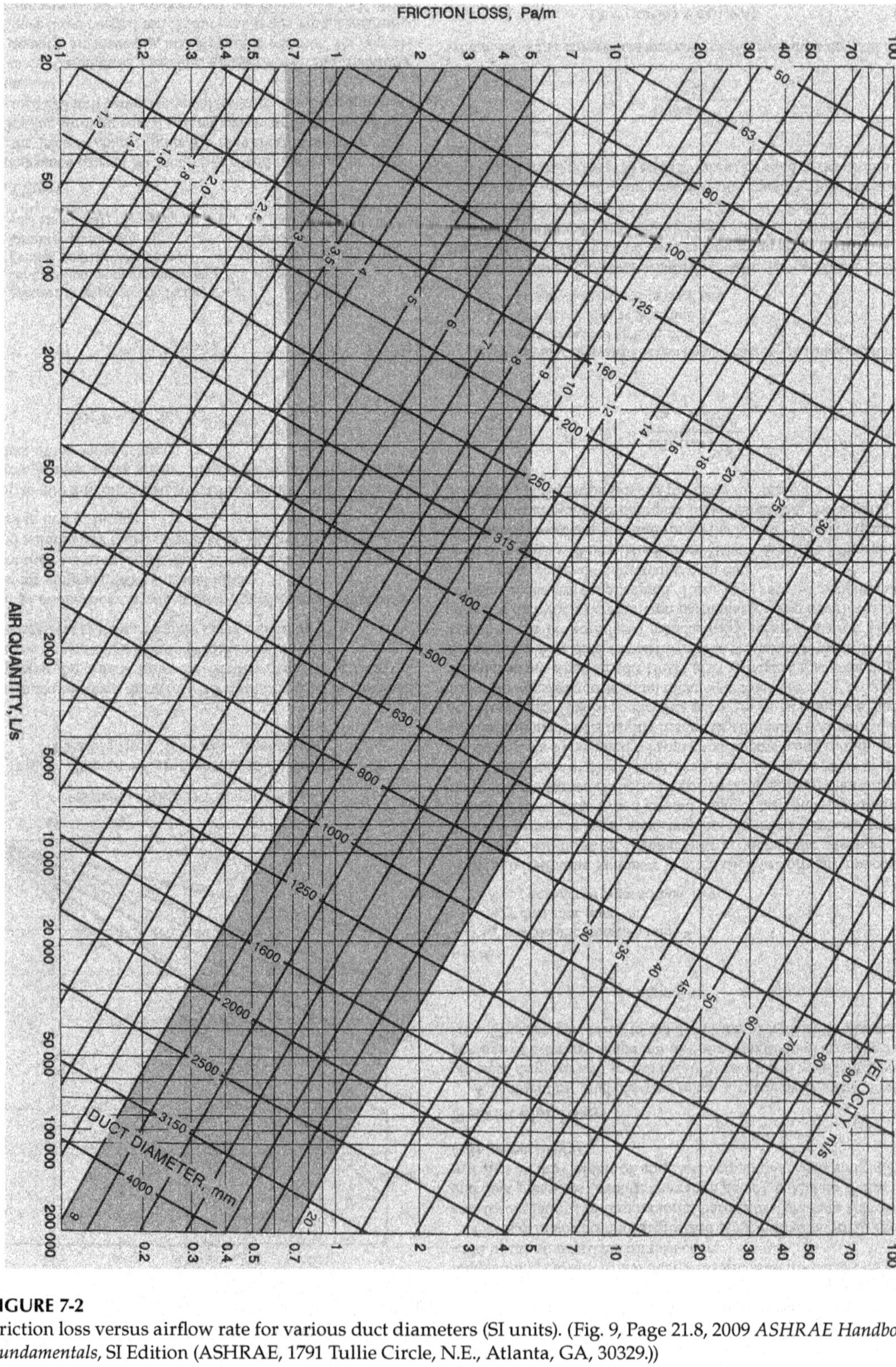

FIGURE 7-2
Friction loss versus airflow rate for various duct diameters (SI units). (Fig. 9, Page 21.8, 2009 *ASHRAE Handbook-Fundamentals*, SI Edition (ASHRAE, 1791 Tullie Circle, N.E., Atlanta, GA, 30329.))

The most efficient duct, in terms of flow capacity at a certain velocity with minimum pressure loss due to friction, is a circular duct. Rectangular, oval, or other shapes will have greater friction losses for the same flow and average velocity. In order to use the charts and tables developed for the friction pressure losses in circular ducts, an *equivalent diameter* may be defined by equating the losses for circular ducts and ducts of hydraulic diameter D_E for equal lengths and equal flow rate, resulting in

$$D_E = 1.3(ab)^{0.625}/(a+b)^{0.25} \tag{7-7}$$

where a and b are the dimensions of the rectangular duct or lengths of the axes of the oval duct. This equivalent diameter, D_E, is therefore the diameter of a circular duct that will have the same pressure loss as the rectangular or oval duct when carrying the same flow rate for the same length.

Tables 7-1 and 7-2 each provide a set of solutions of Eq. (7-7), giving equivalent diameters of rectangular ducts with dimensions specified in each consistent set of units. Duct contractors prefer to fabricate ductwork in increments of 2 in. [50 mm], preferably in even sizes. Ducts over 40 in. [1,000 mm] in any dimension, or with an aspect ratio (long side/short side) greater than 3, add disproportionately to the duct fabrication cost. Tables 7-1 and 7-2 reflect these constraints.

In lieu of using Table 7-1 or 7-2, the Trane "Ductulator" or Carrier's "Duct Calculator" slide rule can be used. Heating, ventilating, and air-conditioning (HVAC) designers can obtain either or both of these devices from sales representatives of each company. Both are simple, easily operated manual devices for quickly determining rectangular duct sizes for a given equivalent round duct size and/or for comparing alternative rectangular dimensions for a given airflow.

TABLE 7-1

Rectangular/Round Duct Equivalent Size for Equal Friction (in.)

Side	6	8	10	12	14	16	18	20	22	24	26	28	30	32	34	36	38	40
6	6.6	7.6	8.4	9.1	9.8	10.1	11.0											
8	7.6	8.7	9.8	10.7	11.5	12.2	12.9	13.5	14.1	14.6								
10	8.4	9.8	10.9	12.0	12.0	13.7	14.5	15.2	15.9	16.5	17.1	17.7	18.3					
12	9.1	10.7	12.0	13.1	14.2	15.1	16.0	16.8	17.6	18.3	19.0	19.6	20.2	20.8	21.4	21.9		
14	9.8	11.5	12.9	14.2	15.3	16.4	17.3	18.2	19.1	19.9	20.6	21.3	22.0	22.7	23.3	23.9	24.5	25.0
16	10.1	12.2	13.7	15.1	16.4	17.5	18.5	19.5	20.4	21.3	22.1	22.9	23.7	24.4	25.1	25.7	26.4	27.0
18	11.0	12.9	14.5	16.0	17.3	18.5	19.7	20.7	21.7	22.7	23.5	24.4	25.2	26.0	26.7	27.4	28.1	28.8
20		13.5	15.2	16.8	18.2	19.5	20.7	21.9	22.9	23.9	24.9	25.8	26.6	27.5	28.3	29.0	29.8	30.5
22		14.1	15.9	17.6	19.1	20.4	21.7	22.9	24.0	25.1	26.1	27.1	28.0	28.9	29.7	30.5	31.3	32.1
24		14.6	16.5	18.3	19.9	21.3	22.7	23.9	25.1	26.2	27.3	28.3	29.3	30.2	31.0	32.0	32.8	33.6
26			17.1	19.0	20.6	22.1	23.5	24.9	26.1	27.3	28.4	29.5	30.5	31.5	32.4	33.3	34.2	35.1
28			17.7	19.6	21.3	22.9	24.4	25.8	27.1	28.3	29.5	30.6	31.7	32.7	33.7	34.6	35.6	36.4
30			18.3	20.2	22.0	23.7	25.2	26.6	28.0	29.3	30.5	31.7	32.8	33.9	34.9	35.9	36.8	37.8
32				20.8	22.7	24.4	26.0	27.5	28.9	30.2	31.5	32.7	33.9	35.0	36.1	37.1	38.1	39.0
34				21.4	23.3	25.1	26.7	28.3	29.7	31.0	32.4	33.7	34.9	36.1	37.2	38.2	39.4	40.3
36				21.9	23.9	25.7	27.4	29.0	30.5	32.0	33.3	34.6	35.9	37.1	38.2	39.4	40.4	41.5
38					24.5	26.4	28.1	29.8	31.3	32.8	34.2	35.6	36.8	38.1	39.3	40.4	41.5	42.5
40					25.0	27.0	28.8	30.5	32.1	33.6	35.1	36.4	37.8	39.0	40.3	41.5	42.6	43.7

(Maximum 3:1 Aspect Ratio).

TABLE 7-2

Rectangular/Round Duct Equivalent Size for Equal Friction (mm)

Side	150	200	250	300	350	400	450	500	600	700	800	900	1,000
150	164	189	210	229	245	260	274						
200	189	219	244	266	286	305	321	337	365				
250	210	244	273	299	322	343	363	381	414	443	470		
300	229	266	299	328	354	378	400	420	457	490	520	548	
350	245	286	322	354	383	409	433	455	496	533	567	597	626
400	260	305	343	378	409	437	464	488	533	573	609	643	674
450	274	321	363	400	433	464	492	518	567	610	649	686	719
500	287	337	381	420	455	488	518	547	598	644	687	726	762
600	310	365	414	457	496	533	567	598	656	708	755	799	840
700	331		443	490	533	573	610	644	708	765	818	866	911
800	350		470	520	567	609	649	687	755	818	875	927	976
900	367			548	597	643	686	726	799	866	927	984	1,037
1,000	384				626	674	719	762	840	911	976	1,037	1,101

(Maximum 3:1 Aspect Ratio).

7.3.2 Calculation of Friction Losses in Fittings

Friction losses in fittings are typically a much greater part of any duct system's losses than the straight duct. These losses are expressed in terms of the velocity pressure since the losses vary as the square of the velocity. Using an empirical loss coefficient, C, the pressure loss due to friction in a fitting can be expressed by Eq (7-8).

$$P_t = C_o P_{v.0} \tag{7-8}$$

In this equation, $P_{v,0}$ is the velocity pressure at section 0 in. wg. Loss coefficients for over 220 rectangular, round, and round-oval fitting types are available in the 2016 American Society of Heating, Refrigerating, and Air-Conditioning Engineers (ASHRAE) *Duct Fitting Database*, which is the best source of this information currently available. These coefficients are utilized by almost all duct design software for computing total static pressure losses in a duct system.

The losses represented by Eq. (7-8) are the dynamic losses, the losses associated with the currents caused by changes in direction of the airflow. Strictly, if the fitting is of substantial length, the loss due to linear movement through the length of the duct, referred to as *simple* friction losses, should be added to the dynamic losses. Simple friction losses are found by treating the flow path length through the duct as a straight section.

To compute the Reynolds number for standard air, Eqs. (7-9) and (7-9a) can be utilized.

$$\text{For IP units,} \quad \text{Re} = 8.56 D_E V_o \tag{7-9}$$

$$\text{For SI units,} \quad \text{Re} = 1.10 D_E V_o \tag{7-9a}$$

where D_E is the duct equivalent diameter in inches [mm] and V_o is the velocity in fpm [m/s].

The relationship between the airflow velocity and the resulting velocity pressure in ductwork is represented by Eqs. (7-10) and (7-10a) at standard air conditions.

$$\text{For IP units,} \quad P_v = (V/4{,}005)^2 \tag{7-10}$$

$$\text{For SI units,} \quad P_v = 0.642\,V^2 \tag{7-10a}$$

where

P_v = velocity pressure, in. wg [Pa]

V = airflow velocity, fpm [m/s]

7.4 Methods of Duct Design

There are four methods used for duct design and sizing: (1) *constant velocity,* (2) *velocity reduction,* (3) *static regain,* and (4) *equal friction.* Each of these is discussed in the following sections, but primary emphasis is given to the equal-friction method, since it is certainly the method in widest use.

7.4.1 Constant Velocity Method

The *constant velocity* method is normally used for exhaust systems conveying particulates and consists simply of sizing each section of the duct system so as to maintain a constant flow velocity. The ducts are sized to operate with a minimum constant velocity so that the materials being conveyed remain in motion. The simplicity of this approach has the additional advantage of lending itself to analytical expressions for minimum operating power cost as a function of design velocity.

7.4.2 Velocity Reduction Method

The *velocity reduction* method involves the selection of some appropriate velocity at the fan outlet, then progressively reducing the velocity at each junction or branch duct. For commercial and public buildings, typical recommended velocities are 1,000–1,300 fpm [5–6.6 m/s] with a maximum of 1,600 fpm [8 m/s] for main ducts, and 600–900 fpm [3–4.5 m/s] with a maximum of 1,300 fpm [5 m/s] for branches. The pressure drops in each duct circuit are calculated, and the circuit with the greatest pressure drop determines the static pressure required to be produced by the fan.

This circuit is called the *critical path,* and its damper will be completely open, whereas the dampers in the other paths will be partially closed. Sometimes the procedure is refined by a trial-and-error technique to find the velocity in each duct section that will result in a pressure loss equal to or less than the pressure available at the entrance to the duct section. Table 7-3 presents typical design velocities for HVAC components.

7.4.3 Static Regain Method

In the *static regain* method of duct design, each duct section is sized so that the increase in static pressure after each branch exactly equals the pressure loss of the next section of duct. Thus, approximately the same static pressure exists at the entrance to each branch, simplifying the balancing of the system. A possible disadvantage of this method, however, is that very large ducts could result at the ends of long duct runs if the velocity near the fan is not large enough. A major drawback of this method is that the process is necessarily iterative,

TABLE 7-3

Typical Design Velocity for HVAC Duct Elements

Duct Element	Design Velocity (fpm)[m/s]
Intake louvers	400–500 [2.0–2.5]
Exhaust or relief louvers	700–800 [3.5–4.0]
Filters	250–500 [1.3–2.5]
Heating coils	500–700 [2.5–3.5]
Cooling coils	400–500 [2.0–3.5]
Air washers	300–500 [1.5–2.5]

requiring several trial-and-error steps to determine the duct size that will produce an acceptable pressure loss, *which is waste of time for VAV systems.*

7.4.4 Equal Friction Method

In the *equal friction* method of duct design, the ducts are sized so as to cause equal pressure losses per unit length of duct.

After the ducts are sized and the pressure losses of the ducts and fittings are calculated, the total loss in each duct circuit is determined. The circuit with the highest pressure loss is designated as the critical path and that pressure loss is used for fan sizing. The differences in pressure losses between duct circuits must be compensated by damping those circuits with lower losses. An additional design iteration to decrease the duct sizes on these circuits may be worthwhile to increase pressure losses, reduce first costs, and reduce the amount of flow damping required. However, the effects of noise at high velocities must be considered and velocities may be limited on that basis.

In medium and low-velocity duct systems, flexible duct is typically utilized as the final connection between the metal duct and terminal units and air distribution outlets. And, occasionally, it may be advantageous to use materials such as cement pipe, plastic pipe, or fiberglass as duct materials. In each of these cases, to utilize Figures 7-1 and 7-2, correction is required. These correction factors, as shown in Table 7-4, are based on "relative roughness," with typical galvanized sheet metal having a relative roughness factor of 1.00.

On a comparative basis, PVC or aluminum ductwork is considered "smooth," glass fiber ductwork (or duct liner) is considered "medium rough," while flexible ductwork and concrete ducts are considered "rough." Some judgment on the part of the designer is required for selecting the K_m factor to apply.

$$P_{da} = P_d \times K_m \tag{7-11}$$

TABLE 7-4

Relative Duct Resistance Roughness Correction Factors

	Correction Factor (K_m)			
Velocity (fpm) [m/s]	**Smooth Duct**	**Average Duct**	**Medium Rough Duct**	**Rough Duct**
1,000 [5.0]	0.90	1.0	1.25	1.80
2,000 [10.0]	0.87	1.0	1.30	1.90
3,000 [15.0]	0.85	1.0	1.35	2.00
4,000 [20.0]	0.80	1.0	1.40	2.10

where

P_{da} = actual pressure drop, in. wg [Pa]

P_d = pressure drop determined from Table 7-1 or 7-2 and/or ASHRAE *Duct Fitting Database*, in. wg [Pa]

K_m = roughness correction factor from Table 7-4

Air velocity in a duct can be calculated using Eq. (7-12), (7-12a), or (7-12b).

$$V = q/A \tag{7-12}$$

$$= 4q/(\pi d^2) \tag{7-12a}$$

$$= q/(a \times b) \tag{7-12b}$$

where

V = air velocity, fpm [m/s]

q = airflow, cfm [m^3/s]

A = area of duct, ft^2 [m^2]

D = diameter of duct, in. [m]

a = width of duct, in. [m]

b = width of duct, in. [m]

7.4.5 Duct Construction Criteria

Duct construction standards utilized in the HVAC industry are those defined in *HVAC Duct Construction Standards, Metal and Flexible* by the Sheet Metal and Air-Conditioning Contractors National Association Inc. (SMACNA). These standards utilize maximum operating pressure, positive or negative, to define the duct construction details (metal gauge [mm], types of horizontal and vertical seams, reinforcing, etc.) required. Table 7-5 defines these construction classifications. (In general, the industry also uses the terms "high, medium, and low" to describe the SMACNA pressure ranges.)

To ensure that the duct contractor provides the required class of duct construction for each section of the duct system under design, SMACNA recommends that the required

TABLE 7-5

Ductwork Construction Classifications

Approximate Pressure Classification	Maximum Internal Static Pressure (in. wg [Pa])	Pressure Type
High	10 [2,500]	Positive
	6 [1,500]	Positive
Medium	4 [1,000]	Positive
	3 [750]	Positive or negative
Low	2 [500]	Positive or negative
	1 [250]	Positive or negative
	1/2 [125]	Positive or negative

TABLE 7-6

Recommended Duct Construction Pressure Class Applications

Application	Construction Pressure Class (in. wg [Pa])
Return and general exhaust ductwork	–2 [–500]
Supply ductwork downstream of air terminal units	+1 [+250]
Supply ductwork with fan static pressure less than 2.5″ wg	+2 [+500]
Supply ductwork with fan static pressure greater than 2.5″ wg	+3 [+750]
Supply ductwork with fan static pressure greater than 5.0″wg	+4 [+1,000]
Smoke control system ductwork	1.5 times the design fan static pressure listed above, but not less than ±3 [±750].

TABLE 7-7

Duct Sealing Classifications

			Leakage Class (cfm/100 ft at 1 in. wg) [L/s at 250 Pa]	
Duct Construction Class (in. wg) [Pa]	Seal Class	Sealant Application Required	Rectangular Duct	Round or Oval Duct
± 1 [± 250] or less	C	Transverse joints only	24 [11]	12 [6]
± 2 [± 500]	B	Transverse joints and seams	12 [6]	6 [3]
± 3 [± 750] and greater	A	All joints and seams and all wall penetrations	6 [3]	3 [1.5]

pressure classification be noted on each section on the design drawings. In general, though, this tedious requirement has been found from experience to be less effective that including "blanket" construction classifications based on the type of air system and fan-rated static pressure. Table 7-6 is a representative of the application requirements for typical HVAC air systems.

HVAC ductwork must be sealed to minimize air leakage in accordance with SMACNA "Sealing Classifications" shown in Table 7-7.

However, experience has shown that Class C sealing is essentially worthless and all ductwork below±3″ wg [750 Pa] pressure construction class should be sealed to meet minimum Class B requirements.

7.4.6 Duct Design Guidelines

Designers must route ductwork, both horizontally and vertically, square and plumb with building walls. *Wherever possible, go straight* … the straighter the duct system, both the energy and first costs will be lower. When laying out a duct system, reduce the number of bends, turns, and transitions to an absolute minimum.

Use round spiral duct wherever it can be fitted within the available space. Round duct is less expensive than oval or rectangular duct, especially when routed in long, straight sections, and round spiral duct requires less sealing than rectangular duct or longitudinally seamed round duct.

With rectangular duct, if at all possible, *avoid the use of square throat elbows with or without turning vanes … use full radius elbows.* Never add turning vanes to a full radius elbow. Where space limitations require the use of square throat elbows, provide airfoil-configured turning vanes that are sized, spaced, and constructed in accordance with the SMACNA standards.

Do not use glass fiber duct or duct liner insulation. Glass fiber, unless provided with foil-faced jacket on the airflow side, will ultimately erode and, in the interim, provides an ideal mold breeding ground. If duct liner must be incorporated into the project, it is recommended that a closed-cell flexible elastomeric material, specifically designed and approved for the application, be used.

Provide opposed blade balancing dampers (with locking quadrants) in low pressure CAV or VAV ductwork for air balancing. *Do not include air splitter controls or "air scoops"... these generate noise and provide very little balancing effect.* Balancing dampers are required in medium- and high-pressure CAV systems but are not needed in VAV systems since the air terminal units serve as "self-balancing" devices as airflows change in response to the imposed loads.

For supply, return, relief, and general exhaust systems, ductwork should be sized using the "equal friction" method and the following criteria:

1. High pressure ductwork: 0.3 in. wg per 100 ft [24.5 Pa/m]
2. Medium pressure ductwork: 0.2 in. wg per 100 ft [16.3 Pa/m]
3. Low pressure: 0.1 in. wg per 100 ft [8.2 Pa/m]

After making the initial duct size selection, "up size" the duct as required to limit duct velocities and potential noise in accordance Table 7-8.

Finally, do not reduce the duct section downstream of a branch or takeoff unless (1) the flow rate in that section is reduced by 25% or more and (2) any dimension changes by at least 4 in. [10 mm]. This reduces fitting pressure losses and costs and generates at least some static pressure regain, which works as a built-in safety factory.

As stated, better duct design attempts to minimize the use of fittings and use only fittings that minimize pressure loss. Figure 7-3 provides recommended duct design criteria to minimize both static pressure losses and potential noise problems.

TABLE 7-8

Velocity to Room Criteria (RC)

		Maximum Allowable Velocity (fpm) [m/s]			
		Trunk/Main Ducts		Branch Ducts	
Location	Design RC	Rectangular	Round or Oval	Rectangular	Round or Oval
In duct shaft or above a hard gypsum wall board ceiling	41+	3,500 [18]	5,000 [25]	2,800 [14]	4,000 [20]
	31–40	2,500 [13]	3,500 [18]	2,000 [10]	2,800 [14]
	≤30	1,700 [9]	2,500 [13]	1,400 [7]	2,000 [10]
Above suspended acoustical tile ceiling	41+	2,500 [13]	4,500 [23]	2,000 [10]	3,600 [18]
	31–40	1,700 [9]	3,000 [15]	1,400 [7]	2,400 [12]
	≤30	1,200 [9]	2,000 [10]	1,000 [5]	1,600 [8]
Exposed ductwork within an occupied space	41+	2,000 [10]	3,900 [20]	1,600 [8]	3,100 [16]
	31–40	1,400 [7]	2,600 [13]	1,200 [6]	2,100 [11]
	≤30	1,000 [5]	1,700 [9]	800 [4]	1,400 [7]

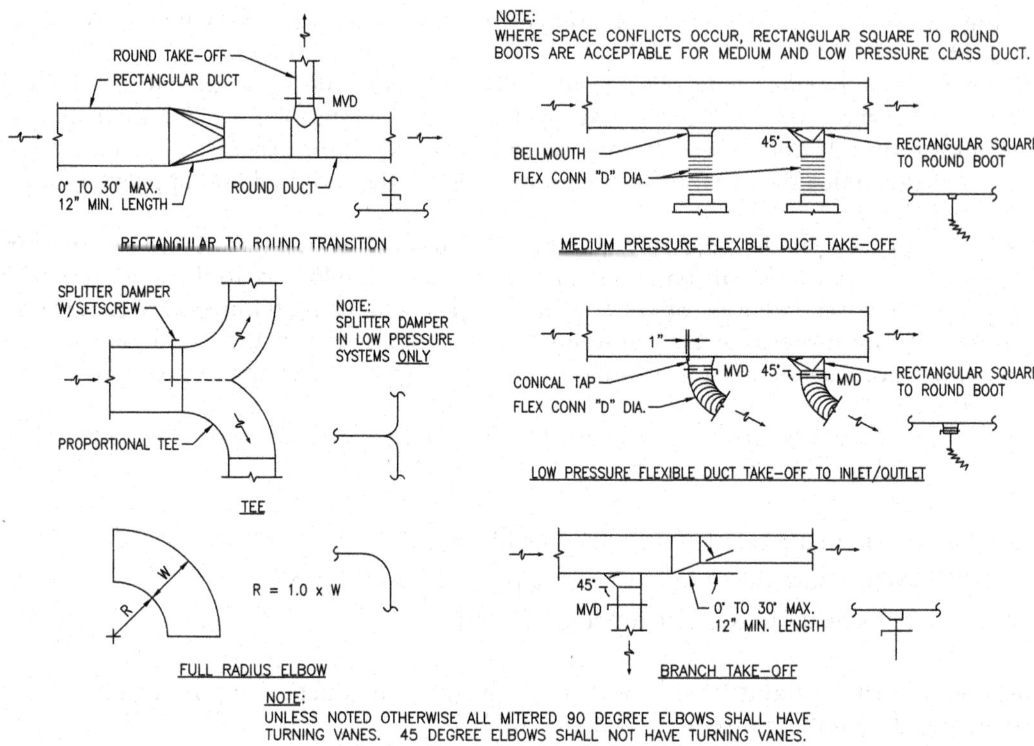

FIGURE 7-3
Recommended duct construction. (SW Standard Detail D01.)

Computing static pressure losses in any duct system is a time-consuming process, and most designers utilize software for this purpose. For preliminary estimates, however, the total length of straight duct, plus 25%–50% to account for fittings, may be multiplied by the static pressure used for duct sizing to obtain an approximate duct pressure loss.

7.5 Special Exhaust Systems Design

7.5.1 Laboratory Exhaust Systems

There are as many approaches to designing laboratory exhaust systems as there are engineers designing them. And, owners will often have specific design criteria that must be met. However, fume hood exhaust systems can be generally classified as "individual exhaust" or "manifold exhaust." A careful *hazard analysis* must be undertaken by the owner to determine which design approach is more appropriate.

The individual exhaust concept results in a separate exhaust fan for each fume hood. Generally, this concept is applied when there are relatively few hoods and/or the hoods are widely dispersed in the building. Under these conditions, the cost of ductwork to manifold the hoods would be cost-prohibitive. Some hoods, due to their use (perchloric acid, radioisotope, special research, etc.), *must* have individual exhaust fans.

With individual hood fans, "on-off" fan control can be activated from a fan switch at the hood.

With individual hood fans, exhaust airflow is considered to be "constant volume." However, dirty exhaust ductwork, filter loading, etc. can change the pressure loss and the resulting air airflow. The designer must select individual fans for the worse case operation to insure that the minimum required hood exhaust airflow is maintained at all times.

The manifold approach has numerous advantages over the individual fan approach, including lower ductwork cost, fewer fans to maintain, fewer roof penetrations and discharge stacks, and providing an opportunity for exhaust air heat recovery. Most important, however, these systems can take advantage of any exhaust system "diversity" that may exist. The design of this type system usually incorporates two to four fans, at least one of which is redundant and operates in a standby mode.

In a manifold system, each hood is equipped with a pressure-independent air valve, and "on-off" hood control can be activated from a switch at the hood to open and close the air valve. (Thus, it is possible that a "usage factor" or diversity may apply in a manifold system and the total exhaust fans sized for a total airflow lower than the sum of the hoods' airflows. However, the prudent engineer will design for *no diversity*, since it is difficult and expensive to increase exhaust capacity in an undersized system.)

Any manifold system must be *pressure independent* and include fan volume control that responds to system pressure changes due to air valves' operation.

With manifold exhaust systems, the *air balance* in the laboratory can be maintained through constant volume or variable volume control modes. The total exhaust requirement for a laboratory is normally dictated by the fume hoods and other exhaust elements. Overall, ventilation airflow equivalent to 6–10 air changes/h (or higher as dictated by the owner or applicable standards) during occupied is recommended for chemical, biological, and biomedical laboratories. But, if the sum of the hood exhaust airflows does not meet this criterion, additional *general exhaust* will be required.

The *constant volume air balance concept* incorporates the use of a pressure independent constant volume airflow regulator on the supply to the laboratory and a separate pressure independent constant volume airflow regulator on the general exhaust (if any), in addition to the regulator(s) on the hood(s) exhaust. Each regulator maintains its required airflow at all times.

A variation to this scheme involves the application of "occupied/unoccupied" modes to the system control. The hood exhaust remains constant, but the overall ventilation airflow can be reduced to as low as 6 air changes/h when the laboratory is unoccupied. Thus, using an occupancy sensor, the laboratory supply and general ventilation airflow rates can be reduced during unoccupied periods as long as (1) sufficient makeup air to offset hood exhaust is provided and (2) the overall room pressure relationship is maintained.

The *variable volume air balance concept* incorporates the use of a pressure independent variable volume airflow regulator on the supply to the laboratory and a separate pressure independent variable volume airflow regulator on the general exhaust (if any), all in addition to the variable volume regulator(s) on the hood(s) exhaust. Under this scheme, the variable volume regulator(s) on the hood(s) reduce the amount of exhaust air in direct proportion to the sash position, down to a fixed minimum airflow of about 20%. The room supply and general exhaust airflow regulators then "track" with the hood(s) regulator(s) to maintain the required room air balance. The exhaust fans' volume controls, then, can respond to the hoods' reduced airflow by monitoring the negative static pressure condition in the exhaust manifold.

The variable volume fume hood exhaust concept is promoted on the basis of energy savings since, as the exhaust volume is reduced, (1) the amount of conditioned makeup air is also reduced and (2) the exhaust fan energy is reduced. However, the complexity and initial cost of these systems is significantly higher than for constant volume systems. Also, since the protection provided by the hood depends on more than simply the sash face velocity, there are concerns (and even a number of law suits) that variable volume hoods may not provide the same level of protection as constant volume hoods. The energy savings claim is also suspect, since researchers tend to leave the hood sash open most of the time, anyway. *The use of variable volume hoods must be discussed with the owner (and the safety officer) to insure that their use is desirable and the necessary level of maintenance can be provided.*

General exhaust systems are always manifold systems since these systems are exhausting room air, along with air from toilets, storage rooms, etc. Their design parameters are the same as for any building exhaust system.

Exhaust fans for fume hood exhaust systems are normally backward inclined or airfoil blade single width single inlet centrifugal fans or vaneaxial fans. The static pressure losses may be very high (5″–8″ wg not being unusual), dictated by the hood loss, ductwork friction, filters, heat recovery coils, and discharge stacks. The designer must evaluate each of these components to determine the fan design static pressure requirements.

The designer must, with the owner's input, evaluate the potential corrosion hazards and select duct and fan materials accordingly:

1. Fans must be chemical resistant and be of non-sparking construction throughout. (While all-stainless steel fans are often used, these fans are not non-sparking.) Motors must be located outside of the air stream. Belt drive fans should use multiple belts rated for at least 150% of motor rating and be nonconducting.
2. Ductwork should be designed in accordance with *Industrial Ventilation: A Manual of Recommended Practice* published by the American Conference of Governmental Industrial Hygienists. The materials for exhaust ductwork must be corrosion resistant for the intended use and the most common materials are Type 304 and 316 stainless steel. However, stainless steel is subject to acid and chloride compound attack and another material, or a coating applied to the stainless steel, may be required. Joints must be butt-welded throughout to eliminate leakage.
3. Perchloric acid hood exhaust must be provided with an internal wash system throughout (including the hood, ductwork, and fan). The fan should be a stainless steel bifurcated straight flow-through fan. Ductwork must be straight, with no elbows or bends.

Filtering of exhaust air is required for the following applications:

1. Biohazard exhaust from biological safety cabinets.
2. Exhaust from radioisotope hoods.

Rooftop exhaust should discharge at 3,000-fpm minimum velocity in order to promote the dilution of contaminants and the discharge point must be at least 10 ft above the roofline.

Architectural screening around exhaust fans is known to adversely affect exhaust dispersion and should not be used. If they are used, stack height must be increased to compensate, so the aesthetic value of the screening becomes questionable.

For variable volume exhaust systems, it is necessary to provide a supplemental outdoor intake with a control damper to maintain the minimum required discharge velocity as the system exhaust volume decreases. Thus, as the hood exhaust is reduced, the outdoor intake damper will modulate open to maintain fan total airflow and the discharge velocity at 3,000 fpm.

To avoid entrainment of exhaust airflow, the stack plume must rise above the *recirculation height*, H_c, computed as shown by Eq. 7-13.

$$H_c = 0.22 \times R \tag{7-13}$$

where

$$R = \left(B_s^{\ 0.67} \times B_L^{\ 0.33}\right)$$

B_s = smaller of upwind building face dimensions (height or width), ft [m]

B_L = larger of upwind building face dimensions (height or width), ft [m]

To account for changes in roof levels, penthouses, and rooftop equipment, the upwind height must be increased by the maximum height of these elements.

Outside air intakes must be located so that discharge from exhaust stacks, even if diluted, does not enter the building. The best location is on the windward side of the building. Intakes should be located in the lower third of the wall, with the bottom of the intake approximately 2 ft [600 mm] above grade.

Most research standard laboratory modules are designed to house two fume hoods. But, in many individual laboratories, only one (or even none) will be installed initially. The designer, however, must design the supply and hood exhaust systems to support the standard module requirement since, in all probability, the additional fume hood(s) will be added in the future.

With the individual fan approach, the provision for adding hoods in the future is fairly simple, since the exhaust fan can be installed when the hood is installed. The supply airflow must, however, be designed to provide make-up for any future hoods. In multi-story buildings, installing exhaust risers in the future will be difficult and expensive and, therefore, the prudent engineer will design for the installation of these risers initially.

With the manifold approach to hood exhaust, the exhaust fans and manifold ducting must be designed to support future hoods.

Exhaust duct sizing for laboratory exhaust systems is typically done using the *constant velocity method*, with 1,000–2,000 fpm [5.0–10.0 m/s] as design minimum velocity.

7.5.2 Industrial Exhaust Systems

All industrial exhaust systems are comprised of local exhaust hoods, duct segments, fittings, etc. leading to an exhaust fan. These systems are generally referred to as *tapered* since duct sizes get gradually larger as additional flows are merged together and airflow velocities are kept essentially constant. The flow velocity must be maintained at or above the minimum velocity required to convey the exhaust materials, in accordance with Table 7-9.

Ductwork should be designed in accordance with *Industrial Ventilation: A Manual of Recommended Practice* published by the American Conference of Governmental Industrial

TABLE 7-9

Range of Minimum Exhaust Duct Design Velocities

Type of Exhaust	Examples	Design Velocity (fpm) [m/s]
Vapors, gases, smoke	All vapors, gases, and smoke	1,000–2,000 [5–10]
Fumes	Welding	2,000–2,500 [10–13]
Very fine, light dust	Cotton lint, wood flour, litho powder	1,500–3,000 [8–15]
Dry dusts and powders	Fine rubber dust, bakelite molding powder dust, jute lint, cotton dust, light wood shavings, soap dust, leather shavings	3,000–4,000 [15–20]
Average industrial dust	Grinding dust, buffing lint, wool just dust, coffee beans, shoe dust, granite dust, silica flour, general material handling, brick cutting, clay dust, foundry (general), limestone dust	3,500–4,000 [18–20]
Heavy dusts	Sawdust, metal turnings, sand blast dust, wood blocks, hog waste, lead dust	4,000–4,500 [20–23]
Heavy or moist materials	Metal chips, moist cement dust, quicklime dust, sticky buffing lint	4,500+ [23+]

Hygienists. The following procedure can be used for design of any industrial exhaust system:

1. Select hood and determine the airflow rate for each exhaust application.
2. Determine the hood static pressure entry loss.
3. Layout ductwork route for smooth airflow, minimizing direction changes, elevation changes, etc. Select fittings to minimize turbulence and pressure loss.
4. Size each duct section based on the recommended minimum conveying velocities. Select commercial duct sizes and undersize as necessary to keep velocity high enough.
5. Determine the duct and fittings pressure losses.
6. Sum all pressure losses.
7. Select the exhaust fan based on the total airflow and total pressure loss.

Air always takes the path of least resistance. A natural balance at any junction will occur and the total exhaust airflow will distribute itself throughout the various branches of the duct system according to the pressure losses of each individual flow path. Therefore, some method of *balancing* the system to attain the correct flows from each branch is required.

While it is possible to design a duct system to be *self-balancing*, this is rarely workable in the field because of construction variations in the duct installation, changes over time, and worker intervention at hoods, etc. Thus, almost all industrial exhaust systems utilize *blast gates* for balancing the system.

Blast gates are "slide dampers" (much like gate valves in piping systems) that can be used at each branch junction to add static pressure loss to obtain the correct airflow through the branch. Once the system is placed in operation, each damper is manually adjusted so that pressure drop through each branch is equal and the proper flow rates are obtained.

7.6 Duct Insulation

Supply air ductwork (and outdoor air intake ductwork) must be insulated to minimize heat gain or loss and, for cold air supply, prevent surface condensation. Indoor supply ducts are typically insulated with *glass fiber insulation* that is manufactured with an external vapor barrier of laminated kraft paper and aluminum foil, reinforced with glass fiber threads.

In the normal range of HVAC supply air temperature (55°F–95°F [13°C–34°C]), this is an adequate type of insulation. But, if supply air temperatures below 55°F [13°C] are utilized, glass fiber insulation runs the risk of absorbing enough moisture from the ambient air to lose enough insulation value that insulation surface condensation can occur, resulting in increased heat gain and supporting mold growth. In these applications, the ductwork must be insulated with closed-cell foam or other non-hydroscopic insulation.

Return air and exhaust air ductwork routed through air-conditioned spaces does not require insulation. However, return air and exhaust air ductwork in ceiling plenums or chases, and when routed through non-air-conditioned spaces, must be insulated.

When concealed above ceilings or in chases, indoor ductwork is typically insulated with blanket insulation complying with American Society for Testing and Materials (ASTM) C 553, Type II, with factory-applied jacket. Insulation density should be 1.00 lb/cf [16.0 kg/m^3] and conductivity, k, tested in accordance with ASTM C 518 or C 177 at 75°F [24°C] mean temperature shall not exceed 0.27 Btu-in./(h-sf-°F) [W/m^2 K]. In exposed areas, such as in mechanical equipment rooms, to reduce the potential for mechanical damage, ducts should be insulated with board insulation complying with ASTM C 612, Type IA or IB, with factory-applied jacket. Insulation density should be 3.0 lb/cf [48.0 kg/m^3]or greater and conductivity, k, tested in accordance with ASTM C 518 or C 177 at 75°F [24°C] mean temperature shall not exceed 0.25 Btu-in./(h-sf-°F) [W/m^2 K].

All indoor composite duct insulation (including insulation, jackets, coverings, sealers, mastics, and adhesives) with flame spread index of 25 or less, and smoke developed index of 50 or less, as tested in accordance with the ASTM E 84 or Underwriters Laboratory (UL) 723 method.

Outdoor insulation, due to the much high ambient humidity levels to which it may be exposed, must be insulated with a closed-cell, non-hydroscopic insulation material, the best of which is a rigid board material manufactured from closed-cell, polyisocyanurate foam between two aluminum foil facers, complying with ASTM C 1289 Type 1, Class 2, with Grade 2 minimum compressive strength. Insulation on outdoor ductwork can be protected by a covering of minimum 20-gauge aluminum complying with ASTM B 209, Alloy 3003, Temper H-14, arranged so that sides are flashed under the top and the bottom is flanged, turned down, and mechanically attached to the sides. All transverse joints are made with a minimum 1/2″ [13 mm] standing seam, mechanically closed, and caulked watertight.

For some applications, the use of double-wall duct may be a good solution to providing better sound attenuation in supply ducts and/or to minimize the potential for insulation damage, such as in gymnasiums. Factory-fabricated double-wall duct and fittings have galvanized sheet metal outer walls and a layer of fiberglass insulation surrounding the inner liner. The insulation is glass fiber insulation with an acrylic coating that resists air erosion of fibers and the growth of fungus and bacteria. A perforated metal inner liner is typical to provide sound attenuation, but solid-wall inner liners are available.

It is recommended that minimum duct insulation thickness be 2 in. [50 mm] for indoor ductwork and 3 in. [75 mm] for outdoor ductwork. *In every case, confirm that the insulation thickness complies with ASHRAE Standard 90.1 for the application.*

7.7 Fire and Smoke Control in Duct Systems

Fire and smoke control for occupant safety is a major design consideration for any building and is particularly important as part of duct systems design. Since duct systems penetrate fire and/or smoke walls and floors in buildings, it is necessary that the duct system be designed to maintain the integrity of the building construction.

Fire rating of walls and partitions is based on independent laboratory testing, the largest and most prominent of which is UL. UL rates wall and floor assemblies for an "F-rating" that defines the number of hours the assembly will delay the fire. *Penetration* of fire rated wall must be fire-stopped by a system or device that has been tested and listed as complying with ASTM E-814 as having a fire rating at least equal to the wall rating.

Systems that comply with ASTM E-814 are generally referred to as *through-penetration firestop systems.* These systems each should be tested and rated by an independent testing laboratory for a specific application or range of applications. UL publishes a listing annually of through-penetration fire-stop systems and every architect and engineer should become familiar with this publication.

Duct penetrations of rated walls, partitions, and floors must be protected with a fire damper rated, listed, and installed in accordance with UL Standard 555. There are two types of fire dampers: *static* dampers that are designed to be installed in duct systems that automatically shut down in the event of fire and *dynamic* dampers that are installed in duct systems designed to continue to operate during a fire. Since most systems are required by code to shut down, static dampers are far more common and dynamic dampers are used only in systems designed to vent the products of combustion or are part of an engineered smoke control system.

Access to each fire damper fusible link must always be provided. An access door is required in the duct to service and reset the fire damper, and access to the duct access door must be provided. Where the duct is above a solid ceiling or a ceiling that is part of a rated roof-ceiling or floor-ceiling assembly, a service opening with the equivalent fire rating must be provided. Each damper access opening must be identified on the architectural plans.

Kitchen exhaust or other conveying exhaust ductwork cannot contain fire dampers but must be fully enclosed within a fire-rated construction or protected by a fire-rated insulation when concealed. If the exhaust duct must go up through several floors of a building to the roof, a fire-rated masonry shaft enclosure will be required.

Sometimes, to achieve the fire-resistance rating required by building codes, *floor-ceiling or roof-ceiling assemblies* may be selected by the architect. Each assembly includes a ceiling membrane specifically designed to remain in place and resist passage of heat and flames into the concealed space above a ceiling. HVAC duct openings penetrating the ceiling membrane need to be provided with thermal protection against radiant heat during the fire exposure. UL Standard 555C classified that *ceiling radiation dampers* provide this protection. Unfortunately, these dampers create numerous problems for the HVAC designer: increased static pressure losses, increased noise, installation detailing required to ensure proper damper operation, added maintenance requirements, etc. *These problems can be avoided simply by not using a fire-resistant rated assembly* (see Chapter 4).

Some types of occupancies (such as hospitals) require the use of *smoke barriers* to compartmentalize areas within a building. Duct penetrations of smoke barriers must be protected by *smoke dampers*. A smoke damper is a single or multibladed damper tested and listed in accordance with UL Standard 555S to impede smoke in a duct system. The UL Standard tests and rates the dampers based on their *leakage rate* and a low leakage (Class I) damper is generally required by codes. The damper is installed inside the duct, typically, within 24 in. [600 mm] of the smoke partition. The actual duct penetration of the wall must be sealed smoke tight in a manner similar to pipe.

Smoke dampers must have a *damper operator* that will close the damper when it is activated by the fire/smoke alarm system. To operate safely and function in the event of electrical power loss during a fire event, these dampers are designed to "fail closed" and stop the spread of smoke. Thus, the damper operator must hold the damper open the vast majority of the time.

Bibliography

HVAC Duct Construction Standards, Metal and Flexible, Sheet Metal and Air-Conditioning Contractors National Association. Inc. (Vienna, VA, 2005).

Fire, Smoke and Radiation Damper Installation Guide for HVAC, Sheet Metal and Air-Conditioning Contractors National Association. Inc. (Vienna, VA, 2002).

ASHRAE Handbook-Fundamentals, American Society of Heating, Refrigerating, and Air-Conditioning Engineers, Inc. (Atlanta, GA, 2017).

Industrial Ventilation: A Manual of Recommended Practice, American Conference of Governmental Industrial Hygienists (Cincinnati, OH, 2013).

8

Piping Design

8.1 Introduction

The solution to any flow problem requires knowledge of the physical properties of the fluid being handled.

8.1.1 Viscosity

Viscosity defines the readiness with which a fluid flows when it is acted upon by an external force. The absolute viscosity of a fluid is the measure of its resistance to internal deformation or shear. Molasses is a highly viscous fluid; water is comparatively less viscous; and the viscosity of gases is very small compared to that of liquids.

The IP units for absolute viscosity are "slugs per foot second" or "pound force seconds per square foot"; however, "pound mass per foot second" or "poundal seconds per square foot" may also be encountered. The absolute viscosity of water at a temperature of 68°F [20°C] is

$$\mu = 0.000672\,\text{lbm/ft-sec}\left[0.10\,\text{gr/cm-s}\right]$$

Kinematic viscosity is the ratio of the absolute viscosity to the mass density as shown by Eq. (8-1).

$$v(\text{sf/sec})\left[\text{m}^2/\text{s}\right] = \mu(\text{lbm/ft-sec})\left[\text{gr/cm-s}\right]/\text{d}(\text{lbm/cf})\left[\text{kg/m}^3\right] \tag{8-1}$$

8.1.2 Weight Density, Specific Volume, and Specific Gravity

The mass density, or specific weight of a substance, is its weight per unit volume expressed in pounds per cubic foot (lbm/cf) [kg/m³].

The specific volume, v, being the reciprocal of the weight density is expressed by Eq. (8-2).

$$v = 1/d \tag{8-2}$$

where d is the weight density.

In the normal range of heating, ventilating, and air-conditioning (HVAC) applications, the effect of pressure on the weight of liquids is of no practical importance in flow problems. The weight densities of gases and vapors, however, are greatly altered by pressure changes. For "perfect" gases, the weight density can be computed from Eq. (8-3).

$$d = 144P/RT \tag{8-3}$$

where P is the pressure, psia [Pa], and T is the absolute temperature, K [R].

The individual gas constant, R, is equal to the universal gas constant divided by the molecular weight of the gas, M, as shown by Eq. (8-4):

$$R = 1{,}544/M \tag{8-4}$$

Specific gravity, S, is a relative measure of weight density. Since pressure has an insignificant effect on the weight density of liquids, temperature is the only condition that must be considered in designating the basis for specific gravity. Thus, the specific gravity of any liquid is its weight density at 60°F [16°C] divided by that of water at standard temperature, 60°F [16°C], as shown by Eq. (8-5)

$$S = d(\text{any liquid at } 60°\text{F}[16°\text{C}])/d(\text{water at } 60°\text{F}[16°\text{C}]) \tag{8-5}$$

8.1.3 Mean Velocity of Flow

The term "velocity," V, (fpm [m/s]) refers to the mean or average velocity at a given cross section, as determined by the continuity equation for steady-state flow, Eq. (8-6).

$$V = q/A \tag{8-6}$$

where A is the cross-sectional area of flow (sf [m^2]) and q is the flow quantity (cf/min [m^3/s]).

The nature of flow in pipe, that is, whether it is laminar or turbulent, depends on (1) the pipe diameter, (2) the density and (3) viscosity of the flowing fluid, and (4) the velocity of flow. The numerical value of a dimensionless combination of these four variables, known as the *Reynolds number*, is the ratio of the dynamic forces of mass flow to the shear stress due to viscosity, all in consistent units. The Reynolds number, R_e, is computed *via* Eq. (8-7).

$$R_e = DVd/\mu \tag{8-7}$$

where D is the diameter of the pipe.

Normally, flow in pipes is considered to be "laminar" if the Reynolds number is less than 2,000 and "turbulent" if the Reynolds number is greater than 4,000. Between these two values lies the "critical zone" where the flow may be laminar, turbulent, or in the process of change and is basically unpredictable. Typically, water and gas flows are fully turbulent in normal flow conditions where R_e is in excess of 10,000.

Flow in a pipe is always accompanied by the friction created by fluid molecules rubbing against one another and against the pipe and, consequently, by loss of energy available for work. Thus, there is always a pressure drop in the direction of flow.

The general equation for friction pressure drop, PD, known as *Darcy's formula* and expressed in feet of fluid, is expressed by Eq. (8-8).

$$\text{PD} = f(LV^2)/D^2g \tag{8-8}$$

where

f = Friction factor (dimensionless)

L = Pipe equivalent length (ft [m])

g = Acceleration of gravity (32.2 ft/s^2) [9.8 m/s^2]
V = Flow velocity (fpm [m/s])
D = Diameter of the pipe (ft [m])

The Darcy equation is valid for laminar or turbulent flow of any liquid in a pipe.

The friction factor for laminar flow conditions is a function of Reynolds number only. However, when flow is turbulent, the friction factor depends not only on the Reynolds number but also on the relative roughness, e/D, where e is the roughness of the interior pipe walls and D is the inside diameter of the pipe.

Since the character of the internal surface of commercial pipe is practically independent of the diameter, the roughness of the walls has a greater effect on the friction factor in the small sizes. Consequently, the pipe with small diameter will approach the *very rough* condition and, in general, will have higher friction factors than will larger pipe of the same material.

8.1.4 Piping Basics

In HVAC design, the engineer is normally dealing with commercially available piping materials (most typically, steel and copper). The design of a piping system must begin with an understanding of pipe, fittings, and common valves.

Steel pipe. This is the most common piping type and is defined by its wall thickness, called "schedule," and finish (i.e., "black" or "galvanized"). For HVAC applications, most steel piping will be "standard weight," that is, Schedule 40 up through 12″ *nominal pipe size* (NPS) [*diameter nominal* (DN)], corresponding to mm] and 0.375″ [9.53 mm] wall thickness for larger pipe.

Where consistent pressures in excess of 250 psig [1,724 kPa] are encountered, for steam condensate, designers will typically specify Schedule 80 or "extra heavy" pipe. Table 8-1 lists the dimensions for steel piping in the sizes commonly used in HVAC applications.

Steel piping is used for water, steam, steam condensate, compressed air, fuel oil, gas, and so on, in HVAC applications and is manufactured two ways. For small pipe sizes that may experience high temperatures and pressures and/or must withstand the stresses of coiling and bending, such as boiler tubes, steel piping is extruded and designated *seamless.* For general hydronic and steam applications, steel piping is manufactured by a rolling process and, thus, has a welded longitudinal seam. Typically, HVAC piping 2-1/2″ NPS [DN65] and larger should comply with American Society for Testing and Materials (ASTM) A53, Type E, Grade B, while piping 2″ NPS [DN50] and smaller should comply with ASTM A53, Type F, Grade A. For high-pressure steam applications, piping should comply with ASTM A106.

Copper tubing is the second most common piping material and is used extensively for water piping and refrigeration piping. For smaller pipe sizes, copper tubing is often used in HVAC systems in lieu of steel piping because of its lower installation labor costs, even though the material cost is higher. Table 8-2 lists the physical properties for copper tubing for the sizes commonly used in HVAC applications.

Copper tubing larger than 2-1/2″ [DN65] is rarely used due to its cost and the lack of availability of qualified installers (large copper tubing is very difficult to heat evenly to create a tight solder joint).

Copper tubing is fabricated in accordance with ASTM B 88, while fittings may be cast copper in accordance with American National Standards Institute (ANSI) B16.18 or wrought copper in accordance ANSI B16.22, with the second type being more common.

TABLE 8-1

Dimensions of Steel Pipe

NPS (in.) [DN]	OD (in.) [mm]	Wall Thickness Schedule (in.) [mm]
1/2 [15]	0.840 [21.3]	Std. (Sch.40) 0.109 [2.77]
3/4 [20]	1.050 [26.7]	Std. (Sch.40) 0.113 [2.87]
1 [25]	1.315 [33.4]	Std. (Sch.40) 0.133 [3.38]
1-1/4 [32]	1.660 [42.2]	Std. (Sch.40) 0.140 [3.56]
1-1/2 [40]	1.900 [48.3]	Std. (Sch.40) 0.145 [3.68]
2 [50]	2.375 [60.3]	Std. (Sch.40) 0.154 [3.91]
2-1/2 [65]	2.875 [73.0]	Std. (Sch.40) 0.203 [5.16]
3 [80]	3.500 [88.9]	Std. (Sch.40) 0.216 [5.49]
4 [100]	4.500 [114]	Std. (Sch.40) 0.237 [6.02]
6 [150]	6.625 [168]	Std. (Sch.40) 0.258 [6.55]
8 [200]	8.625 [219]	Std. (Sch.40) 0.280 [7.11]
10 [250]	10.75 [273]	Std. (Sch.40) 0.322 [8.18]
12 [300]	12.75 [324]	Std. (Sch.40) 0.365 [9.27]
14 [350]	14.00 [356]	Std. (Sch.30) 0.375 [9.53]
16 [400]	16.00 [406]	Std. (Sch.30) 0.375 [9.53]
18 [450]	18.00 [457]	Std. (Sch.30) 0.375 [9.53]
20 [500]	20.00 [508]	Std. (Sch.20) 0.375 [9.53]
24 [600]	24.00 [610]	Std. (Sch.20) 0.375 [9.53]
1/2 [15]	0.840 [21.3]	XH (Sch.80) 0.147 [3.74]
3/4 [20]	1.050 [26.7]	XH (Sch.80) 0.154 [3.91]
1 [25]	1.315 [33.4]	XH (Sch.80) 0.179 [4.55]
1-1/4 [32]	1.660 [42.2]	XH (Sch.80) 0.191 [4.85]
1-1/2 [40]	1.900 [48.3]	XH (Sch.80) 0.200 [5.08]
2 [50]	2.375 [60.3]	XH (Sch.80) 0.218 [5.54]
2-1/2 [65]	2.875 [73.0]	XH (Sch.80) 0.276 [7.01]
3 [80]	3.500 [88.9]	XH (Sch.80) 0.300 [7.62]
4 [100]	4.500 [114]	XH (Sch.80) 0.337 [8.56]
6 [150]	6.625 [168]	XH (Sch.80) 0.432 [11.0]
8 [200]	8.625 [219]	XH (Sch.80) 0.500 [12.7]
10 [250]	10.75 [273]	XH (Sch.80) 0.593 [15.1]
12 [300]	12.75 [324]	XH (Sch.80) 0.687 [17.4]

TABLE 8-2

Dimensions of Copper Tubing

NPS (in.) [DN]	OD (in.) [mm]	Wall Thickness (in.) [mm]		
		Type K	Type L	Type M
1/2 [15]	5/8 [15.9]	0.049 [1.24]	0.040 [1.02]	0.028 [0.71]
3/4 [20]	7/8 [22.2]	0.065 [1.65]	0.045 [1.14]	0.032 [0.81]
1 [25]	1-1/8 [28.6]	0.065 [1.65]	0.050 [1.27]	0.035 [0.89]
1-1/4 [32]	1-3/8 [34.9]	0.065 [1.65]	0.055 [1.40]	0.042 [1.07]
1-1/2 [40]	1-5/8 [41.3]	0.072 [1.83]	0.060 [1.52]	0.049 [1.24]
2 [50]	2-1/8 [54.0]	0.083 [2.11]	0.070 [1.78]	0.058 [1.47]
2-1/2 [65]	2-5/8 [66.7]	0.095 [2.41]	0.080 [2.03]	0.065 [1.65]

Special piping has limited HVAC applications that includes the following:

1. Cast iron and ductile iron pressure pipe used for underground water systems.
2. Plastic (including polyvinyl chloride [PVC], chlorinated PVC, high density polyethylene) is used for underground hydronic piping, cooling condensate drain lines, makeup (service) water piping, and in some applications with systems with small temperature gradients.

Typical types of fittings encountered in HVAC piping systems include the following:

1. *Elbow (ell).* 90° or 45° change in direction. Ells may be "standard" or "long" radius. Typically, standard radius ells are used in NPS 2-1/2" [DN65] and smaller, while long radius ells are used in larger pipe sizes.
2. *Tee.* Tees split or combine flows and may be line size (e.g., 2" × 2" × 2") [50 mm × 50 mm × 50 mm] or "reducing" tees (e.g., 2" × 1" × 2") [50 mm × 25 mm × 50 mm].
3. *Reducer.* Reduces (or increases) the pipe size in the direction of flow. Reducers may be "concentric," implying that the outlet is centered with the inlet or "eccentric," where the outlet and inlet are offset to yield a flat side on the fitting.
4. *Coupling.* Designed to join together two sections of equal-sized screw pipe.
5. *Return bend.* A 180° change in direction.
6. *Bushing.* An insert to create an abrupt change in pipe size. A bushing is usually installed in a coupling, elbow, or tee to change pipe sizes without using a smooth reducer. But, it is a poor piping practice due to the high pressure losses that result.
7. *Union.* A threaded three-piece coupling that allows pipe connections to be easily disconnected. Unions are utilized at equipment connections, valves, and so on to facilitate piping removal for maintenance.
8. *Flange.* A bolted connection device for equipment, valves, and piping specialties in larger pipe sizes. A gasket is inserted between the flanges, which are bolted together for a tight joint.

Piping ends are typically either "threaded" for *screw-together joints* or "plain" for *welded joints.* Labor and material costs for steel pipe dictate that screwed fittings and pipe are normally utilized in steam and water systems up to 2 in. NPS, while larger piping is welded. Copper tubing joints are soldered ("sweated").

Other joining methods encountered in HVAC applications may include *grooved* piping joints for steel piping that utilizes factory-fabricated fittings and *pressed* joints for both steel and copper piping, again using factory-fabricated fittings.

Grooved joints are created in one of two ways: *machine cut* or *rolled and swaged* and may be used in lieu of welded flanged joints for steel piping 2-1/2" NPS [DN65] and larger at HVAC equipment located in mechanical equipment rooms to facilitate piping removal for equipment maintenance. The use of cut groove end joints is not acceptable for HVAC pressure piping since each groove represents a thinning of the pipe wall thickness, reducing its strength and creating a point for stress corrosion to occur. Rolling and swadging the pipe ends creates a thicker wall at the joint, eliminating the thinning problem when grooves are cut, but is a much more expensive process. Couplings and fittings used with grooved piping joints should be forged steel.

In lieu of screwed or soldered joints in piping 2" NPS [DN50] and smaller, the use of pressed joints and copper fittings may be allowed. With this system, factory-fabricated

TABLE 8-3

Valve Types and Applications

Duty	Flow Medium	Description
Shutoff	Hydronic	1/2″-2″ NPS [15–50 DN], ball valve, 2-1/2″+ NPS [65+ DN], butterfly valve.
Check		All sizes, swing or lift check valve.
Balancing (modulated flow)		1/2″-2″ NPS [15–50 DN], globe valve or ball valve, 2-1/2″+ NPS [65+ DN], butterfly valve.
Shutoff	Steam	1/2″-2″ NPS [15–50 DN], ball valve, 2-1/2″+ NPS [65+ DN], gate valve
Check		All sizes, swing or lift check valve.

copper fittings are hydraulically pressed onto the plain ends of either copper or steel pipe. Press copper fittings are fabricated of a copper alloy and comply with ASME B16.22. Sealing elements integrated into the fittings are ethylene propylene diene terpolymer O-rings and may be reinforced with additional stainless steel grip rings.

Valves are an important part of any HVAC piping system and generally are applied as shown in Table 8-3.

HVAC valves are generally applied as follows:

1. *Gate valve.* It opens by lifting a round or rectangular gate/wedge out of the path of the fluid. The distinct feature of a gate valve is that the sealing surfaces between the gate and seats are planar, so gate valves are often used when a straight-line flow of fluid and minimum restriction is desired. The gate faces can be parallel but are most commonly wedge-shaped. Gate valves are primarily used to permit or prevent the flow of liquids and should not be used for regulating flow. "Rising stem" gate valves have threads external to the valve body and the operator can observe that the valve stem is moving up/down when the valve is opening/closing. "Non-rising stem" gate valves have threads that contained within the valve body and are exposed to the fluid being controlled. Non-rising stem gate valves require a bigger disc, hence their cost is higher.
2. *Ball valve.* It is a "quarter-turn" valve which uses a ported, rotating ball to control flow. It is open when the ball's port is in line with the flow and closed when it is pivoted 90° by the valve handle. The handle aligns with the flow when open and is perpendicular to it when closed, making for easy visual confirmation of the valve's status. Valves may be "full port" type when the port through the ball is the same diameter as the interior valve body or "reduced port" when the port has a smaller diameter that the valve body. Full port valves are recommended to "shut off" duty in order to minimize the valve pressure drop when open. Reduced port valves are recommended for flow balancing to make the valve's flow characteristic more linear.
3. *Globe valve.* It is a type of valve used for regulating flow and consists of a movable disk-type element and a stationary ring seat in a generally spherical body. Globe valves are named for their spherical body shape with the two halves of the body being separated by an internal baffle. This has an opening that forms a seat onto which a movable plug, called a disc or disk, can be screwed in to close the valve. Large globe valves are very expensive and, thus, butterfly valves are used in piping 2-1/2″ NPS [250 DN].

4. *Butterfly valve.* A butterfly valve is another type of quarter-turn valve. The "butterfly" is a metal disc mounted on a rod. When the valve is closed, the disc is turned 90° to the direction of flow so that it completely blocks off the passageway. When the valve is fully open, the disc is rotated a quarter turn so that it allows an almost unrestricted passage of the fluid. Because of their lower cost and high reliability, butterfly valves are used in hydronic piping 2-1/2" NPS [250 DN] and larger for both shutoff and balancing duty. Butterfly valves are not rated for steam duty.
5. *Swing check valve.* A swing check valve is a check valve in which the disc, the movable part to block the flow, swings on a hinge or trunnion, either onto the seat to block reverse flow or off the seat to allow forward flow. The seat opening cross section may be perpendicular to the centerline between the two ports or at an angle. Although swing check valves can come in various sizes, large check valves are often swing check valves. A common issue caused by swing check valves is known as water hammer. This can occur when the swing check closes and the flow abruptly stops causing a surge of pressure resulting in high velocity shock waves that act against the piping and valves, placing large stress on the metals and vibrations in the system.
6. *Lift check valve.* It is a check valve in which the disc, sometimes called a lift, can be lifted up off its seat by higher pressure of inlet or upstream fluid to allow the flow to the outlet or downstream side. A guide keeps motion of the disc on a vertical line, so the valve can later reseat properly. When the pressure is no longer higher, gravity or higher downstream pressure will cause the disc to lower onto its seat, shutting the valve to stop reverse flow.
7. *Spring check valve.* These valves come is several configurations. The key benefit of a spring in a check valve is that it largely eliminates the effect of gravity on check valve function. Without a spring, gravity and, thus, the valve's vertical orientation can become a critical factor for a check valve to work properly. The spring's position applies a constant pressure holding the valve sealing surfaces together. Fluid flow in the correct direction and at the right pressure (the downstream pressure) overcomes the force of the spring and pushes the sealing surface of the plunger away from the seat. This allows flow to take place. When the upstream pressure drops low enough, the spring pushes the sealing surface of the stopper back against the sealing surface of the valve seat, which closes the valve and prevents reverse flow. Among other things, the effect of the spring is to eliminate the potential for water hammer.

8.2 Hydronic Piping

The majority of piping design for HVAC systems is water piping: hot water for heating and chilled water and condenser water for cooling. Using Eq. (6-1) (repeated here), the overall heat transferred by a given water flow rate can be computed.

$$Q = m \times C_p \times \mathrm{TD} \tag{6-1}$$

where m is the fluid mass flow rate, C_p is the specific heat, and TD is the change in temperature.

Typically, to use Eq. (6-1) for a given volumetric flow rate (in gpm [m³s]), the properties of water at 60°F [16°C] are utilized, resulting in a C_p of 1.0 Btu/lbm-°F [277.8 kWs/kJ]. Utilizing the conversion factor of 7.48 cf/gal. [998 kg/m³] and converting for the time rate yields Eqs. (8-9) and (8-9a).

In IP units, Eq. (6-1) can be restated as follows using the properties of water at 60°F:

$$Q = \left[(62.4\ \text{lbm/cf})/(7.48\ \text{cf/gal.})\right] \times 60\ \text{min/h} \times 1.0\ \text{Btu/lbm °F} \times W \times \text{TD}$$

or

$$Q = 500 \times W \times \text{TD} \tag{8-9}$$

where

Q = Heating/cooling capacity, Btu/h
W = Volumetric flow rate, gpm
TD = Temperature difference, °F

In SI units, Eq. (6-1) can be restated as follows using the properties of water at 16°C:

$$Q = 4.1868(\text{kJ/kg °C}) \times 1{,}000\,\text{kg/m}^3 \times W \times \text{TD}$$

or

$$Q = 4{,}187 \times W \times \text{TD} \tag{8-9a}$$

where

Q = Heating/cooling capacity, kW (kJ/s)
W = Volumetric flow rate, m³/s
TD = Temperature difference, °C

The operating pressure of a pumped system is calculated in the SI units of *meters of water column* (m). To maintain dimensional consistency, any pressure values used within the calculations are therefore converted from kPa into meters on the basis of 1 kPa = 0.102 m of water column. In IP units, the pressure in a pumped system is based on psig, which can be converted on the basis of 1 psig is equal to 2.54 ft of water column.

8.2.1 Pipe Routing and Sizing

For constant flow HVAC hydronic piping, it is necessary that balancing in each flow circuit be done to ensure adequate water flow to each load. Typically, due to cost constraints, most piping systems are designed on the basis of a "direct return" configuration (i.e., the first load supplied is the first load returned), which creates pressure imbalances in the system that must be addressed by utilizing manual balancing valves. The alternative "reverse return" configuration (i.e., the first load supplied is the last load returned) requires more piping, but since each flow circuit is essentially the same length, it is more "self-balancing."

For variable flow systems, manual balancing of the overall system is not required since the action of control valves makes the system inherently "self-balancing." However, to minimize the pressure loss requirements imposed on control valves, piping systems should be designed, insofar as feasible, as follows:

1. Oversize piping "mains"... size piping for lower pressure drops through long supply and return runs.
2. Use a reverse return piping configuration for sub-loops in larger systems (i.e., each building of a campus system, each wing of a large building, and each floor of a multistory building)

For all piping systems, piping should be routed and configured using the following guidelines that can be specified by the designer for use in building information modeling software:

1. Route piping requirements, both horizontally and vertically, square and plumb with building walls.
2. Branches should have top, 45°, or side outlets from the main, never bottom outlets.
3. Do not use "bull headed tees" to make the final branches from a main. Use a tee for the first branch and then a short nipple and elbow for the second.
4. Remember that piping must slope downward in the direction of flow to allow the system to be drained. *Specify hanging height(s) to avoid interferences.*

Using the Figures 8-1a and b or 8.2a and b, select pipe sizes using the following procedure:

1. Select each NPS on the basis of friction losses of 1.0–4.0 ft of water/100 lf of pipe [0.1–0.4 kPa/m], with a maximum allowable flow rate resulting in flow velocity of 10 ft/s (fps) [3 m/s], as summarized in Table 8-4.
2. *For short runs,* select for higher pressure drops for reduced pipe sizes and costs. *For long runs,* select lower pressure drops to minimize pumping energy requirements.
3. Where piping is to be installed adjacent to noise-sensitive areas (rated for RC 30 or lower), piping velocity should not exceed 4 fps [1.2 m/s] for pipe 2″ [DN50] and smaller or 6 fps [2.0 m/s] for pipe 2-1/2″ [DN65] and larger.
4. Do not select 1-1/4″ [DN32], 3-1/2″ [DN90], or 5″ [DN125] pipe sizes except when needed to match existing piping. Although these sizes exist, 1-1/2″ [DN40] costs essentially the same as 1-1/4″ [DM32] pipe, 3-1/2″ [DN90] pipe is rarely even manufactured, and the cost of 5″ [DN125] pipe generally equals or exceeds that of 6″ [DN150] pipe.

Since each circuit in a *constant flow hydronic piping system* will typically have a different pressure loss, some "balancing" may be required to ensure design flow rates in each circuit in constant flow systems. While manual balancing valves (globe, ball, plug, or butterfly type) are adequate, the use of calibrated flow balancing stations is acceptable since these stations provide ready-made pressure taps for flow readings.

However, for *variable flow systems,* no balancing valves are required... the control valves that regulate flow in direct proportion to imposed load make these systems "self-balancing" over their full range of flows. *Manual balancing valves, manual set point automatic*

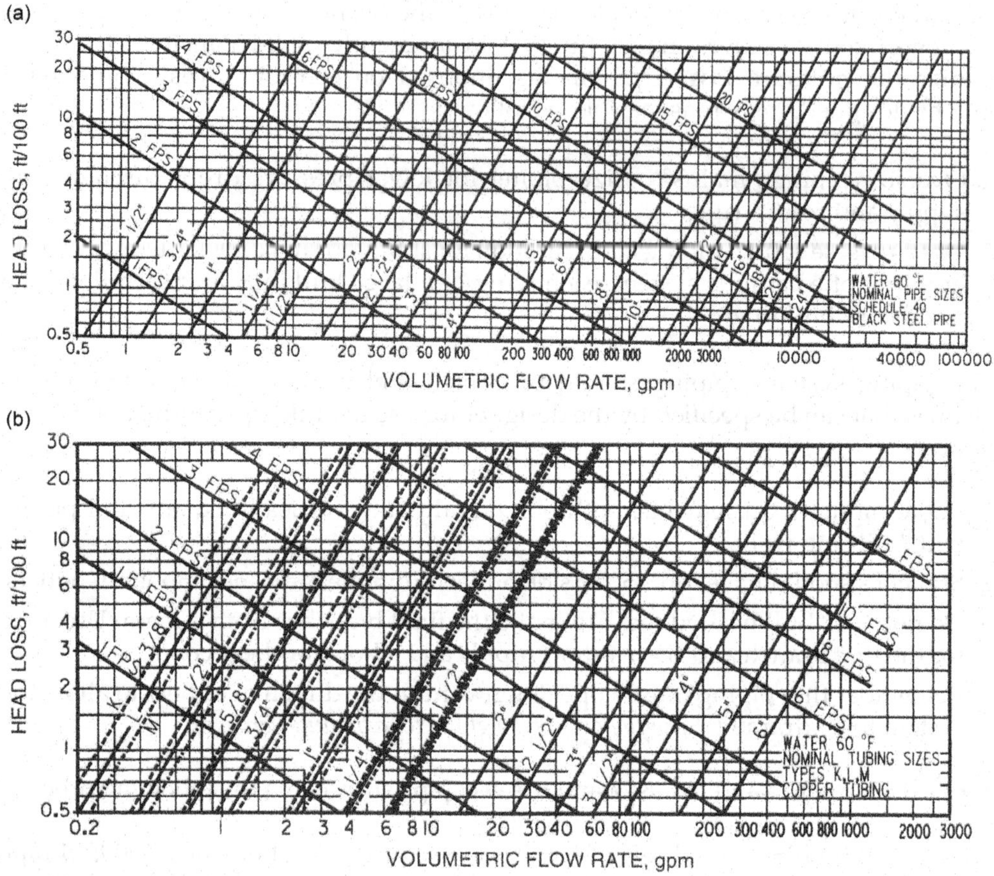

FIGURE 8-1
(a) Friction loss for water in steel pipes (IP units). (Fig. 8-2a, First Edition; ASHRAE, 1791 Tullie Circle, N.E., Atlanta, GA 30329.) (b) Friction loss for water in copper tubes (IP units). (Fig. 8-2b, First Edition; ASHRAE, 1791 Tullie Circle, N.E., Atlanta, GA 30329.)

flow control valves, and three-way bypass control valves should never be included in any variable flow hydronic system… these valves simply add to system pressure losses and conflict with the proper action of the two-way control valves. However, for any variable flow system, ensure that each "loop" has a low limit differential pressure sensor installed to allow override of the pump speed control, ensuring that adequate flow is provided to the remote part of any loop in the system under light load conditions.

Valves are an important part of any hydronic system and are required for balancing, isolation, and service of equipment and components throughout the system. There are predominately three types of valves used: shutoff or isolation valves, balancing valves, and check valves.

Shutoff or isolation valves are designed to do just that, cut off flow in a section of the piping system, primarily when that section, or the equipment within that section, requires service. The most commonly used valves for shutoff duty in piping 2" NPS [DN65] is the full port *ball valve*. In larger piping, the butterfly valve is used for this duty. While the gate valve is operated by a hand wheel and requires several rotations to fully open or close, both the ball valve and the butterfly valves are "quarter turn vales," requiring only a

(a)

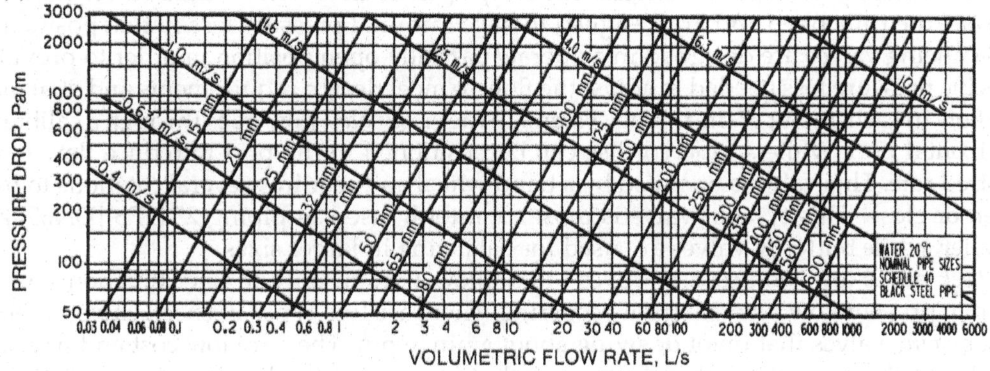

Friction Loss for Water in Commercial Steel Pipe (Schedule 40)

(b)

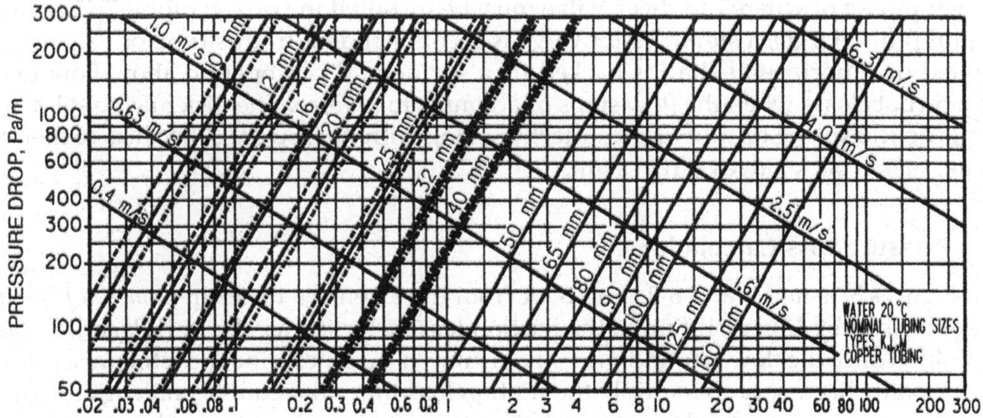

Friction Loss for Water in Copper Tubing (Types K, L, M)

FIGURE 8-2
(a) Friction loss for water in steel pipes (SI units). (Fig. 4, Page 22.7, 2009 *ASHRAE Handbook-Fundamentals*, SI Edition; ASHRAE, 1791 Tullie Circle, N.E., Atlanta, GA 30329.) (b) Friction loss for water in copper (SI units). (Fig. 5, Page 22.7, 2009 *ASHRAE Handbook-Fundamentals*, SI Edition; ASHRAE, 1791 Tullie Circle, N.E., Atlanta, GA 30329.)

TABLE 8-4
Maximum Recommended Flow Rates in Hydronic Piping Systems

NPS (in.) [DN]	Max. Flow (gpm) [L/s]	NPS (in.) [DN]	Max. Flow (gpm) [L/s]
1/2 [15]	9 [0.6]	3/4 [20]	18 [1.1]
1 [25]	28 [1.8]	1-1/4 [32]	49 [3.1]
1-1/2 [40]	63 [4.0]	2 [50]	100 [6.3]
2-1/2 [65]	160 [10]	3 [80]	240 [34]
4 [100]	400 [25]	6 [150]	900 [57]
8 [200]	1,500 [95]	10 [250]	2,500 [158]
12 [300]	3,500 [220]	14 [350]	4,200 [303]
16 [400]	5,500 [347]	18 [450]	7,000 [442]
20 [500]	8,700 [549]	24 [600]	12,500 [789]

90° movement of the operating handle to fully open or close. All of these valves normally operate the fully open and are designed to have minimal pressure loss in that position.

Balancing valves are designed in a operate partially open position in order to provide a specific pressure loss needed to adjust the flow to meet design requirements and "balance" the flow in each circuit of the piping system. In the *globe valve*, flow goes through a calibrated horizontal port, while a tapered disc can be positioned in the port to reduce flow to the desired rate. This valve is operated by a hand wheel and requires several rotations to fully open or close. This type of valve is most commonly used in piping 2″ NPS [DN65] and smaller, while butterfly valves are used for balancing in larger sizes.

Check valves are "non-return valves" that allow flow in only in one direction and are used to prevent flow reversal from occurring. There are two types of check valves applied: *lift* or *swing* valves that pivot or swing about a hinge pin. They are low cost and have low head loss characteristics when fully open, but because of the long stroke and inertia of the disc, may slam in vertical pipe installations and are not recommended for that duty. In larger piping or where the check valve must be installed in vertical piping (particularly at a pump discharge), a *spring loaded* valve is recommended. In this type of valve, a disk mounted on a horizontal shaft is backed by a spring and can open to allow flow in one direction but closes when the flow stops or attempts to reverse. The disk maybe solid, with a single helical spring or split like a butterfly's wing with dual springs. Because of their fast closure, these check valves rarely slam.

8.2.2 Pressure Loss Calculation

From Figures 8-1a and b and 8-2a and b (or from the extensive tables in *Cameron Hydraulic Data* published by Ingersoll Rand), the friction loss factor (ft. wg/100 ft-[kPa/m]) is easily determined once the flow rate and pipe size is known. To determine the total pressure loss, then, it is only necessary to multiply the friction loss times the equivalent length of pipe.

There are several methods for calculating the pressure loss through standard pipe fittings and valves. The *3K-method* is the most accurate, and most piping system analysis computer programs use this method. But, if hand calculations are required, then the *equivalent length method* or the *K-method* is generally used.

While these latter two methods are not as accurate as the 3K-method, both are sufficiently accurate for most HVAC hydronic piping systems, especially when it is well understood that the piping arrangement shown of the drawings may differ from the installed arrangement and designers often include a 5%–10% "safety factor" at the end to ensure adequate pump head and capacity (see Chapter 9).

The American Society of Heating, Refrigerating, and Air-Conditioning Engineers (ASHRAE) *Handbook-Fundamentals* presents the K-method is some detail and that reference may be used at the designer's option. However, for day-to-day HVAC piping design, the equivalent length method is acceptable. Table 8-5 shows the equivalent length, L_{eq}, for standard fittings and valves and is based on steel pipe.

To compute the total head loss for a piping system, Eq. (8-10) is used.

$$H = F \times \left[L + \Sigma(E \times L_e)\right]/100 \tag{8-10}$$

where

H = Total pressure ("head") loss, ft. wg [kPa]

F = Average friction factor, ft. wg/100 ft of length [kPa/m], used for pipe sizing

TABLE 8-5

Equivalent Length of Pipe Fittings and Valves

Pipe Size	Equivalent Length, L_e (ft.), for Types of Fittings and Valves[a]															
(NPS) [DN]	Type 1	Type 2	Type 3	Type 4	Type 5	Type 6	Type 7	Type 8	Type 9	Type 10	Type 11	Type 12	Type 13	Type 14	Type 15	Type 16
3/4 [20]	0.96	2.06	1.10	—	—	1.40	1.90	2.00	4.12	—	0.48	22.9	—	0.48	1.13	6.86
1 [25]	1.22	2.62	1.40	—	—	1.70	2.30	2.60	5.24	—	0.61	29.1	—	0.61	1.50	8.74
1-1/4 [32]	1.61	3.45	1.84	—	—	2.30	3.10	3.30	6.90	—	0.81	38.3	—	0.81	1.88	11.5
1-1/2 [40]	1.88	4.02	2.15	—	—	2.60	3.70	4.00	8.00	—	0.54	44.7	—	0.54	2.25	13.4
2 [50]	2.41	5.17	2.76	—	—	3.30	4.30	5.00	10.1	—	1.21	57.4	—	1.21	3.00	17.2
2-1/2 [65]	2.88	6.16	3.29	4.12	2.47	4.10	5.60	6.00	12.3	4.15	1.44	68.5	9.26	1.44	3.75	20.6
3 [80]	—	—	—	5.11	3.07	—	—	—	—	5.00	1.79	—	11.5	—	—	25.5
4 [100]	—	—	—	6.71	4.03	—	—	—	—	6.67	2.35	—	15.1	—	—	33.6
6 [150]	—	—	—	10.1	6.07	—	—	—	—	10.0	3.54	—	22.7	—	—	50.5
8 [200]	—	—	—	13.3	7.98	—	—	—	—	13.3	4.65	—	29.9	—	—	33.3
10 [250]	—	—	—	16.7	10.0	—	—	—	—	16.7	5.85	—	29.2	—	—	41.8
12 [300]	—	—	—	19.9	11.9	—	—	—	—	20.0	6.96	—	34.8	—	—	49.7
14 [350]	—	—	—	21.8	13.1	—	—	—	—	23.3	7.65	—	38.3	—	—	54.7
16 [400]	—	—	—	25.0	25.0	—	—	—	—	26.7	8.75	—	31.3	—	—	62.5
18 [450]	—	—	—	28.1	16.9	—	—	—	—	30.0	9.85	—	35.2	—	—	70.3
20 [500]	—	—	—	31.4	18.8	—	—	—	—	33.3	11.0	—	39.2	—	—	78.4
24 [600]	—	—	—	37.7	22.6	—	—	—	—	40.0	13.2	—	47.1	—	—	94.4

Type 1: 45° Elbow, screwed, standard radius
Type 2: 90° Elbow, screwed, standard radius
Type 3: 90° Elbow, screwed, long radius
Type 4: 90° Elbow, welded, standard radius
Type 5: 90° Elbow, welded, long radius
Type 6: Tee, screwed, straight through flow, no size reduction
Type 7: Tee, screwed, straight through flow, 25% size reduction
Type 8: Tee, screwed, straight through flow, 50% size reduction
Type 9: Tee, screwed, flow through branch, no size reduction
Type 10: Tee, welded, straight through flow, no size reduction
Type 11: Gate valve, 100% open
Type 12: Globe valve, 100% open
Type 13: Butterfly valve, 100% open
Type 14: Ball valve, 100% open
Type 15: Plug valve, 100% open
Type 16: Check valve, lift or swing

[a] Multiple values in Table 8.5 by 0.3048 to obtain equivalent length in meters.

L = Length, ft. [m], of straight pipe
E = Quantity of each fitting or valve type
L_e = Equivalent length, ft. [m], for each type of fitting or valve

To obtain the total pressure loss in the piping system, the designer must also include the anticipated head loss of control valves, coils, heat exchangers, and other equipment items included in the system. The pressure loss for coils, heat exchangers, and other equipment (including strainers, air separators, and spring loaded check valves) must be obtained from manufacturer's data. And, finally, open control valve pressure losses are defined by the valve sizing criteria presented in Chapter 13.

While most HVAC hydronic systems are closed systems, cooling tower and closed circuit cooler systems are open systems (see Chapter 11) and require that *static lift* be added to the piping pressure losses as computed previously (see Chapter 9).

8.2.3 Expansion and Air Removal

Water expands when heated and contracts when cooled in almost direct proportion with the temperature change. Since water is not compressible, a lack of expansion space in a closed piping system means that any volume increase will cause a pressure increase that can result in mechanical damage. The most common way to accommodate water volume changes is to use an air cushion *compression tank*.

Note that the term *expansion tank* refers to an open tank, installed at the highest elevation in the piping system. This type of tank is no longer used, though the term "expansion tank" is still often misapplied to describe a compression tank.

For each closed hydronic system, it is necessary to provide an *air separator* to collect any residual air remaining in the piping system after it has been filled and "bled" and a compression tank to compensate for water volume changes that occur as fluid temperatures change.

For air separation, a tangential air separator, *without a strainer*, is recommended for piping 2-1/2" NPS [DN65] and larger. On smaller piping, an *air separation fitting* is typically satisfactory. On primary pumping systems, locate the air separator or separation fitting at or near the pump(s) suction. For primary/secondary piping systems, locate the air separator or separation fitting at the secondary pumps.

Generally, the air separation and removal components must be installed in accordance with Figure 8-3.

There are several types of compression tanks that may be used:

1. *Standard compression tanks* have no diaphragm or bladder. Since the water and air in the tank is compressed from atmospheric pressure, the size of the tank tends to be larger since it has to be sized on the basis of the thermal expansion of the fluid, plus leaving space for the air "cushion" above the water. Poor sizing/selection of a standard compression tank can to lead to "waterlogging" or loss of the air cushion, but that problem is easily solved by simply adding an additional tank in parallel.
2. *Diaphragm compression tanks* have a flexible membrane separating the liquid from the air. These tanks are designed to be charged with compressed air (or, if desired, nitrogen) to 15–30 psig [103–207 kPa] and, therefore, result in a much smaller tank being required. However, the diaphragm is not replaceable, and when it fails, the

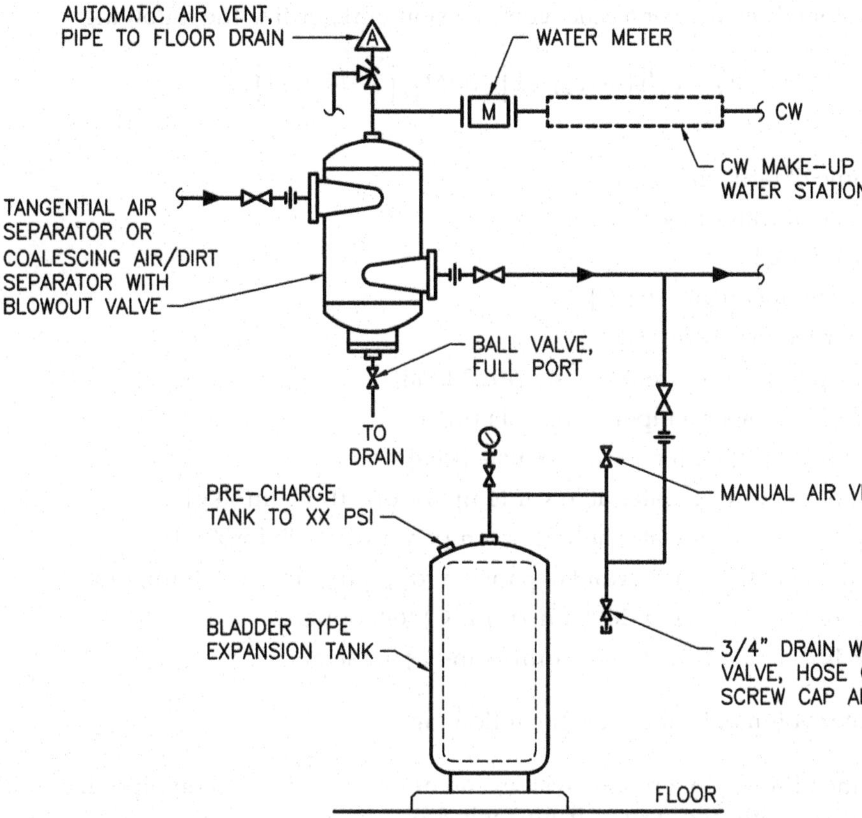

FIGURE 8-3
Air separation and removal components installation. (SW Standard Detail W01.)

entire tank must be replaced. In small systems, this type of tank saves money... both first cost is lower than a standard compression tank and replacement cost is low if proper provisions for removing the tank in the future have been made.

3. *Bladder compression tanks* are essentially the same as diaphragm tanks except that the flexible membrane is designed to be replaceable. This type of tank would be appropriate for large systems since replacement of a bladder that has failed would be far less expensive than a new tank.

Compression tank sizing is far more critical in hot water systems than in chilled water systems. For hot water systems, the water volume in the system can expand significantly as the water is heated from its ambient condition when the system is filled to its normal operating temperature. For chilled water systems, the water volume actually contracts slightly when the system is started and the tank needs to only compensate for the temperature rise that occurs when the system shuts down and the water warms up again.

The formula used for determining the tank volume, V_t, will depend on the type of tank being applied. For a closed system *standard* compression tank with an air/water interface, utilize Eq. (8-11).

$$V_t = V_s\left\{\left[\left(v_2/v_1\right)-1\right]-3\sigma\Delta t\right\}/\left[\left(P_a/P_1\right)-\left(P_a/P_2\right)\right] \tag{8-11}$$

For a closed system *diaphragm* or *bladder* compression tank, utilize the following:

$$V_t = V_s\left\{\left[\left(v_2/v_1\right)-1\right]-3\sigma\Delta t\right\}/\left[1-\left(P_1/P_2\right)\right] \quad (8\text{-}12)$$

where

V_t = Volume of tank, gal. [m³]
V_s = Volume of water in system, gal. [m³]
$\Delta t = (t_2 - t_1)$ °F [°C]
t_1 = Lower temperature, °F [°C]
t_2 = Higher temperature, °F [°C]
P_a = Atmospheric pressure, 14.7 psia [101.3 kPa]
P_1 = Pressure at lower temperature, psia [kPa]
P_2 = Pressure at higher temperature, psia [kPa]
v_1 = Specific volume of water at lower temperature, ft³/lb [m³/kg]
v_2 = Specific volume of water at higher temperature, ft³/lb [m³/kg]
σ = Linear coefficient of thermal expansion for piping, in/in °F [mm/mm °C]
= 6.5×10^{-6} in/in °F [17×10^{-6} mm/mm-°C] for copper
= 9.5×10^{-6} in/in °F [10.8×10^{-6} mm/mm-°C] for steel

The system water volume, V_s, includes the following:

1. The volume of water in a piping system can be estimated based on pipe sizes and their lengths within the system from Table 8-6.
2. The volume of water in a tangential air separator can be estimated using Table 8-7.
3. While heat exchangers (coils, chillers, and so on) are a common part of closed water systems, there are far too many types of heat exchangers to attempt to tabulate water volumes here. *It is necessary for the designer to obtain the water volume data for each heat exchanger in the system from manufacturer's data.*

In almost all hydronic systems, the low pressure in the system is selected to hold a positive pressure at the highest point in the system, usually about 10 psig [479 Pa], equivalent

TABLE 8-6
Water Volumes in Piping

NPS (in.) [DN]	Water Volume(gal./lf) [L/m]	NPS (in.) [DN]	Water Volume (gal./ft) [L/m]
1/2 [15]	0.0158 [0.0182]	4 [100]	0.660 [0.762]
3/4 [20]	0.0276 [0.0318]	6 [150]	1.501 [1.732]
1 [25]	0.0449 [0.0518]	8 [200]	2.597 [2.996]
1-1/4 [32]	0.0774 [0.0893]	10 [250]	4.098 [4.728]
1-1/2 [40]	0.106 [0.122]	12 [300]	5.974 [6.893]
2 [50]	0.174 [0.201]	14 [350]	7.168 [8.270]
2-1/2 [65]	0.255 [0.294]	16 [400]	9.506 [10.968]
3 [75]	0.383 [0.442]	18 [450]	12.156 [14.026]

TABLE 8-7

Water Volumes in Tangential Air Separators

Separator Size, NPS (in.) [DN]	Water Volume (gal.) [L]	Separator Size, NPS (in.) [DN]	Water Volume (gal.) [L]
2-1/2 [65]	3 [16]	10 [250]	150 [568]
3 [75]	7 [26]	12 [300]	291 [1,101]
4 [100]	13 [45]	14 [350]	472 [1,787]
6 [150]	34 [129]	16 [400]	723 [2,737]
8 [200]	90 [341]	18 [450]	1,149 [4,349]

TABLE 8-8

Pressure and Specific Volume of Water

Temperature (°F) [°C]	Absolute Pressure (psia) [Pa]	Specific Volume (ft³/lb) [m³/kg]
40 [4.4]	0.1217 [5.8]	0.01602 [0.2566]
60 [15.6]	0.2564 [12.3]	0.01603 [0.2568]
80 [26.7]	0.5074 [24.3]	0.01607 [0.2574]
100 [37.8]	0.9504 [45.5]	0.01613 [0.2584]
120 [48.9]	1.6949 [81.1]	0.01620 [0.2595]
140 [60.0]	2.8929 [139]	0.01629[0.2609]
160 [71.1]	4.7472 [227]	0.01639 [0.2625]
180 [82.2]	7.5196 [360]	0.01651 [0.2645]
200 [93.3]	11.5376[553]	0.01663 [0.2664]

to 24.7 psia [1,183 Pa]. The high pressure is usually set by the maximum pressure allowable at the safety relief valve(s) without opening, usually 150 psig [7.2 kPa] (164.7 psia [7.9 kPa]), though sometimes lower.

For hot water systems, the low temperature at the tank connection occurs when the system is cold (i.e., $t_1 = 50°F$ [10°C]) and the high temperature occurs during operation (i.e., t_2 = hot water supply temperature).

In the case of a chilled water system, the high temperature will occur when the system is off (i.e., $t_2 = 90°F$ [32°C]) and the low temperature will occur when the system is at operating temperature (i.e., t_1 = chilled water supply temperature).

The high and low pressures and specific volumes for water in the system can be determined from Table 8-8.

Finally, a commercially available compression tank size, with at least the required volume V_t, may be selected using Eq. (8-12) and Table 8-9. To save time, the online application at http://westank.com/calculator/ can be used to compute the minimum compression tank size… open the application and select the "Expansion" option. Where more than 300 gal. tank capacity is required, select multiple tanks to be piped in parallel.

Size piping between the air separator and compression tank using Table 8-10.

8.2.4 Freeze Protection

To protect outdoor HVAC piping from freezing, *heat trace tape* is wound around the pipe before it is insulated and controlled by an outdoor thermostat or temperature controller that energizes the heat trace tape and keeps the pipe from freezing. Heat trace tape provides electrical resistive heating for the pipe and is fabricated with a heat trace heating

TABLE 8-9

Compression Tank Sizes[a]

	Conventional Tank			Bladder Tank		
Tank Volume (gals)	Diameter (in.)	Length (in.)	Water-Logged Wgt. (lb)	Diameter (in.)	Height (in.)	Water-Logged Wgt. (lb)
15	12	33	175	14	24	150
24	12	51	275	16	37	290
30	14	48	330	20	37	375
40	14	63	440	N/A	N/A	N/A
50	16	72	635	24	43	630
80	20	62	830	24	55	895
100	20	78	1,020	30	49	1,134
120	24	65	1,220	N/A	N/A	N/A
135	24	72	1,355	30	57	1,335
175	30	62	1,780	N/A	N/A	N/A
220	30	77	2,205	32	76	2,310
240	30	84	2,425	36	75	2,715
300	30	106	2,985	36	87	3,225

[a] IP units only are shown here. See manufacturers' data for dimensions and weights for tanks manufactured for specific SI tank volumes.

TABLE 8-10

Pipe Sizes between Air Separator and Compression Tank

System Capacity (1,000 Btu/h) [W]	Chilled Water, NPS (in.) [DN]	Hot Water, NPS (in.) [DN]
Up to 3,000 [up to 879]	3/4 [20]	1 [25]
3,001–6,000 [880–1,758]	1 [25]	1-1/4 [32]
6,001–12,000 [1,759–3,517]	1-1/4 [32]	1-1/2 [40]
12,001+ [3,518+]	1-1/2 [40]	2 [50]

wire (typically 16 AWG) inside a conductive core. The conductive core is surrounded by a thermoplastic insulation that has an outer copper over shield that is encased in thermoplastic insulation as an outer jacket to protect the inner layers.

Generally, the criteria from Table 8-11 can be used to prevent pipe freezing. In Climate Zones 1 and 2, and coastal regions of Climate Zone 3, where the outdoor winter design temperature is 20°F [–7°C] or higher and there are no prolonged periods at ambient temperatures below freezing, heat tracing may be safely eliminated for piping larger than 4".

8.3 Steam and Condensate Piping

Steam piping systems for HVAC applications are generally classed as *low pressure* (0–15 psig [0–103 kPa]), *medium pressure* (16–100 psig [104–690 kPa]), and *high pressure* (101+ psig [695+ kPa]). Rarely are pressures in excess of 150 psig [1,030 kPa] utilized in HVAC systems.

TABLE 8-11

Electric Heat Tracing for Outdoor Piping

Pipe Size NPS (in.)[DN]	Watts/Linear Foot [W/m] of Piping
1/2 [15]	1.6 [5.25]
3/4 [20]	1.7 [5.58]
1 [25]	1.9 [6.23]
1-1/4 [32]	2.0 [6.56]
1-1/2 [40]	2.3 [7.55]
2 [50]	2.9 [9.51]
2-1/2 [65]	3.4 [11.2]
3 [75]	4.5 [14.8]
4 [100]	4.9 [16.9]
6 [150]	7.1 [23.3]
8+ [200+]	0.68/in. [0.027/mm] of pipe diameter

8.3.1 Pipe Routing and Sizing

Steam piping systems consist of two distinct parts:

1. Steam piping that carries (primarily) a vapor or gas.
2. Condensate return piping that carries a liquid (condensed steam). And, there are two parts to that system: initially, condensate flows from the steam traps to a condensate receiver by gravity and then pumped from the condensate receiver to the boiler feedwater tank.

As for hydronic piping, the designer should route steam and condensate piping, both horizontally and vertically, square and plumb with building walls, with the appropriate slope:

1. Steam mains must be installed with at least 1/2% downward slope in the direction of steam flow, and branch piping and riser offsets require 1% downward slope in the direction of condensate return. Run out piping to terminal coils, and so on requires 2% downward slope in direction of condensate return.
2. Condensate gravity return piping requires 1/4% downward slope in direction of flow.

Piping must be "drained" of condensate that forms due to heat loss in the system to prevent water hammer and trap damage. High and medium pressure *drip legs* are required at each elevation change and at minimum 300 ft [90 m] intervals along piping mains. Low pressure drip legs are required at each elevation change and at minimum 100 ft intervals along piping mains.

Branches should have top or 45° top outlets from the main, *never side or bottom outlets.*

Avoid the use of "bull headed tees" in steam piping when making the final branch connections from a main. Rather, use a side outlet tee for the first branch and then a short nipple and elbow for the second.

There are several special considerations relative to condensate return pipe routing:

1. A steam trap can "lift" condensate only a short distance and, even then, only if the trap is very carefully selected. Therefore, always route trap condensate piping for downward, gravity flow.

2. Steam condensate at or above 25 psig [172 kPa] should be routed through a *flash tank* to reduce the pressure to atmospheric pressure before entering the condensate receiver. While low pressure "flash" steam will not damage a receiver, higher pressures may. Also, the flash steam may be recovered for use in low pressure applications.

A flash tank is designed to allow flash steam from medium or high pressure condensate to be reduced to a lower pressure before being introduced into a low-pressure condensate return system. In a boiler room, condensate from high pressure traps can be collected, flashed to a lower pressure in a flash tank, and that low pressure steam used for a deaerator or feedwater heating. In the latter case, a back pressure regulator is required to control the low-pressure steam pressure.

The amount of flash steam generated depends on the upstream steam pressure and the final downstream condensate pressure, which is also the pressure of the flash steam and is equal to the sensible heat of the condensate at the higher pressure before discharge minus the sensible heat in the condensate at the lower pressure to which discharge takes place, divided by the latent heat in the steam at the lower pressure to which the condensate has been discharged, as summarized in Table 8-12.

Flash tanks are specified on the basis of the condensate load entering the vessel, the corresponding amount of flash steam that is generated, and the disengaging area as well as the volume of the tank. The disengaging area is the surface through which the steam must flow to separate from the liquid and is the rectangle that is formed by the intersection of the water surface and the inside of the tank when the tank is exactly half-full. For a vertical cylindrical tank, the disengaging area is simply the area of the circle formed by the inside diameter of the tank, while for horizontal tanks, it is the area computed as the tank diameter multiplied by its length. The required flash tank disengaging area can be determined from Table 8-13.

The minimum tank volume can be determined from Table 8-14.

A flash tank must be located below any equipment or steam lines being drained. And when multiple return lines are fed into a flash tank, check valves are required on each line to prevent reverse flow of condensate and resultant flash steam in the condensate lines.

A properly sized safety relief valve must be installed on the flash tank. It should be set for the flash tank's maximum allowable working pressure or the maximum allowable working pressure of equipment being supplied by the low pressure steam.

TABLE 8-12

Percent (%) of Flash Steam Generated

Steam/Condensate Pressure, psig [kPa]	Back Pressure, psig [kPa]			
	0 [0]	10 [69]	20 [138]	30 [207]
15 [103]	3	1	—	—
25 [172]	5	3	—	—
50 [345]	7	5	4	2
75 [517]	11	8	6	5
100 [690]	13	10	8	7
150 [1,034]	15	12	11	9

TABLE 8-13

Flash Tank Disengaging Area Requirements

Steam Pressure Entering Tank (psig) [kPa]	Disengaging Area Required (ft²/1,000 pph) [m²/kW × 10⁻⁵] — Flash Tank Back Pressure		
	0 psig [0 kPa]	5 psig [34.4 kPa]	15 psig [103 kPa]
20 [138]	0.21 [0.55]	0.10 [0.26]	0.05 [0.13]
40 [276]	0.32 [0.85]	0.18 [0.48]	0.07 [0.18]
60 [414]	0.35 [0.92]	0.23 [0.61]	0.10 [0.26]
80 [552]	0.39 [1.03]	0.28 [0.74]	0.14 [0.37]
100 [690]	0.47 [1.24]	0.30 [0.79]	0.17 [0.45]

TABLE 8-14

Flash Tank Volume Requirements

Steam Pressure Entering Tank (psig) [kPa]	Flash Tank Volume Required (ft³/1,000 pph) [m³/kw × 10⁻⁶] — Flash Tank Back Pressure, psig [kPa]		
	0 [0]	5 [34.4]	15 [103]
20 [138]	0.40 [3.22]	0.18 [1.45]	0.05 [0.40]
40 [296]	0.60 [4.83]	0.35 [2.82]	0.14 [1.13]
60 [414]	0.75 [6.04]	0.48 [3.86]	0.22 [1.77]
80 [532]	0.88 [7.08]	0.58 [4.67]	0.30 [2.42]
100 [690]	0.97 [7.81]	0.66 [5.31]	0.37 [2.98]

3. A small amount of condensate discharged from a steam trap may be piped directly into a pumped condensate return line (such as the discharge from line traps on steam distribution piping) in accordance with Figure 8-4.

Steam piping may be sized based on the recommended allowable steam pressure losses from Table 8-15.

The recommended steam flow velocity range for quiet operation is 6,000–8,000 fpm [30–55 m/s].

Based on these criteria, steam piping may be sized using the ASHRAE *Handbook-Fundamentals* methodology, the recommended maximum steam flow rates provided by Table 8-16, or the online pipe sizing tool at https://www.tlv.com/global/TI/ calculator/ steam-pipe-sizing-by-pressure-loss.html.

In Table 8-16, the unshaded areas represent flows in which the velocity is less than 6,000 fpm [30 m/s], while the shaded areas represent flows in which the velocity exceeds 6,000 fpm [30 m/s]s but is less than 8,000 fpm [55 m/s].

To estimate the available steam supply pressure at any point in the system, multiply the straight pipe length by 1.5 (to account for fittings and valves), then multiply by pressure loss, psig/100 ft [kPa/30 m]. Subtract the result from starting pressure (i.e., boiler or *pressure-reducing valve* discharge pressure.

Size *gravity* flow condensate piping based on 1/4% slope is given in Table 8-17.

Pumped flow condensate piping may be sized on the same basis as hydronic piping on the basis of three times the steam flow rate, since condensate pumps "cycle" on/off with higher flow rates than those entering the receiver. Conversion of steam flow (lbm/h) [kW] to water (condensate) flow (gpm) [kg/h] can be done using Eq. (8-13) or (8-13a).

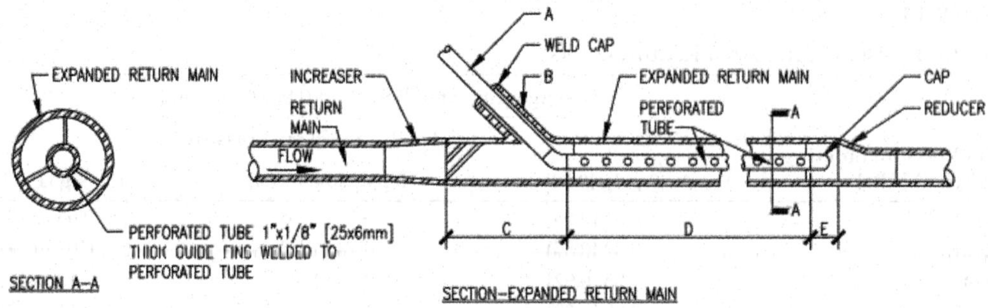

A	SIZE, TRAP DISCHARGE LINE	15mm [1/2"]	20mm [3/4"]
B	SIZE, 45° WELDING NIPPLE	25mm [1"]	32mm [1-1/4"]
C	LENGTH OF EXPANDED MAIN AHEAD OF TRAP DISCHARGE PIPE	175mm [7"]	175mm [7"]
D	LENGTH OF PERFORATED PIPE	415mm [16-1/2"]	415mm [16-1/2"]
E	LENGTH OF EXPANDED MAIN FOLLOWING PERFORATED PIPE	50mm [2"]	50mm [2"]

RETURN MAIN SIZE	UP TO 40mm [1-1/2"]	50mm [2"]	75mm [3"] & OVER
EXPANDED RETURN MAIN SIZE		65mm [2-1/2"]	SAME SIZE

NOTES:

1. 15mm [1/2"] PERFORATED TUBE SHALL HAVE 40 - 16mm [1/8"] DIAMETER HOLES SPACED 40mm [1-1/2"] O.C. IN 4 ROWS.
2. 20mm [3/4"] PERFORATED TUBE SHALL HAVE 78 - 6mm [1/8"] DIAMETER HOLES SPACED 40mm [1-1/2"] O.C. IN 6 ROWS.
3. HOLES IN TUBE SHALL BE SPACED EQUALLY AROUND PERIMETER.

FIGURE 8-4
Gravity condensate connected to pumped condensate return. (Public Domain; Veterans Administration.)

TABLE 8-15
Recommended Steam Piping Design Allowable Pressure Loss

System Pressure (psig) [kPa]	Allowable Pressure Loss (psig/100 lf) [kPa/m]
0–15 [0–1–3]	0.125 [0.86]
16–59 [104–384]	0.25 [1.72]
60+ [385+]	0.50–0.75 [3.451–5.71]

$$\text{For IP units, gpm} = (\text{lbm/h})/500 \tag{8-13}$$

$$\text{For SI units, kg/h} = 1.8 \times \text{kW} \tag{8-13a}$$

8.3.2 Steam Trap Application and Sizing

Steam traps perform two functions in a steam distribution system: (1) hold steam in the heat transfer side of the system until it fully condenses, giving up all its latent heat and (2) release the liquid condensate to the return system. *In other words, traps are designed to pass liquids but stop gases.*

TABLE 8-16

Maximum Recommend Steam Flow Rate, PPH [kW]

Pipe Size, NPS (in.) [DN]	Steam Supply Pressure, psig [kPa]/Maximum Pressure Drop, psig per 100 ft [kPa per 30 m]					
	5/0.125 [73/0.86]	15/0.125 [103/0.86]	30/0.25 [207/1.72]	60/0.25 [414/1.72]	100/0.50 [690/3.45]	125/0.75 [862/5.17]
3/4 [20]	10 [2.84]	13 [3.70]	22 [6.25]	28 [7.96]	50 [14.2]	66 [18.8]
1 [25]	24 [6.82]	27 [7.68]	45 [12.8]	57 [16.2]	100 [28.4]	134 [38.1]
1-1/4 [32]	52 [14.8]	59 [16.8]	99 [28.1]	127 [36.1]	220 [62.5]	296 [84.2]
1-1/2 [40]	81 [23.0]	91 [25.9]	156 [44.4]	197 [56.0]	340 [96.7]	460 [131]
2 [50]	160 [45.4]	180 [51.2]	316 [89.8]	382 [109]	690 [273]	930 [264]
2-1/2 [65]	270 [76.8]	300 [85.3]	520 [148]	660 [188]	1,140 [342]	1,535 [436]
3 [80]	400 [114]	560 [159]	953 [271]	1,205 [343]	2,090 [594]	2,815 [800]
4 [100]	1,040 [296]	1,180 [335]	2,020 [574]	2,550 [726]	4,420 [1,257]	5,930 [1,686]
6 [150]	3,160 [898]	3,580 [1,018]	6,100 [1,734]	7,725 [2,196]	13,420 [3,815]	18,070 [5,137]
8 [200]	6,500 [1,848	7,450 [2,118]	12,700 [3,611]	16,100 [4,577]	27,930 [7,941]	37,550 [10,675]
10 [250]	12,020 [3,417]	13,000 [3,696]	23,200 [6,596]	29,400 [8,358]	50,970 [14,491]	68,500 [19,475]
12 [300]	19,290 [5,484]	21,830 [6,199]	37,250 [10,690]	47,100 [13,390]	81,810 [23,259]	110,200 [31,330]

TABLE 8-17

Pipe Sizing for Gravity Flow Condensate

Maximum Condensate Flow Rate (lb/h) [kW]	Minimum Pipe Size, NPS (in.) [DN]
1,000 [284]	3/4 [20]
2,240 [637]	1 [25]
4,880 [1,387]	1-1/4 [32]
7,480 [2,127]	1-1/2 [40]
13,900 [3,952]	2 [50]
37,400 [10,633]	2-1/2 [65]
79,700 [22,659]	3 [80]

The more common types of traps are:

1. *Float traps* operate by the rise and fall of a float in response to the condensate level in the trap cavity. When the trap is empty, the float is in its lowest position and the discharge valve is closed. As the condensate accumulates, the float rises, mechanically opening the valve, and the pressure difference forces the condensate out. The discharge is generally continuous, since the opening of the valve is directly proportional to the flow of condensate through it.
2. *Thermostatic traps* consist of thin metal bellows, or a disk capsule, which contains a small amount of volatile liquid such as alcohol. This liquid vaporizes when steam comes in contact with it, and the pressure created expands the element and closes the trap valve. When condensate (or air) comes in contact with the element, it cools and contracts, opening the valve and allowing the condensate and/or air to escape. The discharge is intermittent, varying with load.
3. *Float and thermostatic traps* are a combination of the float trap and the thermostatic trap and are used in draining condensate from unit heaters, air heating coils, and so on, which have variable loads and where there is apt to be both air and a large

volume of condensate. The float element of this trap handles the condensate, and the thermostatic element permits the venting of the air.

4. *Inverted bucket traps* normally fill as they accumulate a "load" of condensate, and for proper operation, it must remain flooded. When steam or non-condensables reach the trap, they collect in the top of the inverted bucket and impart buoyancy. The bucket then floats and mechanically lifts the valve toward its seat. Air passes through the vent hole in the top of the bucket, and the steam passing through the vent is condensed. Condensate entering the trap fills the bucket, which sinks, opening the trap valve. The condensate discharges through the open valve until enough steam enters to float the bucket, and the cycle is repeated. In any application that requires condensate *lift*, that is, raising the condensate from one elevation to a higher elevation, an inverted bucket trap is required.
5. *Thermodynamic traps* are traps that use the evaporation of saturated liquid, when released into a lower pressure, to produce the flow-restricting or trapping action. *Phase-change* traps rely on the vaporizing action to choke the flow through a continually open orifice, whereas *impulse* traps impart an "impulse" that operates a valve to provide the trapping action.

The correct trap for the application is often a difficult decision for the designer, but Table 8-18 provides recommendations for the more common steam trap applications.

Sizing of steam traps is based on empirical data derived from experience and is summarized as follows:

TABLE 8-18

Recommended Steam Trap Applications

Application	Recommended Trap Type
Air heating coils	
Low and medium pressure	Float-and-thermostatic
Hot water heater	
Instantaneous	Float-and-thermostatic
Storage	Float-and-thermostatic
Shell-and-tube exchangers	
Small, high pressure	Thermodynamic
Steam-jacketed vessels	
High pressure	Thermodynamic
Low pressure	Float-and-thermostatic
Steam line drip traps	
0 15 psig	Float-and-thermostatic
16–125 psig	Thermodynamic
126–600 psig	Thermodynamic or inverted bucket
Steam coils	Thermostatic or float-and-thermostatic
Steam tracer lines	Thermodynamic
Unit heaters	Float-and-thermostatic
Sterilizers and autoclaves	Thermodynamic or inverted bucket

1. Low- and medium-pressure traps:
   a. Traps for equipment such as coils, converters, unit heaters, fan coil units, and humidifiers should be sized for three times the maximum unit capacity at 1 psi [6.0 Pa] pressure drop.
   b. Traps for end-of-main drips shall be sized for 50–130 lbm/h [42–111 kW] at a 1 psi [6.0 Pa] pressure drop.
2. High-pressure traps for end-of-main drips shall be sized for 300 lbm/h [256 kW] at 5 psi [34.0 Pa] pressure drop.

8.3.3 Control of Steam Piping Expansion

Steam piping can have significant thermal expansion, resulting in pipe movement that must be addressed by the HVAC designer since expansion creates stress and stress may produce failure.

For steel piping, the linear coefficient of thermal expansion for piping is 9.5×10^{-6} in/in-°F [10.8×10^{-6} mm/mm-°C]. To deal with this movement and relieve stress, one or more of the following measures are typically required:

Expansion loops: U-shaped piping configurations

Expansion bends: L- or Z-shaped piping configurations

Pipe expansion devices: Constrained, linear expansion devices may be used only where space limitations preclude the use of bends or loops.

Eq. (8-14) can be used to compute the thermal expansion in any section of piping.

$$\Delta L = \sigma \times L \times \Delta T \quad (8\text{-}14)$$

where

ΔL = Length change due to expansion, in [mm]

σ = Coefficient of linear expansion for steel, in/in [mm/mm]

L = Length of piping section between piping anchors, in [mm], as shown in Figure 8-5.

ΔT = Temperature difference between the operating temperature and the initial temperature, °F [°C]

Typically, the initial temperature is based on a winter installation at 30°F [0°C]. The final temperature is the system maximum operating temperature. For steam, the following pressure/temperature relationships can be used to determine the final temperature:

Low pressure steam	250°F [120°C]
Medium pressure steam	300°F [150°C]
High pressure steam	350°F [175°C]

Use Tables 8-19 and 8-20 to determine the piping lengths required for either expansion loops or bends. Limit the expansion that must be relieved to less than 6″ [150 mm]. Generally, size loops for total length required equal $W+2H$, where W equals 20% of the required length and H equals twice W.

For linear expansion devices, utilize manufacturer's data for selection.

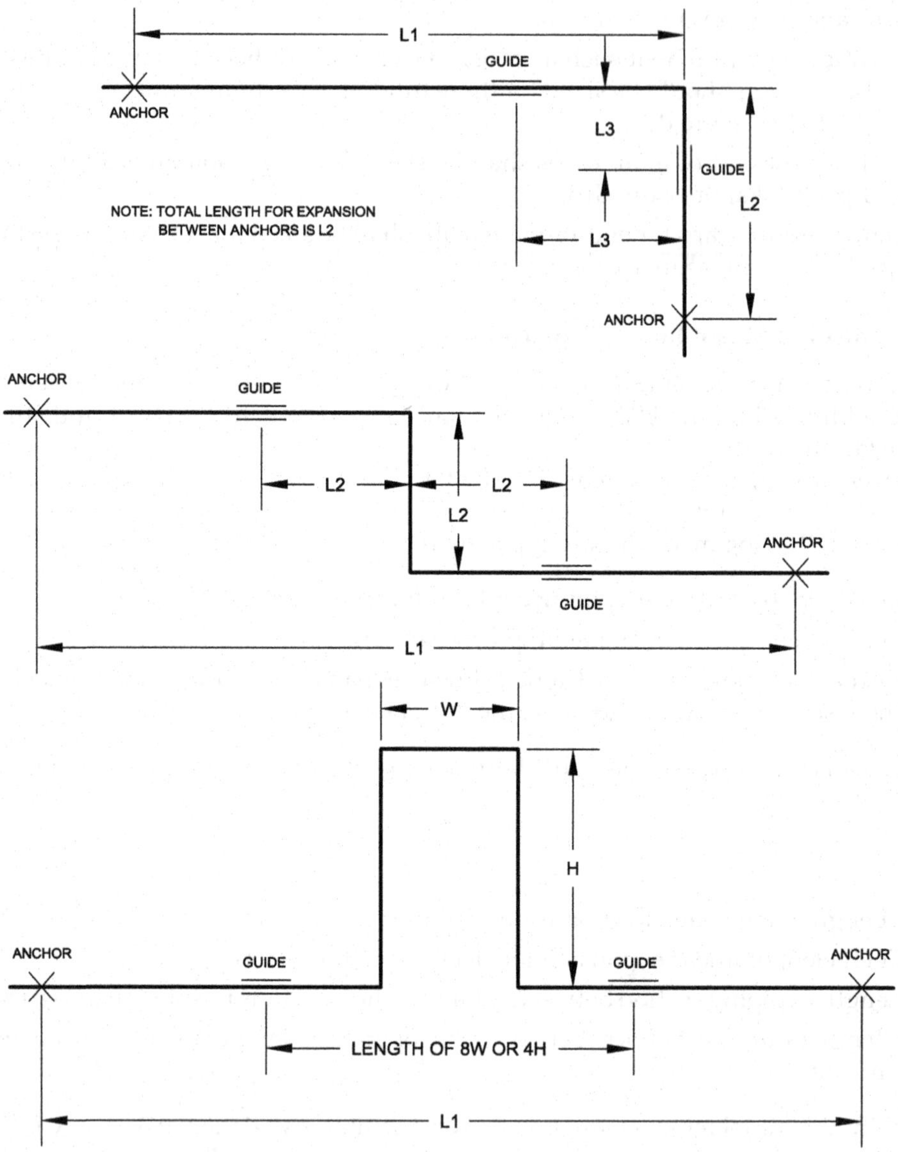

FIGURE 8-5
Typical steam system thermal expansion control.

8.4 Refrigerant Piping

8.4.1 Design Considerations

Engineer-designed, field-installed refrigerant piping is almost nonexistent in today's HVAC applications, except for the suction and liquid lines between indoor refrigerant cooling coils (*evaporators*) and outdoor factory-piped condensing units (which include the compressor and condenser) in a split system configuration (see Chapter 3). Even in this case, the manufacturers

TABLE 8-19

Thermal Expansion Control with Loop

Expansion Loop Length Required, *W*+2*H* (feet) [m]						
	Total Expansion, Δ*L* (in) [mm]					
NPS (in) [DN]	**1″ [25]**	**2″ [50]**	**3″ [75]**	**4″ [100]**	**5″ [125]**	**6″ [150]**
2 [50]	10 [250]	15 [380]	19 [480]	23 [580]	26 [660]	29 [740]
2-1/2 [65]	10 [250]	15 [380]	19 [480]	23 [580]	26 [660]	29 [740]
3 [75]	13 [330]	18 [460]	22 [560]	27 [690]	30 [760]	34 [860]
4 [100]	13 [330]	20 [510]	26 [660]	31 [790]	35 [890]	39 [990]
6 [150]	15 [380]	24 [610]	32 [810]	39 [990]	44 [1,120]	50 [1,270]
8 [200]	16 [410]	27 [690]	35 [890]	43 [1,090]	50 [1,270]	57 [1,450]
10 [250]	17 [430]	28 [710]	38 [970]	46 [1,170]	55 [1,400]	62 [1,570]
12 [300]	18 [460]	30 [760]	40 [1,020]	50 [1,270]	60 [1,520]	69 [1,750]

TABLE 8-20

Thermal Expansion Control with L- or Z-Bend

Expansion Bend: Length Required, H (feet) [m]						
	Expansion of Longer Leg, Δ*L* (in)					
NPS (in) [DN]	**1″ [25]**	**2″ [50]**	**3″ [75]**	**4″ [100]**	**5″ [125]**	**6″ [150]**
2 [50]	8 [200]	13 [330]	16 [380]	18 [460]	20 [510]	22 [560]
2-1/2 [65]	9 [230]	14 [360]	17 [430]	19 [480]	22 [560]	24 [610]
3 [75]	10 [250]	15 [380]	18 [460]	20 [510]	23 [580]	26 [660]
4 [100]	11 [280]	16 [380]	19 [480]	22 [560]	25 [640]	28 [710]
6 [150]	13 [330]	19 [480]	23 [580]	27 [690]	31 [790]	34 [860]
8 [200]	18 [460]	22 [560]	27 [690]	31 [790]	35 [890]	40 [1,020]
10 [250]	20 [510]	26 [660]	30 [760]	35 [890]	40 [1,020]	45 [1,020]
12 [300]	20 [510]	29 [740]	34 [860]	49 [1,240]	45 [1,020]	50 [1,270]

will provide specific sizing information for their equipment. But, every HVAC designer should have a basic understanding of refrigerant piping configurations and requirements.

Refrigerant piping is divided into four segments, each with different sizing criteria.

1. *Suction piping*: The piping between the evaporator and the compressor inlet. This piping transports low-pressure (cold) refrigerant gas.
2. *Discharge (or hot gas) piping*: The piping between the compressor discharge and the condenser. This piping transports high-pressure, high-temperature refrigerant gas. Discharge piping is normally factory-installed in condensing or packaged units to minimize the pressure loss between the compressor and the condenser.
3. *Liquid piping*: The piping between the receiver and the evaporator. This piping transports high-pressure liquid between the receiver and the expansion valve and low-pressure liquid between the expansion valve and the evaporator.
4. *Condensate piping*: The piping between the condenser and the receiver when a receiver is installed. This piping transports liquid (hot) from the condenser to the receiver. Condensate piping is sized the same as the liquid line size and is normally factory-installed.

A refrigerant piping system requires the same general design considerations as any fluid flow system. However, there are additional factors that have significant effect on refrigerant piping design, as follows:

1. Pressure loss at full load must be kept to within reasonable limits to avoid cooling capacity loss and increased energy consumption.
2. Oil return must be maintained at minimum load conditions.
3. The design must minimize the possibility of liquid refrigerant draining or migrating to the compressor during shutdown and prevent oil and refrigerant from returning to the compressor as liquid during operation.

The flow of refrigerant vapor in the hot gas and suction lines should be accomplished with minimum pressure loss to avoid penalizing compressor capacity and increasing power input. Since designers typically think in terms of saturated suction, discharge, and condensing temperatures, it is common to express friction pressure loss in refrigerant lines in terms of the equivalent temperature difference.

Each 1°F [0.6°C] drop in saturated temperature also reduces compressor capacity by about 2% and increases power input by about 2%. Therefore, good practice calls for suction lines to be sized for a maximum 2°F [1°C] temperature difference and liquid lines to be sized for maximum 1°F [0.6°C] temperature difference. Discharge lines are sized for maximum 2°F [1°C] temperature rise.

A second design consideration is the return of lubricating oil to the compressor. A small amount of oil will be carried by the refrigerant vapor throughout the refrigerant system. This oil must be returned to the compressor or it will fail due to lack of lubrication (and lots of other problems will occur, also).

Finally, the design of refrigerant piping must protect the compressor during both operation and upon shutdown. During shutdown, the oil in the compressor is capable of causing refrigerant to "migrate" to the crankcase where it is absorbed by the oil. These refrigerant rich oils tend to come into equilibrium at the ambient temperature. Then, upon start-up, the pressure in the crankcase is reduced to essentially the suction pressure condition. This rapid reduction in pressure causes an almost explosive foaming of the oil-refrigerant mixture. With hermetic compressors, the foam vents back through the compressor motor and into the suction line. In open compressors, foam vents into the suction manifold of the compressor. In either case, the oil in the compressor suction enters the cylinders. If the oil enters as a liquid, this is called *slugging* and can result in severe compressor damage as the compressor attempts to compress an incompressible liquid.

Even if slugging is avoided, oil entrained with refrigerant is carried through the compressor and out the hot gas line. This may very quickly "rob" the compressor of oil and result in compressor failure due to the lack of lubrication. In less severe cases, there may be only a partial loss of oil upon start-up, but this condition can result in lubrication problems that causes premature compressor failure.

To prevent this problem, the compressor must be provided with controls to minimize refrigerant migration to the crankcase during shutdown. One method is to use a liquid line solenoid valve and *pump down control*. A variation of this method is a combination of single pump down at shutdown, plus a crankcase heater. A third method is to energize a crankcase heater whenever the compressor is inoperative. The crankcase heater, by keeping the oil-refrigerant temperature at an elevated level, reduces the ability of the oil to attract and absorb refrigerant.

All refrigerant lines should be routed to be as short and level as possible. Horizontal lines should be installed level and hangers installed at close intervals to prevent sagging and the creation of "pockets" that will collect oil. Every effort should be made to prevent creating "traps" in the piping that will collect oil.

Allowance should be made for expansion and vibration isolation. Copper tubing expands at the rate of approximately 0.6″ per 100 linear feet [150 mm per 30 m] for each 50°F [28°C] increase in temperature. Compressors are normally mounted on vibration isolators and, at shutdown or start-up, compressor movement may be significant. To account for these two factors, the piping must include offsets, loops, and/or manufactured flexible connectors to absorb line movement and avoid rupture.

Copper tubing and solder-type wrought copper fittings are used for refrigerant piping. Copper tubing is available in three standard wall thicknesses, designated as K, L, and M, as shown in Table 8-2.

Refrigerant tubing size is always specified on the basis of outside diameter (OD). Type L copper has a working pressure up to 250 psig [1.7 kPa] and is normally used for refrigeration duty. While any copper tubing can be used, the tubing must be free of dirt and scale. Since field cleaning is both difficult and expensive, copper tubing for refrigerant duty is provided by the manufacturers cleaned, deoxidized, and capped to prevent contamination prior to use. This type of copper tubing is designated as "ACR" tubing.

8.4.2 Suction Lines

Suction lines must be sized to assure a reasonable pressure drop, not to exceed 2°F [1.1°C]. When suction lines are short and direct, with suction temperatures between 35°F [1.7°C] and 45°F [7.2°C], single suction lines can usually be used. But, for longer runs or lower suction temperatures, *double suction risers* will be required.

As the length of the suction line increases, a larger diameter line is required to keep the pressure drop within design limits. But, a single line selected to ensure oil transport at minimum load may be so small that the resulting pressure drop is excessive. This conflict is overcome by using a double suction riser shown by Figure 8-6. The smaller line is sized

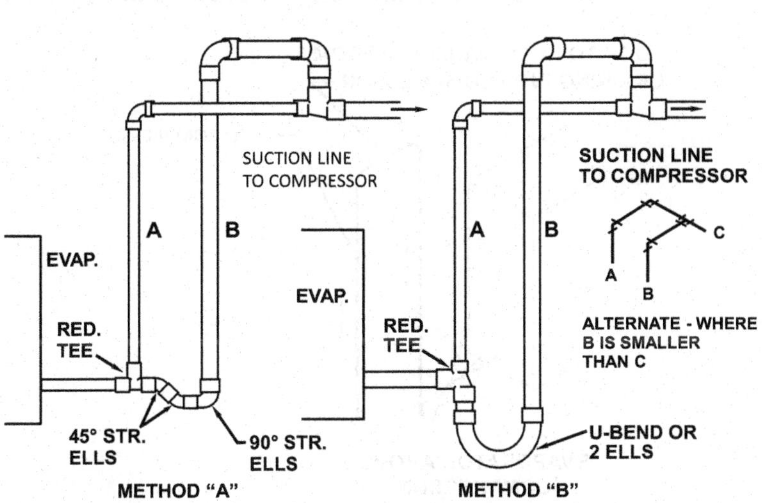

FIGURE 8-6
Typical double suction riser connections.

on the basis of the system minimum load, while the larger line is sized so that the pressure loss with maximum flow through *both* lines does not exceed the design limit.

A "trap" is provided at the base of the risers, as shown. During light load periods when the refrigerant gas velocity is insufficient to return oil up both risers, the trap gradually fills with oil until the larger riser is sealed off. When this occurs, the gas travels up the smaller riser with sufficient velocity to carry the entrained oil with it.

The fittings at the bottom of the riser must be close-coupled so that the oil holding capacity of the trap is minimized. A large trap can accumulate enough oil to seriously lower the compressor oil level and may increase the risk of slugging as the gas flow rate increases. The loop at the top of the riser prevents oil drainage into riser "B" when it is idle during light load operation.

The refrigerant that is trapped in an evaporator located above the compressor during shutdown will migrate to the compressor and be absorbed by the lubricating oil in the crankcase. On start-up, crankcase foaming and loss of oil can occur. In many design manuals, a suction line loop at the evaporator, as shown in Figure 8-7 is recommended. The purpose of the loop is, supposedly, to prevent liquid refrigerant and oil from draining to the compressor through the suction piping.

In most cases, however, this loop is ineffective. If a small quantity of liquid refrigerant lies in the evaporator, the oil in the crankcase will cause the refrigerant to migrate to the oil and be absorbed, regardless of the loop. After shutdown, if the evaporator temperature exceeds the compressor temperature, the resulting pressure difference can "lift" the liquid refrigerant up into the loop (approximately 2 ft. [610 mm] of rise for each psig difference in pressure) and, again, the loop serves no real purpose.

The best way to prevent liquid refrigerant form draining or migrating to the compressor is to use automatic pump down control. The *automatic pump down* "system" consists of a normally closed liquid-line solenoid valve, a normally closed electric shutoff valve, and a thermostat located in the conditioned space. When the desired space temperature is reached, the thermostat will open and de-energize the liquid-line solenoid valve, closing it. The compressor will continue to run to evacuate any refrigerant on the solenoid valve's outlet side, up to and including the compressor. The refrigerant will be stored in the condenser and receiver on the high pressure side of the refrigeration system.

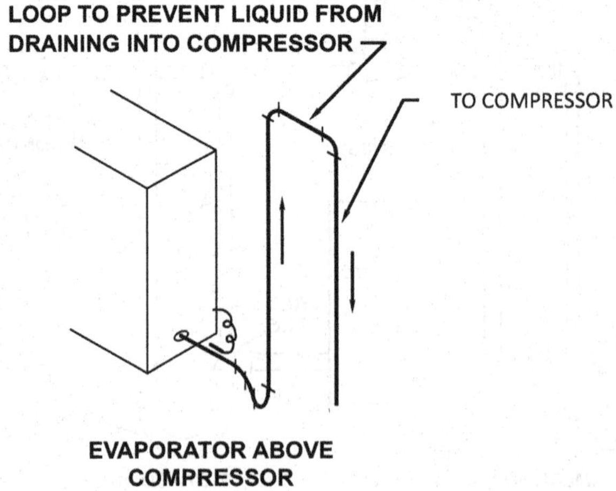

FIGURE 8-7
Evaporator suction loop piping.

The compressor will then be shut off by a low-pressure switch set to open at about 5–10 psig [34–68 kPa]. This pressure setting will ensure that no refrigerant will migrate to the compressor while it is off.

On a call for cooling, the thermostat will close and energize the liquid-line solenoid valve, opening it. Liquid refrigerant will flow to the expansion valve and into the empty evaporator, thus increasing the evaporator pressure. Once the "cut-in" pressure of the low-pressure control is reached, the compressor will easily start and resume a normal refrigeration cycle. The cut-in pressure must be set at a high enough pressure to ensure that the system will not *short-cycle,* damaging motor windings and starters due to overheating.

8.4.3 Liquid Lines

Liquid lines have few design problems simply because the liquid refrigerant and liquid oil mix well, and there is no oil return concern.

It is usual to find solenoid valves (thermal expansion valves), filter dryers, strainers, and/or sight glasses in liquid lines, each of which adds pressure loss to the circuit. Since thermal expansion valves are rated and selected on the basis of 100% liquid at the valve inlet, the presence of "flash gas" in the supply reduces the valve capacity and results in loss of system capacity. To ensure that little or no flashing occurs due to pressure reduction resulting from pressure losses in the liquid line, *subcooling* of the liquid is utilized. Thus, pressure losses will result in some heat gain, but the refrigerant will remain in the liquid state.

8.5 Piping Insulation

Piping that conveys heated or cooled fluids (vapor or liquid) must be insulated to limit heat losses or gains to minimize energy waste. For "cold piping," carrying fluids below 60°F and including pipe, fitting, valves, unions, flanges and all equipment cold surfaces, must be insulated and a continuous vapor barrier must be maintained. "Ambient" piping, carrying fluids between 61°F and 105°F, does not require insulation. Finally, "hot piping," conveying fluids at greater than 105°F and including pipe, fitting, valves, unions, flanges and all equipment hot surfaces, must be insulated, but no vapor barrier is required. However, hot gas refrigerant lines do not require insulation unless the heat dissipation is anticipated to be objectionable or there is a danger of personnel contact.

8.5.1 Hot Piping Insulation

For hot water, steam, condensate, and boiler feedwater piping, preformed mineral fiber insulating materials, including fiberglass, is typically used, with a minimum recommended thickness as shown in Table 8-21.

All service jacket (ASJ), provided by the insulation manufacturer as part of preformed piping insulation, is commonly used for hot piping, along with preformed PVC fitting covers.

8.5.2 Cold Piping Insulation

Cold piping includes chilled water, cooling condensate drains, liquid and cold gas refrigerant piping, and condenser water piping incorporated in a waterside economizer.

TABLE 8-21

Hot Piping Minimum Insulation Thickness, in. [mm]

	Fluid Temperature (F) [°C]				
	Hot Water, Low Pressure Steam, and Condensate		Medium and High Pressure Steam, Boiler Feedwater		
NPS (in.) [DN]	105–140 [40–60]	141–200 [61–90]	201–250 [91–120]	251–350 [121–175]	351+ [175+]
<1 [25]	1 [25]	1-1/2 [40]	2-1/2 [65]	3 [75]	4-1/2 [115]
1–1-1/4 [25–32]	1-1/2 [40]	1-1/2 [40]	2-1/2 [65]	4 [100]	5 [125]
1-1/2–3 [40–75]	1-1/2 [40]	2 [50]	2-1/2 [65]	4-1/2 [115]	5 [125]
4+ [100+]	1-1/2 [40]	2 [50]	3 [75]	4-1/2 [115]	5 [125]

Historically, chilled water and other cold piping have been insulated with mineral preformed fiber insulating materials with ASJ. These are open-structured materials that rely largely on entrapped still air for much of their insulating properties. They have little long-term resistance to water and no resistance to vapor flow. *Thus, the potential for moisture absorption can be very high if the factory-applied vapor retarder as part of the ASJ is either damaged or inadequately sealed.*

Research has been undertaken into the effect of moisture on mineral fiber insulants by Achtziger and Cammerer of FIW in Germany. Their research concluded that 1% moisture content by volume could increase the thermal conductivity of the material by 36%–107% with four of the five samples tested falling within the 95%–107% increase range. (*Forschungsvorhaben* Nr. 815-80.01.83-4 contained within CEN TC 88 WG 4-N484). *Tests by the National Insulation Manufacturers Association (NIMA) in 2000 showed that fiberglass insulation with ASJ can gain as much as 50% in weight from water absorption when installed in high humidity locations.*

Thus, the key to keeping fiberglass efficient on cold piping is keeping it dry.

Typical ASJ vapor barriers used on HVAC piping insulation have a rated permeance of 0.02 perm-inches, which is very low. Some tests have shown that the ASJ material may have an actual perm rating as high as 0.13 perm-inches, but even this rating is still acceptable for cold piping applications. However, all of these ratings are based on testing samples of ASJ in the laboratory, and there is no data that define ASJ performance as installed by the typical insulation contractor in the field.

There are numerous opportunities for the contractor to fail in the installation of ASJ. First, the longitudinal sealing strip must be kept clean and be carefully aligned to provide a vapor tight seal. The butt joints of the jacket must be carefully lapped and sealed with vapor barrier mastic, and all joints at fittings, valves, equipment, wall penetrations, and so on must also be 100% sealed with vapor barrier mastic or tape. At hangers, rigid insulation inserts and sheet metal shields must be installed to protect the integrity of the vapor barrier. *The chances of all of this happening on the typical project are very poor.* Thus, every cold piping system insulated with fiberglass offers opportunities for vapor diffusion and the introduction of low-level moisture content.

The conventional wisdom for selecting the thickness of insulation on cold piping is to (1) prevent surface condensation and/or (2) maintain heat gain to a fixed maximum. *Tables provided by fiberglass insulation manufacturers show that the minimum thickness required to prevent surface condensation is typically 1".*

However, mold will grow on most surfaces if the relative humidity (RH) at the surface is above 70% and the surface temperature (above 40°F [4°C]) is conducive to growth. The longer the RH remains above 70%, the more likely visible mold growth; and the higher the humidity or temperature, the shorter the time needed for germination. The surface RH is a complex function of material moisture content, material properties, and local temperature and humidity conditions. In addition, mold growth depends on the type of surface.

Based on the maximum 70% RH ambient humidity condition, analysis shows that the minimum required mineral fiber insulation thickness is 2″ [50 mm] for pipe smaller than 2″ NPS [DN50] and 2-1/2″ [65 mm] for larger pipe, *assuming dry insulation.* If 1% moisture content, by volume, has diffused into the insulation due to poor ASJ installation or damage to the ASJ over time, insulation thermal conductivity increases by 100% and the required minimum insulation thickness also increases by 100% to 3–4″ [75–100 mm] for all pipe sizes. *No designer currently requires that thickness of fiberglass insulation on chilled water piping* and, thus, mold growth on chilled water piping in hot, humid climates is almost universal.

Mineral fiber insulation that is of insufficient thickness allows a boundary layer of high humidity air to be created and the mold immediately begins to grow on the surface. If the ASJ does allow moisture to invade, as time goes by the insulation moisture content rises to a level where the thermal conductivity increases enough to allow the surface temperature to drop to or below the ambient dew point and surface condensation forms. This liquid water further augments mold growth on the insulation surface and condensation that drips onto ceiling and wall materials below to create additional habitats for mold growth.

While secondary vapor barriers (usually welded PVC jacket or vapor proof mastic coatings) can be added to ASJ, the only real solution to this condition is to *avoid the use of mineral insulation and, instead, to use a non-hydroscopic insulation, insulation that will not absorb moisture,* on cold piping. The insulation effectiveness, then, is not dependent on the quality of the installation of the ASJ vapor barrier to maintain a surface temperature that results in humidity boundary layer below 70% RH.

For indoor applications, cellular glass (which is a closed cell material) is the recommended material for insulating cold piping systems. For exterior applications, polyisocyanurate or expanded polystyrene can be used. But, because both of materials are combustible, their use indoors should be avoided (and can be a building code violation).

Recommended thickness of cellular glass insulation installed on indoor chilled water piping and condenser water piping used in a waterside economizer is shown in Table 8-22. For outdoor applications, cellular glass insulation thickness should be increased by 1″ [25 mm] simply to reduce heat gain, while the recommended minimum thickness for outdoor polyisocyanurate or polystyrene insulation is 3″ [75 mm] for all pipe sizes.

For refrigerant piping, both indoor and outdoor, the liquid and cold gas lines require a minimum of 1/2″ [13 mm] *closed cell flexible foam insulation* to minimize heat gain. Suction and liquid lines should be insulated where the two lines are clamped together during the installation.

TABLE 8-22

Chilled Water Piping Minimum Insulation Thickness, in. [mm]

Pipe Size, in. NPS [DN]	Thickness, in. [mm]
≤3 [75]	2 [50]
4–10 [100–50]	2-1/2 [65]
12+ [300+]	3 [75]

Bibliography

ASHRAE Handbook-Fundamentals, American Society of Heating, Refrigerating, and Air-Conditioning Engineers, Inc. (Atlanta, GA, 2017).
Cameron Hydraulic Data, Ingersoll-Rand (Woodcliff Lake, NJ, 1988).

Section III

Systems and Components

9

Pumps and Fans

9.1 Introduction

A very large part of heating, ventilating, and air-conditioning (HVAC) design involves the transport of fluids: air, water, steam, and so on to accomplish heat transfer and satisfy heating/cooling loads imposed on the system(s). Two of these fluid transport applications require that energy be expended by mechanical devices to create fluid movement:

1. Water movement requires *pumps* (except for a very limited number of applications where gravity may be used).
2. Air movement requires *fans*.

9.2 Pumps and Applications

9.2.1 Pump Types and Characteristics

Typical HVAC systems utilize pumps to circulate water through piping systems. With only minor exceptions, these pumps are *centrifugal pumps* of several available configurations.

A pump converts the energy provided by a prime mover, such as an electric motor, steam turbine, or gas engine, to energy within the liquid being pumped. This energy within the liquid is present as velocity energy, pressure energy, static elevation energy, or a combination of these. The rotating element of a centrifugal pump, which is turned by the prime mover, is called the *impeller.* The liquid being pumped surrounds the impeller and, as the impeller rotates, its motion imparts a rotating motion to the liquid, as shown in Figure 9-1. The other parts of the centrifugal pump are indicated in Figure 9-2.

For any centrifugal pump impellor, curves can be developed to show the full range of various flow/pressure relationships. When all these characteristics are plotted on one coordinate system, typically with *flow rate* as the ordinate and *pressure* (typically referred to as "head") as the abscissa, the capabilities of the pump are completely defined. Every pump performance curve is based on a particular speed, casin geometry, impeller shape and diameter, and the viscosity of water.

Figure 9-3 indicates the relationship between head developed by the pump versus flow through the pump. As flow rate or capacity increases, the total head that the pump is capable of developing is reduced. In general, the highest head is produced at a point where there is no flow, referred to as the *cutoff point*.

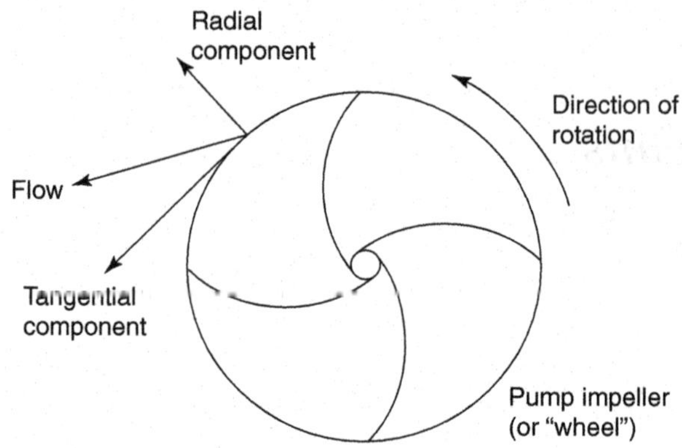

FIGURE 9-1
Impellor/flow relationship.

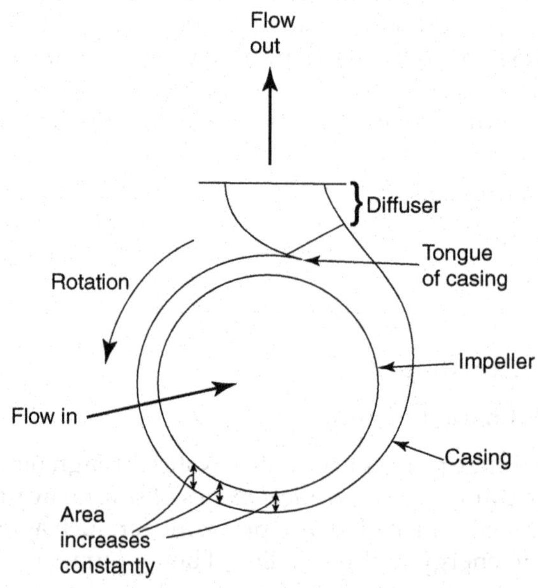

FIGURE 9-2
Centrifugal pump configuration.

For the centrifugal pump to deliver the desired capacity, a certain *break horsepower* (bhp) or *load kW* must be supplied. The bhp or load kW requirement for any pump can be *estimated* using Eq. (9-1) or (9-1a).

In IP units,

$$\text{BHP} = \text{gpm} \times \text{head} \times \text{specific gravity} / (3{,}960 \times \text{pump efficiency}) \quad (9\text{-}1)$$

where *head* is defined in terms of feet of water column (or gauge), *specific gravity* is found in Table 9-1, and typical *pump efficiency* is found in Table 9-2.

In SI units,

$$\text{load kW} = \text{m}^3/\text{h} \times \text{head} \times \text{specific gravity} / (367.2 \times \text{pump efficiency}) \quad (9\text{-}1a)$$

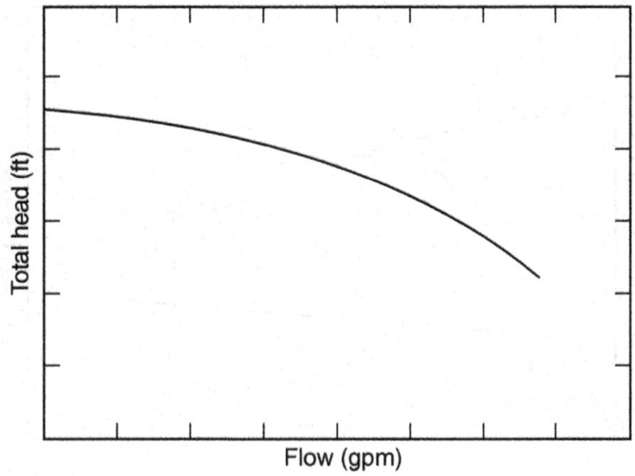

FIGURE 9-3
Typical pump flow/pressure relationship.

TABLE 9-1
Specific Gravity of Water

Fluid Temperature (°F) [°C]	Specific Gravity
≤100 [≤38]	1.000
120 [49]	0.990
140 [60]	0.985
160 [71]	0.979
180 [82]	0.972
200 [93]	0.964

TABLE 9-2
Typical Pump Efficiency

Pump Type	Typical Efficiency (%)
Inline circulator (fractional HP)	50
Inline pump, horizontal (1–5 HP)	60
Inline pump, vertical (5–20 HP)	75
Base-mounted, single (end) suction pump	80
Base-mounted, double suction pump	85
Vertical turbine	78

where *head* is defined in terms of meters of water column (or gauge).

Finally, all of the performance characteristics will be given by an *overall pump curve* similar to Figure 9-4. In Figure 9-4, NPSH is "net positive suction head" and is the minimum required absolute fluid head at the pump inlet for proper pump performance.

To select the pump properly, the required system head must be determined. This head is the sum of two different system pressure losses, as discussed in Chapter 8:

1. *Friction head.* Press losses due to pipe and fittings friction and other components in the system.

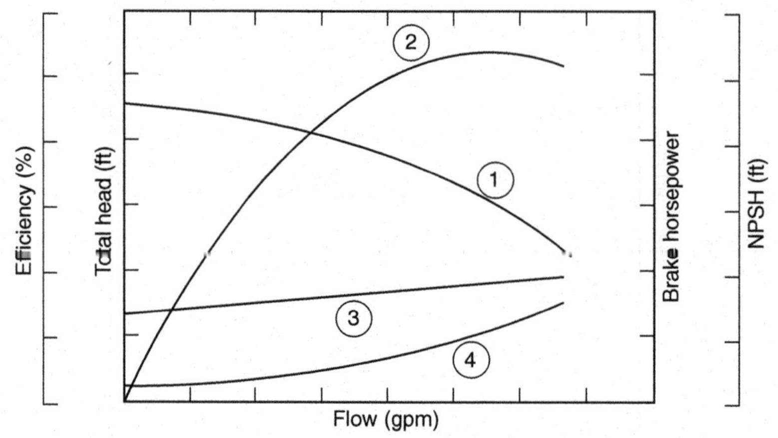

FIGURE 9-4
Typical overall pump curve.

2. *Static head.* In an open piping system (such as condenser water systems), the difference in elevation between the lowest reservoir connect to the pump suction and the highest reservoir on the pump discharge, as illustrated by Figure 9-5.

For friction head loss, the loss increases as the flow rate increases, as shown by Eqs. (9-2) and (9-3).

$$H_1^{0.5}/W_1 = H_2^{0.5}/W_2 \qquad (9\text{-}2)$$

or

$$W_2^2/W_1^2 = H_2/H_1 \qquad (9\text{-}3)$$

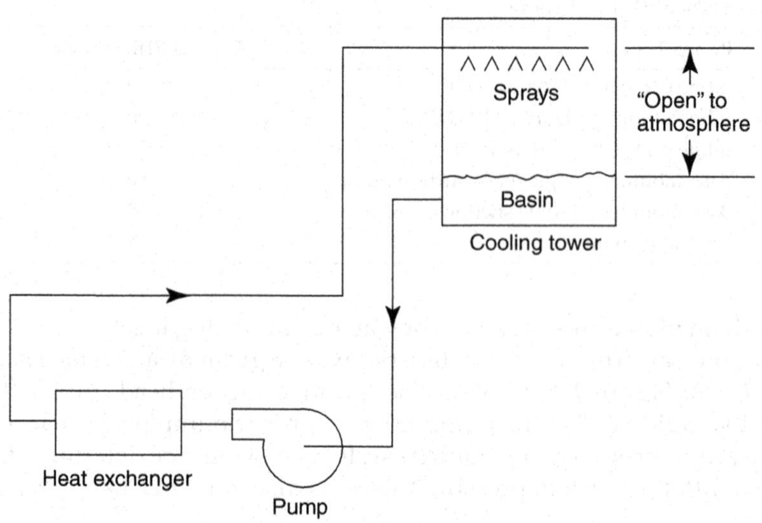

FIGURE 9-5
Static lift or head in open hydronic piping systems.

where H_1 and H_2 equal two different heads (ft wg) [m wg] at the two flows rates W_1 and W_2 (gpm).

Thus, if the design or "known" flow and head is W_1 and H_1, the head at any other flow, W_2, is computed using Eq. (9-4).

$$H_2 = (W_2^2 / W_1^2) \times H_1 \tag{9-4}$$

Since we are concerned about flow/head relationships, it is obvious that the head varies with the square of the flow ratios. Thus, to construct a *system curve* to represent graphically the flow/head relationships in any piping system, it is convenient to plot the conditions at four points: (1) the design flow, (2) zero flow, (3) half of design flow, and (4) twice design flow. The head condition at each different flow can be estimated using Eq. (9-4). For example, for a hydronic system designed for 250 gpm [16 L/s] flow at 40 ft [12 m] head, the four points for constructing a system curve would be as shown in Table 9-3.

Centrifugal pumps are classified as *end suction* and *double suction* (often called *split case*, again either horizontally or vertically) pumps. An end suction pump is configured so that the fluid enters on one side of the impeller only. For a double suction pump, water enters on both sides of the impeller. Normally, end suction pumps are utilized for lower flow ranges and heads (e.g., less than 300 gpm [20 L/s] and 125 ft [40 m] total head), while double suction pumps are available up to 15,000 gpm [950 L/s] and 800 ft [250 m] total head. End suction pumps may also be used where there are significant space constraints.

Double suction pumps, of either *horizontal split case* or *vertical split case* construction, are preferred for base-mounted pump applications. They are more efficient than end suction pumps and do not require an inlet diffuser (in fact, a long radius elbow at the pump suction works better than an inlet diffuser).

Figure 9-6 shows two manufacturer published pump curves.

On both graphs, a number of performance or head/capacity curves are given, each for a different diameter impeller. For example, the 2 [DN50] × 1-1/2 [DN40] × 6 [DN150] pump size has a nominal 6 in. diameter [DN150] impeller. However, ratings are given for impeller sizes ranging from 4.4 in. [112 mm] in diameter to 6.5 in. [165 mm].

Note also that the pump curves terminate short of the full scale. A pump will not operate "off the end of the curve" without *cavitating* and performing erratically. A pump must never be subjected to operating conditions beyond the end of its rating curve.

9.2.2 Pump Arrangements

Multiple pumps are used in HVAC systems. *Parallel pumping*, shown in Figure 9-7, is often used to reduce first costs when standby capacity is needed. Often, a designer, to ensure against possible pump failure problems, will specify the installation of two pumps, each capable of handling the entire load. With this configuration, one pump or the other

TABLE 9-3

Example System Curve Points

Flow (gpm) [L/s]	Head (ft-wg) [m-wg]
0	0
125 [8]	10 [3]
250 [16]	40 [12]
500 [32]	160 [48]

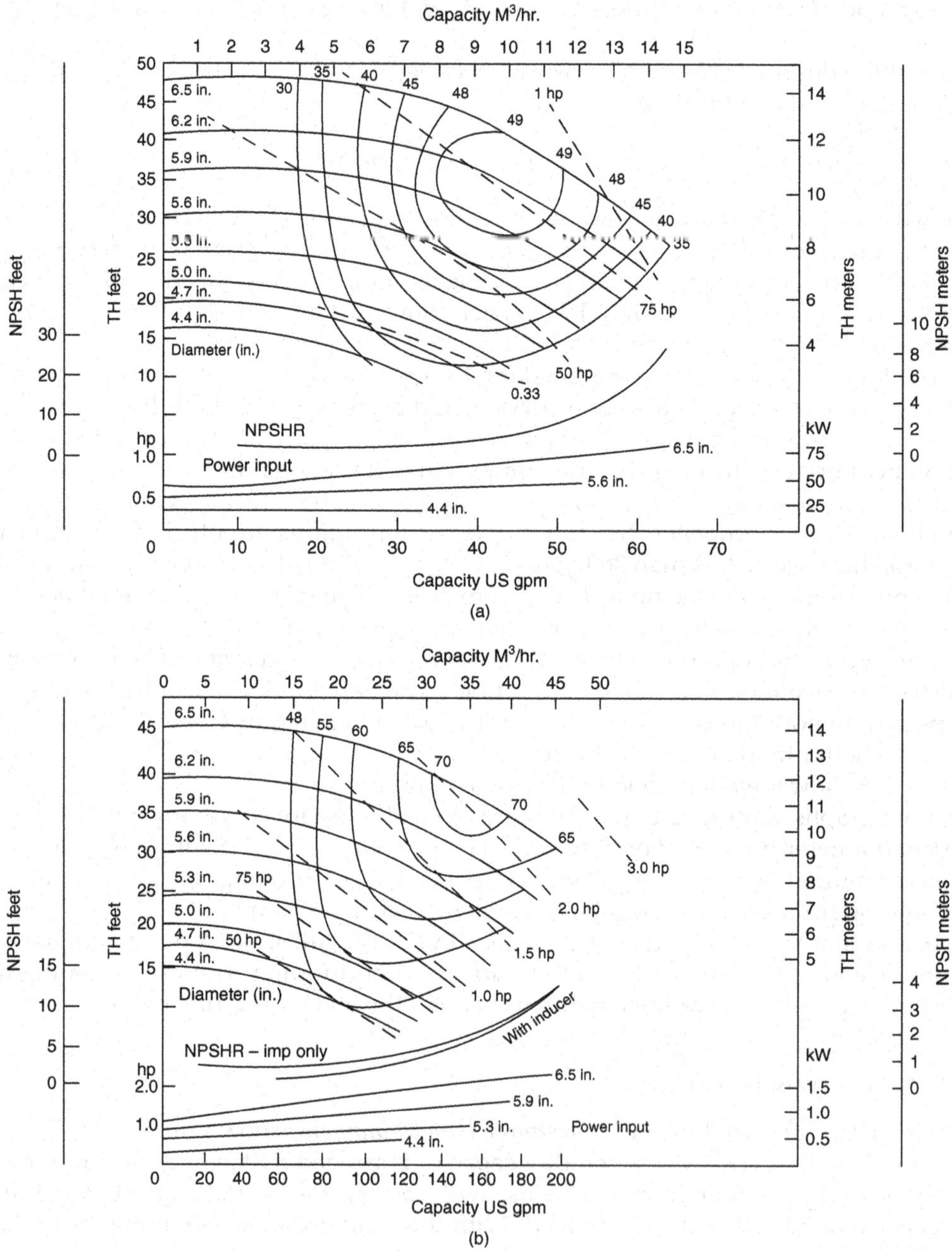

FIGURE 9-6
Typical actual pump curve.

operates, but not both at the same time. For most installations, however, the question of cost precludes any thought of full standby protection. Parallel pumping, wherein both pumps run at the same time, can provide very high (in most cases around 70%) standby protection with only marginally increased cost.

It can be seen from Figure 9-7 that each pump is pumping one-half of the total (design) flow rate.

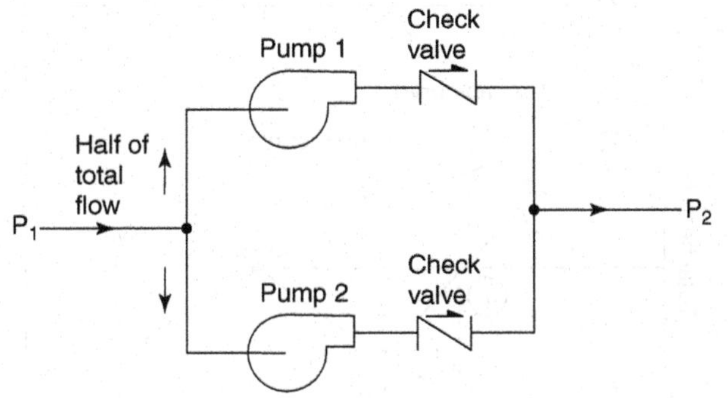

FIGURE 9-7
Typical parallel pump configuration.

Since the pumps are identical and are each pumping the same flow rate, it follows that each pump will produce the same pressure head. Each pump operates at the same point on its pump curve and then produces one-half of the total flow rate at the total system head.

To analyze the system accurately, construction of a *parallel pump curve*, a graph of the total pumping condition as it will actually exist on the system must be made. To construct a parallel pump curve, the flow produced by a single pump at any given head is simply doubled. This procedure is shown in Figure 9-8.

By overlaying the system curve on the pump curves, as shown by Figure 9-9, the overall system performance can be determined. When both pumps are operating, the system head/capacity condition is indicated by point 2, as each individual pump operates at point 1 (one-half the total flow). However, if one pump fails, the remaining pump will "ride its curve" to operate at point 3, actually delivering more flow at lower head than it did when both pumps operated. (As a "rule of thumb," one pump typically will deliver about two-thirds of the required system flow when both pumps operate.)

Series pumping, as shown in Figure 9-10, is often applied when exceptionally high heads are required.

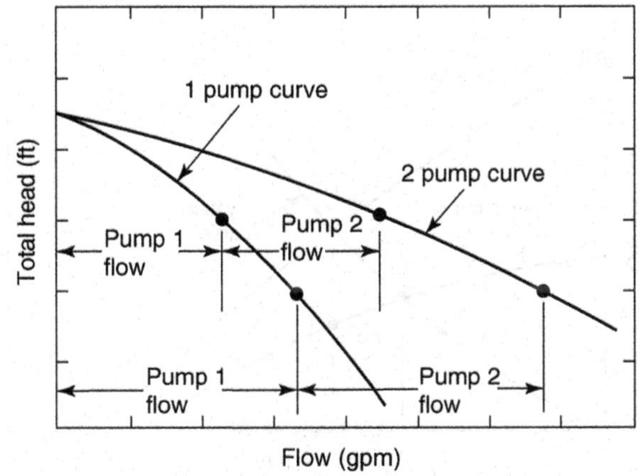

FIGURE 9-8
Parallel pump curve construction.

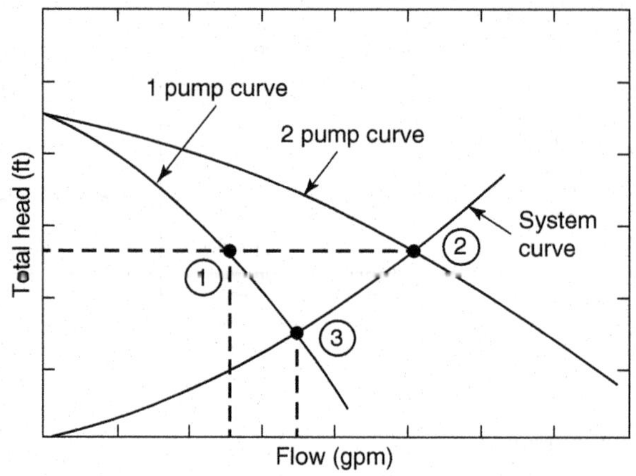

FIGURE 9-9
Parallel pump system operation.

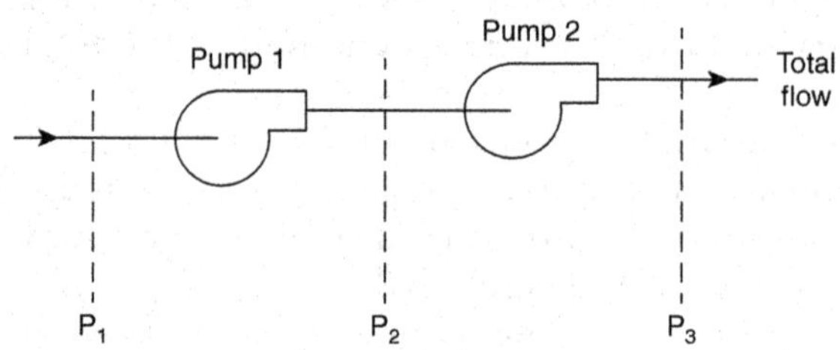

FIGURE 9-10
Typical series pump configuration.

To analyze a series pumping application, it is necessary to construct a series pump curve. Assuming identical pumps (e.g., each pump rated at full flow and one-half of the required system head), the series pump curve would be as shown by Figure 9-11.

As shown by Figure 9-12, with both pumps operating, the system head/capacity condition is indicated by point 1, as each pump operates at point 2 (one-half the head). However,

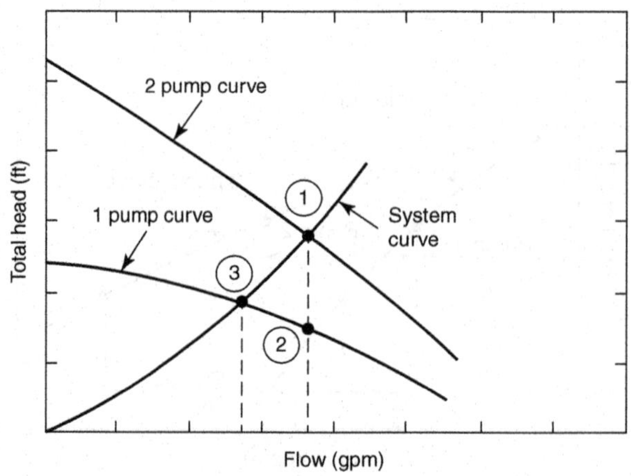

FIGURE 9-11
Series pump curve construction.

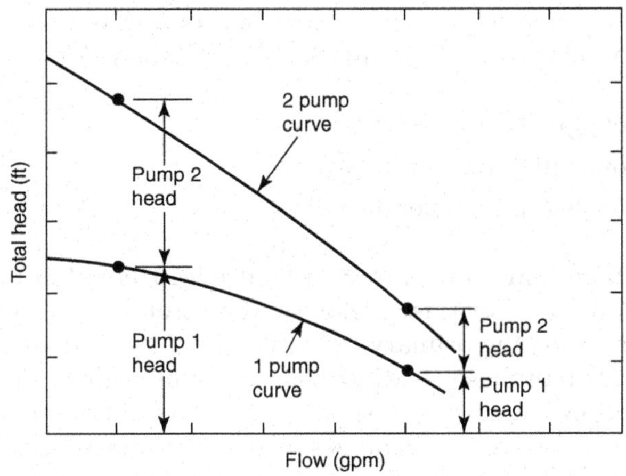

FIGURE 9-12
Series pump system operation.

if one pump fails, the remaining pump will now operate at point 3, delivering less flow (at a higher head) than it did when both pumps operated.

A common application in HVAC piping systems is the *primary–secondary pumping circuit*. With this arrangement, the flow rates in the secondary loops are independent of the primary loop flow rate. The basic primary–secondary connection is shown in Figure 9-13.

Common piping is defined as a length of piping common to both the primary and secondary circuit flow paths. The common piping length is made quite short, producing minimum pressure drop and ensuring hydraulic isolation of the two circuits. Thus, flow in the primary circuit will not cause flow in the secondary and vice versa. The secondary circuit pump is sized to provide design flow rate through the secondary circuit with the secondary circuit pressure drop only.

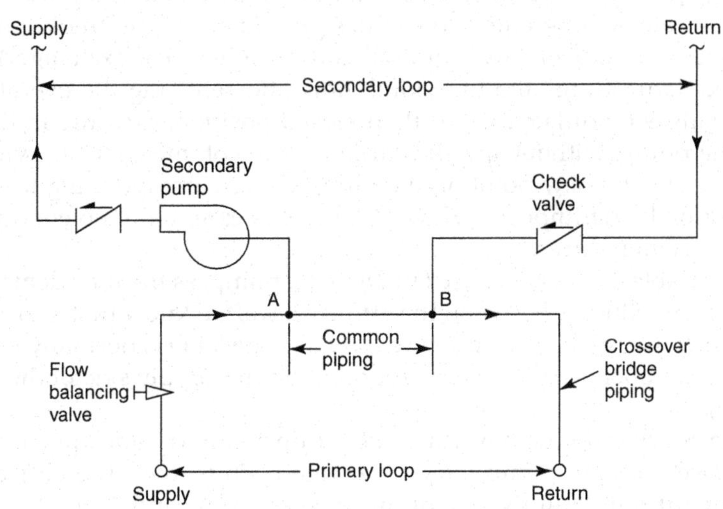

FIGURE 9-13
Primary/secondary pumping system configuration.

The amount and direction of flow in the common piping will vary based on which of three relationships exist between the primary and secondary flow rates:

Case 1: primary flow greater than secondary flow

Case 2: primary flow equal to secondary flow

Case 3: primary flow less than secondary flow

For Case 1, there will be flow from A to B in Figure 9-13, equal to the primary flow rate *less* the secondary flow rate. For Case 3, flow will be from B to A in a quantity equal to the secondary flow rate *less* the primary flow rate. For Case 2, there would be no flow in the common piping. Each of these conditions is easily determined by applying the *tee law*, which states that the flows into a tee must equal the flows out of the tee.

There are a number of specific advantages to using primary–secondary pumping and typical examples include the following:

1. Chilled water systems, where refrigeration machine performance may depend on a constant flow rate, yet there is a desire to save pumping energy by using variable flow control at the cooling coils served.
2. Systems where temperature drops or rises in the primary loop are different from the secondary loops. Large chilled and hot water distribution systems are often designed on this basis to save piping and pumping costs.
3. Systems where the operating times for the secondary loops are not coincident. For example, a hot water system may serve several areas of a building with a single primary loop, but each building area's secondary loop will operate on a different time schedule.

9.2.3 Water Flow Volume Control

Variable flow is commonly used (and is highly recommended) in all but the smallest HVAC hydronic piping systems. When the peak heating or cooling load is imposed on a hydronic system, the flow rate will be equal to the design value. But, as the imposed loads are reduced, the two-way control valves at each coil (or other heat exchanger) will begin to modulate closed, "throttling" the individual flow and reducing the overall system flow. As the control valves modulate toward their closed position, pressure in the system will increase and the pump, without specific variable flow volume controls, will simply "ride its curve" to operate at a new point on the pump curve corresponding to each new flow and head condition being imposed. However, there is essentially no energy reduction as the flow in the system reduces.

By adding a variable frequency drive (VFD) to the pump, as the flow demand is reduced, pump speed can be reduced. Based on the affinity laws, we know that a corresponding 3:1 reduction in pump energy consumption will result. The VFD is normally programmed to maintain a specific "operating pressure" set point on the supply side of the piping system (see Chapter 13).

Since a pump curve is defined by a specific pump geometry and speed, as the speed is reduced, effectively a new pump curve is created. Thus, over the entire speed range of the VFD, an infinite "family" of pump curves exist, and since the system pressure is maintained at a fixed set point, *the system curve becomes essentially a straight horizontal line.*

However, pump speed cannot be reduced to zero. First, *differential pressure sensors* are required in each sub-circuit of the piping system with set points that ensure a minimum pressure in each remote part of the system such that no part of the system is "flow starved" at light load conditions.

The second limiting factor on flow rate for the pump is the *turndown ratio* of the VFD, typically 5:1. When total flow demand drops to about 20% of design flow, the system returns to "constant flow" operation. Then, as two-way control valves throughout the piping system throttle flow further, the pump will "ride" its 20% rpm curve to, ultimately, "cutoff" or zero flow. At zero flow, the heat added by motor and pump inefficiencies will increase the water temperature until it flashes to steam and the pump will cavitate, damaging both the impellor and the casing.

To protect the pump, a pump pressure relief valve must open to maintain a minimum flow through the pump. The minimum flow required is computed using Eq. (9-5) based on a temperature rise of 15°F [8°C] that is quickly dissipated in the piping system by convection.

$$q = P / 2.95 \times c_p \times \text{SG} \tag{9-5}$$

where

q = minimum required flow rate (gpm) [L/s] through the pump that equals the pump pressure relief control valve rated flow.

P = power input (BHP) [kW] at minimum flow and head. The pump horsepower at cutoff can be obtained from the family of pump curves. However, an approximation is that the cutoff brake horsepower at 20% of design rpm will be only the losses in the pump and motor itself, amounting to less than 5% of the pump design input energy.

c_p = specific heat capacity (1 Btu/lb °F) [4.1868 kJ/kg °K].

SG = specific gravity, the ratio of density of water at a given temperature to the density of water is determined from Table 9-2.

9.3 Fans

Fans are devices designed to impart energy to the air in HVAC systems in order to effect the distribution of the air through the duct system. This energy is in the form of static pressure and kinetic energy.

9.3.1 Fan Energy Relations

It is useful to examine the operation of a fan from a thermodynamic standpoint. For the fan shown in Figure 9-14 in operation at steady-state conditions, applying the first law of thermodynamics stipulates that the rate of energy input equals the rate of energy output as defined by Eq. (9-6).

$$mh_1 + m\left(V_1^2/2g_c\right) + P_i = mh_2 + m\left(V_2^2/2g_c\right) \tag{9-6}$$

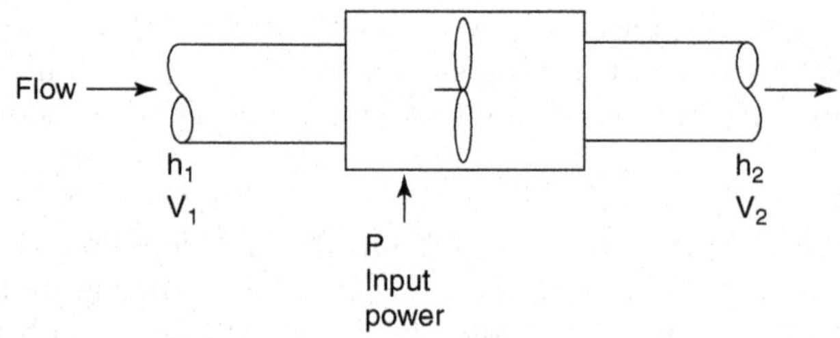

FIGURE 9-14
Schematic of fan energy relationships.

where P_i is input power to the fan expressed as a positive number, or

$$P_i = m(h_2 - h_1) + (m/2g_c)(V_2^2 - V_1^2) \quad (9\text{-}7)$$

and, since $h = u + P_v$,

$$P_i = m(u_2 - u_1) + m(P_2 v_2 - P_1 v_1) + (m/2g_c)(V_2^2 - V_1^2) \quad (9\text{-}8)$$

If the specific volume is taken to be constant at v, then $m = V/v$, and if $(u_2 - u_1)$ is taken to be $C_v (t_2 - t_1)$, and if appropriate conversion factors are used, the expression may be modified to

$$P_i = F_a(V_0/v)\, C_v (t_2 - t_1) + F_b V_0 (P_2 - P_1) + F_c (m/2vg_c)(V_2^2 - V_1^2) \quad (9\text{-}9)$$

Input power	=	internal energy difference due to friction	+	static pressure energy difference	+	kinetic energy difference

where
P_i = input power, bhp [kW]
V_0 = volumetric flow rate, cfm [L/s]
C_v = specific heat of air at constant volume, 0.171 Btu/(lbm-°F) [0.716 kJ/kg °K]
t = inlet or outlet air temperature, °F [°C]
v = specific volume of air, cf/lbm [m³/kg]
g_c = conversion factor, 32.2 ft/(s²) [9.8 m/s²]
P = static pressure, in. water [249 Pa]
V = velocity of air, ft/s [0.3 m/s]
F_a = conversion factor, 1 bhp/(42.4 Btu/min)
F_b = conversion factor, 1 bhp/(6,346 cfm/in. wg)
F_c = conversion factor, 1 bhp/(33,000 ft-lbf/min)

In Eqs. (9-8) and (9-9), the internal energy due to friction represents the losses inside the fan. If the fan were 100% efficient, all the input power would be used either to increase the static pressure or to increase the kinetic energy (velocity) of the output airflow. Any actual fan, of course, has friction losses that tend to increase the temperature of the outlet air. We may, therefore, define the efficiency of the fan as follows:

$$E = 1/P_i \times V_0[F_b(P_2 - P_1) + F_c(V_2^2 - V_1^2)/(2vg_c)] \tag{9-10}$$

where E is the efficiency of the fan and other variables are defined as before. If the velocity term is omitted in Eq. (9-10), the resulting efficiency is defined as the *static efficiency* defined by Eq. (9-11).

$$E(\text{static}) = 1/P_i \times [V_0 F_b(P_2 - P_1)] \tag{9-11}$$

Typical fan efficiencies are shown in Table 9-4.

9.3.2 Fan Types and Characteristics

The most general types of fans are centrifugal and axial flow. In the *centrifugal fan*, the air enters through the center portion of an impeller wheel and exits through a scroll and outlet at right angles to the inlet path.

Figures 9-15 and 9-16 illustrate the two types of centrifugal fans.

In the *double-width, double-inlet* (DWDI) centrifugal fan of Figure 9-15, the air enters from both sides, as opposed to the *single-width, single-inlet* (SWSI) fan of Figure 9-16. The impeller blades may be inclined or curved in the direction of the flow, in the opposite direction to the flow, or radially, as shown in Figure 9-17.

Centrifugal fans are widely used because they are capable of moving air over a wide range of flow rates and pressures. Efficiencies for fans using inclined, airfoil-shaped blades are relatively high, and these fans are generally less noisy.

Two particular characteristic differences between *backward-inclined* and *forward-curved* centrifugal fan blades should be noted. First, the mechanical efficiency curves for backward-inclined blades peak at a somewhat higher efficiency and at higher fractions of maximum flow rates than do the forward-curved fans. Second, the power input continues to increase through full load power as the flow rate is increased to maximum capacity in the forward-curved fan, whereas in the backward-curved blade fan, the power input peaks at between 60% and 80% of flow capacity. The practical consequence of this difference is that the forward-curved blade fan may overload its motor if allowed to operate without restriction, whereas the backward-curved blade fan is self-limiting.

The shaded area of Figure 9-18 represents the normal selection range for a selection of a centrifugal fan.

TABLE 9-4

Typical Fan Efficiency

Fan Type	Typical Efficiency (%)
Centrifugal, airfoil blade	88
Centrifugal, backward curved blade	84
Centrifugal, backward inclined blade	80
Centrifugal, forward curved	70
Vaneaxial	86
Tubeaxial	75

FIGURE 9-15
DWDI centrifugal fan. (Widely circulated with a Creative Commons license.)

FIGURE 9-16
SWSI centrifugal fan. (Widely circulated with a Creative Commons license.)

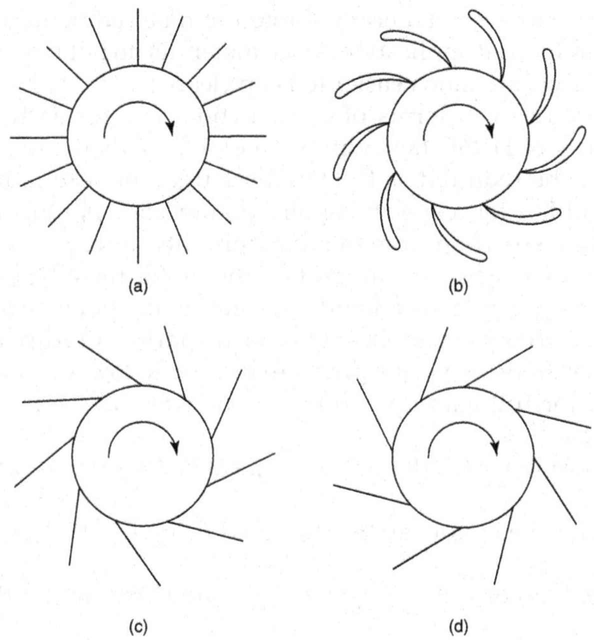

FIGURE 9-17
Centrifugal fan blade configurations.

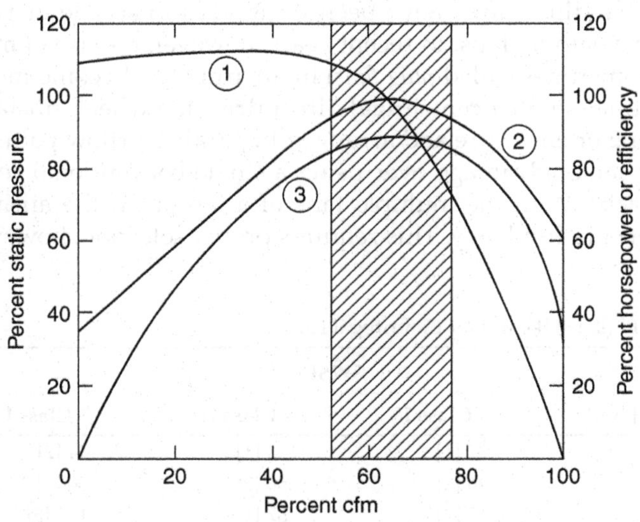

FIGURE 9-18
Typical performance characteristics of centrifugal fans.

Utilizing manufacturers' published fan curves, the designer should select a fan for the required operating condition at the highest point of fan efficiency, thus minimizing fan losses and energy requirements.

Except for fancoil or small blower coil units, designers should always use backward-curved or backward-inclined blade centrifugal fans to maximize energy efficiency. For fans requiring 4,000 cfm or more, *backward-inclined airfoil* blade centrifugal fans are recommended.

Radial blade configurations are generally limited to material handling, such as sawdust collection and other industrial application, since materials do not tend to accumulate at the blade root and the wheels are much easier to keep clean.

Centrifugal fans are rated in terms of construction *class* for each range of pressures. While the vast majority of HVAC fans will be Class I fans, there are applications where a Class II or III fan may be required as the fan discharge pressure rating increases. When selecting a centrifugal fan for a specific application, designers must always confirm the required class of fan construction from manufacturer literature.

When selecting a motor for a centrifugal fan, the motor must be capable of both driving the fan at its rated operating speed and accelerating the fan wheel to that speed upon start-up. For some centrifugal fan applications, particularly for return air fans in large air-handling systems, the fan design static pressure may be so low, yet the fan wheel so large, that the motor sized for full load operation may be *undersized* for overcoming the fan's initial *moment of inertia.*

Moments of inertia, WR^2 (lb ft^2) [kg m^2], for typical HVAC centrifugal fans can be determined from Table 9-5.

The WR^2 required for the motor can be computed using Eq. (9-12).

$$\text{Required motor WR}^2 = 1.1 \times \text{fan WR}^2\ (\text{fan RPM / motor RPM})^2 \tag{9-12}$$

Motor torque and starting inertia ratings are given Table 9-5.

The data of Table 9-6 are based on use of a 1,800 rpm motor. Correction factors for other motor speeds can be computed as 1,800/motor rpm.

Historically, air-handling units have been designed to use one or two (parallel) conventional centrifugal supply fans. In recent years, however, the use of multiple unhoused centrifugal fans (sometimes called *plug fans*) in an array has become more common. This *fan array* uses multiple, smaller centrifugal direct-drive fan wheels, installed with an inlet bell without a casing or enclosure and arranged in parallel airflow paths.

A fan array has some advantage over using a single conventional centrifugal fan, but it also has drawbacks. With the multiple fans of a fan array, the air-handling unit can provide some level of redundancy. This requires proper selection, however, to ensure that

TABLE 9-5
Moments of Inertia, WR^2 (lb ft^2) [kg m^2] for Centrifugal Fans

	SWSI		DWDI	
Wheel Diameter (in.) [mm]	**Class I**	**Class II**	**Class I**	**Class II**
24 [610]	32 [1]	34 [1]	44 [2]	49 [2]
27 [686]	46 [2]	50 [2]	66 [3]	79 [3]
30 [762]	72 [3]	87 [4]	105 [5]	136 [6]
33 [838]	103 [4]	125 [5]	152 [6]	198 [8]
36 [914]	179 [8]	200 [8]	260 [11]	303 [13]
40 [1,016]	300 [13]	333 [14]	435 [18]	505 [21]
44 [1,118]	434 [18]	541 [23]	642 [27]	863 [36]
49 [1,245]	663 [28]	821 [35]	971 [41]	1,310 [55]
54 [1,372]	1,120 [47]	1,350 [57]	1,760 [74]	2,120 [89]
60 [1,524]	1,920 [81]	2,250 [95]	2,850 [120]	3,480 [147]
66 [1,676]	3,000 [126]	3,200 [135]	4,650 [196]	5,070 [214]
73 [1,854]	4,480 [189]	4,780 [201]	6,950 [293]	7,570 [319]

TABLE 9-6

Motor Torque and Starting Inertia Ratings

Motor (hp) [kW]	Full Load Torque (ft-lbs) [J]	Maximum Starting Inertia, WR^2 ($lb\text{-}ft^2$) [kg m^2]
3 [2.2]	8.9 [12]	44 [1.85]
5 [3.7]	14.9 [20.2]	70 [2.95]
7-1/2 [5.5]	22.6 [30.6]	100 [4.71
10 [7.5]	30.1 [40.8]	150 [6.32]
15 [11]	44.7 [60.6]	200 [8.43]
20 [15]	59.7 [80.9]	250 [10.5]
25 [18]	74.4 [101]	300 [12.6]
30 [22]	89.5 [121]	350 [14.7]
40 [30]	118.6 [161]	425 [17.9]
50 [37]	149.0 [202]	525 [22.1]
60 [45]	178.0 [241]	650 [27.4]
75 [55]	223.0 [302]	800 [33.7]
100 [75]	297.0 [403]	1,025 [43.2]

when one fan fails, the remaining fans are able to provide the required airflow. *Typically, an array of three or four fans is able to provide the same level of redundancy as an array that consists of many, smaller fans.*

When considering the use of a fan array in an air-handling system, the designer needs to consider several factors, as follows:

1. When multiple smaller fans are used in an array, individual fans and motors are lighter and easier to replace, if necessary.
2. Using a single fan (rather than a fan array) will typically result in the lowest cost air-handling unit. When a fan array is desired, using fewer, larger fans will typically cost less than using many, smaller fans.
3. Using a single fan will typically require less input power than an array of multiple fans. When a fan array is desired, using fewer, larger fans will typically require less power than using many, smaller fans.
4. Since the fans are operating at a slower speed and larger fans are more efficient, the fewer the number of fans, typically the better the acoustics.
5. Fewer moving parts (and, therefore, fewer fans) generally results in better reliability. However, the redundancy provided by using multiple fans in an array may offset concerns over a single fan failing for critical applications.
6. The need for a backdraft damper for each fan in the array is required to prevent the short cycling of air in the event of a fan failure.

9.3.3 Fan and System Characteristics

The relationships between a fan and the duct system it serves are illustrated in Figures 9-19 through 9-21.

First, consider the static pressure required to produce a certain volumetric flow rate of air through a given duct system. As the volumetric flow rate increases, the static pressure

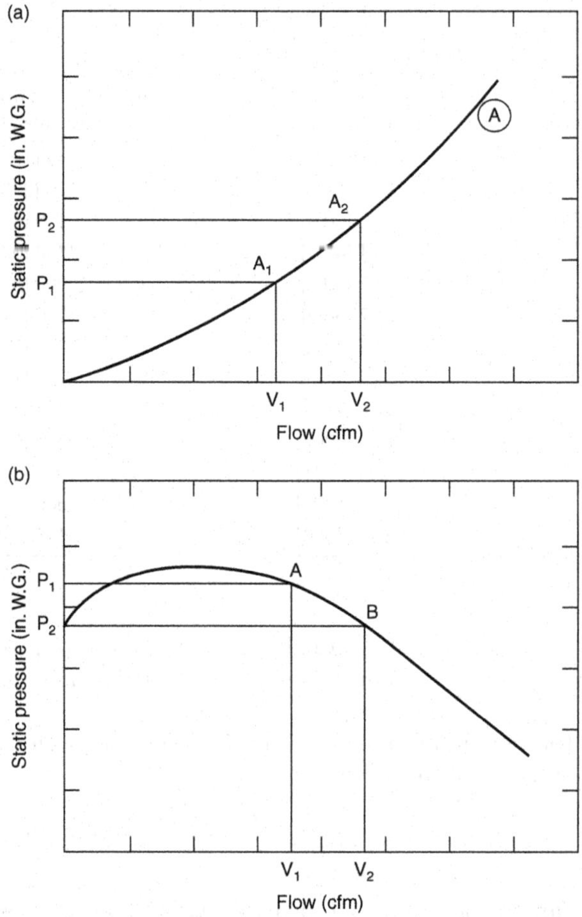

FIGURE 9-19
Duct system and fan system characteristics.

required also increases but increases approximately as the square of the increase in flow rate (i.e., $P_2/P_1 = V_2^2/V_1^2$).

Figure 9-19 illustrates changes in the system and fan characteristics. Any time the resistance of the duct system changes, a new system characteristic curve will be generated, whether the change is caused by a change in duct size, opening or closing a damper, or because a filter has become clogged. The same general shape of the curve will be retained, however. Similarly, if the fan speed of a given fan is changed, a new fan curve is generated. Obviously, the shape of the curve depends on the type of fan, but the general shape of the curve is the same for a particular fan at different speeds.

When a particular fan operating at a particular speed is placed in a particular duct system, the only stable operating point is the one that satisfies both the fan and the system characteristic curves (i.e., at the intersection of the two curves), as illustrated in Figure 9-21.

9.3.4 System Effects

In general, system effects on fan performance result from errors in estimating system losses or system changes and from deviations in the installed condition of the fan from

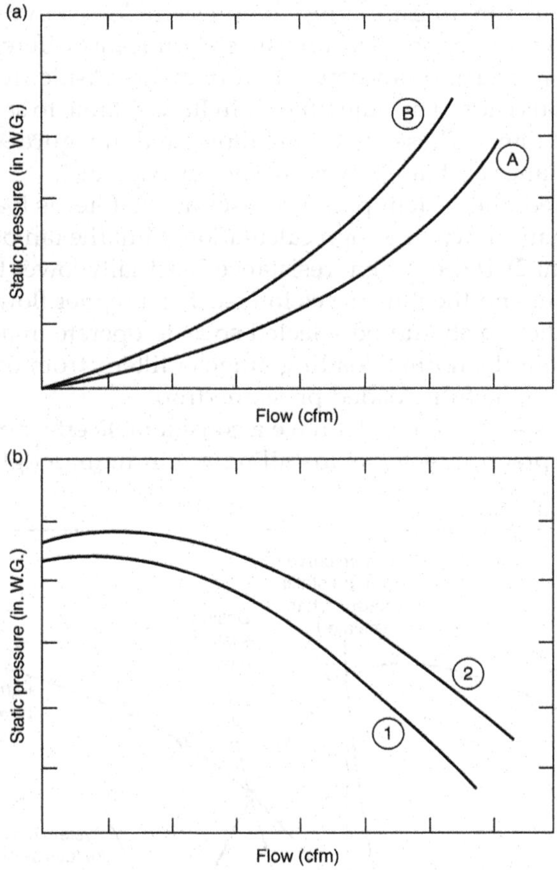

FIGURE 9-20
Changes in duct system and fan system characteristics.

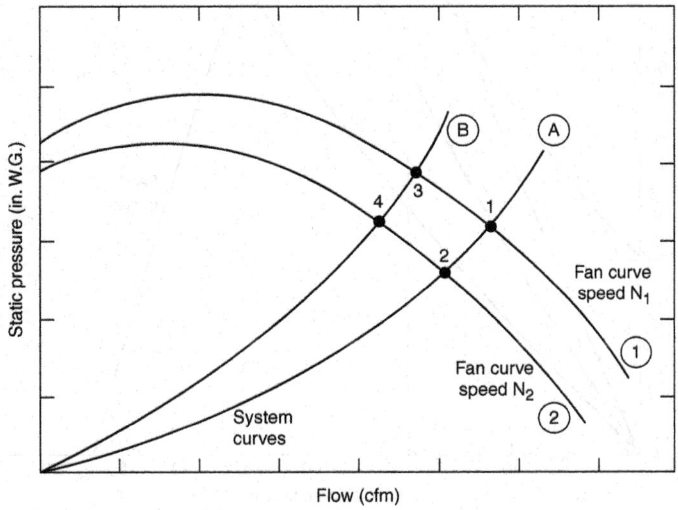

FIGURE 9-21
Fan and duct systems characteristic curves interposed to show determination of operating point.

ideal (test) conditions. Errors in estimating the pressure losses in duct systems or deviations in installed duct sizes or shapes can cause significant differences in actual versus expected values of flow rate and fan power requirements. Aside from these errors, there will always be variations caused by adjustments in flow volume to selected spaces and due to gradual clogging of filters. These effects are illustrated in Figure 9-22, where the design values of system resistance and fan power are shown as point 1.

As would be expected, the system pressure rises as the filter is loaded (or if the system has more resistance than shown in design calculations), but the fan power decreases as the flow rate falls off (point 2). If the system resistance is actually lower than shown in design calculations, the system and the fan curves intersect at a higher flow rate, requiring more fan power (point 3). The fan should be selected so as to operate in an acceptable range of performance throughout the normal loading range of filters, from the least pressure drop in the clean condition to the fully loaded pressure drop.

The duct connections at the fan outlet have a considerable effect on the performance of a fan. Recommended practice includes installing a minimum length of straight duct for

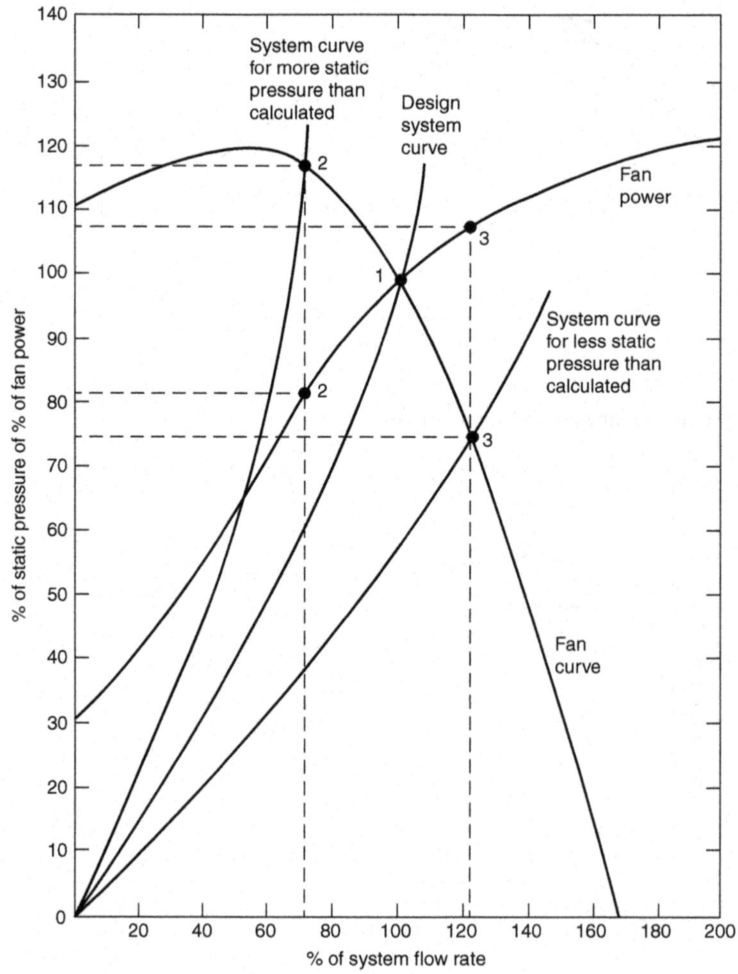

FIGURE 9-22
Effects of changes in system resistance on fan performance.

controlled diffusion between the fan outlet and the first elbow, branch, or damper. The Air Moving and Control Association, Inc. publishes system effect factors, which are additional pressure losses beyond those for the fan under ideal, or as tested, conditions for various fan outlet and inlet connections which may be necessary for particular installations. These factors are velocity dependent, increasing with the volumetric flow rate of the fan.

The 100% minimum effective straight duct length recommended for fan outlets is 2.5 *equivalent duct diameters* for velocities of 2,500 fpm [12.7 m/s] or less, with one additional duct diameter for each additional 1,000 fpm [5.1 m/s]. If the fan outlet duct is shorter than this and if the fan outlet area (to which the duct is connected) is greater than the fan blast area (the exit area created by the fan housing scroll cutoff), there will be losses due to poor diffusion to the larger duct. In effect, there will be less static regain because of the poor transition, and some of the kinetic energy due to the air velocity will be lost to heating of the air. For example, for an outlet velocity of 2,000 fpm [10.2 m/s] and a blast area-to-outlet area ratio of 60%, the static pressure loss penalty that must be added to the system losses varies from 0.1 in. wg [24.9 Pa] for a straight outlet duct of 25% of a recommended minimum length to 0.25 in. wg [62.3 Pa] for no straight duct at all.

The direction of outlet duct elbows also affects fan performance. Figure 9-23 shows various fan outlet connections rated with respect to losses.

The fan inlet conditions are also major factors affecting fan performance. For the fan to achieve rated performance, the air must enter the fan distributed uniformly over the inlet area and without prerotation. Best performance is achieved with smooth, rounded surfaces, such as with a *bellmouth* at the inlet. Use of turning vanes in elbows in inlet ducts near the inlet is recommended to avoid swirl in inlet air. Fan inlets should be at least 0.6 wheel diameter from the walls of inlet plenums, and there should be 1.2 diameters between inlets of adjacent tandem fans.

It is important to note that the results of poor inlet conditions are not simply increases in system resistance but are equivalent to changes in the fan performance curves themselves and may severely degrade fan performance.

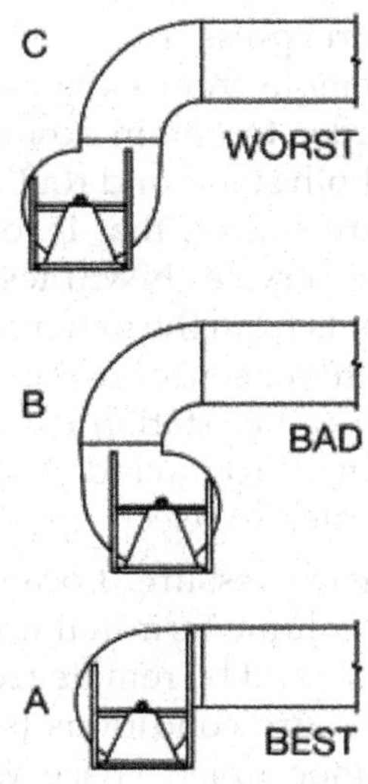

FIGURE 9-23
Outlet configurations for centrifugal fans. (Widely circulated with a Creative Commons license.)

9.3.5 Variable Air Volume Fan Control

In variable air volume (VAV) systems, the volumetric flow rate of a fan or fan array may be changed through the use of *speed control* through the use of a VFD. A change in the rotational speed of a centrifugal fan via a VFD results in a new *fan curve* and produces the greatest fan power reduction as the airflow rate is reduced.

The VAV supply air fan speed is typically controlled to maintain a fixed pressure condition at a control point normally about two-thirds of the downstream distance from the fan discharge. As VAV terminal units vary, the supply airflows are delivered to each conditioned space, the total airflow need will change accordingly, and the supply fan responds by slowing or speeding up to maintain the operating pressure in the duct system set point. For all practical purposes, then, the *system curve* is dictated by the pressure set point as a straight line over the entire airflow range.

Speed control of the return or relief fans is of more complex due to the need to maintain the minimum outdoor ventilation airflow to meet code requirements. The more common method of return/relief fan speed control, called *fan tracking*, can be implemented by either of the following methods:

1. Provide an airflow monitoring array in the outdoor air intake to continuously measure the ventilation airflow rate. Vary the speed of the return/relief fan as required to maintain the ventilation airflow rate at the required set point condition.
2. Provide an airflow monitoring array in both the supply fan inlet and the return/relief air inlet to continuously measure each fan's airflow rate. Vary the speed of the return/relief fan as required to maintain the difference between the supply airflow rate and the return/relief airflow rate at the required set point condition, equivalent to the minimum required outdoor airflow rate.

A less common method for controlling return or relief airflow is to monitor overall *building pressure* and to control return/relief fan speed based on building pressure at a fixed set point condition. To implement this method, the following is required:

1. Measure the outdoor barometric pressure. Locate the outdoor sensor at least 8 ft [1.5 m] above the roof and remote from exhaust fans and install the sensor in a wind-proof enclosure. In order to obtain accurate outdoor pressure measurements, the sensor needs to be both stable and routinely calibrated against a known reference atmospheric pressure source, that is, one that is professionally maintained. The National Weather Service (NWS) tests their pressure gauges annually using calibrated instruments from the national pressure standards laboratory. Routine recalibration of the outdoor sensor using the air pressure measurement of the nearest NWS automated weather station is required. To prevent fluctuations in outdoor air pressure attributed from wind gusts "exciting" the control system, digital filters must be incorporated to reject input spikes.
2. Measure the indoor barometric pressure. Locate the indoor sensor in interior space served with a constant volume terminal unit, with no direct exhaust from the space. The selected space should be remote from elevators, stair towers, adjacent spaces with variable pressure conditions (such a toilets and laboratories), and, ideally, should be an unoccupied space with minimum door opening/closing.

3. Average the monitored barometric pressures over a sliding 5-min period to establish differential pressure between the indoor and outdoor values to "smooth" control of return/relief air fan speed to maintain a set point of 0.05 in. wg [12.44 Pa] differential pressure.

Obviously, the building pressure control method is more complex and much more prone to error than the fan tracking method and this method is not recommended.

9.4 Pump and Fan Affinity Laws

Approximate relationships for assessing effects of changes in fan or pump speed, volumetric flow rate, pressure, and power are known as the *affinity laws*. There are three essential laws, with many variations for specific cases, as follows:

1. Volumetric flow rate, V, is directly proportional to the rotational speed N, in rpm:

$$V_2 = V_1 \times (N_2 / N_1) \tag{9-13}$$

2. Pressure, P, is proportional to the square of the rotational speed.

$$P_2 = P_1 \times (N_2 / N_1)^2 \tag{9-14}$$

3. Power, P_i, is proportional to the cube of the rotational speed.

$$P_{i2} = P_{i1} \times (N_2 / N_1)^3 \tag{9-15}$$

The affinity laws can be used to make an estimate of the performance of any pump or fan under different speeds.

Bibliography

ASHRAE Handbook-HVAC Systems and Equipment, American Society of Heating, Refrigerating, and Air-Conditioning Engineers, Inc. (Atlanta, GA, 2016).

Training manuals TEH 1109A, *Parallel and Series Pump Application*; TEH 775A, *Primary Secondary Pumping Application*; TEH 910A, *Variable Primary Flow Systems*; and TEH 1166B, *Principals of Centrifugal Pump Construction*, Bell and Gossett/Xylem Inc. (available online at http://bellgossett.com/training-education/training-manuals/).

"Part 2 - Air Distribution," *Carrier System Design Manual*, Carrier Corporation (Syracuse, NY, 1960).

10

Terminal Systems and Components

10.1 Systems Types and Applications

10.1.1 Selecting the Right System for the Application

In Chapter 3, we introduced the basic *secondary* or terminal system types utilized in building energy systems. The next step (beyond simply understanding how each system functions) is to understand those aspects about each system that in any given design situation will allow the designer to select the right system for the application.

Table 10-1 shows the basic design aspects for each type of secondary system. "Typical applications" defines the type(s) of occupancy or heating, ventilating, and air-conditioning (HVAC) conditions that are most readily satisfied by each type of system, based on (1) the design conditions, load conditions, and load variations imposed on the system and (2) the ability of the system to respond to and satisfy these conditions. "Control tolerance" defines the ability of the system to satisfy a wide range of design indoor temperature and humidity conditions. "Response to load variations" is a measure of the system's ability to maintain space design temperature and humidity set points as heat gains/losses in the space change. This is a function of both the time and the degree of response. "Maintenance requirements" defines the amount (e.g., cost) to maintain the system, considering both preventive and breakdown maintenance. "Maintenance sophistication" is a measure of the requirements for maintenance resources, both in personnel and in spares. It addresses the complexity of the system and the burden imposed on the operating personnel to maintain proper performance from the system.

These data can be used as a "reference point" for the designer when attempting to select a secondary system(s) for a particular application. Coupled with the "systems design rules" from Chapter 4, the data of Table 10-1 should enable the designer to select systems resulting in a successful design.

10.1.2 Space Needs and Other Requirements for Terminal Systems

While smaller buildings may use outdoor rooftop air-handling units (AHUs), larger buildings will require "fan rooms" to house this equipment. The fan room dimensions are based on the size of the AHU. From Table 10-2, the length (L) and width (W) of a typical AHU required to serve a given building area can be obtained. The overall room dimensions, however, must be increased to allow for service access, additional support equipment, and so on. Therefore, the typical room dimensions can be established as follows:

$$\text{Room length} = 1.1 \times L$$

$$\text{Room width} = 2.5 \times W$$

$$\text{Room height} = 12 - 14\,\text{ft}\left[4\,\text{m}\right](\text{minimum to underside of structure})$$

To obtain L and W for an AHU from Table 10-2, it is necessary to adjust the actual floor area served by the unit by a factor that represents the typical supply airflow ratio between office buildings (the basis for Table 10-2) and other types of buildings. To obtain the equivalent

TABLE 10-1

Typical Secondary System Applications

System Type	Typical Applications	Control Tolerance	Response to Load Variation	Maintenance	
				Requirements	Sophistication
Single zone (including packaged DX systems)	Restaurants, shops, assembly areas, and other single-zone applications, including any one zone in a multiple-zone building	Good (fair for packaged systems)	Good (fair for packaged systems)	Low	Average
Multi-zone (bypass)	Small office buildings, classroom buildings, perimeter zones of larger office buildings	Good	Good	Low–moderate	Average
Variable volume (cooling only)	Core zones of buildings	Good	Good (cooling only)	Low–moderate	Above average
Variable volume terminal reheat	Office buildings, hospitals, laboratories, computer rooms	Excellent	Excellent	Moderate	Above Average
Variable volume fan box (series)	Perimeter zones in office buildings	Very good	Excellent	Moderate	Above average
Dual duct (sequenced air)	Office buildings, hospitals, classroom buildings	Good	Excellent	Moderate	Above average
Terminal reheat (constant volume)	Hospital and laboratory spaces requiring positive room pressures	Excellent	Excellent	Moderate	Average
Incremental units	Hotel/motel rooms, nursing home patient rooms, apartments, dormitories	Fair	Fair	High	Below average
Fan coil units (four pipes)	Hotel/motel rooms, hospital or nursing home patient rooms, apartments, classrooms, dormitories	Good	Fair	Moderate	Below average
Water source heat pump	Classroom buildings, nursing homes, small office buildings, apartments, large homes	Good	Good	High	Average

TABLE 10-2

Representative AHU Dimensions

Building Area Served (SF) [m²]	*L* = Unit Length (Ft) [m]	*W* = Unit Width (Ft) [m]
1,000–1,600 [93–149]	19.7 [6.0]	2.6 [0.8]
1,601–2,900 [150–269]	20.8 [6.3]	3.7 [1.1]
2,901–3,800 [270–353]	22.3[6.8]	4.0 [1.2]
3,801–4,800 [354–446]	22.5 [6.9]	5.0 [1.5]
4,801–6,000 [447–557]	22.8 [7.0]	5.3 [1.6]
6,001–7,000 [558–650]	23.0 [7.0]	5.7 [1.7]
7,001–8,400 [651–780]	23.3 [7.1]	6.2 [1.9]
8,401–10,400 [781–966]	23.5 [7.2]	6.3 [1.9]
10,401–12,200 [967–1,133]	26.3 [8.0]	6.5 [2.0]
12,201–14,500 [1,134–1,347]	26.3 [8.0]	7.6 [2.3]
14,501–17,000 [1,348–1,579]	30.0 [9.0]	8.0 [2.4]
17,001–19,600 [1,580–1,820]	30.0 [9.0]	9.1 [2.8]
19,601–25,000 [1,821–2,323]	30.5 [9.3]	10.1 [3.1]
25,001–33,000 [2,324–3,066]	34.9 [10.6]	11.4 [3.5]
33,001–39,000 [3,067–3,623]	37.7 [11.5]	11.4 [3.5]
39,001–50,000 [3,624–4,645]	40.5 [12.3]	12.7 [3.9]

building area, multiply the actual area by the appropriate factor from Table 10-3 and use this equivalent area to obtain the dimensions from Table 10-2.

Table 10-2 is based on the following data and assumptions:

1. Unit dimensions are based on blow-through AHU configuration with a mixing plenum, cartridge filter section (with prefilter), heating coil with access section, fan section, fan discharge diffuser, cooling coil with access section, and discharge plenum.
2. Add 10% to *L* if a preheat coil is required.
3. Subtract 20% from *L* if the unit is variable air volume (VAV) cooling-only type.

The designer must ensure that *flood control* is provided in all fan rooms to avoid water damage in the event drain pan overflows, piping leaks, and so on. All floor openings should be protected with 2″–4″ [50–100 mm] high curbs or pipe sleeves to "dam" loose water on the floor. Floor drains must be provided and the floors sloped to these drains near each AHU.

Sufficient ceiling space for HVAC air distribution (ductwork); HVAC, plumbing, and fire protection piping; and electrical raceway is required. These space requirements are in addition to the minimum space required for the ceiling itself and recessed lighting fixtures.

Table 10-4 reflects typical minimum commercial and institutional building ceiling space requirements when air secondary systems are used:

Thus, the minimum floor-to-floor height in a multistory building is 44″ [1,100 mm] + floor-to-ceiling height + structure depth.

When water-type secondary systems are used, the ductwork space requirement can be eliminated in single-story buildings and reduced to only 12″ in multistory buildings.

The space requirements listed previously must be increased for special occupancies such as laboratories, hospitals, and kitchens because of the much higher "density" of distribution systems in these types of facilities.

TABLE 10-3

Adjustment to Building Area in Table 10-2

Building Type	Adjustment Factor
Residential	
Single family housing	1.2
Condominiums and apartments (high rise)	1.2
Retail facilities	
Strip shopping centers	0.8
Regional shopping centers	0.9
Discount and outlet stores	1.0
Supermarkets	0.8
Department stores	0.9
Multiple use complexes	1.0
Commercial and public buildings	
Correctional facilities	1.0
Court houses	0.8
Office buildings	1.0
Dining and entertainment facilities	0.7
Libraries and museums	0.9
Warehouses	1.4
Assembly buildings	
Houses of worship	1.0
Auditoriums	0.7
Arenas and stadiums	0.7
Convention and exhibition centers	0.7
Community centers	0.7
Motels, hotels, and dormitories	
Hotels and motels	1.2
Dormitories	1.2
Nursing homes	1.2
Education facilities	
Preschools and elementary schools	1.2
Middle schools	1.1
High schools	1.0
Colleges and universities (classrooms, admin)	0.9
Colleges and universities (laboratories)	0.8
Gymnasiums	1.0
Health care facilities	
Hospitals	0.7
Outpatient facilities	0.9
Research and development laboratories	0.5

The architect must evaluate ceiling space requirements and coordinate between the plumbing, fire suppression, HVAC, and electrical systems to establish a reasonable clear cavity height and to prevent the common (and embarrassing) need to lower ceilings during the construction phase to provide adequate space. *While building information modeling (see Chapter 16) has gone a long way toward avoiding this type of problem, it still exists.*

TABLE 10-4

Minimum Ceiling Cavity Space Requirements

Lighting	8″ [200 mm]	6″ [150 mm] for fixture plus 2″ [50 mm] to lift and move fixture above the ceiling.
Ductwork	24″ [600 mm]	Ducts with high aspect ratio, width to height, increase installation costs.
Piping and Electrical Raceway	12″ [300 mm]	May increase to incorporate fire protection and/or roof drainage systems.
Total	44″ [1100 mm]	

There are two types of building exhaust that may be present: general exhaust from toilets, copy rooms, and so on and kitchen exhaust. Each of these exhausts have different requirements:

1. General exhaust fans are typically low profile fans that discharge downward against the roof surface and can be located anywhere convenient so long that the discharge is at least 10 ft [3 m] away from any ventilation air intake.
2. Kitchen exhaust fans are upblast fans designed to discharge the grease-laden air from range hoods and dishwashers at 2,000 fpm [10 m/s] or higher exit velocity. These fans are typically large and may be 5 ft [1.5 m] or greater in height. Again, the discharge cannot be located any closer than 10 ft [3 m] from any ventilation air intake.

In multistory buildings where AHUs are located on the roof or in upper-story fan rooms, vertical piping and duct risers must be routed in fire-rated chases. The chases are normally required by building codes to be constructed for a 1-h fire rating for three stories of less in height and for a 2-h rating if greater than three stories. Because the chase walls are rated, fire dampers or combination fire/smoke dampers must be installed in the ductwork penetrations of these walls. Thus, access through the adjacent ceilings must be provided for inspection and service of the fire dampers.

Rectangular chases are more useful than square chases and at least three sides of the chase must be available for duct and pipe egress to and from the adjacent ceiling cavity. Care must be taken to avoid placing chases adjacent to stairs and/or elevators where these vertical elements effectively block access to the chase.

The amount of chase area will vary with the type of secondary system and the area of the air-conditioned building that must be served from the chase. Air systems require space for supply and return duct, exhaust duct, and piping. Water systems require space only for exhaust duct and piping.

10.2 AHUs and Components

10.2.1 AHUs

AHU is the general term for several types air-treating and delivery systems, each of which consists of one or more supply fans, cooling and/or heating coils, air filters, and so on, all contained within a common enclosure or "casing" and factory-assembled.

1. *Blower coil units* (BCUs) are factory-fabricated "light commercial" duty AHUs typically used for smaller (up to about 4,000 cfm [2,000 L/s]) systems applied in schools, offices, retail, or other less critical; comfort air-conditioning applications; and/or where cost is an issue. These units are fabricated with 1" thick, double wall casings. For higher capacity and better quality, use commercial-grade modular AHUs.
2. *"Modular"* AHUs are factory-fabricated "standard commercial" duty AHUs rated for up to 60,000 cfm [28,000 L/s] supply airflow and fabricated with 2" [50 mm] thick double-wall casings. For higher airflow capacity, or where applications are more critical (e.g., hospitals, laboratories, and so on), utilize custom AHUs.
3. *"Custom"* AHUs are available with a capacity of 5,000 cfm [2,400 L/s] to 140,000 cfm [66,000 L/s] (or greater) and are "one-off" fabricated to meet specific project capacity, dimensions, and/or materials requirements. These units are *required* for air systems exceeding about 60,000 cfm [28,000 L/s] and are *recommended* for critical applications (hospitals, laboratories, and so on) with an airflow requirement of 5,000 cfm [2,400 L/s] or greater.

Field-erected, custom AHUs are applied in industrial applications and for large facilities to deliver airflows up to 500,000 cfm [240,000 L/s]. However, the use of field-erected units today is very rare, even in industrial applications, due to their higher cost.

BCUs are typically available with only forward curved fans up to about 1,000 cfm [470 L/s] but should be specified to have backward inclined centrifugal fans in larger sizes. AHUs, modular, custom, or field-fabricated, should always be specified to utilize backward included, airfoil blade centrifugal fans to maximize efficiency.

The following design considerations must be addressed for <u>any</u> AHU:

1. If the *minimum* ventilation airflow results in a mixed air temperature 40°F [4°C] or lower at the winter outdoor design temperature, provide a preheat coil upstream of the cooling coil (see Chapter 6). Table 10-5 may provide some guidance to when preheating is required.

   Also, with every VAV system, the mixed air temperature can drop to 40°F [4°C] or less when the supply airflow is reduced due to lower cooling load requirements when the outdoor air temperature is below 40°F [4°C]. This condition will almost

TABLE 10-5

Preheat Coil Requirement

Percent Outdoor Air	Winter Outdoor Design Temperature at Which a Preheat Coil Is Required (at 70°F [21°C] Return Air Temperature)
100	≤40°F [≤4°C]
90	≤38°F [≤3°C]
80	≤35°F [≤2°C]
70	≤30°F [≤–1°C]
60	≤22°F [≤–6°C]
50	≤10°F [≤–12°C]
40	≤–5°F [≤–20°C]
30	≤–30°F[≤–34°C]
20 or less	Never

always occur, since the percentage of outdoor air steadily increases as the supply airflow rate and corresponding return airflow rate decreases. *Thus, it is prudent to include an outdoor air preheat coil as part of any VAV system.*

2. The *blow-through* AHU configuration is preferred for units exceeding 5,000 cfm [2,400 L/s] airflow for the following reasons:
   a. Since the fan is located upstream of the coil, the *fan heat* contribution is added to the entering air condition of the coil and does not affect the required leaving air condition. This, in turn, means that the chilled water supply temperature does not have to be lowered to compensate any downstream fan heat temperature rise in the supply air temperature.
   b. The fan is not connected directly to the duct system. Having the cooling coil downstream of the coil can provide some noise attenuation, reducing potential for transferring fan noise to occupied spaces. Additionally, the fan is isolated within the AHU and vibration problems can be effectively addressed.
   c. Changes over time may require that additional airflow and unit cooling capacity be provided. Since the fan heat release goes up with the cube of airflow, increasing airflow in a draw-thru configuration can significantly increase the unit's leaving air temperature. Since the chilled water supply temperature cannot normally be lowered significantly, this increased discharge temperature will offset the increased airflow, effectively eliminating the possibility of increasing the unit's cooling capacity.
3. In a hospital or laboratory where a high-efficiency final filter is installed at or near the discharge of the AHU, a blow-through AHU configuration may result in moisture condensing on the filter, rendering it useless.

   Typically, high efficiency filters have a pressure drop of 0.5–2.0 in. wg [125–500 Pa]. Following the perfect gas law $PV = RT$ (where P = pressure, V = volume, T = temperature, and R = universal gas constant), the reduction in pressure as the air passes over a high pressure drop filter will also result in a reduction in temperature (approximately 0.5 F [0.3°C] for each 1.0 in wg [250 Pa] of pressure change). Since the air leaving the cooling coil is already saturated, any further reduction in temperature will result in condensation, depositing moisture on the filter.

   By inserting the fan between the cooling coil and the final filter in a *draw-through* AHU configuration, the fan heat raises the coil discharge air to a non-saturated condition, eliminating the condensation problem within the final filter. Note, however, that care must be taken in this configuration to prevent high velocity impingement on the final filters.
4. If an airside economizer cycle is used, *always* utilize a return or relief fan ... *barometric relief does not work* unless the relief dampers are located within the area served, which is rarely possible except for gymnasiums, manufacturing plants, and so on.
5. Select modular and custom AHUs with access sections upstream and downstream of each major component (filter section, each coil section, fan(s), and so on).
6. For fans and coils in AHUs installed at an elevation exceeding 2,000 ft [600 m] above sea level, their performance must be corrected for elevation (see Chapter 14).
7. Rooftop AHUs:
   a. Rooftop AHUs are not recommended for use in high wind or hurricane areas or in locations where winter snow and ice load are heavy.

b. Select either *full curbs* or *structural steel stands* for installation. If structural steel stands are selected, the steel must be galvanized and the stand designed with a work platform the full length of the unit on at least one side, complete with handrails and steps. Also, the stand must provide sufficient elevation to allow for roof maintenance below the rooftop unit.
c. Ensure that the roof design incorporates walk pads to the equipment and, for curb-mounted equipment, walk pads completely around the unit to prevent damage during maintenance operations.
d. Minimize exposed piping on the roof to serve the AHUs. Where rooftop piping must be used, make sure the piping is sufficiently elevated to allow for roof maintenance.

For modular air-handling units, it is necessary for the designer to select each individual section and component of the air handler: preheat coil, filter(s), heating coil, cooling coil, fan (capacity, size, motor size, and location), type and location of dampers, and so on, so that a complete unit is defined for the manufacturer. The manufacturers' catalogs typically provide a series of standard sections that the designer "assembles" to obtain the AHU configuration and performance that are required. Consult the manufacturer's product options in the selection of factory-fabricated components that can be assembled to create an AHU for a specific need.

Factory-fabricated AHUs impose three limitations on the designer:

1. Because the units are assembled from standard sections with fixed dimensions, equipment spaces must "fit" the unit rather than the unit fitting the space.
2. Standardized manufacturing limits, to some extent, the types of components, particularly fans and filters, available for use.

For large capacity requirements, then, it may become cost-effective for the designer to design an AHU to be field-fabricated by selecting each individual component and arranging these components within a field-erected housing (casing) to suit the specific project needs. Again, units may be basically horizontal or vertical, draw-through or blow-through. But, the fan(s), coil(s), and filters can be arranged in an infinite variety to satisfy specific constraints of space, duct routing, and so on.

Whereas the manufacturers test and guarantee the performance of their factory-fabricated units, there are no guarantees (beyond the designer's own ability) for the performance of field-fabricated units, even though there are adequate test data and guarantees associated with each component. Thus, whereas filters and coils rarely present problems, there are large numbers of field-fabricated units whose fan performance is below expectations.

As discussed in Chapter 9, fans are rated for a specific set of conditions, normally an unobstructed (or "free") inlet and an outlet with a minimum of 2.5 diameters of straight duct. Thus, the conditions for fans installed with the air-handling plenum, particularly if multiple fans are utilized, will rarely match the rating conditions, and the fan performance will be negatively affected. To account for the effects of plenums and multiple fans, Air Movement and Control Association (AMCA) has developed data that allow the determination of the impact of fan spacing arrangement on fan performance, defined as "system effects."

Based on these data, and knowing the specifics of the fan installation, the additional static pressure loss imposed on the fan(s) can be determined. Failure to account for the plenum system effect will result in selecting a fan for a static pressure loss less than what the fan will actually encounter.

10.2.2 Particulate Air Filtration

Particulate air filters (*air cleaners*) are applied in air-handling systems to remove impurities such as dust, pollen, and smoke from the air. American Society of Heating, Refrigerating, and Air-Conditioning Engineers (ASHRAE) *Standard 52.2* defines a methodology of rating and applying filters in HVAC systems by establishing *minimum efficiency reporting values* (MERVs) for filters and a test method for determining the MERV. Table 10-6 summarizes the application guidelines for each type of filter and establishes a range of recommended MERVs. This table can be used by the designer to select and specify air cleaner performance required for most HVAC applications.

Electrostatic filters in Table 10-6 are ranked at the low end of the MERV scale, despite claims by the manufacturers that this type of filter is both effective and reliable. The standard recognizes that these claims are, at best, exaggerated and the filters do not perform well or reliably in HVAC systems, often due to poor maintenance.

TABLE 10-6

Particulate Filter Application Guidelines

	Approximate Standard 52.1 Results		Application Guidelines		
Standard 52.2 MERV	**Duct Spot Efficiency**	**Arrestance**	**Typical Controlled Contaminant**	**Typical Applications and Limitations**	**Typical Air Filter/Cleaner Type**
20	N/A	N/A	**0.30 μm particle or smaller**	Clean rooms	**HEPA/ULPA filters**
19	N/A	N/A	Virus (unattached)	Radioactive materials	≥99.999% efficiency on 0.10–0.20 μm particles, IEST Type F
18	N/A	N/A	Carbon dust	Pharmaceutical manufacturing	≥99.999% efficiency on 0.30 μm particles, IEST Type D
17	N/A	N/A	Sea salt	Carcinogenic materials	≥99.99% efficiency on 0.30 μm particles, IEST Type C
			All combustion smoke Radon progeny	Orthopedic surgery	≥99.97% efficiency on 0.30 μm particles, IEST Type A
16	N/A	N/A	**0.30–1.0 μm particle**	Hospital inpatient care	**Bag filters** Nonsupported (flexible) microfine fiberglass or synthetic media. 300–900 mm (12–36 in.) deep, 6–12 pockets.
15	>95%	N/A	All bacteria	General surgery	
14	90%–95%	>98%	Most tobacco smoke	Smoking lounges	
13	80%–90%	>98%	Droplet nuclei (sneeze) Cooking oil Most smoke Insecticide dust Copier toner Most face powder Most paint pigments	Superior commercial buildings	

(Continued)

TABLE 10-6 (*Continued*)

Particulate Filter Application Guidelines

	Approximate Standard 52.1 Results		Application Guidelines		
Standard 52.2 MERV	**Duct Spot Efficiency**	**Arrestance**	**Typical Controlled Contaminant**	**Typical Applications and Limitations**	**Typical Air Filter/Cleaner Type**
12	70%–75%	>95%	**1.0–3.0 μm particle**	Superior residential	**Box filters** Rigid style cartridge filters. 150–300 mm (6–12 in.) deep may use lofted (air laid) or paper (wet laid) media.
11	60%–65%	>95%	Legionella	Better commercial buildings	
10	50%–55%	>95%	Humidifier dust	Hospital laboratories	
9	40%–45%	>90%	Lead dust Milled flour Coal dust Auto emissions Nebulizer drops Welding fumes	Schools	
8	30%–35%	>90%	**3.0–10.0 μm particle**	Commercial buildings	**Pleated filters** Disposable, extended surface, 25–125 mm (1–5 in.) thick with cotton–polyester blend media, cardboard frame.
7	25%–30%	>90%	Mold	Better residential	**Cartridge filters** Graded density viscous-coated cube or pocket filters, synthetic media
6	<20%	85%–90%	Spores	Industrial workplaces	**Throwaway** Disposable synthetic media panel filters
5	<20%	80%–85%	Hair spray Fabric protector Dusting aids Cement dust Pudding mix Snuff Powdered milk	Paint booth inlet air	
4	<20%	75%–80%	**>10.0 μm particle**	Minimum filtration	**Throwaway** Disposable fiberglass or synthetic panel filters
3	<20%	70%–75%	Pollen	Residential	**Washable** Aluminum mesh, latex coated animal hair, or foam rubber panel filters
2	<20%	65%–70%	Spanish moss	Window air conditioners	**Electrostatic** Self charging (passive) woven polycarbonate panel filter
1	<20%	<65%	Dust mites Sanding dust Spray paint dust Textile fibers Carpet fibers		

TABLE 10-7

Typical Initial Static Pressure Loss at 500 fpm [2.5 m/s] Face Velocity

Filter Efficiency (MERV)	Initial Static Pressure Loss (in. wg) [Pa]
17–20	0.50–0.80
13–16	0.40–0.60
9–12	0.25–0.30
6–8	0.15–0.20

ASHRAE *Standard 62.1* requires a filter MERV not be less than 6 (and that filters be installed upstream of the cooling coils, which conflicts with the final filter requirements for some hospital and laboratory applications). *But, in general, to reduce particulates to acceptable levels, analysis by Lawrence Berkeley National Laboratory (LBNL) indicates that filters with an MERV of 9 are the minimum required. This same analysis indicates that, for comfort air-conditioning applications, filters with an MERV greater than 13 provide only minor improvement in indoor air quality (IAQ).* However, higher MERV levels may be required by regulatory bodies for special applications such as some hospital areas, research facilities, and clean rooms.

Typical initial or "clean" filter static pressure losses may be estimated from Table 10-7. But, the design static pressure loss to be imposed on the supply fan(s) is always the final or "dirty" filter static pressure loss which is typically 2–3 times the initial static pressure loss. Consult manufacturer literature for recommended filter initial and final static pressure losses.

10.2.3 Gas-Phase Filtration

For buildings located where the ambient air does not meet clean air quality standards or for those that may be considered potential terrorist targets, *gas-phase air cleaning* may be required.

Gas-phase air cleaning devices are sorbent filters, designed to remove pollutant gases and vapors from the building environment. Sorbents use one of two mechanisms for capturing and controlling gas-phase air contaminants, physical adsorption, and chemisorption. Both the capture mechanisms remove specific types of gas-phase contaminants from indoor air. Unlike particulate filters, sorbents cover a wide range of highly porous materials, varying from simple clays and carbons to complexly engineered polymers. Many sorbents, not including those that are chemically active, can be regenerated by application of heat or other processes.

Understanding the precise removal mechanism for gases and vapors is often difficult due to the nature of the adsorbent and the processes involved. While knowledge of adsorption equilibrium helps in understanding vapor protection, sorbent performance depends on such properties as mass transfer, chemical reaction rates, and chemical reaction capacity. Some of the most important parameters of gas-phase air cleaning include the following:

1. *Breakthrough Concentration*: the downstream contaminant concentration, above which the sorbent is considered to be performing inadequately. Breakthrough concentration indicates the agent has broken through the sorbent, which is no longer giving the intended protection. This parameter is a function of loading history, relative humidity (RH), and other factors.
2. *Breakthrough Time*: the elapsed time between the initial contact of the contaminant at a reported challenge concentration on the upstream surface of the sorbent bed, and the breakthrough concentration on the downstream side of the sorbent bed.

3. *Challenge Concentration*: the airborne concentration of the contaminant entering the sorbent.
4. *Residence Time*: the length of time that the contaminant spends in contact with the sorbent. This term is generally used in the context of superficial residence time, which is calculated on the basis of the adsorbent bed volume and the volumetric flow rate.
5. *Mass Transfer Zone or Critical Bed Depth*: interchangeably used terms, which refer to the adsorbent bed depth required to reduce the chemical vapor challenge to the breakthrough concentration. When applied to the challenge chemicals that are removed by chemical reaction, mass transfer is not a precise descriptor but is often used in that context. The portion of the adsorbent bed not included in the mass transfer zone is often termed the capacity zone.

Before selecting an air-cleaning strategy, the designer must evaluate potential contaminants, ventilation and particulate filtration requirements, and IAQ. The answers to these questions will guide in making decisions about what types of filters and sorbents (if any) should be installed in an HVAC system, how efficient those filters and/or sorbents must be, and what procedures must be developed to maintain them after installation.

For gas-phase air cleaners, the integrity of the filter rack or frame system has a major impact on the installed filtration efficiency. When specifying gas-phase air cleaners, careful specification of the frame system is required to minimize air bypass and efficiency degradation.

10.2.4 Ultraviolet Air and Surface Treatment

Ultraviolet germicidal irradiation (UVGI) generally refers to UV wavelength of 254.7 nm. The wavelength is near optimal for damaging nucleic acid (DNA, RNA). UV irradiation by itself does not clean air. The microorganisms are still there and, in the case of some microorganisms, may still contain the ability to cause noninfectious (e.g., allergenic) disease. While there is potential for UV to destroy allergenic sites on the surface of a bioaerosol, this ability has not been documented or quantified.

UV irradiation is commonly used in the disinfection of drinking water. Although to a lesser extent, UV has also been used in the disinfection of air and is required in some health care settings. Anecdotal health information as well as laboratory research suggests it is an effective technology for inactivating airborne viruses and bacteria and their spores. However, there is a significant difference between addressing disease-causing organisms in a laboratory setting and ensuring that deadly biological warfare pathogens are completely removed from an airstream. Installation, construction, and design of UV systems must be of paramount importance, and a focus of air decontamination must now include securing airstreams in ducts and HVAC systems from outside infiltration. Since the threat of this kind of terrorism has only recently been realized, independent scientific investigators have begun to do fundamental research into the biological decontamination of airstreams to update this science.

A major research study, funded in part by the U.S. Department of Energy, attempted to determine the effectiveness of UV germicidal radiation in inactivating bacteria and spores in a "typical" HVAC duct environment. The results were encouraging, with inactivation effectiveness greater than 90% for bacteria, although efficacy was less so for more resistant bacterial and fungal spores. Viruses were not tested in the study, but research in the laboratory indicates that a highly UV-resistant virus (bacteriophage MS2) is less resistant to

UV radiation (in air) than bacterial *Bacillus* spores (regarded as the most resistant bacteria to UV radiation).

Laboratory research has determined that air temperature, RH, flow rate, lamp design, and ballast engineering all have significant impacts on the effectiveness of these systems. Furthermore, the location of the decontamination UV array is of crucial importance because of the potential for air eddies and corners of ducting to allow some of the flow to pass through untreated. HVAC system designers have to factor these components into their plans and base decontamination system designs on tested parameters in order to ensure maximum effectiveness. Because of the variable airflow in modern ventilation systems, making this work in an HVAC system requires coordination between dampers (in order to slow the passage of the air enough to allow all particles to receive a sanitizing ray of UV light) and the UV light itself. While this kind of attention to detail might not be necessary in installations primarily concerned with improving the day-to-day healthiness of air, in order to control the spread of a potential bioterrorist pathogen (such as anthrax or smallpox), the design of an UVGI system is of critical importance.

The EPA has issued several warnings concerning UVGI companies that make unsubstantiated claims as to the effectiveness of their technology. For example, reports by sterilization and decontamination companies making claims such as "the simultaneous emission of ultrasound or ultrasonic waves and ultraviolet light complement each other and can effectively sterilize either organic or inorganic items in a non-liquid environment." No peer-reviewed research has ever been published confirming these claims.

10.2.5 Airside Economizer Cycle

Interior lighting, people, appliances, and so on represent heat gains that may produce a need for cooling, in at least part of the building, on a year-round basis. Thus, to eliminate the need to provide mechanical refrigeration during cold weather and to comply with code requirements, *economizer cycles* are often added to all-air and air–water systems. There are basically three options for controlling the operation of an integrated airside economizer cycle:

Option 1: Fixed Outdoor Air Dry Bulb Temperature. The optimal strategy for minimizing cooling coil energy use by adjusting damper position is as follows:

If outdoor air temperature > return air temperature, use minimum outside air.

If outdoor air temperature < high limit temperature set point, use 100% outside air to minimize the imposed mechanical cooling load.

If outdoor air temperature ≤ required supply air temperature, blend outside air with return air so that the mixed air temperature equals the supply air temperature but maintain outside airflow at or above the minimum ventilation requirement, preheating as needed.

Option 2: Differential Enthalpy, Outdoor Air Enthalphy versus Indoor Air Enthalpy. Although temperature-based economizer control leads to reduced cooling coil energy use, it can actually increase the load on the cooling coil when the outdoor air is cool and humid. There are three distinct regions that define the operating modes of an enthalpy economizer cycle:

1. Region 1 is any condition at which the outdoor air temperature is equal to or less than the required supply air temperature and represents the dry-bulb economizer region. Any time the outdoor conditions fall within this region, outdoor air can be utilized to satisfy the system cooling requirements fully.

2. Region 2 is any condition at which the outdoor air temperature is greater than the required supply air temperature, while the outdoor air enthalpy is less than or equal to the indoor design enthalpy. This represents the enthalpy economizer region. When outdoor air conditions are within this region, its total heat content is lower than the room conditions, and 100% outdoor air, with supplemental mechanical cooling, would be utilized in the system.
3. Region 3 is any condition at which the outdoor air enthalpy is greater than the indoor design enthalpy. Thus, outdoor conditions would impose additional cooling load on the system. In this region, only enough outdoor air to satisfy minimum ventilation requirements should be introduced into the system.

Option 3: Fixed Outdoor Air Dry-Bulb Temperature and Fixed Outdoor Air Enthalpy. To reduce the higher control costs and complexity of Option 2, a third approach to airside economizer control is to add outdoor enthalpy measurement to the temperature-based control mode of Option 1, as follows:

If outdoor air temperature > return air temperature *or* if outdoor air enthalpy > high limit enthalpy set point, use minimum outside air.

If outdoor air temperature < high limit temperature set point *and* if outdoor air enthalpy < high limit enthalpy set point, use 100% outside air to minimize the imposed mechanical cooling load.

If outdoor air humidity ≤ required supply air temperature, blend outside air with return air so that the mixed air temperature equals the supply air temperature, but maintain outside airflow at or above the minimum ventilation requirement, preheating as needed.

The use of enthalpy determination is a difficult process that is prone to error, while temperature measurement is easy and generally reliable, as follows:

1. Since enthalpy is a thermodynamic property of air, it cannot be directly measured. An enthalpy transmitter requires a temperature sensor, a humidity sensor, and firmware to compute enthalpy based on the two measured conditions to produce a 4–20 mA linear output signal.
2. There are two types of humidity sensors utilized for HVAC applications. Typical indoor humidity sensors are the resistive type, using a bulk polymer element that expands and contracts based on moisture level of the air, resulting in a linear change in resistance. These sensors are relatively accurate (± 1%–2% RH) within the range of 20%–95% RH. However, they have very high drift and/or failure rate when used in outdoor air measurement applications. A study reported in a paper entitled "Economizer Control Using Mixed Air Enthalpy" by University of Nebraska researchers indicates that humidity sensors of this type, installed in outdoor air applications, have an accuracy drift of 3.5% after only 4 months and high failure rates within 7 months after installation. Humidity in the outdoor air exceeding 95% RH, along with sensor wetting due to rain or surface condensation, saturates the polymer element and fairly quickly results in its failure.
3. Capacitive-type humidity sensors that utilize ceramic sensor elements are less accurate (± 2%–3% RH) and have even poorer accuracy at high humidity levels and elevated temperatures. Temperature compensation firmware is typically incorporated by sensor manufacturers to improve sensor accuracy at higher

temperatures. The important advantage to capacitive sensors is that they are not damaged by high humidity or wetting in outdoor locations and long-term drift is limited to about 1% per year (again, requiring recalibration only about every 5 years).

The November 2010 issue of the *ASHRAE Journal* presented a lengthy article by Steven Taylor, PE describing a detailed analysis of each of the three airside economizer cycle control modes. Based on this analysis, the article reaches the following conclusions:

> Including sensor error, the *best* (or very close to the best) option in all climates is simply fixed dry-bulb control [Option 1], assuming the setpoint is optimized by climate.
>
> Including sensor error, the *worst* (or very close to the worst) option in all climates is the differential enthalpy control [Option 2]. This control logic is considered the "best" anecdotally among many design engineers and is required for some climate zones by [ASHRAE] Standard 189.1. *Yet, in practice, with realistic (even optimistic) sensor error, it performs among the worst of all options* … it will be almost impossible for sensors to be accurate enough for differential enthalpy control to beat a simple dry-bulb switch; and, certainly impossible for differential enthalpy control to be life-cycle cost effective versus a dry-bulb switch given the significant added first costs and maintenance costs.
>
> Fixed enthalpy control when combined with fixed dry-bulb control [Option 3] also performs well. The error in the enthalpy sensor is buffered by the addition of the dry-bulb limit, and the dry-bulb limit resolves the inefficiency problems the fixed enthalpy sensor has in dry climates. *But, it performs only slightly better than fixed dry bulb alone, even in humid climates, so it is not likely to be cost effective given the added first costs and maintenance (calibration) costs of the outdoor air humidity sensor.*

Based on this information, the best airside economizer design remains the fixed dry-bulb temperature option (Option 1), assuming that the designers establish upper limit temperature set points based on local climate data. If an owner insists on having enthalpy control, then Option 3 should be applied using capacitive-type humidity sensors that utilize ceramic sensor elements for outdoor air humidity sensing. *Option 2 should never be used.*

Dampers used for controlling ventilation airflow rate and/or as part of an airside economizer cycle shall be sized on the basis of ASHRAE *Guideline No. 16*, as follows:

1. *Return fan system* (consisting of a supply fan, return fan, and modulating control dampers for airside economizer control):
   a. Select the relief air damper first. An opposed blade damper is recommended since it will have lower pressure drop than a smaller parallel blade damper when selected for linear throttling characteristics, thus reducing the cost of operation.

      The damper selection is based upon the maximum design relief air (design supply airflow rate minus mechanical exhaust minus exfiltration and pressurization air). Determine the relief air path pressure drop without the relief damper, including ductwork, plenums, and discharge louver. For a linear response opposed blade damper, the ratio between the pressure drop in the relief air path and the pressure drop through the open relief air damper should be 8–15. This means that the damper will have a wide-open pressure drop of 7%–15% of the relief path pressure drop at design flow. The resulting wide-open damper face velocity will frequently be around 1,000–1,500 fpm.

Once this damper is sized, the pressure in relief/return plenum relative to ambient can be estimated as the relief air path pressure drop plus the pressure drop across the relief air damper.

b. For constant air volume systems, select a two-position minimum outdoor air damper. This damper may be either parallel or opposed blade. It should be sized for the minimum outdoor airflow rate at the same face velocity as the maximum (or "economizer") outdoor air damper.

c. For VAV systems, the minimum outdoor air damper must be modulating for capacity control. It should be selected using the procedure similar to that described above for the relief air damper.

d. Select the maximum (or "economizer cycle") outdoor air damper. The size of this damper is not critical since it does not directly control outdoor airflow. The total flow of outdoor air through maximum and minimum outdoor air dampers is always the difference between the supply airflow and the return airflow, and therefore, the damper size and type have no effect on the amount of outdoor air. Dampers with a low pressure drop are generally used to minimize the cost of operating the supply fan. Damper face velocities in the range of 800–1,500 fpm are typical. Because this damper does not control flow, either opposed blade or parallel blade action can be used.

e. Finally, select the return air damper. This damper selection is based upon the maximum return airflow rate, which is the difference between the design supply air and the minimum outdoor airflow rate. The return air damper should be selected so that the pressure drop across the damper is equal to the differential pressure available between the return air (+ pressure) plenum and the outdoor air (– pressure) plenum.

The pressure in the outdoor air plenum is equal to the pressure drop at the minimum outdoor airflow rate from the outdoors through the outdoor air damper when the damper is at minimum position. The minimum position is up to the designer, but it should be well below full open so that the damper can open further when the VAV supply fan airflow rate is low and the plenum pressure approaches ambient pressure.

2. *Relief air fan system* (consisting of a supply fan, relief fan, and control dampers for airside economizer control):

a. First, size the relief air damper. This damper is typically interlocked with the relief air fan start/stop control and with relief airflow volume control (variable speed drive and/or fan staging). As such, it should be selected for low pressure drop to minimize fan energy and may be either parallel or opposed blade. Damper face velocities from 400 to 1,000 fpm at design flow are typical.

b. Select the minimum outdoor air damper in accordance with the procedure for systems with return air fans.

c. The maximum (or "economizer") outdoor air damper should be selected for low pressure drop to minimize the cost of operating the supply fan. Damper face velocities from 800 to 1,500 fpm at design flow are typical.

d. Finally, select the return damper. Return air damper sizing is based upon the maximum return airflow rate, which is the difference between the design primary supply air and the minimum required outdoor airflow rate. With a relief fan, the pressure in outdoor air plenum must be below the pressure in the relief air plenum.

The pressure in the relief air plenum is pressure drop from the space to the plenum through the return air ductwork. This will usually be more negative than outdoor air plenum pressure. Therefore, the pressure drop between the two plenums (including the return damper pressure loss) should be as low as possible to minimize the negative pressure required in the return air plenum and minimize supply fan energy.

Damper face velocities from 800 to 1,500 fpm at design flow are typical.

e. Where outdoor and return air dampers are controlled so that one opens when the other closes, both outdoor and return air dampers should be parallel blade dampers. Tests show that a relatively constant pressure drop across the return and outdoor air paths throughout the economizer cycle can be achieved by using parallel blade dampers. Where outdoor and return air dampers are sequenced independently, either parallel or opposed action may be used.

3. To ensure good mixing of the two airstreams when the outdoor and return airstreams enter the AHU mixed air plenum at 90° to each other, parallel blade dampers should be used. The two dampers should be arranged so that the two airstreams impinge with each other at 180°.

10.2.6 Cooling Coils Condensate Drainage

Moisture removed by latent cooling must be collected and removed from the AHU. In a blow-through configuration, an AHU creates a *positive pressure condition* at the cooling coil and condensate drain pan. The condensate drain, therefore, must be provided with an air seal (a "trap") of sufficient height to account for the positive static pressure condition that occurs under normal operating conditions. This pressure is typically equal to the downstream duct, box, and air distribution losses, plus any plenum losses in the AHU. As a safety factor, add an additional 1" to this total.

In a draw-through configuration, the AHU creates a *negative pressure* condition at the cooling coil and condensate drain pan. The condensate drain, therefore, must be provided with an air seal (a "trap") of sufficient height to account for the negative static pressure condition that occurs under "worst case" conditions. This pressure is typically equal to the sum of the AHU losses with dirty filters (including damper, coil, filter, and AHU losses).

The amount of condensate produced depends on the amount of outdoor air introduced and the climatic conditions that exist at the tune.

Condensate drain design calls for a properly sized drain line and a trap, as shown in Figure 10-1, designed specifically for the pressure condition expected.

There is no "scientific" method for sizing condensate drain lines, but Table 10-8 lists that the recommended pipe sizes have been developed from experience over the years.

And, improper use and maintenance of the trap can cause condensate drainage problems:

1. The trap must be cleaned periodically (least once per year at the beginning of the cooling season) to avoid blockage by algae formation, sediment, and so on in the line.
2. It is common for the water seal to evaporate during the non-cooling season. It is typically necessary for the trap to be manually filled at the beginning of the cooling season.

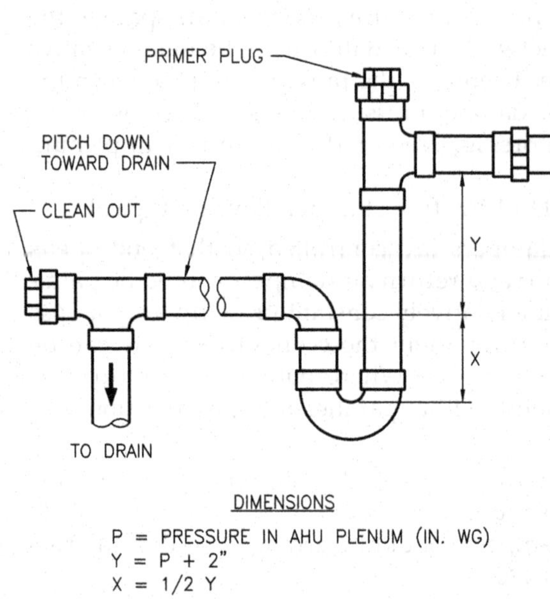

SIZE

PIPE SIZE	MAX COIL LOAD (TONS)
3/4"	2
1"	5
1-1/4"	30
1-1/2"	50
2"	160
3"	300
4"	430
6"	1000

FIGURE 10-1
Cooling coil condensate drain piping. (SW Standard Detail C02.)

TABLE 10-8
Recommended Condensate Drain Pipe Sizes

Drain Size Nominal Pipe Size [DN]	Maximum Coil Load (Tons) [kW]
¾″ [19]	2 [7]
1″ [25]	5 [18]
1-1/4″ [32]	30 [106]
1-1/2″ [40]	50 [176]
2″ [50]	160 [563]
3″ [75]	300 [1,055]
4″ [100]	430 [1,512]
6″ [150]	1,000 [3,500]

10.3 Humidification

As discussed in Chapter 2, humidification is the process of adding moisture to air. This may be accomplished by any one of the following methods:

1. *Water pan humidifiers* are duct-mounted devices, usually used only in residential applications, which can add limited amounts of moisture by direct evaporation. Basically, water is passed over a water pan that may be heated and/or agitated to increase the evaporation rate. The performance of these devices is usually poor.
2. *Water spray humidifiers* are available in two varieties: *air atomizers* that use compressed air to break water into fine droplets to increase the air–water contact area and thus enhance evaporation and *pressure atomizer*, generally called *foggers*, which

use high pressure pumps (1,000 psig [7,000 Pa]) and special nozzles to create a fine water mist that evaporates rapidly. Air atomizers are most often used as space humidifiers, while spray humidifiers are typically installed within an AHU and include a "de-misting" pad downstream of the nozzles to capture any water droplets that may be formed.

3. *Direct steam injection humidifiers* are duct-mounted units that use a tubular grid of nozzles, along with a modulating steam flow control valve, to distribute steam into the supply airstream. Since steam is already water in a vapor state, evaporation is required and the moisture is rapidly absorbed by the airflow.
4. *Indirect steam injection humidifiers* use steam from boilers to generate low pressure steam at the point of injection, thus eliminating the potential for introducing water treatment chemicals or other contaminates that can be carried by the boiler steam supply. These humidifiers use no water treatment (and sometimes are supplied with distilled water supply) and require periodic replacement of the heater reservoir due to solids buildup. These humidifiers are used only in the most critical of applications such as some hospitals, biological research, and pharmaceutical manufacturing.

Both pan and water spray humidifiers introduce the potential for Legionella contamination (see Chapter 14) and are not recommended.

The design humidification requirement for a space is best determined using the psychrometric chart to find the two points that define the required space temperature and humidity condition, along with the supply air temperature and humidity condition. There are basically four different conditions that may occur:

1. Space conditions are defined by the minimum design RH level required. For typical comfort air-conditioning applications, the design humidity level is typically 30% RH. For laboratories, hospitals, and so on, higher levels may be required. Based on the design space temperature and RH, the specific humidity requirement for the space can be obtained from the psychrometric chart.
2. For constant air volume systems with a fixed outdoor ventilation airflow, the supply air humidity will be the same as the mixed air humidity. By locating the mixed air point on the psychrometric chart, the specific humidity of the supply air can be obtained.
3. For constant air volume systems with an economizer system, the outdoor airflow will be sufficient to create a mixed air condition that is dictated by the supply air temperature set point for the system. By locating the mixed air point on the psychrometric chart, the specific humidity of the supply air can be obtained.
4. For VAV systems with an economizer system, the outdoor airflow will be sufficient to create a mixed air condition that is dictated by the supply air temperature set point for the system but at a reduced airflow dictated by the terminal units being served. By locating the mixed air point on the psychrometric chart, the specific humidity of the supply air can be obtained.

The design humidification requirement for a space is determined with IP units using Eq. (10-1).

$$W = 4.45V(W_2 - W_1) \tag{10-1}$$

where

W = humidifier rating, lb water/h

V = supply airflow rate, cfm (based on standard specific air volume of 13.5 cf/lb of dry air)

W_2 = room design specific humidity, lb water/lb dry air

W_1 = supply air specific humidity, lb water/lb dry air

The design humidification requirement for a space is determined with SI units using Eq. (10-1a).

$$W = 4.10V(W_2 - W_1) \tag{10-1a}$$

where

W = humidifier rating, kg/h

V = supply airflow rate, L/s (based on standard specific air volume of $0.88\,\text{m}^3/\text{kg}$ of dry air)

W_2 = room design specific humidity, kg water/kg dry air

W_1 = supply air specific humidity, kg water/kg dry air

Humidifier installation in the supply ductwork is critical to ensure that the airflow absorbs all of the moisture introduced and liquid water does not occur inside the duct (resulting in corrosion, mold and mildew growth, and so on). The following general rules apply to humidifier installation:

1. Locate the humidifier at least four duct diameters upstream of a turn or branch.
2. Construct ductwork 4 ft [1.2 m] upstream and 8 ft [1.4 m] downstream of either aluminum or stainless steel to minimize corrosion potential.
3. Provide a drain (with trap) just downstream of the humidifier. Slope the bottom of the ductwork to the drain.
4. Interlock the humidifier controls with the AHU so that the humidifier cannot operate when the supply fan is not in operation.

10.4 Ventilation and Building Pressurization

Ventilation air is required for a building space when it is occupied. But, most buildings are not occupied around the clock. Therefore, the ventilation system(s) design must address unoccupied periods, ranging from overnight to weekends to summer vacations, to maintain low humidity level and help prevent mold growth.

The recommended approach to preventing high humidity from occurring during these periods is to establish both high limit temperature and high limit humidity space set points, monitored by space temperature and humidity sensors, that will cause the HVAC controls to cycle the HVAC systems on to maintain these set points. Ideally, this would be done for each air handler. But, because central chiller and boiler plants may have a minimum load requirement, care must be taken by the designer to ensure that enough systems are cycled on to provide this minimum load or there is no benefit to the unoccupied period control.

10.4.1 Outdoor Air Intakes

Dilution ventilation is a process by which some percentage of "dirty" indoor air is continuously replaced by "clean" outdoor air to control odors and reduce pollutant concentrations. Thus, each HVAC must intake outdoor air. The best location is on the windward side of the building, located in the lower third of the wall, with the bottom of the intake at least 2 ft [600 mm] above grade level ... *never install outdoor intakes in wells below exterior grade level.* These wells tend to fill with trash, leaves, and snow that can impede outdoor air intake flow.

For a facility that may be a potential terrorist target, Chapter 14 provides additional guidance relative to outdoor air intake location and protection.

Outside air intakes must be separated from any potential pollutant source, as defined in Table 10-9.

For industrial applications (and for hospitals and research laboratories), where hazardous exhaust discharge is required, special care must be taken to ensure that reentry of pollutants does not take place. This is discussed in more detail in Chapter 7.

Outdoor air intake louvers to protect wall openings from rainwater penetration must be selected on the basis of the following criteria:

a. Size each louver on the basis of a maximum face velocity at least 25% below the first point of *water penetration rated face velocity* as cataloged by the louver manufacturer, not exceeding 500 fpm [2.5 m/s] face velocity. Louvers should be rated for maximum rain intrusion/entrainment of 0.01 oz/sf of louver free area at the maximum design intake velocity, determined for a minimum 15-min test duration when subjected to a water flow rate of 0.25 gal/min as described under the "Water Penetration Test" in AMCA *Standard 500-L-9913.*

TABLE 10-9
Minimum Separation between Pollutant Source and Outdoor Air Intake

Object	Minimum Distance (ft) [m][a]
General building exhaust	10 [3]
Significantly contaminated exhaust or vents	15 [4.6]
Noxious or dangerous exhaust or vents[b]	30 [9]
Garage entry, automobile loading area, or drive-in queue	15 [4.6]
Truck loading area or dock, bus parking/idling area	25 [7.6]
Driveway, street or parking place	10 [3]
Thoroughfare with high traffic volume	25 [7.6]
Roof, grade, or other surface directly below intake	1 [0.3]
Garbage storage/pick-up area, dumpsters	15 [4.6]
Cooling tower intake or basin	15 [4.6]
Cooling tower discharge	25 [7.6]

[a] Minimum separation for health care facilities for any object is 30 ft [9 m].
[b] Laboratory fume hood exhaust air outlets shall be in compliance with NFPA 45-1991 and ANSI/AIHA Z9.5-1993.

The International Building Code defines *hurricane-prone regions* of the United States where the ultimate design wind speed is greater than 115 mph [51.4 m/s], plus all of Hawaii, Puerto Rico, the Virgin Islands, Guam, and Samoa. Louvers in these areas must be selected to avoid wind-driven rain penetration.

This same code defines *wind-borne debris regions* as areas within hurricane-prone regions if located within 1 mile [1.6 km] of the coastal mean high water line where the wind speed is 130 mph [58 m/s] or greater. For HVAC designers, this second definition is a fine differentiation that can be ignored ... *all facilities within hurricane-prone regions should also be considered to be within wind-borne debris regions.*

Thus, louvers, intake or exhaust, within hurricane-prone regions should be specified to comply with AMCA *Standard 540* and *Standard 550.* Additionally, louvers must be licensed to bear the AMCA listing label for high velocity rain resistance and impact resistance in accordance with AMCA *Standard 512.*

b. Select louvers to have a pressure loss below 0.2 in. wg for most applications. To achieve this, increase the opening size or select a louver model with higher free area, higher first point of water penetration, lower pressure drop, or a combination of these factors.

c. Provide an indoor plenum at least a 12″ deep downstream of the louver, with its floor sloped to one or more drains piped to waste, in order to trap and collect any rainwater that does penetrate the louver. The plenum must include adequate access for damper maintenance.

Each louver must include a removable and cleanable wire mesh *birdscreen* to prevent birds and airborne debris from entering.

In "snowy" locations, the outdoor intake must be arranged to avoid snow or ice clogging. Louvers provide some protection by separating snow particles from the airflow. But, this may result in snow buildup on the louver blades, ultimately partially or fully blocking intakes. Some louver manufacturers offer integral electric heating devices that are used to melt the snow and increase the louver's snow penetration efficiency, but these become worthless during a power outage. The most effective approach to avoiding snow or ice clogging of outdoor air intakes is to prevent snow from impinging on the intakes by installing *wind and snow shelters*, open only at the bottom so that wind cannot directly blow into the intake, as illustrated by Figure 10-2.

The bottom intake opening can be sized for the required airflow at 1,000 fpm [5 m/s] face velocity.

1. If the top of the wind shelter is located close to the roof line, a snow fence along the roof, extending well beyond the edges of the shelter in both directions, may be required to prevent snow buildup and structural loading on the shelter. Alternatively, the top of the shelter can be sloped at 60° or more to ensure that impinging snow slides off.
2. Since the shelter serves as a rain shield, no louver is required in the intake opening ... only a birdscreen.

Louvers for relief or exhaust airflows may be sized for 750 fpm [4 m/s] face velocity. For projects in coastal regions, select louvers for high wind conditions as discussed above for intake louvers.

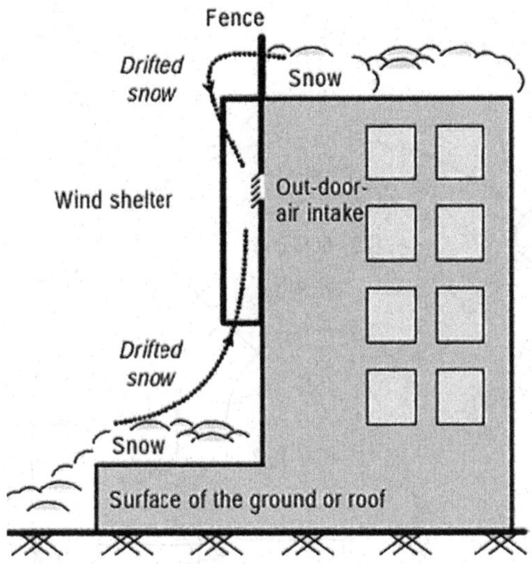

FIGURE 10-2
Wind shelter for outside air intake. ("The Prevention of the Snow Entrance to the HVAC-Systems," Vesa Asikainen and Pertti Pasanen, University of Kuopio, Department of Environmental Science, *Proceedings of Clima 2007 Well Being Indoors*.)

Rooftop outdoor intakes and discharges may utilize either factory-fabricated penthouses or shop-fabricated "goosenecks." Factory-fabricated penthouses are available as a low silhouette basic four-way hood design or, where aesthetics are important, a fully louvered configuration. For industrial applications, goosenecks, as shown by Figure 10-3, are commonly applied. Again, these intakes/discharges must be sized to prevent rainwater intrusion and designed for high wind conditions, as applicable.

10.4.2 Understanding and Applying ASHRAE Standard 62.1

The design criteria for dilution ventilation airflow rates has changed over the years as shown in Table 10-10.

Establishing design values and methods of controlling minimum ventilation airflow in HVAC systems is becoming more and more complex. There are numerous codes and/or standards that apply to establishing and controlling ventilation rates:

ASHRAE *Standard 62.1*
ASHRAE/IESNA *Standard 90.1*
Applicable state and/or local building codes
ASTM *Standard D6245, Standard Guide for Using Indoor Carbon Dioxide Concentrations to Evaluate Indoor Air Quality and Ventilation*

The "gold standard" for ventilation design is ASHRAE Standard 62.1. However, the minimum airflow requirements of this standard may result in lower ventilation rates than dictated by the applicable state and/or local building codes and vice versa. Therefore, *the designer must evaluate both sets of requirements and select ventilation rates that meet the more stringent of the two.*

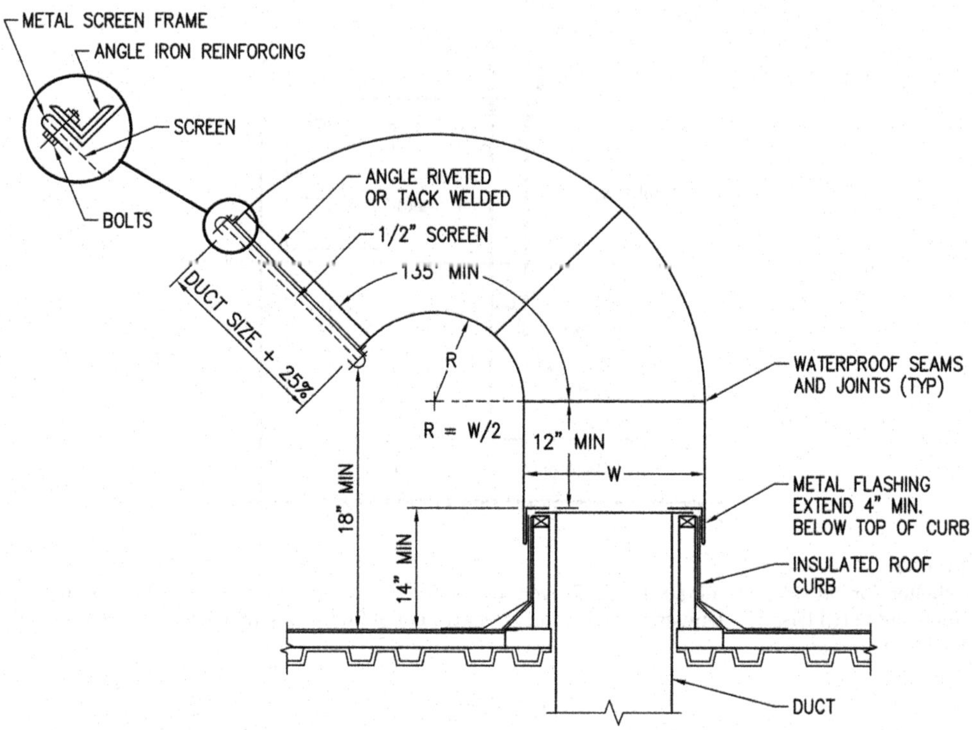

FIGURE 10-3
Gooseneck outdoor air intake/exhaust air discharge. (SW Standard Detail D04.)

TABLE 10-10
Minimum Ventilation Rates

Time Period	Source	Baseline Minimum Ventilation Rate
1875–1922	ASHVE guidelines	30 cfm/person [14.2 L/s per person]
1923–1946	ASHVE standards	30 cfm/person [14.2 L/s per person] in 1923, reduced to 10 cfm/person [4.7 L/s per person] in 1946 [4.7 L/s per person]
1946–1979	ASHRAE (ASHRAE Standard 62 issued in 1973)	10 cfm/person [4.7 L/s per person] (though, common design practice during this period was to set the outdoor airflow to 25% of total airflow)
1979–1989	ASHRAE Standard 62-1979	5 cfm/person [2.4 L/s per person], nonsmoking; 20 cfm/person [9.4 L/s per person], smoking
1989–2004	ASHRAE Standard 62-1989	15 cfm/person [7.1 L/s per person]
2004+	ASHRAE Standard 62.1	0.06–0.30 cfm/sf [0.33–1.52 L/s per m^2] + 5–10 cfm/person [2.4–4.7 L/s per person]

Minimum ventilation airflow requirements in the breathing zone of each occupied space are mandated by ASHRAE *Standard 62.1* or by the applicable state and/or local building codes. However, the total outdoor air supply delivered by the AHU must be determined by further calculation.

While ventilation rates must comply with codes and standards above, rates must also be sufficient to maintain an overall positive building pressure in order to eliminate infiltration and reduce humidity control problems. *Select the design ventilation airflow to be at least 10% (or greater) than the total building exhaust airflow.*

ASHRAE *Standard 62.1* requires testing of ambient air conditions to determine whether they meet National Ambient Air Quality Standards (NAAQS) standards. If not, to remove externally introduced volatile organic carbons and other pollutant gases or vapors, gaseous filtration is required.

This standard defines ventilation air requirements with each space's "breathing zone," the volume between 3" [76 mm] and 72" [1,800 mm] above the floor and 24" [600 mm] and greater inside each wall. Designers must become familiar with applicable ASHRAE *Standard 62.1* requirements, as follows:

1. The minimum ventilation airflow rate delivered by an HVAC to each zone is determined not only by the type of occupancy and population of the zone but also by the area of the zone. Table 6.1 of the standard introduces two measures that are *additive* for determining the required zone ventilation rate: a *people outdoor air rate,* defined in terms of cfm/person, and an *area outdoor rate,* defined in terms of cfm/sf. Indoor air contaminates can originate from occupants and their activities or from the building and its furnishings. If both sources are present, the ventilation rate required for adequate dilution is the sum of the rates needed to handle each source separately.
2. The actual amount of outdoor air that must be brought into the HVAC system must account for the *zone air distribution effectiveness* of the system and the system's *ventilation efficiency.* This first factor is based on the idea that not all of the air supplied through outlets will reach the zone occupants' breathing zone and represents the ratio of airflow that actually enters the breathing zone to the total supply airflow. The second factor defines the effectiveness of an air system in distributing the outdoor air from the outdoor air intake to the breathing zone of the most critical ventilation zone.

To apply ASHRAE *Standard 62.1,* the following steps are required:

1. The first step in the application of this method is to determine the minimum required outdoor airflow to each zone using Table 6.1 of the standard and the specific zone area and population data.

   Then, each zone's minimum ventilation airflow rate (V_z) is computed via the following equation:

$$V_z = R_p P_z + R_a A_z \tag{10-2}$$

   where

   R_p = People outdoor air rate required, cfm/person [L/s per person]

   R_a = Area outdoor air rate required, cfm/sf [L/s/m^2]

   P_z = Zone peak population number

   A_z = Zone area, sf [m^2]

   For zones that have varying populations, the standard allows for the zone population to be "averaged" over a *time period, T,* in minutes, computed by Eq. (10-3), to determine the zone design population:

$$T = 3v / V_z \tag{10-3}$$

where

v = zone volume, cf [m³]

2. Each zone's total ventilation rate must then be "corrected" to account for the *ventilation zone air distribution effectiveness*, E_z, of the air distribution configuration used, as shown in Table 10-11.

   The amount of outdoor air that must be brought into the HVAC system for each zone, V_{oz}, is computed by Eq. (10-4).

$$V_{oz} = V_z / E_z \qquad (10\text{-}4)$$

   ASHRAE defines a zone as "one … or several occupied spaces with similar occupancy category, occupant density, zone air distribution effectiveness, and zone primary airflow." Ventilation zones may or may not be the same as temperature/humidity control zones.

3. The *outdoor air intake airflow* must be determined on the basis of the type of HVAC system used, as follows:
   a. For systems that serve only one zone, the outdoor intake airflow is simply equal to the zone airflow.
   b. For 100% outdoor air systems, the outdoor intake airflow is simply equal to the sum of zone airflows.
   c. For systems that serve multiple zones with a mix of outdoor and return air, the outdoor intake airflow is computed as follows:
      1. For each zone, the *primary outdoor air fraction*, Z_p, must be computed by dividing the zone outdoor air rate, V_{oz}, by the total supply airflow to the zone delivered by the air handler.
      2. Then, the *uncorrected outdoor air intake* airflow, V_{ou}, can be computed by the following equation:

$$V_{ou} = \left[D \times \sum (R_p P_z)\right] + \sum (R_a A_z) \qquad (10\text{-}5)$$

TABLE 10-11

Zone Air Distribution Effectiveness

Air Distribution Configuration	E_z
Ceiling supply of cool air	1.0
Ceiling or floor supply of warm air and floor return	1.0
Floor supply of warm air and ceiling return	0.7
Ceiling supply of warm air (at least 15°F [8°C] above space temperature) and ceiling return provided that the 150 fpm [0.8 m/s] supply air jet does not reach within 4.5 ft [1.4 m] of floor level.	0.8
Ceiling supply of warm air (at least 15°F [8°C] above space temperature) and ceiling return provided that the 150 fpm [0.8 m/s] supply air jet reaches within 4.5 ft [1.4 m] of floor level.	1.0
Floor supply of cool air and ceiling return provided that the 150 fpm [0.8 m/s] supply air jet reaches at least 4.5 ft [1.4 m] above the floor (which is met by most underfloor air distribution systems).	1.0
Floor supply of cool air and ceiling return provided low-velocity displacement ventilation achieves unidirectional flow and thermal stratification (almost impossible to achieve in occupied spaces).	1.2

TABLE 10-12

System Ventilation Efficiency

Max (Z_p)	E_v
£0.25	0.9
0.26–0.35	0.8
0.36–0.45	0.7
0.46–0.55	0.6
>0.55	See ASHRAE *Standard 62.1* for Computation Method

where D is the *occupant diversity* factor (0.0–1.0), computed by dividing the sum of each zone's peak population by the system's peak population.

4. Finally, the actual *design system outdoor intake* airflow rate, V_{ot}, may be determined through use of Eq. (10-6).

$$V_{ot} = V_{ou} / E_v \tag{10-6}$$

where E_v is the *system ventilation efficiency* shown in Table 10-12.

These requirements may seem "computationally intensive" because ASHRAE Standard 62.1, by design, does not provide a way in which the HVAC designer can avoid taking into account the ventilation effectiveness of a system serving multiple zones.

10.4.3 Demand Control Ventilation

Demand controlled ventilation (DCV) is considered a key approach to help minimize building energy consumption by HVAC systems, as it allows the amount of outdoor introduced for dilution ventilation to be reduced in proportion to any reduction in the number of people occupying a space.

DCV may be controlled on the basis of an occupancy schedule, occupancy sensors, and/or scheduled or manual "event" switching for spaces such as gymnasiums and auditoriums. Since sensing of the number of occupants in a space is required, but direct "people counting" is not yet a fully developed technology, DCV is most commonly applied using an indirect indicator of population density, *carbon dioxide concentration.*

Carbon dioxide (CO_2) sensors are used to compare the outdoor level of CO_2 with each indoor space's CO_2 concentration. The typical set point for the difference between outdoor and indoor concentrations, based on ASHRAE *Standard* 62.1, is typically 650–700 ppm since research shows that at that concentration at least 80% of "unadapted" persons (i.e., visitors) will find the level of body odor released by occupants to be acceptable.

The National Institute for Occupational Safety and Health (NIOSH) recommends an indoor high limit concentration set point of 1,000 ppm. NIOSH reports that this concentration indicates inadequate ventilation and complaints of headaches, fatigue, and eye and throat irritation will be more widespread. Since outdoor CO_2 concentration may be much higher in urban or industrial areas, this indoor high limit is often the controlling factor for DCV applications.

The rationale for using CO_2 as a sensed control variable is that (1) CO_2 concentration generally reflects occupancy level since people are typically the only source of CO_2 in buildings and (2) its concentration indicates whether there is a need for more or less outside air. *When implementing a DCV system based on CO_2 sensing, CO_2 control is a method, not a goal.*

The key to successful DCV performance is being able to accurately measure CO_2 concentrations, both indoors and outdoors. In HVAC applications, *nondispersive infrared* (NDIR) sensors are spectroscopic sensors that detect CO_2 in air by its characteristic infrared (IR) light absorption pattern that is most commonly applied.

The essential components of an NDIR sensor include one or two IR radiation sources (lamps), one or more IR detectors, band-pass optical filters, and an optical path between the source and the detector which is open to the air sample. IR sources emit radiation over a broad range of wavelengths; therefore, band-pass filters are used to limit the IR spectrum to specific wavelengths used in the particular sensor design.

Sensors are calibrated using gas mixtures with a known CO_2 concentration. Often the calibration gas is an N_2/CO_2 mixture. Calibration gases are available in various CO_2 concentrations for testing sensors. The accuracy of the CO_2 gas concentration in the mixture is expressed either as deviation in CO_2ppm (i.e., ±10 ppm) or as a percentage of the CO_2 concentration (i.e., 1%–2%).

For HVAC designers, there are two major performance requirements for CO_2 sensing … initial sensor accuracy and long-term sensor stability without the need for frequent recalibration by the facility owner or manager.

Unfortunately, there is no current national standard that defines a uniform test method for sensor "accuracy," and studies by LBNL and Iowa Energy Center indicate there are significant accuracy problems with CO_2 sensors on the market today.

1. The LBNL studies found that, at 1,010 ppm CO_2 concentration, 40% of sensors tested had errors greater than ±75 ppm and 31% of sensors had errors greater than ±100 ppm. At 760 ppm, 47% of sensors had errors greater than ±75 ppm and 37% of sensors had errors greater than ±100 ppm. Even more significantly, at 1,010 ppm 19% of sensors had an error greater than 200 ppm and 13% of sensors had errors greater than 300 ppm. Statistical analyses indicated that there were significant differences between the average accuracies of sensors from different manufacturers using the same technology, so the inaccuracy problems seem endemic.
2. Iowa conducted a laboratory-based study of the accuracy of 15 models of new CO_2 sensors. Their report shows that most of these tested sensors had errors greater than 75 ppm and errors greater than 200 ppm were not unusual. Of the 15 models tested by the Iowa Energy Center, only two met their manufacturer's accuracy statement for two out of the three sensors tested … no manufacturer had results where all sensors of the same model met manufacturer's accuracy statements and about half of all sensor models did not meet the manufacturer's accuracy statement at all.

Thus, based on the LBNL and Iowa studies, it is obvious that most CO_2-based DCV systems will, because of poor sensor accuracy, fail to perform as designed.

Long-term stability refers to how sensor accuracy degrades over time and defines the required frequency of recalibration … a stable sensor will require less frequent recalibration. Long-term stability can be specified as "percent full scale per year" or "ppm per year," with the second used more frequently in the literature.

In an effort to "improve" long-term stability, some sensor manufacturers include firmware logic for "automatic baseline adjustment" in an attempt to reduce long-term sensor accuracy degradation. Typically, this "self-adjustment" feature is based on an arbitrary baseline of 400 ppm. However, establishing a routine 400 ppm condition for the sensor to use as a baseline ranges from difficult to impossible in normal building operations. And, most sensors with this feature cannot be field-calibrated and actual sensor accuracy can never be verified. *Therefore, this feature is counter-productive to accurate CO_2 measurement.*

Real-world experience has shown that twice a year verification and potential calibration of the CO_2 sensors is required to maintain proper DCV operation. Further support for this level of sensor maintenance is provided in ASHRAE *Standard 62.1*:

> Sensors whose primary function is dynamic minimum outdoor air control … shall have their accuracy verified as specified…. This activity shall occur *at a minimum of once every six months* [emphasis added]…. A sensor failing to meet the accuracy specified … shall be recalibrated or replaced.

As a potential way to reduce the cost of recalibration, many CO_2 sensors offer an *autocalibration feature,* and vendors purport that their sensors do not need to be calibrated for 5 years. Autocalibration assumes that in the middle of the night or early morning hours, the building will have been flushed to outdoor background levels, which should be constant at about 400 ppm. Based on this, the sensor averages the nighttime values and periodically recalibrates its offset, using one point calibration, to this averaged nighttime value.

This autocalibration function often causes more accuracy problems than it solves, due to three major issues. First, in many buildings, particularly during the week, nighttime CO_2 building levels do not reach background or outdoor levels, particularly if fans are turned off during unoccupied times. Consequently, the CO_2 sensor will read a level higher as the background level, say 600 ppm, and operate as if it is the assumed outdoor background level of 400 ppm. Over time, this will cause a sensor error of 200 ppm or 33%.

A second issue is that outdoor background levels themselves can vary by more than 100 ppm. As a result, if in the middle of the night the building did reach outdoor background levels, this level is likely not to be the assumed 400 ppm and will probably vary over time. If the nighttime level is 500 ppm and the assumed background level is 400 ppm, this alone will create an error of 100 ppm or 25% error.

CO_2 sensor location in the space also contributes to the accuracy of DCV control. ASHRAE requires that the sensor be located in the occupied zone or in the return airstream. In-room sensors are normally wall-mounted but should not be located close to doorways, locations that are impinged by direct supply air, or locations that are close to where people regularly sit or stand (within 6 ft [2 m] or so) must also be avoided.

Duct-mounted sensors may be located in the return air of single-zone air systems that utilize ducted returns. Sensors should never be located within return air plenums or at the face of return air inlets. Finally, any duct-mounted sensor must be "bias" calibrated to reflect CO_2 concentration in the occupied zone of the space.

Even if everything is done correctly, the true performance of any DCV system design depends on numerous, typically unquantifiable, factors. So, when should DCV be utilized? As yet, there is no definitive answer to this question. However, the following guidelines appear reasonable and may be applied until more research is performed and better information becomes available:

1. *Minimize the use of CO_2 sensing as a control mode for DCV.* Utilize space occupancy schedules and/or space occupancy/vacancy sensors to turn space ventilation airflows on and off to ensure that DCV control will not be active when the space is unoccupied.
2. Obviously, apply DCV where dictated by standards or codes. *But, DCV will be most effective in spaces that must not only be designed for high occupancy, but when that high occupancy occurs only at infrequent intervals … auditoriums, gymnasiums, cafeterias or*

dining halls, conference rooms, and so on. These types of spaces are good candidates for DCV using *multilevel occupancy schedules* and ventilation airflow rates that switch on/off for each level of occupancy.

3. *Finally, never apply DCV using CO_2 sensing as the control mode in a facility whose owner is unable or unwilling to commit to implementing a CO_2 sensor test/recalibration program as part of an ongoing annual maintenance program.*

10.4.4 Dedicated Outdoor Air Systems

One approach to meeting ventilation air requirements is to use *dedicated outdoor air systems* (DOAS) that condition the outdoor ventilation makeup air separately from the return air from the conditioned space. This approach to handling ventilation air results in superior humidity control by dealing with the primary source of humidity in most buildings, the ambient humidity carried into the building by the ventilation air, directly at its source.

When the DOAS removes enough extra moisture from the ventilation air to handle the building interior latent cooling load, energy savings can be obtained by running the separate, sensible cooling only, interior cooling system at higher evaporating temperature, improving the energy efficiency. Further energy savings are realized by providing only the amount of ventilation air necessary and by using enthalpy recovery for the building exhaust air to precool the makeup air.

Handling the treatment and distribution of ventilation air and return air from the occupied space with separate, parallel systems offers a number of potential advantages over conventional VAV systems that help to overcome the problems discussed earlier. Many of these advantages directly result in significant energy savings:

1. The ventilation makeup air system can be sized and operated to provide the ventilation airflow rate required by code and provide this flow rate regardless of the interior temperature, without any need to oversize the ventilation rate. The ventilation rate can be constant, or it can be varied based on the building operating/occupancy schedule or in response to the actual occupancy (on a real-time basis). Moreover, a DOAS allows easy verification that the system supplies the minimum OA quantities to different portions of a building. Energy recovery heat exchange between the makeup air and exhaust is readily implemented in this configuration, reducing peak cooling and heating loads to condition makeup air.
2. The predominant humidity load in most buildings in hot, humid climates is the humidity brought in with the ventilation makeup air in hot weather. Thus, the entire humidity load for the building can be handled efficiently by separately conditioning the makeup air via a DOAS so that excess ambient humidity is removed, along with an additional amount to offset internal moisture gain.
3. With the ventilation air separately conditioned, and with the entire building humidity load handled in the process, the main air-conditioning system can be operated to maintain temperature control by addressing sensible cooling loads only. Consequently, the cooling system can be operated at a higher than normal evaporating temperature, increasing the compressor coefficient of performance (COP).

In addition to providing independent temperature and humidity control, a DOAS is ideal for use with VAV air-conditioning systems. The conditioned airflow rate is varied in proportion to the net cooling or heating load, saving significant amounts of fan power during the large proportion of the year when full heating or cooling capacity is not required. Meanwhile, the parallel ventilation air system continues to deliver the appropriate amount of air to each space for IAQ purposes without introducing the control complexity required for a VAV system under ASHRAE *Standard 62.1*.

These advantages can be realized in either a single- or a multiple-zone HVAC system configuration. A DOAS can realize efficiency gains by greatly reducing the introduction of excess outdoor air required to achieve minimum outside air levels in a multi-zone system, thus reducing the amount of outside air-conditioning (both heating and cooling) required.

Thus, DOAS achieves energy savings via three primary factors: optimal use of the ventilation air provided; ready use of heat recovery to precool the outdoor air; and allowing interior cooling load to be handled at higher refrigerant temperature and COP since the DOAS can take care of all latent cooling requirements. And, combining DOAS with a sensible cooling only VAV system saves additional energy by reducing total ventilation airflow and by handling sensible cooling loads more efficiently.

But, DOAS has its disadvantages:

1. An additional air-handling unit and duct system is required, which may present serious space limitation problems and will certainly increase costs.
2. For the ventilation air to offset each individual zone's latent cooling requirements, the zonal distribution of latent loads must be essentially the same as the distribution of ventilation airflows. This will be the case when the only source of latent loads in the building are people. But, in zones where there are latent loads from process equipment or other sources, ventilation airflow may be insufficient to offset the moisture gains and these zones will develop high humidity.
3. *High aspiration diffusers* are often required when the design calls for distribution of ventilation air directly into each space. A diffuser design with a high aspiration ratio induces room air toward the supply diffuser, providing good mixing with relatively low ventilation airflow through the diffuser. If standard diffusers are used, air velocities will be too low and drafts or dumping will result.
4. Ventilation air dew-point temperatures of 40°F–45°F [4°C–7°C] are normally required for the ventilation air to offset all building latent cooling loads. This will result in a cooling coil sensible heat ratio of 0.30–0.40, which is almost impossible (or, at least, very expensive) to achieve with DX equipment. *Therefore, the use of chilled water and/or desiccant dehumidification is recommended.*

Bibliography

ASHRAE *Handbook-HVAC Systems and Equipment*, American Society of Heating, Refrigerating, and Air-Conditioning Engineers, Inc. (Atlanta, GA, 2016).

"Economizer High Limit Controls and Why Enthalpy Economizers Don't Work," Steven T. Taylor, P.E., Fellow ASHRAE, and C. Hwakong Cheng, Member ASHRAE, *ASHRAE Journal* (November 2010).

11

Refrigeration Systems and Components

11.1 Refrigeration Basics

11.1.1 Refrigerant Issues

A refrigerant is any chemical compound that is capable of going through phase changes in order to absorb or release heat.

Early refrigerants, developed in the 1920s and 1930s for use in vapor compression cycle refrigeration applications (e.g., R-11, R-12, and R-503), were predominately chemical compounds made up of *chlorofluorocarbons* (CFCs). While stable and efficient in the range of temperatures and pressures required for heating, ventilating, and air-conditioning (HVAC) use, any escaped refrigerant gas was found to be long-lived in the atmosphere. In the lower atmosphere, the CFC molecules absorb infrared radiation and, thus, contribute to atmospheric warming. Once in the upper atmosphere, the CFC molecule breaks down to release chlorine that destroys ozone and, consequently, damages the atmospheric ozone layer that protects the earth from excess ultraviolet (UV) radiation.

The manufacture of CFC refrigerants in the United States and most other industrialized nations was eliminated by international agreement in 1996.

Researchers found that modifying the chemical compound of CFCs by substituting a hydrogen atom for one or more of the chlorine or fluorine atoms resulted in a significant reduction in the life of the molecule, thereby reducing the negative environmental impact it may have. These new compounds, called *hydrochlorofluorocarbons* (HCFCs), were used in HVAC refrigeration systems such as R-22 and R-123.

While HCFCs have reduced the potential environmental damage by refrigerants released into the atmosphere, the potential for damage has not been totally eliminated. Again, under international agreement, this class of refrigerants is slated for phaseout for new equipment installations by 2020, with total halt to manufacturing and importing mandated by 2030/2040, as summarized in Table 11-1.

A third class of refrigerants, *hydrofluorocarbons* (HFCs), are not regulated by international treaty, are considered, at least for the interim, to be the most environmentally benign compounds, and are now widely used in HVAC refrigeration systems. However, HFCs can contribute to increased greenhouse gas levels, and both the Kyoto Protocol (1997) and the Paris Agreement (2016) call for reducing use of these refrigerants.

HVAC refrigeration equipment has undergone a transition in the use of non-HCFC refrigerants, including R-410A, R-407C, and R-134A, and further transitions are expected as the search for an environmental benign refrigerants continues.

American Society of Heating, Refrigerating, and Air-Conditioning Engineers (ASHRAE) *Standard 34* classifies refrigerants according to their toxicity (A = nontoxic and B = evidence of toxicity identified) and flammability (1 = no flame propagation, 2 = low flammability,

TABLE 11-1

Implementation of HCFC Refrigerant Phaseout in the United States

1 January of Year	Phaseout of HCFC-22	Phaseout of HCFC-123
2020	All production/import of HCFC-22 stops. Significant recycling required to provide refrigerant for servicing.	Refrigerant production reduced to 0.5% of 1989 level… end of new equipment using HCFC-123. Significant recycling required to provide refrigerant for servicing.
2030	By this date, it is anticipated that no HCFC-22 equipment will remain in service.	All production/import of HCFC-123 stops. Significant recycling required to provide refrigerant for servicing.
2040	Not applicable.	By this date, it is anticipated that no HCFC-123 equipment will remain in service.

and 3 = high flammability). Thus, all refrigerants fall within one of the six "safety groups": A1, A2, A3, B1, B2, or B3. *For HVAC refrigeration systems, only A1 refrigerants should be considered.* Table 11-2 lists the safety group classifications for refrigerants commonly used in HVAC applications.

11.1.2 Vapor Compression Refrigeration Cycle

The term *refrigeration*, as part of a building energy system, generally refers to a *vapor compression system* wherein a chemical substance alternately changes from liquid to vapor (gas) and back again. The first phase change is *evaporation*, during which the refrigerant absorbs heat and provides a cooling effect, and the second is *condensation*, where the refrigerant releases heat to some "heat sink," usually the outdoor air or water that is cooled by a cooling tower. This "cycle" actually has four distinct steps:

1. *Compression*: Low-pressure refrigerant gas is compressed, thus raising its pressure by expending mechanical energy. (Thermodynamically, there is a corresponding increase in temperature along with the increased pressure.)
2. *Condensation*: The high-pressure gas is cooled by outdoor air or water and condenses into a liquid at high pressure.
3. *Expansion*: The high-pressure liquid flows through an orifice in the expansion valve, thus reducing the pressure. (A small portion of the liquid "flashes" to gas due to the pressure reduction.)
4. *Evaporation*: The low-pressure liquid absorbs heat from the indoor air or water and evaporates into gas. The low-pressure gas returns to the compressor, and the process repeats.

TABLE 11-2

Current HVAC Refrigerants Safety Groups

Refrigerant	Type	Safety Group
R-134A	HFC	A1
R-407C	HFC blend	A1
R-410A	HFC blend	A1
R-717	Ammonia	B2
R-718	Water	A1

As shown in Figure 11-1, the vapor compression refrigeration system consists of four components that perform the four steps of the refrigeration cycle.

The *compressor* raises the pressure of the initially low-pressure refrigerant gas. The *condenser* is a heat exchanger that cools the high-pressure gas so that it changes phase to liquid. The *expansion valve* controls the pressure ratio, and thus flow rate, between the high- and the low-pressure regions of the system. The *evaporator* is a heat exchanger that heats the low-pressure liquid, causing it to change phase to gas.

Thermodynamically, the most common representation of the basic refrigeration cycle is made using a *pressure–enthalpy chart*. For each refrigerant, the phase-change line represents the conditions of pressure and total heat content (enthalpy) at which it changes from liquid to gas and vice versa. Thus, each of the steps of the vapor compression cycle can easily be plotted to demonstrate the actual thermodynamic processes at work, as shown in Figure 11-2.

Point 1 represents the condition of the refrigerant *entering* the compressor. Compression of the gas raises its pressure from P_1 to P_2. The "work" that is done by the compressor adds heat to the refrigerant, raising its temperature and slightly increasing its heat content. Point 2 represents the condition of the refrigerant *leaving* the compressor and *entering* the condenser. In the condenser, the gas is cooled, reducing its enthalpy from h_2 to h_3.

Point 3 to point 4 represents the pressure reduction that occurs in the expansion process. Due to a small percentage of the liquid evaporating as a result of the pressure reduction, the temperature and enthalpy of the remaining liquid is also reduced slightly. Point 4, then, represents the condition *entering* the evaporator. Point 4 to point 1 represents the heat gain by the liquid, increasing its enthalpy from h_4 to h_1, completed by the phase change from liquid to gas at point 1.

For any refrigerant whose properties are known, a pressure–enthalpy chart can be constructed and the performance of a vapor compression cycle analyzed by establishing the high and low pressures for the system. Note that Figure 11-2 represents an "ideal" cycle, and in actual practice, there are various departures dictated by second-law inefficiencies.

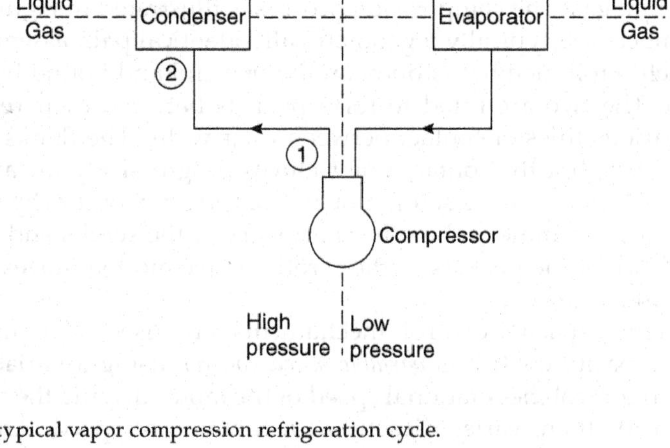

FIGURE 11-1
Components of the typical vapor compression refrigeration cycle.

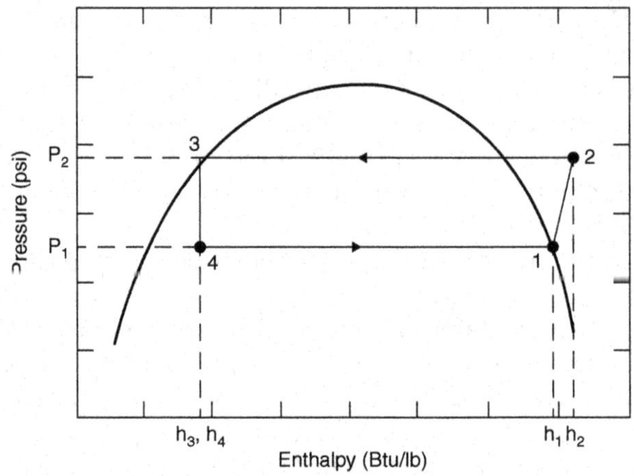

FIGURE 11-2
Ideal refrigeration cycle imposed over pressure versus enthalpy diagram.

Any substance that absorbs heat may be termed a *refrigerant*. Secondary refrigerants, such as water or brine, absorb heat but do not undergo a phase change in the process. *Primary refrigerants, then, are those substances that possess the chemical, physical, and thermodynamic properties that permit their efficient use in the vapor compression cycle.*

In the vapor compression cycle, a refrigerant must satisfy several (and sometimes conflicting) requirements:

1. The refrigerant must be chemically stable in both the liquid and the vapor state.
2. Refrigerants for HVAC applications must be nonflammable and have low toxicity.
3. Finally, the thermodynamic properties of the refrigerant must meet the temperature and pressure ranges required for the application.

There are three main types of compressors used for HVAC applications: *scroll, screw,* and *centrifugal*. The first two are *positive displacement* compressors, whereas the third raises the pressure of the refrigerant gas by increasing its kinetic energy.

1. *Scroll compressors*: Scroll compressors are orbital motion compressors that use two spiral-shaped scroll members, one that is fixed and the other that rotates, to compress refrigerant gas. The compression process is illustrated in Figure 11-3.

   Scroll members are typically a geometrically identical pair, assembled 180° out of phase. Each scroll member is open on the one end and bound by a base plant on the other. The two are fitted to form pockets between their respective base plates and various lines of contact between their walls. The flanks of the scrolls remain in contact, but the contact point moves progressively inward, compressing the refrigerant gas, as one scroll moves. Compression occurs by sealing gas in pockets of a given volume at the other periphery of the scrolls and progressively reducing the size of the pockets as the scroll relative motion moves them inward toward the discharge port.

   Two different capacity control mechanisms are used. The most common approach to capacity control is *variable speed control*, using a variable frequency drive (VFD) to control the rotational speed of the motor driving the moving scroll. Cooling capacity, then, varies directly as a function of speed. Another control

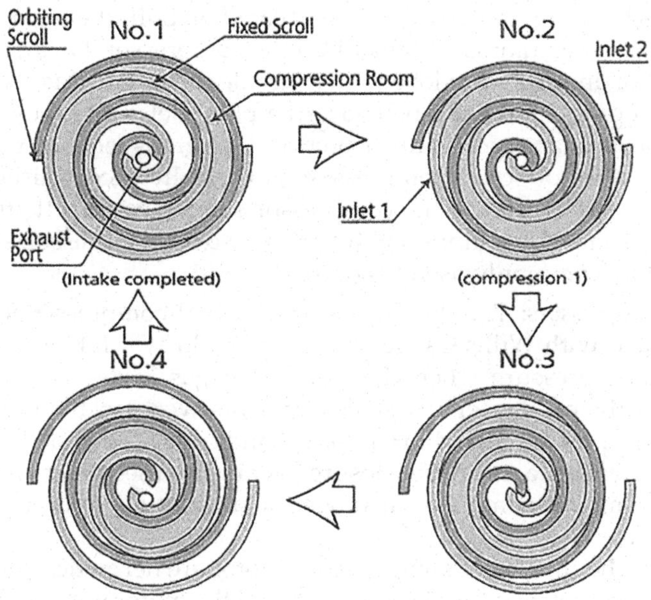

FIGURE 11-3
Scroll compressor operation. [EggEncyclopedia (http://www.enggcyclopedia.com/2012/03/scroll-compressors/). Widely circulated with a Creative Commons license.]

method is called *variable displacement*, which incorporates "porting" holes in the fixed scroll. Capacity control is provided by disconnecting or connecting compression chambers on the suction side by closing or opening these porting holes.

2. *Screw compressors*: Figure 11-4 illustrates the rotary screw compressor.
   Screw compressors typically use double-mating helical grooved rotors with "male" lobes and "female" flutes or gullies within a stationary housing. Compression is obtained by a direct volume reduction with pure rotary motion.

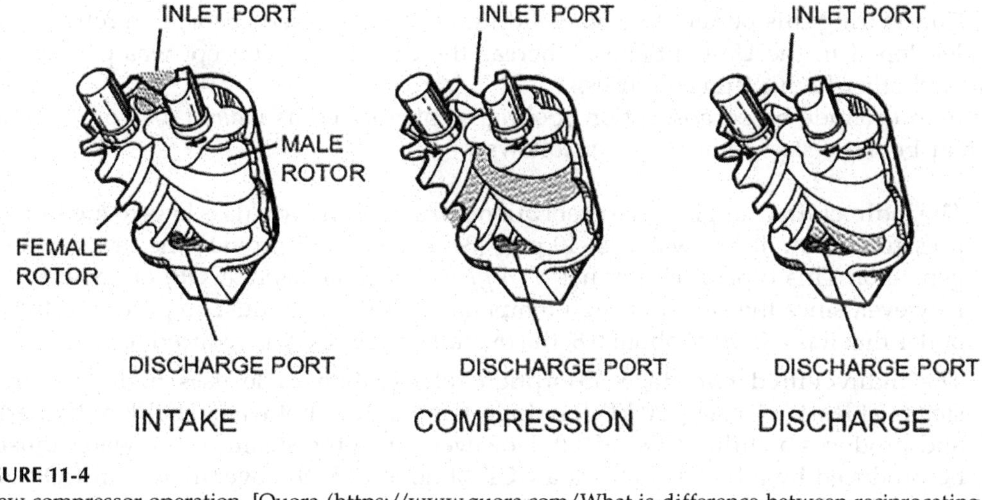

FIGURE 11-4
Screw compressor operation. [Quora (https://www.quora.com/What-is-difference-between-reciprocating-and-screw-compressor). Widely circulated with a Creative Commons license.]

As the rotors begin to unmesh, a void is created on both the male and the female side, allowing refrigerant gas to flow into the compressor. Further rotation starts the meshing of another male lobe with a female flute, reducing the occupied volume, and compressing the trapped gas. At a point determined by the design volume ratio, the discharge port is uncovered and the gas is released to the condenser.

Capacity control of screw compressors is typically accomplished by opening and closing a slide valve on the compressor suction to throttle the flow rate of refrigerant gas into the compressor. In a larger sized compressor, speed control is becoming more commonly used to control capacity.

3. *Centrifugal compressors*: The design of a centrifugal compressor for refrigeration duty originated with Willis Carrier just after World War I. Kinetic energy is converted to static pressure when the refrigerant gas leaves the compressor and expands into the condenser. These compressors work on the same principle as for fans and pumps discussed in Chapter 9 and consist of impellor(s) on a motor-driven shaft within a casing/enclosure. Typically, these compressors are constructed with multiple impellors arranged in a series configuration and rotate at very high rpm.

   Commonly, the compressor and its drive motor are hermetically sealed within a single casing and refrigerant gas is used to cool the motor windings during operation. Low-pressure gas flows from the evaporator to the compressor. Capacity is controlled by a set of *pre-swirl inlet vanes* and/or the compressor rotational speed to regulate the refrigerant gas flow rate through the compressor in response to the cooling load imposed. Normally, the output of the compressor is fully variable within the range 15%–100% of full-load capacity, though the most efficient operating point for the compressor is usually about 80% of full-load capacity.

11.1.3 Vapor Absorption Refrigeration Cycle

Absorption refrigeration cycle is relatively old technology. The concept dates from the late 1700s, and the first absorption refrigeration machine was built in the 1850s. However, by World War I, the use of vapor compression reciprocating compressors had advanced to the point where interest in and development of absorption cooling essentially stagnated until the 1950s. During this period, two-stage lithium bromide-based absorption refrigeration was developed in the United States, whereas the direct-fired concept was perfected in Japan and other Pacific Rim countries.

Lithium bromide-based absorption cooling is categorized as *indirect-fired* or *direct-fired* and may be *single-stage* or *two-stage*, as follows:

1. The indirect-fired single-stage absorption refrigeration cycle uses low- to medium-pressure steam or hot water (5–40 psig [34–276 kPa]) to provide the heat for the generator. This type of chiller has a *coefficient of performance* (COP) of about 0.67. However, since the steam or high-temperature hot water must be produced by a boiler that has a COP of about 0.8, the overall system COP is reduced to 0.54.
2. The indirect-fired two-stage absorption refrigeration cycle uses high-pressure steam (at least 100 psig [690 kPa]) or high-temperature hot water (400°F or higher) and produces a chiller COP of 1.0. However, since the steam or hot water must be produced by a boiler that has a COP of about 0.8, the overall system COP is reduced to 0.80.

3. The direct-fired cooler, as its name implies, does not use steam or hot water but uses a natural gas and/or fuel oil burner directly to provide heat. These chillers are of two-stage, with a resulting overall COP of 1.0–1.1. The elimination of the additional production of high-pressure steam in a separate boiler and the resulting increase in efficiency is the primary advantage to direct-fired equipment. The refrigerant used in absorption cooling is *distilled water.* Water will "boil," changing from a liquid to a gas, at lower temperatures as its pressure is reduced.

For water to act as a refrigerant, the inside of absorption coolers must be maintained at low pressures: about 0.15 psia [1.03 kPa] in the evaporator and 1.5 psia (3.054 in. Hg) in the condenser. Both of these conditions are well below the standard atmospheric pressure of 14.7 psia [101 kPa].

The absorption cycle uses a second fluid, called an *absorbent,* which is used to "carry" the refrigerant from the low-pressure evaporator to the higher-pressure condenser. The most common absorbent for large coolers is *lithium bromide,* a nontoxic salt that has a very high affinity for water, whereas the most common absorbent for small coolers is *ammonia.*

As shown in Figure 11-5, with a single-effect absorption refrigeration cycle, an absorber, generator, pump, and recuperative heat exchanger replace the compressor used in the vapor compression refrigeration cycle.

The cycle "begins" when high-pressure liquid refrigerant from the condenser passes through a metering device (1) into the lower-pressure evaporator (2) and collects in the evaporator pan or sump. The "flashing" that occurs at the entrance to the evaporator cools the remaining liquid refrigerant. Similarly, the transfer of heat from the comparatively warm system water to the now-cool refrigerant causes the latter to evaporate (2), and the resulting refrigerant vapor migrates to the lower-pressure absorber (3).

There, it is "soaked up" by an absorbent lithium bromide solution. This process not only creates a low-pressure area that draws a continuous flow of refrigerant vapor from the evaporator to the absorber, but also it causes the vapor to condense (3) as it releases the heat of vaporization picked up in the evaporator. This heat, along with the heat of dilution

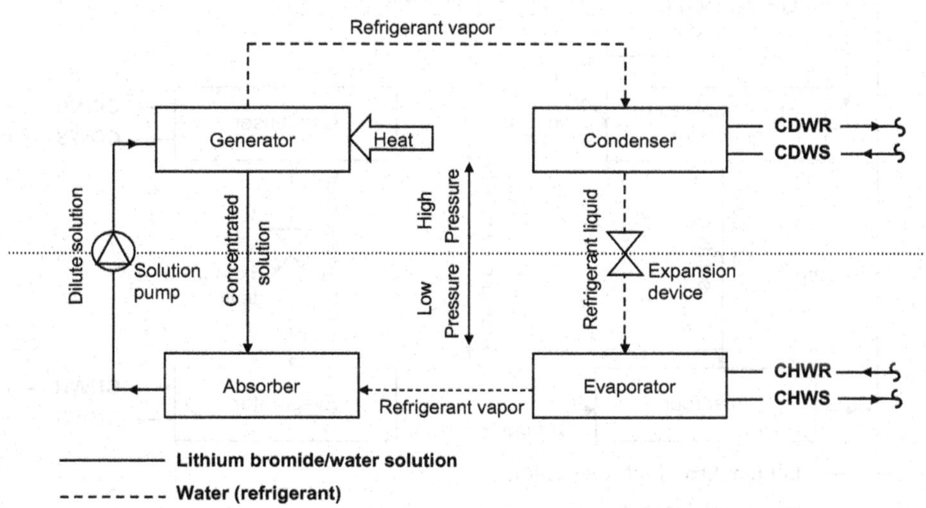

FIGURE 11-5
Lithium bromide single-effect absorption refrigeration cycle.

produced as the refrigerant condensate mixes with the absorbent, is transferred to the cooling water and released through the condenser.

Assimilating refrigerant dilutes the lithium bromide solution and reduces its affinity for refrigerant vapor. To sustain the refrigeration cycle, the solution must be reconcentrated. This is accomplished by constantly pumping (4) dilute solution from the absorber to the generator (5), where the addition of heat boils the refrigerant from the absorbent. Once the refrigerant is removed, the reconcentrated lithium bromide or ammonia solution returns to the absorber, which is ready to resume the absorption process.

Meanwhile, the refrigerant vapor "liberated" in the generator migrates to the cooler condenser (6). There, the refrigerant returns to its liquid state as the cooling water picks up the heat of vaporization carried by the vapor. The liquid refrigerant's return to the metering device (1) completes the cycle.

The double-effect absorption refrigeration cycle, illustrated in Figure 11-6, takes absorption to the next level.

The easiest way to picture a double-effect cycle is to think of two single-effect cycles stacked on top of each other. The cycle on top is driven either directly by a natural gas or oil burner, or indirectly by steam. Heat is added to the generator of the topping cycle (primary generator), which generates refrigerant vapor at a relatively higher temperature and pressure. The vapor is then condensed at this higher temperature and pressure, and the heat of condensation is used to drive the generator of the bottoming cycle (secondary generator), which generates refrigerant vapor at a lower temperature and pressure.

If the heat added to the generator is thought to be equivalent to the heat of condensation of the refrigerant, it becomes clear where the efficiency improvement comes from. For every unit of heat into the primary generator, two masses of refrigerant are boiled out of solution, or generated: one in the primary generator and the other in the secondary generator. In a single-effect cycle, only one mass is generated.

Therefore, in a double-effect system, twice the mass flow of refrigerant is sent through the refrigerant loop per unit of heat input, so twice the cooling is delivered per unit of

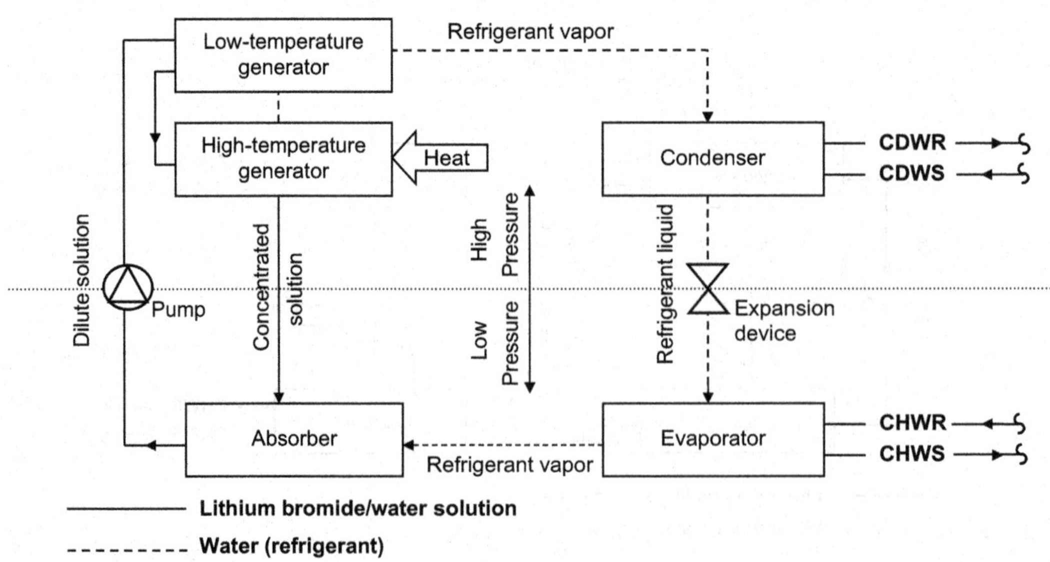

FIGURE 11-6
Lithium bromide double-effect absorption refrigeration cycle.

heat input. Using this approach, a double-effect system has a COP that is roughly twice that of a single-effect cycle. However, this simplifying assumption does not account for cycle inefficiencies and losses, and the net result is a chiller that is only about 50% more efficient.

11.2 Direct Expansion Refrigeration Systems

The basic vapor compression cycle, when applied directly to the job of space cooling, is referred to as a *direct expansion* or "DX" *refrigeration system*. This reference comes from the fact that the building indoor air that is cooled passes "directly" over the refrigerant evaporator without a secondary refrigerant being used.

All residential and the majority of commercial HVAC systems use one DX piece equipment or split systems to provide cooling, with or without a heating function. The Air-Conditioning, Heating, and Refrigeration Institute (AHRI) standard ratings for packaged equipment are defined as follows:

Evaporator entering air temperature	80°F DB/67°F WB [26.7/19.4°C]
Outdoor air temperature for air-cooled condensers	95°F DB/75°F WB [35.0/23.9°C]
Water conditions for water-cooled condensers	85°F EWT/95°F LWT [29.4/35.0°C]

AHRI defines seven different configurations of packaged equipment for rating purposes:

1. *Single-package cooling units*: Electric-drive DX cooling with optional electric or hydronic heating
2. *Single-package cooling/heating units*: Electric-drive DX cooling with gas heating
3. *Single-package cooling units for variable air volume systems*: Electric-drive DX cooling with optional electric heat, with evaporator variable airflow
4. *Multizone cooling/heating units*: Electric-drive DX cooling with gas, electric, or hydronic heating and multizone air distribution
5. *Split systems*: Electric-drive DX cooling provided by a compressor/condenser assembly with gas, electric, or hydronic heating provided by a separate furnace or air handler assembly
6. *Single-package heat pumps*: Electric-drive DX cooling/heating with (optional) electric supplemental heating in a single assembly
7. *Large-capacity split system (air-source) heat pump*: Electric-drive DX cooling/heating provided by an outdoor unit (consisting of compressor, outdoor coil and fan, and reversing valve) and a separate indoor unit (blower coil units [BCU] or air-handling unit [AHU])

Table 11-3 shows the typical performance for a nominal 10-ton [35 kW] single-package unit.

In Table 11-3, "TC" represents the total cooling capacity of the unit (Btu/hr), "SC" is the sensible cooling capacity (Btu/h [kW]), and "kW" is the input power requirement to the

TABLE 11-3

Performance Data for Typical Nominal 10 ton [35 kW] Single-Package Unit

Outdoor Air Temperature (°F) [°C]		Evaporator Airflow (CFM [L/s]/Coil Bypass Factor)								
		3,000 [1,416]/0.153			4,000 [1,888]/0.187			5,000 [2,360]/0.216		
		Evaporator Entering Air WB Temperature (°F)								
		72 [26.2]	67 [19.4]	62 [16.7]	72 [26.2]	67 [19.4]	62 [16.7]	72 [26.2]	67 [19.4]	62 [16.7]
85 [29.4]	TC	132 [39]	123 [36]	110 [32]	136 [40]	128 [38]	118 [35]	138 [40]	131 [38]	124 [36]
	SC	65 [19]	81 [24]	95 [28]	70 [20]	91 [27]	109 [32]	75 [22]	99 [25]	120 [35]
	kW	12.2	11.7	11.1	12.4	11.8	11.4	12.5	12.1	11.7
95	TC	128 [38]	116 [34]	100 [29]	133 [39]	123 [36]	110 [32]	135 [40]	136 [40]	117 [34]
	SC	63 [18]	79 [23]	91 [27]	70 [21]	90 [26]	106 [31]	75 [22]	99 [29]	117 [34]
	kW	13.0	12.3	11.9	13.2	12.3	11.9	13.4	12.9	12.3
105	TC	122 [36]	107 [31]	90 [26]	127 [37]	114 [33]	100 [29]	130 [38]	119 [35]	109 [32]
	SC	61 [18]	75 [22]	85 [25]	68 [20]	87 [25]	100 [29]	73 [21]	97 [28]	109 [32]
	kW	13.6	13.1	12.6	13.9	13.1	12.9	14.1	13.5	13.1

unit at the defined conditions. All of these ratings are based on 80°F [26.7°C] DB air entering the evaporator.

To select a packaged unit to satisfy a given building or zone cooling requirement, both the total cooling load and the sensible cooling load must be evaluated. From Chapter 2, we know that the sensible heat ratio (SHR) line defines the room conditions for the specific sensible and total cooling requirements and a given supply air condition. For DX cooling, the supply air temperature is dictated by the evaporator *saturated suction temperature* (typically about 40°F [4.4°C]) and the evaporator coil bypass factor. Neither of these conditions can be controlled by the designer.

Therefore, to apply packaged equipment to satisfy a specific set of space conditions and loads, the actual performance of the unit must be determined, as opposed to the cataloged performance under specific rating conditions. Rarely will actual evaporator coil entering air conditions and actual SHR match the rating conditions.

Manufacturers provide both printed catalog data (much like Table 11-3) and computer-based product selection software for the designer to use to select packaged equipment to satisfy actual space load and condition requirements. *It is strongly recommended that designers avoid making any specific product selections using manufacturer-supplied data or software.* First, a "nominal" capacity unit from one manufacturer may or may not match the performance of a "nominal" capacity unit by another manufacturer. Second, interpolation of printed data tables or use of a software package may introduce errors on the part of the designer due to him or her not being familiar with the material or simply making a mistake. *It is much more prudent for the designer to specify the actual conditions and loads required of the packaged equipment and let each individual manufacturer offer an equipment selection to satisfy these requirements. Then, if the unit does not perform, the manufacturer is liable, not the designer.*

Packaged equipment is designed to keep the costs of HVAC systems low. Therefore, it is critical that packaged equipment be applied and installed in accordance with the manufacturer's requirements and ratings to ensure that the resulting actual performance meets the manufacturer's advertised performance.

To provide a consistent method of evaluating the energy efficiency of heating and cooling equipment, the U.S. Department of Energy (USDOE), as required by the *National Appliance*

Energy Conservation Act, has established test methods and terminology to define "energy conservation efficiency," as follows:

1. *Annual fuel utilization efficiency* (AFUE): The measure of seasonal or annual efficiency of a residential heating furnace or boiler. It takes into account the cyclic on/off operation and associated energy losses of the heating unit as it responds to changes in the load. (Note: AFUE is stated as "percent," and these values must, therefore, be divided by 100 to obtain the decimal value.)
2. *Seasonal energy efficiency ratio* (SEER): The measure of seasonal or annual efficiency of an electric-drive central air conditioner or air-conditioning heat pump during the cooling season. It takes into account the variations in temperature that can occur within a season and is the average number of Btu's of cooling delivered for every Watt-hour of electricity used by the air conditioner or heat pump over a cooling season.
3. *Heating season performance factor* (HSPF): The measure of heating seasonal efficiency of an electric-drive heat pump. The HSPF is the ratio of seasonal heating output in Btu's [Watt-hours] divided by the seasonal heating power consumption in Watt-hours.

All AFUE, SEER, and HSPF ratings are based on standardized weather profiles and cycling factors dictated by USDOE.

The Federal seasonal heating and cooling efficiency rating methods and terms were designed to allow comparison between competing equipment of the same type; that is, different air conditioners could be compared on the basis of SEER, whereas different furnaces could be compared on the basis of AFUE. However, these measures do not allow comparison between different types of systems, there is no direct correlation between AFUE and HPSF that allows the comparison of a furnace to the heating cycle of a heat pump simply by comparing the AFUE of the furnace to the HPSF of the heat pump.

Designers must be aware that USDOE minimum efficiency requirements change over time and that new requirements are expected in 2023. As these requirements are used to update energy conservation building codes, due diligence by designers is required.

To reduce seasonal efficiency measurement terms to a common denominator, engineers use the COP to evaluate the performance of heating and cooling equipment. The COP is the ratio of output Btu's divided by the input Btu's, and the COP can be computed on a peak-load basis, seasonal basis, or any other operating basis that is defined. The AFUE, SEER, and HSPF ratings can be converted to rated *seasonal COP* (SCOP) as follows:

Furnace or boiler heating seasonal COP ($SCOP_h$) = AFUE/100
Air conditioner or heat pump Seasonal cooling COP ($SCOP_c$) = SEER/3.413
Heat pump heating COP ($SCOP_h$) = HSPF/3.413

Electric-drive compressor-based packaged equipment ratings require that the equipment be tested in accordance with American National Standards Institute (ANSI)/ASHRAE *Standard 37* and be rated on the basis of two criteria:

1. *Performance rating*: Standard rating conditions used to establish "cooling steady state" performance are based on 95°F [35°C] outdoor temperature at indoor conditions of 80°F [27°C] dry bulb (DB) temperature and 67°F [19°C] wet bulb (WB)

temperature (50% relative humidity). Standard rating conditions used to establish high- and low-temperature heating performance for heat pumps are 47°F [8°C] and 17°F [–8°C], respectively, outdoor temperature and 70°F [22°C] indoor temperature. These values are used by both AHRI and manufacturers to define the performance of equipment.

2. *Energy efficiency rating.* The AHRI standards contain detailed calculation procedures to be used, utilizing data from specified cooling condition tests and heating condition tests, to determine seasonal cooling and heating energy efficiency ratings, SEER and HPSF. Specific calculation requirements vary based on the type of compressor operation control used, that is, single speed, two-speed, three-stage, and variable speed.

11.3 Chilled Water Refrigeration Systems

DX systems have application, capacity, and performance constraints that limit their use in larger, more complex HVAC applications. For these applications, the use of chilled water systems is dictated. Typical applications for chilled water systems include large buildings (offices, laboratories, etc.) or multibuilding campuses where it is desirable to provide cooling from a central facility.

The typical water-cooled HVAC system has three heat transfer loops, which are as follows:

Loop 1: Cold air is distributed by one or more AHUs to the spaces within the building. Sensible heat gains, including heat from temperature-driven transmission through the building envelope; direct solar radiation through windows; infiltration; and internal heat from people, lights, and equipment, are "absorbed" by the cold air, raising its temperature. Latent heat gains, including moisture added to the space by air infiltration, people, and equipment, are also absorbed by the cold air, raising its specific humidity. The resulting space temperature and humidity condition is an exact balance between the sensible and latent heat gains and capability of the entering cold air to absorb those heat gains.

Loop 2. The distributed air is returned to the AHU, mixed with the required quantity of outdoor air for ventilation, and then directed over the cooling coil where chilled water is used to extract heat from the air, reducing both its temperature and moisture content so it can be distributed once again to the space.

As the chilled water passes through the cooling coil in counterflow to the air, the heat extraction process results in increased water temperature. The chilled water temperature leaving the cooling coil (chilled water return) will be 8°F–16°F warmer than the entering water temperature (chilled water supply) at the design load. This temperature difference (*range*) establishes the flow requirement *via* the relationship shown in Eq. (8-9) or (8-9a).

The warmer return chilled water enters the water chiller where it is cooled to the desired chilled water supply temperature by transferring the heat extracted from the building spaces to a primary refrigerant. This process, obviously, is not "free" since the compressor must do work on the refrigerant for cooling to occur and, thus, must consume energy in the process. Since most chillers are *refrigerant-cooled*, the compressor energy, in the form of heat, is added to the building heat and both must be rejected through the condenser.

Loop 3. The amount of heat that is added by the compressor depends on the efficiency of the compressor. This *heat of compression* must then be added to the heat load on the chilled water loop to establish the amount of heat that must be rejected by the condenser to a *heat sink*, typically the outdoor air.

11.3.1 Determining Chilled Water Supply Temperature

The first step in evaluating a chilled water system is to determine the required chilled water supply temperature. For any HVAC system to provide simultaneous control of space temperature and humidity, the supply air temperature must be low enough to simultaneously satisfy both the sensible and the latent cooling loads imposed. Thus, the required temperature of the supply air is dictated by two factors:

1. The desired space temperature and humidity set point
2. The SHR that is defined by dividing the sensible cooling load by the total cooling load

The intersection between the "room" line and the saturation line on the psychrometric chart represents the required *apparatus dew-point (ADP) temperature* for the cooling coil. However, since no cooling coil is 100% efficient, the air leaving the coil will not be at a saturated condition but will have a temperature of about 1°F–2°F [0.6°C–1°C] above the ADP temperature.

Based on typical coil conditions, the required supply air temperature can be determined by plotting on a psychrometric chart the room conditions point and a line having a slope equal to the SHR passing through the room point, defining the ADP temperature intersection point, and then selecting a supply air condition on this line based on a 95% coil efficiency. Table 11-4 summarizes the results of this analysis for several different typical HVAC room design conditions and SHRs.

For a chilled water cooling coil, *approach* is defined as the temperature difference between the entering chilled water and the leaving (supply) air. While this approach can range as low as 3°F [1.7°C] to as high as 10°F [5.6°C], the typical value for HVAC applications is approximately 7°F [4°C]. Therefore, to define the required chilled water supply temperature, it is only necessary to subtract 7°F [4°C] from the supply air DB temperature determined from Table 11-4.

11.3.2 Establishing Chilled Water Temperature Range

Once the required chilled water supply temperature is determined, the desired temperature range must be established.

TABLE 11-4

Typical Supply Air Temperature Required to Maintain Desired Space Temperature and Humidity Conditions

Space Conditions		Supply Air DB/WB Temperature Required (°F) [°C]	
DB Temperature (°F) [°C]	**Relative Humidity (% RH)**	**0.90 + SHR**	**0.80–0.89 SHR**
75 [24]	50	54/53 [12]	52/51 [11]
70 [21]	50	50/49 [10]	44/43 [7]
65 [18]	50	44/43 [7]	41/40 [5]

From Eq. (8-9) or (8-9a), the required chilled water flow rate is dictated by the imposed cooling load and the selected temperature range. The larger the range, the lower the flow rate and, thus, the less energy consumed for the transport of chilled water through the system. However, if the range is too large, chilled water coils and other heat exchangers in the system require increased heat transfer surface and, in some cases, the ability to satisfy latent cooling loads is reduced.

A 10°F [5°C–6°C] range has typically been used for chilled water systems, resulting in a required flow rate of 2.4 gpm/ton [0.43 L/s·kW] of imposed cooling load. For smaller systems with relatively short piping runs, this range and flow rate are acceptable. However, as systems get larger and piping runs get longer, the use of larger ranges will reduce pumping energy requirements. Lower flow rates can also result in economies in piping installation costs since smaller sizes of piping may be used.

At a 12°F [7°C] range, the flow rate is reduced to 2.0 gpm/ton [0.36 L/s·kW], and at a 14°F [8°C] range, it is reduced to 1.7 gpm/ton. For very large campus systems, a range as great as 16°F [9°C] (1.5 gpm/ton [0.27 L/s·kW]) to 20°F [11°C] (1.2 gpm/ton [0.22 L/s·kW]) may be used (though these low flow rates may introduce problems in selecting cooling coils).

11.3.3 Vapor Compression Cycle Water Chillers

Chilled water has the advantage that fully modulating control can be applied, and thus, closer temperature tolerances can be maintained under almost any load condition.

In the HVAC industry, the refrigeration machine that produces chilled water is generally referred to as a *chiller* and consists of the compressor(s), evaporator, and condenser, all packaged as a single unit. The condensing medium may be water or outdoor air.

The evaporator, called the *cooler,* consists of a shell-and-tube heat exchanger with refrigerant in the shell and water in the tubes. Coolers are typically designed for 3–11 fps [0.9–3.3 m/s] water velocities (though some may operate satisfactorily at flows as low as 1.5 fps [0.5 m/s]) when the chilled water flow rate is selected for a 10°F [6°C] to 20°F [11°C] range.

For air-cooled chillers, the condenser consists of an air-to-refrigerant heat exchanger and fans to provide the proper flow rate of outdoor air to transfer the heat rejected by the refrigerant. For water-cooled chillers, the condenser is a second shell-and-tube heat exchanger with refrigerant in the shell and condenser water in the tubes. Condenser water is typically supplied at 70°F–85°F [21°C–29°C], and the flow rate is selected for a 10°F–15°F [6°C–9°C] range. A cooling tower is typically used to provide condenser water cooling, but other cool water sources (heat sinks), such as wells and ponds, can be used.

Chiller efficiency is typically defined in terms of its COP. The COP, in the IP system, is the ratio of output Btu divided by the input Btu. If the nominal rating of the chiller is 1 ton of refrigeration capacity, equivalent to 12,000 Btu/hr output, and the input energy is equivalent to 1 kW, or 3,413 Btu/hr, the resulting COP is 12,000/3,413 or 3.52. In SI units, the COP would be computed on the basis of 1 kW input energy and 3.517 output energy, again resulting in a value of 3.52.

The peak cooling capacity and the corresponding input energy for water chillers are rated in accordance with the AHRI *Standard 550/590.* Chillers must be constructed and installed in accordance with ASHRAE Standard 15, *Safety Code for Mechanical Refrigeration.* Since the evaporator and condenser for R-22 or R-134A chillers operate at pressures exceeding 15 psig, they are defined as *pressure vessels* and must be constructed in accordance with Section 8 of the American Society of Mechanical Engineers *Boiler and Pressure Vessel Code.* And, all chillers must comply with the requirements of Underwriters Laboratory *Standard 465.*

The rated capacity of a water chiller, determined in accordance with AHRI *Standard 550/590*, is based on a "standard" set of operating conditions, which are as follows:

Chilled water supply temperature	44°F
Chilled water temperature range/flow rate	10°F/2.4 gpm/ton [6°C/0.43 L/s·kW]
Evaporator fouling factor	0.0001
Condenser water supply temperature/flow rate (water-cooled)	85°F/3.0 gpm/ton [29°C/0.54 L/s·kW]
Condenser fouling factor	0.00025
Entering air temperature (air-cooled)	95°F [35°C]

If any of these conditions vary for a specific project or location, the manufacturers will provide a specific cooling rating for the required conditions.

To provide data to evaluate chiller part-load performance, AHRI *Standard 550/590* requires chiller manufacturers to determine the part-load operating performance characteristics, which are as follows:

1. *Integrated part-load value* (IPLV): This rating method requires the manufacturer to determine the input energy requirement for each chiller at the four operating conditions defined in Table 11-5. The "weighing factor" defined in Table 11-5 represents the percentage of the hours of the cooling season at which the chiller is expected to operate at the defined rating point.

   Since chiller energy performance will improve with reduced load and reduced condensing temperatures, the IPLV can be used to compare alternative chiller seasonal energy consumption using a common rating method. All other factors being equal, the chiller with the lower IPLV will have lower seasonal operating costs.
2. *Nonstandard part-load value* (NPLV): The IPLV rating is only directly applicable to the single chiller installation where the condenser water supply temperature is allowed to fall as the outdoor WB temperature falls. Anytime multiple chillers are used, particularly if careful control allows the chillers to be used at or near their optimum efficiency range of 40%–80% of rated capacity, the IPLV has little validity. Likewise, if the condenser water temperature control methods or manufacturer requirements do not allow the condenser water supply temperature to go as low as 65°F [18°C], then the IPLV is not valid. *Therefore, to compare the seasonal energy performance of alternative chillers, it is necessary to define the specific weighing factors and condensing conditions to be applied to compute an NPLV.*

Air-cooled chillers, because they are only about 40%–60% as efficient as water-cooled chillers, should be used only for smaller applications (less than 100 tons [35 kW]) *or where*

TABLE 11-5

IPLV Rating Conditions

Rating Point	Percent of Full Load	Weighing Factor (%)	Water-Cooled Condenser Water Temperature (°F) [°C]	Air-Cooled Entering Air Temperature (°F) [°C]
A	100	1	85 [29]	95 [35]
B	75	42	75 [24]	80 [27]
C	50	45	65 [18]	65 [28]
D	25	12	65 [18]	55 [13]

the client insists. In addition to being less efficient, air-cooled chillers have shorter service lives, higher maintenance costs, and poorer reliability than water-cooled chillers.

In cold climates, freeze protection is required for the outdoor piping and the evaporator associated with an air-cooled chiller. Typically, air-cooled chiller evaporators are factory-insulated with flexible closed-cell foam insulation to a thickness of 1/2–3/4″ [13–20 mm] to prevent surface condensation, as required by the guide specifications. But, in cold climates, the thickness of insulation on the evaporator must be increased to at least 1-1/2″ [40 mm], and the evaporator should be protected by electric heat tracing or factory-installed electric heater. However, a power outage will negate the protection offered by electric heat, and the use of glycol propylene in the chilled water circuit should be considered.

All three types of refrigeration compressors, as previously discussed, are applied for water chiller use: *scroll compressors, rotary screw compressors,* and *centrifugal compressors.* Overall, Table 11-6 summarizes the types of chillers and basic plant capacity control methods that should be applied.

Despite these guidelines, designers must be aware that chiller technology is in a constant state of flux and that "due diligence" is required when selecting chillers for any new chilled water system.

11.3.4 Scroll Compressor Water Chillers

Scroll compressor chillers should be applied only in a capacity range up to about 130 tons [450 kW]. Beyond that, better control and efficiency is offered by variable speed rotary screw or centrifugal compressor chillers. For any scroll compressor chiller over 10 tons

TABLE 11-6

Preferred Chiller Configuration and Capacity Control Method(s)

Total Chiller Capacity (Tons) [kW]	Preferred Chiller Configuration and Capacity Control Method(s)
≤10 [35]	Single or tandem (dual) scroll compressor configuration. With one compressor, it should be variable speed type with internal permanent magnet motor and integral invertor, sized for 100% of design cooling load. Minimum chiller capacity should be 25% or less of chiller rated capacity. With two compressors, one of which should be variable speed type with internal permanent magnet motor and integral invertor, sized for approximately 33% of the design cooling load, and one of which should be constant speed, single stage type with on/off control sized for approximately 67% of the design cooling load. Minimum chiller capacity should be 10% or less of chiller rated capacity.
11–40 [36–140]	Tandem (dual) or trio (three) scroll compressor configuration. With two compressors, one of which should be variable speed type with internal permanent magnet motor and integral invertor, sized for approximately 33% of the design cooling load, and one of which should be constant speed, single stage type with on/off control sized for approximately 67% of the design cooling load. Minimum chiller capacity should be 10% or less of chiller rated capacity. With three compressors, one of which should be variable speed type with internal permanent magnet motor and integral invertor, and the other two should be constant speed, single stage type with on/off control. Each compressor should be sized for approximately 33% of the design cooling load. Minimum chiller capacity should be 10% or less of chiller rated capacity.

(Continued)

TABLE 11-6 (*Continued*)

Preferred Chiller Configuration and Capacity Control Method(s)

Total Chiller Capacity (Tons) [kW]	Preferred Chiller Configuration and Capacity Control Method(s)
41–150 [141–525]	Multiple (3–8) scroll compressor configuration, *with no single compressor exceeding about 30 tons [100 kW] capacity*. Where more than three compressors are applied, there should be an equal number of compressors piped on two independent refrigerant circuits. With three compressors, one of which should be variable speed type with internal permanent magnet motor and integral invertor, and the other two should be constant speed, single stage type with on/off control. Each compressor should be sized for approximately 33% of the design cooling load. Minimum chiller capacity should be 10% or less of chiller rated capacity. With four or more compressors, one compressor should be variable speed type with internal permanent magnet motor and integral invertor, and all others should be constant speed, single stage type with on/off control. Each compressor should be sized for approximately 100/*N* percent of the design cooling load, where *N* is the number of compressors. Minimum chiller capacity should be 25% or less of the first-stage variable speed compressor's rated capacity. Currently, variable speed compressors with internal permanent magnet motor and integral invertor are available only from a limited number of chiller manufacturers. To reduce cost and improve bidding competition for chillers with four or more compressors, all compressors may be constant speed, single stage type with on/off staging control. Where compressors are not of equal capacity, the smallest compressor should be used as the first/last stage of cooling.
151–250 [526–875]	Single variable speed rotary screw compressor type driven by an induction motor and using chiller-mounted VFD to control compressor speed/chiller capacity. In the capacity range of 75–400 tons [260–1,400 kW], some manufacturers offer two-stage centrifugal compressors with internal permanent magnet motor and integral invertor that may be considered. *HGBP piping and control valve, integrated with capacity control sequence, is required to maintain chiller operation down to 0.5°F [0.3°C] differential chilled water temperature but is needed only if this is a single chiller application.*
251–500 [876–1,750]	In this capacity range, the type of chiller/compressor should be selected on the basis of life-cycle cost. *Option 1*: Single/multiple variable speed rotary screw compressor type to control compressor speed/chiller capacity. *Option 2*: Centrifugal compressor type driven by an induction motor and using chiller-mounted VFD. Compressor capacity control is based on the use of pre-rotation inlet guide vanes down to about 50% capacity and then compressor speed control down to 10% capacity. Variable speed centrifugal chillers are available with oil-lubricated bearings, ceramic bearings, or magnetic bearings. *HGBP piping and control valve, integrated with capacity control sequence, is required to maintain chiller operation down to 0.5°F [0.3°C] differential chilled water temperature, but this is needed only if this is a single chiller application.*
>800 [1,750]	Single/multiple centrifugal compressor type driven by an induction motor and using chiller-mounted VFD. The compressor capacity control is based on the use of pre-rotation inlet guide vanes down to about 50% capacity and then compressor speed control down to 10% capacity. Variable speed centrifugal chillers are available with oil-lubricated bearings, ceramic bearings, or magnetic bearings.

[3.5 kW] capacity, the designer should require multiple compressors with at least two independent refrigerant circuits (with the exception that one refrigerant circuit must be acceptable for chillers of ≤20 tons [7 kW] capacity since that is all that is typically offered by manufacturers).

The most common scroll compressor configurations are the single-stage scroll compressor and the two-stage scroll compressor, which incorporates bypass ports that are covered/uncovered during operation to allow the compressor to operate at either of two conditions: 100% or 67% of full capacity. Typically, these compressors are applied in single compressor chillers of 10 tons [35 kW] or less capacity.

In larger chillers, with two equal-capacity two-stage scroll compressors, the minimum steps of capacity reduction are given in Table 11-7.

As the number of compressors is increased from two to four in higher capacity chillers, the minimum chiller load rating improves to 17%, and to 12% with six compressors.

The lack of capacity reduction steps at the low end of the total design load range is good reason to require *hot gas bypass* (HGBP) for water chillers. HGBP will allow the chiller to operate with one compressor, unloaded to 67% of its individual capacity, to continue to operate down to (theoretically) 0% load. In actual practice, though, the available accuracy limitation for chilled water supply temperature sensing/control, coupled with limits imposed by manufacturers to ensure proper oil circulation on the refrigerant side, means that a compressor will cycle off at about 10% of load (although Trane limits theirs to about 15% minimum load). *For typical comfort air-conditioning applications, this type of control is adequate.*

The *digital scroll compressor* concept was introduced basically to provide better capacity-to-load operating characteristics in smaller, single compressor applications. The digital scroll compressor is capable of modulating its cooling capacity from full capacity to 10% minimum capacity (though some chiller manufacturers limit the turndown on their scroll compressors to only 25%–40%, *caveat emptor*).

The digital scroll compressor adds a normally closed (de-energized) solenoid valve to control HGBP on a constant speed, conventional single-stage scroll compressor. When the solenoid valve is in its normally closed position, the compressor is "loaded" and operates at 100% capacity. When the solenoid valve is energized and opens, pressure reduction causes the two scroll elements to move apart axially, and thus, there is no compression. In this mode, the compressor is "unloaded," delivering essentially 0% capacity even though the compressor motor continues running and the scroll elements continue to rotate.

The digital compressor loading cycle consists of a loaded state and an unloaded state. By varying the time of the loaded and unloaded states *via* pulse width modulation (PWM) control of the solenoid valve, an "average" resulting capacity over a given time cycle is

TABLE 11-7

Minimum Capacity Reduction Steps for Dual Two-Stage Scroll Compressors

Percent Total Chiller Capacity	Compressor No. 1 Percent Capacity	Compressor No. 2 Percent Capacity
100	100	100
83	100	67
67	67	67
33	0	67
0	0	0

obtained. The lowest achievable capacity is 10%, which equates to 1.5 s of cooling during one 15 s cycle (or 2.0 s during a 20 s cycle used by some manufacturers).

An example for the 15 s controller cycle: In any 15 s cycle, if the loaded time is 10 s and the unloaded time is 5 s, the average capacity is 67%. If the loaded time is 5 s and the unloaded time is 10 s, the average capacity during that 15 s period is 33%.

Effectively, the digital scroll compressor is operating with "on/off" and unloading control, but instead of cycling the compressor motor, the compressor mechanical components are cycled. However, even when unloaded, the compressor still consumes some energy overcoming mechanical and "windage" losses in both the compressor and its motor; thus, more energy is consumed by the digital compressor than would be consumed from the same compressor with on/off cycling capacity control.

For water chillers using two equal-capacity scroll compressors, one that is a single-stage and the other that is digital (the most common configuration offered by manufacturers), the load/capacity operation is given in Table 11-8.

As long as the lead compressor will unload to 10%–20% of its rated capacity with this configuration, there is probably no need for HGBP for typical comfort air-conditioning applications. However, if the lead capacity will unload only to 30%–40% of its rated capacity, overall chiller performance is really no better than with two-stage compressors. The designer must carefully evaluate the low load performance of digital scroll chillers to ensure that minimum imposed loads can be met satisfactorily.

Since about 2010, the application of variable speed scroll compressors has become common. These compressors come in two configurations: (1) induction motor-driven compressors whose speed is varied *via* a PWM VFD and (2) direct current permanent magnet motor-driven compressors using electronic commutation motors, wherein speed is varied by an integral invertor. The part-load characteristics of the variable speed scroll compressor is essentially the same as for the digital scroll compressor, but the energy consumption by the compressor at part-load operation is reduced significantly (by as much as 60% at 25% load).

Even though water chillers with digital or variable speed compressors may be satisfactory for comfort air-conditioning applications without need for HGBP, HGBP is required for cooling applications that demand tight, continuous, thermal control, particularly if large amounts of outdoor air, widely varying loads, or excessive compressor on/off delays

TABLE 11-8

Dual Scroll Compressor Capacity Reduction Capability

Percent Total Chiller Capacity	Compressor No. 1 Percent Capacity (Lead)	Compressor No. 2 Percent Capacity (Lag)
100	100	100
90	80	100
80	60	100
70	40	100
60	20	100
50	100	0
40	80	0
30	60	0
20	40	0
10	20	0
5	10	0

are involved. These applications are characterized as "process" cooling and include most research, medical, laboratory, and similar critical applications. Typically, HGBP is not available on digital scroll or variable speed compressors, and the use of single- or two-stage scroll compressors may be required.

If applied properly, HGBP can (1) prevent excessive compressor cycling, (2) more closely match capacity to load across the entire load range, and (3) allow the system to operate at safe balance points under low load conditions. However, to minimize the negative energy consumption aspect of HGBP, designers must select chillers with multiple-step refrigerant circuits (using multiple two-stage scroll compressors) or that already have significant unloading capability (variable speed scroll compressors, rotary screw compressors with slide valve and/or variable speed control, or centrifugal compressors with inlet vane and/or variable speed control).

It appears that current manufacturers apply scroll compressors more or less consistently as follows:

1. Single compressor systems are offered for chillers producing up to 10 tons [35 kW] of cooling capacity or less (though only a few manufacturers offer a chiller at less than 20 tons [70 kW] capacity).
2. In the range of 10–20 tons [35–70 kW] total capacity, the chiller may have a single compressor or may have tandem (two) compressors, this varies from manufacturer to manufacturer. Where tandem compressors are offered, they are universally piped on a common refrigerant circuit… no manufacturer offers independent refrigerant circuits in this size range.
3. From 20 to 80 tons [70–280 kW] total capacity, chillers will typically have four compressors, piped in pairs on two separate refrigerant circuits.
4. From 80 to 130 tons [280–455 kW] total capacity, chillers will typically have six compressors, piped with three compressors on each of two separate refrigerant circuits.
5. All manufacturers offer HGBP as an option, *but only with two-stage scroll compressors*. HGBP is typically not available when digital scroll or variable speed scroll compressors are provided.
6. Manufacturers offer two types of evaporators. Typically, for a total capacity less than 120 tons [420 kW] or so, the evaporator will be a brazed plate-and-frame heat exchanger. For larger capacities, the evaporator will be a conventional U-tube heat exchanger. For two circuit chillers, each evaporator has two independent refrigerant circuits.

Thus, it is recommended that designers select scroll compressor water chillers based on the following:

1. *Avoid the use of digital scroll compressors*: This is a relatively crude technology that was originally developed to enable low-end, residential grade, and light commercial DX equipment to have better part-load performance. But it achieves this better part-load performance with an energy penalty and dependence on a high-cycling rate solenoid valve and variable geometry compressor, a sure formula for a short service life. While these compressors are perhaps adequate for small DX-packaged systems, it is recommended that designers prohibit their use and, especially where

good load control is needed, require the use of variable speed scroll compressors (preferably with permanent magnet motors and integral inverters).

2. If a single scroll compressor chiller is applied for conventional comfort air-conditioning duty, the scroll compressor should be variable speed (ideally, the permanent bearing motor with invertor option) in order to have adequate capacity versus load control. *For process cooling requirements, a variable speed compressor is required.*

11.3.5 Rotary Screw Compressor Water Chillers

Since the rotary screw compressor is a *positive displacement* compressor, its capacity is dependent on the compressor size and its rotational speed. Capacity control, therefore, may be based on slide valves, which open to "dump" part of the refrigerant load passing through the compressor lobes, or *via* speed control. In smaller capacities, the first approach represents lower cost, while speed control maximizes energy efficiency under part-load conditions.

With capacity ranges of about 150–550 tons [525–1,925 kW], rotary screw compressors are ideal for mid-range capacity requirements. These chillers have found to be reliable, and if selected correctly, they are quieter than centrifugal compressor water chillers, especially under part-load conditions.

11.3.6 Centrifugal Compressor Water Chillers

The energy requirement for a centrifugal compressor chiller at peak load is a function of (1) the required leaving chilled water temperature and (2) the temperature of the available condenser water. As the leaving chilled water temperature is reduced, the energy requirement to the compressor increases. Similarly, as the condenser water temperature increases, the compressor requires more energy. Thus, the designer can minimize the cooling energy input by using a centrifugal compressor chiller selected to operate with the highest possible leaving chilled water temperature and the lowest possible condenser water temperature.

The energy consumption by a centrifugal compressor chiller decreases as the imposed cooling load is reduced. These chillers operate efficiently at between approximately 30% and 100% load and *most* efficiently between 40% and 80% load. Under these conditions, the gas flow rate is reduced, yet the full heat exchange surface of the cooler and condenser is still available, resulting in higher heat transfer efficiency.

Below about 30% load, the refrigerant gas flow rate is reduced to the point where (1) heat pickup from the motor and (2) mechanical inefficiencies have stabilized input energy requirements.

The vast majority of electric-drive centrifugal compressor water chillers use a single compressor. However, if the imposed cooling load profile indicates that there will be a significant chiller usage at or below 30% of peak load, it may be advantageous to use a dual-compressor chiller or multiple single compressor chillers.

The dual-compressor chiller uses two compressors, each sized for 50% of the peak load. At 50%–100% of design load, both compressors operate. But if the imposed load drops below 50% of the design value, one compressor is stopped and the remaining compressor is used to satisfy the imposed load. This configuration has the advantage of reducing the inefficient operating point to 15% of full load (50% of 30%), reducing significantly the operating energy penalties that would result from a single compressor operation.

Selection of centrifugal water chillers has become more complex since most energy codes impose particularly low energy consumption limits on them. Manufacturers have responded to this by making numerous improvements to heat exchangers and impellor designs, but there still remains one significant issue, *surge*.

When the pressure entering the compressor is higher than the compressor outlet pressure, refrigerant flow tends to reverse or even flow back in the compressor. As a consequence, the inlet pressure will decrease, inlet pressure will increase, and the flow reverses again. This phenomenon, called "surge," repeats and occurs in cycles. Thus, the compressor loses the ability to maintain the peak head when surge occurs and the entire system becomes unstable. Ultimately, this condition will cause physical damage to the impellor wheel and bearings.

Historically, refrigerant gas flow into a centrifugal compressor has been controlled by adjustable *inlet guide vanes*, sometimes called *pre-swirl* or prerotation *vanes*, that are essentially *radial blade inlet dampers*. These vanes are arranged at the inlet to the compressor impeller and can be opened and closed by an external operator.

Since each vane rotates around an axial shaft, they affect the direction of the flow entering the impeller. When the inlet vanes are fully open, gas enters the impeller at 90° to the impeller. However, as the inlet vanes begin to close, flow enters the impeller at an increasing angle in the direction of the radial flow along the impeller blades. This "pre-swirl" condition reduces the ability of the impeller to impart kinetic energy to the refrigerant gas, thus reducing the flow rate. *Inlet vanes do not produce a pressure drop or throttling to reduce refrigerant flow through the centrifugal compressor.*

A minimum volumetric rate flow through a centrifugal compressor is required for stable operation. If the volumetric flow rates fall below this minimum, the compressor will become unstable and *surge*. When this happens, the refrigerant flows alternatively backward and forward through the compressor, producing noise and poor operation. Extended operation under surge conditions will cause mechanical damage to the compressor. The surge envelope will vary from compressor to compressor but usually occurs when the volumetric flow rate is reduced by 40%–60%.

To prevent surge from occurring, internal HGBP may be used to allow capacity to be reduced while maintaining sufficient gas flow through the compressor.

Capacity control of centrifugal chillers by speed control is now sometimes applied. Here, a large VFD is applied to the chiller motor and the motor speed modulated to control capacity. Generally, speed control improves efficiency over inlet vane control down to about 55% of rated capacity, and whole inlet vane control is more efficient below 55% of rated capacity.

Compressor speed is directly related to capacity, but the pressure (lift) produced by the compressor is a function of the square of the speed. This may produce surge as the chiller unloads under speed control.

Even if speed control is not used for capacity control, the potential for surge in centrifugal chillers has become more common since the manufacturers have determined that compressor efficiency is at its maximum when it operates just below the pressure conditions that produce surge. However, this leaves little margin for error, and the designer should make chiller selections on the basis of reducing the potential for surge. There are two basic ways of doing this: (1) requiring that the chiller be provided with an internal HGBP valve that opens as the compressor nears surge conditions, effectively reducing compressor outlet pressure, or (2) simply selecting the chiller with a larger compressor than needed for the expected peak load imposed on it. This effectively moves the compressor operating curve away from the surge curve.

Oil-lubricated compressors are conventional, but *magnetic bearing* centrifugal compressors are also available. Magnetic compressor water chillers have energy ratings 40%–50% lower than conventional centrifugal compressors and are much quieter. But there is a higher cost associated with them. There are several manufacturers that offer these chillers, but the configurations vary widely and careful evaluation is required.

11.3.7 Absorption Water Chillers

Since the water/lithium bromide solution in large absorption chillers is a corrosive salt solution, the primary potential for corrosion in these chillers are the ferrous metals used in them. *The generator (or "concentrator") is the most critical location for potential ferrous corrosion since the highest salt concentration and highest temperatures are present in this heat exchanger, along with the potential impact of surface erosion as refrigerant vapor is driven off tube surfaces by boiling.*

Basic ferrous corrosion occurs when iron reacts with water to produce an iron oxide called *magnetite* and hydrogen. Under acidic conditions, this reaction is greatly accelerated. (Also, in an absorption chiller, hydrogen is a "noncondensable" gas; that is, it does not act as a refrigerant, and the performance of the chiller is negatively impacted.)

There are basically two approaches to corrosion protection: (1) choose a metal compatible with the chemical environment in which it has to survive or (2) modify the chemical environment so it is less corrosive to the metal. The first approach can add significantly to the cost of an absorption chiller, and most manufacturers and owners have been unwilling to pay the premium involved. *Since chemical modification is much cheaper, this is the approach most commonly taken.*

Most chiller manufacturers offer optional tube materials and/or tube wall thickness to improve corrosion resistance, and this additional cost is highly recommended. Typically, though, two types of chemical modifications are made to reduce corrosion:

1. To reduce the acidity of the water/lithium bromide solution, a compatible alkaline solution, *lithium hydroxide*, is added. To reduce ferrous corrosion, it is desirable to maintain the solution at alkaline condition. But since copper corrodes readily at higher alkaline levels, it is necessary to maintain the level high enough to help protect the steel, without being too high and accelerating copper corrosion.
2. To additionally protect the steel in a corrosive environment, a compatible *corrosion inhibitor*, such as *lithium chromate*, is added. Lithium chromate "passivates" the iron that it comes into contact (i.e., makes it less reactive) by forming a protective molecular film on the surface. Lithium chromate has the advantage of working well at low alkalinity levels, allowing the solution to maintain at levels more suitable for the copper in the chiller.

The chemistry balance, then, is both complex and requires routine attention to maintain correctly. If the alkalinity is not adjusted properly, the solution will become acidic and accelerate ferrous metal corrosion. If the alkalinity is too high, the copper in the chiller will corrode. If the lithium chromate level is too low, it offers poor protection to the ferrous metals in the chiller. But if the level is too high, it can initiate pitting in the ferrous metals due to scaling and the resulting localized corrosion while also increasing copper corrosion rates. Thus, as the chiller is operated and maintained, careful attention to the chemical balance of the refrigerant/absorbent solution is critical to maintaining proper chiller performance and to attaining its design life.

Lithium bromide absorption chillers typically fail due to corrosion resulting from two major factors:

1. Air, or more specifically, oxygen, is the culprit. As air leaks into any absorption chiller, it introduces oxygen, which in turn provides the corrosive reaction essential for ferrous metal to corrode in aqueous systems. This process is accelerated by the acidic salts in the absorption solution, primarily lithium bromide. An occasional air leak is not sufficient for dissolving enough iron to plug the fine orifices of spray headers and heat exchangers, causing the chiller to fail. Instead, it is the continuous air leak that causes so much damage in lithium bromide-based systems. Finding and stopping an air leak prevents more oxygen from getting into the aqueous system, and the corrosion process merely stops by itself. If air leaks are not stopped, the system ultimately fails since corrosion becomes intense from the continuous supply of oxygen into the lithium bromide charge.
2. Improper solution chemistry is the second cause of failure. All that is needed to produce the refrigeration cycle is lithium bromide and water, but additional chemicals are added, including *lithium hydroxide* and *lithium chromate.* While each chemical helps to reduce internal corrosion, without proper and continuous monitoring of the solution chemistry, the chemical addition can actually do more harm than good.

Lithium bromide corrosion control in absorption refrigeration is a very complex scheme of proper mechanical operation and the balance of chemicals in solution. What complicates the situation still further is that there are not a large number of highly skilled personnel available to service absorption systems, in the real world, absorbers are tricky to maintain, requiring specialized service personnel who are not readily available.

Once an absorption chiller starts to corrode, it can degenerate quickly. Fine particulates of rust and dissolved copper can easily plug spray header nozzles and heat exchangers, reducing efficiency. A large amount of debris can foul pumps and reduce heat transfer. The high-stage generator tubes can develop leaks due to ferrous corrosion, accelerated by scaling and surface boiling factors.

With their limited scope of application and corrosion-based failures resulting from poor maintenance, lithium bromide absorption chillers are rarely applied today except in Pacific Rim nations that have energy-based government incentives (or penalties) that promote their use. But, for facilities that have waste heat or geothermal heat available, they are an option that should be evaluated.

11.3.8 Chilled Water System Configurations

The basic chilled water piping configuration is for a single chiller to provide chilled water to one or more cooling coils using a single chilled water pump, as shown in Figure 11-7.

For small systems, this configuration has the advantage of lower initial cost but does have some basic disadvantages:

1. With the single compressor system, failure of any component (compressor, pump, or condenser) will result in no cooling being available. For most facilities, this is unacceptable and the use of multiple chillers allows at least some cooling (50% or more) to be provided even if one chiller fails. In cases where cooling is critical to the facility (data or server centers, hospital, laboratories, pharmaceutical

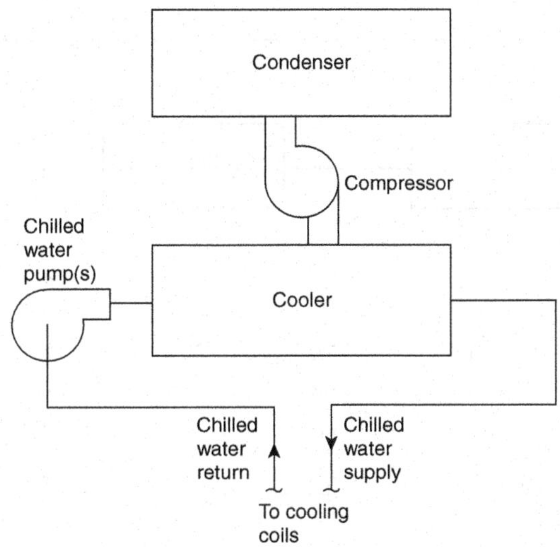

FIGURE 11-7
Constant flow, single chiller piping configuration.

manufacturing, etc.), $N+1$ multiple chillers are typically used, where N is the number of chillers required to satisfy the peak-imposed cooling load. In this case, even if one chiller fails, 100% of the design cooling load can still be met.

2. Once the cooling load imposed on a rotary screw or centrifugal compressor chiller falls below about 30% of the chiller capacity, the efficiency of the chiller begins to decline. Thus, multiple chillers will allow a better overall capacity-to-load ratio and improved low load operating efficiency.

A goal for chiller operation is to produce chilled water as economically as possible. To meet this goal not only must individual chillers be efficient, but also the entire multichiller system must be designed to function efficiently.

The first step in evaluating the requirements for a chilled water system is to determine the peak cooling load requirement. However, the peak load will occur only for a very limited number of hours during the year. On an annual basis, the imposed load will vary based on the time of day due to occupancy patterns, solar heat gains, and diurnal temperature swings, and the time of year due to solar and temperature seasonal variations. Another factor that influences the load imposed on the chilled water plant is the amount of "core" building area that will require cooling year-around and whether an airside economizer cycle is installed to provide "free" cooling during the winter. *With all of these variables, every chilled water system will operate far more hours at a part-load condition than it will operate at the peak-load condition.*

Thus, in addition to determining the peak load to be imposed on the system, it is also necessary to determine the part-load profile, either as a function of time, as a function of outdoor temperature, or as a function of both. Then, multiple chillers can be selected to better match load to capacity at all points of operation.

With multiple chillers, there are two basic configurations that can be used: *series* or *parallel*.

In the series configuration with two chillers, as shown in Figure 11-8, each chiller is selected to produce half of the required cooling at the full system flow rate. Thus, half of

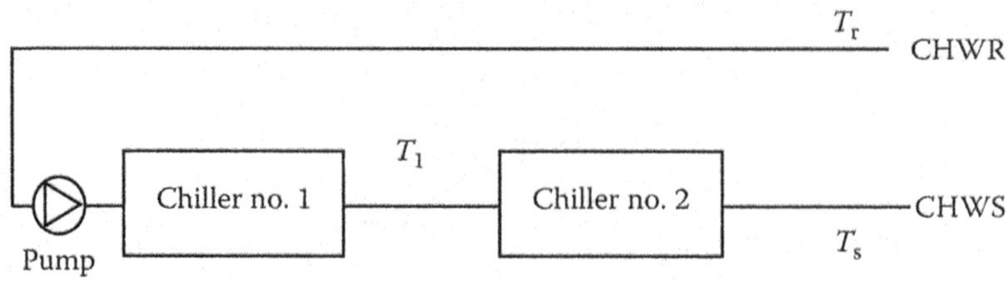

Temp	Full Load	75% Load	50% Load	25% Load
T_s	44	44	44	44
T_r	56	53	50	47
T_1	50	50	50	47

FIGURE 11-8
Series chiller piping configuration. [Fig. 2.2, *HVAC Water Chillers and Cooling Towers*, Herbert W. Stanford III, PE (CRC Press/Taylor & Francis, 2012).]

the total design range is produced by each chiller. Load ratios other than 50/50 are possible, but this is by far the most common condition because of control problems with chillers at very small temperature differences.

Series chiller systems are rarely done any longer because this configuration requires a constant chilled water flow rate at all times, resulting in high pumping costs. But if a relatively large temperature difference is required or if there is a very steady base cooling load, the series configuration may offer some advantages.

The parallel chiller configuration is far more common. In a two-chiller configuration, each chiller is typically selected to operate with the same design range but with only a half of the total system flow requirement. This again results in a 50/50 load split, but other load ratios may be selected if dictated by operational requirements.

The *one-pump parallel chiller* configuration is shown in Figure 11-9, but with this configuration, there is an inherent problem. If both machines were operated for the full-load range (15%–100% of peak capacity), by the time the total system load drops to 30% of full load, each individual chiller would be operating very inefficiently. Thus, most designers use controls to shut off one chiller when the total system load, as evidenced by the return water temperature, falls below 50% of full load.

However, if one chiller is not in operation, chilled water from the operating chiller will mix (blend) with the return water passing through the nonoperating chiller, effectively raising the chilled water supply temperature to the system. In many cases, this may not be a problem, but generally the interior spaces of large buildings still require more-or-less full cooling even when the perimeter spaces require no cooling at all. An increased chilled water supply temperature may not satisfy these interior load conditions.

To attempt to eliminate the blended supply water problem with the one-pump configuration, some designers have used flow isolation valves to shut off flow through each chiller when it is de-energized. This arrangement results in increased flow through the operating chiller but does reduce the blending problem.

To ensure that the blended water condition does not occur, the *multiple-pump parallel chiller* configuration shown in Figure 11-10 is widely used.

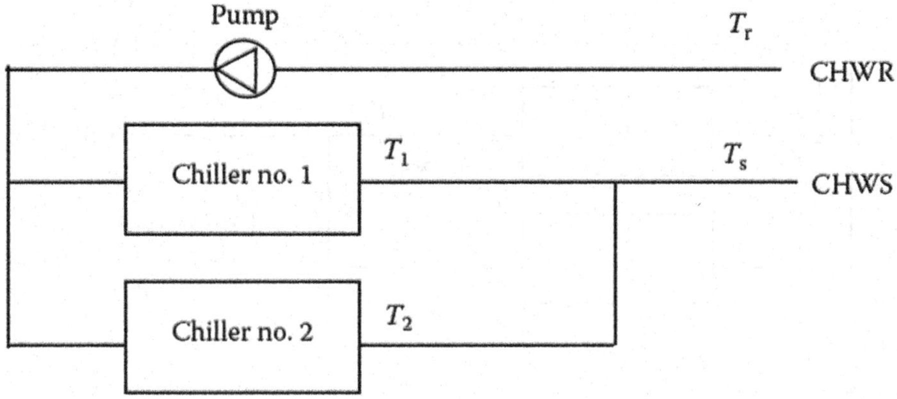

		Chiller No. 1		Chiller No. 2		
% Load	T_r	**Flow (%)**	T_1	**Flow (%)**	T_2	T_s
100	56	50	44	50	44	44
75	53	50	44	50	44	44
50	50/56	50	44	50	56[a]	50
25	50	50	44	50	50[a]	47

[a] Chiller No. 2 off.

FIGURE 11-9
One-pump parallel chiller piping configuration and operation. [Fig. 2.3 and Table 2.2, *HVAC Water Chillers and Cooling Towers*, Herbert W. Stanford III, PE (CRC Press/Taylor & Francis, 2012).]

With this configuration, each chiller has an individual chilled water pump. Thus, when one chiller is not operating, one pump is off, flow through the nonoperating chiller is zero, and no blending results.

Each of the configurations discussed so far is essentially *constant flow* system that uses three-way control valves at the cooling coils. Constant flow systems circulate the same amount of chilled water no matter what the imposed cooling load is, and consequently result in higher pumping energy costs.

To reduce these costs, the *primary–secondary variable flow* piping arrangement illustrated in Figure 11-11 is very common. Here, the *production (primary) loop* through the two chillers is hydraulically isolated from the *distribution (secondary) loop* by a piping *bridge*. The bridge is a short section of piping shared by the two loops and designed to have little or no pressure drop. Thus, flow in either loop is not affected by flow in the other.

On the primary or "production" loop side, the system acts as multiple-pump parallel chiller installation. Flow in this loop varies in "steps" as the chillers are staged on or off and their respective pumps are started and stopped.

However, in the secondary or "distribution" loop, the cooling coils use two-way control valves and the pump is fitted with a variable speed controller that modulates the flow from 0% to 100% as a function of the imposed cooling load. Thus, this loop has variable flow but a constant range. At any load condition, the supply water temperature is the same as the

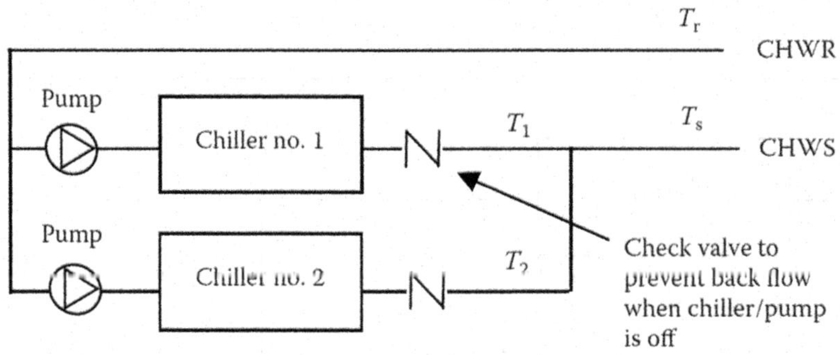

		Chiller No. 1		Chiller No. 2		
% Load	T_r	Flow (%)	T_1	Flow (%)	T_2	T_s
100	56	50	44	50	44	44
75	53	50	44	50	44	44
50	50	50	44	50/0	N/A[a]	44
25	47	50	44	0	N/A[a]	44

[a] Chiller No. 2 off. Thus there is no flow and T_2 does not affect T_s.

FIGURE 11-10
Multiple-pump parallel chiller piping configuration and operation. [Fig. 2.5 and Table 2.4, *HVAC Water Chillers and Cooling Towers*, Herbert W. Stanford III, PE (CRC Press/Taylor & Francis, 2012) Book.]

water temperature leaving the chiller(s), as long as the production loop flow rate equals or exceeds the distribution loop flow rate.

Centrifugal compressor water chillers are manufactured with integral digital control systems. This represents a significant improvement in chiller control and allows the use of the *variable primary flow* piping arrangement shown in Figure 11-12 to save even more in pumping energy (typically, approximately 15%–20%).

The flow through the chillers is varied in response to the system load. A bypass valve opens to maintain a minimum flow equivalent to about 30% of the flow through the largest chiller. This system has certain operational limitations since rapid load fluctuations and flow changes may result in fluctuation in the chilled water supply temperature, but it is a viable option where the critical chilled water temperature control is not required.

Basic design criteria for this system include the following:

1. Flow through each chiller should be maintained in the range of 3–11 fps [0.9–3.3 m/s]; though, some chillers will operate satisfactorily with a flow as low as 1.5 fps [0.5 m/s]. *Consequently, good-quality flow meters, with high turndown ratios, which must have periodic recalibration, are required for each chiller.*
2. If flow rates through the chiller change too rapidly, the chiller controls cannot keep up. Therefore, load fluctuations must be limited to not more than 30% per minute.

One misconception about the variable primary flow configurations is that the chillers in this type of system will operate more efficiently. However, with variable flow through the

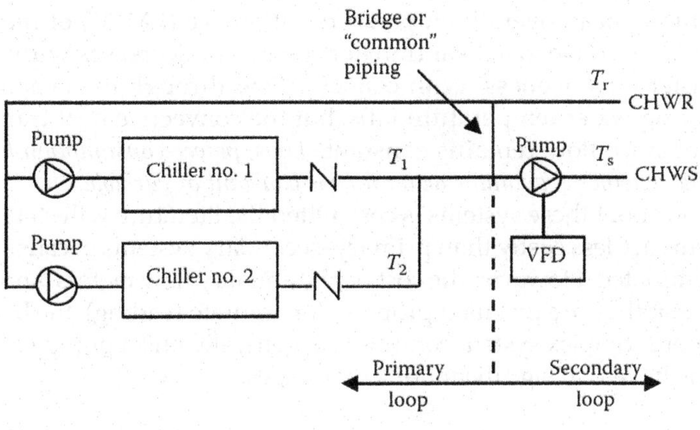

% Load	T_r	% Secondary Chilled Water Flow	% Primary Chilled Water Flow	T_m	Chiller No. 1 Flow (%)	T_1	Chiller No. 2 Flow (%)	T_2	T_s
100	56	100	100	56	50	44	50	44	44
75	56	75	100	53	50	44	50	44	44
50	56	50	50	56	50	44	50/0	N/A[a]	44
25	56	25	50	53	50	44	0	N/A[a]	44

[a] Chiller No. 2 off. Thus there is no flow and T_2 does not affect T_s.

FIGURE 11-11
Primary–secondary flow parallel chiller piping configuration and operation. [Fig. 2.6 and Table 2.5, *HVAC Water Chillers and Cooling Towers*, Herbert W. Stanford III, PE (CRC Press/Taylor & Francis, 2012).]

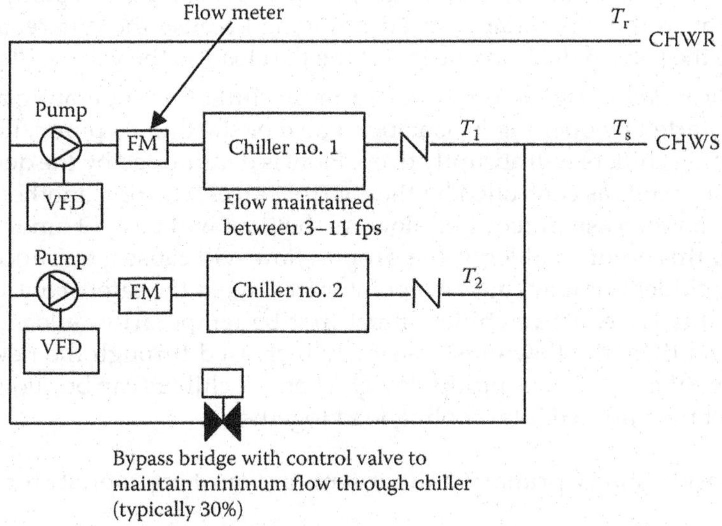

FIGURE 11-12
Primary-only flow parallel chiller piping configuration. [Fig. 2.7, *HVAC Water Chillers and Cooling Towers*, Herbert W. Stanford III, PE (CRC Press/Taylor & Francis, 2012).]

evaporator, the logarithmic mean overall temperature difference (LMTD) of the evaporator remains constant but the convective heat transfer coefficient decreases with flow, thus reducing the heat transfer effectiveness. With constant flow through the evaporator, the LMTD falls as the entering water temperature falls, but the convective heat transfer coefficient remains constant since flow remains constant. *Thus, power consumption by a chiller will be essentially the same whether evaporator water flow is constant or variable.*

Another misconception about these systems is cost. Often the literature will state that variable primary flow systems are less costly than primary–secondary systems because one set of pumps and piping is eliminated. However, the cost of high-quality flow meters (and the additional piping to ensure that they are installed properly for accurate reading), the bypass flow control valve, and the more complex system controls will generally offset pump cost savings.

Primary-only systems have two significant disadvantages:

1. *Complexity of bypass control*: For all but chiller plants with a large number of chillers, some type of bypass valve will be required in order to ensure that minimum flow rates are maintained through operating chillers. This control is somewhat complex.
   a. Selecting the bypass control valve and tuning the control loop are difficult because of the very high differential pressure across it caused by its location near the pumps. This can be mitigated by placing the bypass valve out in the system near the most remote coils. However, this increases flow through distribution piping, which increases pump energy at low loads. Also, it increases piping heat gain if flow is maintained in piping that might otherwise be inactive. On the other hand, placing the valve in a remote location can prevent "slugging" the plant with warm water when a remote coil starts up and may result in more stable plant control. A pressure-independent control valve should be considered for this application.
   b. Control system programming is difficult when there are multiple chillers or stages each requiring different minimum flow set points. The pressure-activated bypass valves commonly used in the past with constant speed pumping systems will not work with variable speed pumping because the differential pressure across the valve will always be less at the part load so the valve will never open.
2. *Complexity of chiller staging*: When one or more chillers are operating and another chiller is started by opening its isolation valve or starting its pump, flow through the operating chillers will abruptly drop. Flow is determined by the demand of the chilled water coils as controlled by the control valves. Starting another chiller will not create an increase in required flow, so the flow will be split among the active chillers. If this occurs suddenly, the drop in flow will cause a nuisance trip in the operating chillers, or may cause evaporator freezing if the safety controls are slow. To avoid this issue, active chillers must first be temporarily unloaded (*demand-limited*), and then, the flow must be slowly increased through the new chiller by ramping up the speed of its pump slowly. Then, all chillers can be allowed to ramp up to meet the imposed total cooling load together.

Given these considerations, primary-only systems are most appropriate for the following:

1. Plants with more than three chillers and with fairly high base loads where the need for bypass is minimal and flow fluctuations during staging are small due to the large number of chillers

2. Plants where future on-site operators will understand the complexity of the controls. Numerous authors note that the sheer complexity of the bypass and staging control makes it likely that the control sequence will fail, since increased complexity typically results in a higher probability of failure, requiring quick intervention by the on-site operators.

So, for most systems, the simpler primary–secondary system configuration remains the recommended approach for the majority of multiple chiller systems, though variable primary flow should be evaluated for larger systems with long distribution runs *and* more sophisticated operators.

11.3.9 Chilled Water Buffer Tanks

Chilled water storage (called "buffer") tanks are required when the total water volume of the distribution network is less than the minimum requirements imposed by chiller manufacturers to insure enough thermal mass to prevent control issues and compressor "short cycling." The volume of water in the chilled water system acts as a damper and "smooths" the effects due to load change. Process loads tend to be the most abrupt and can cause the chiller to shut down on a safety if the system is not designed correctly. As well, when a chiller is close-coupled to a large dominant air-conditioning load (e.g., a chiller connected to a single large AHU), the chiller can "hunt" in response to the control valve position, resulting in unstable operation (and, in fact, with this configuration, it may be better to eliminate the coil control valve and operate with a "wild running coil").

There are other issues that need to be considered, as well. For instance, a small chiller using scroll compressors will change capacity in discrete steps. For example, a four-compressor chiller will have a stepped capacity of 25%, 50%, 75%, and 100% cooling capacity. If the load is between the steps, say 65% capacity, then the chiller will operate at the step above and then the step below the actual load. The chiller will require the chilled water fluid volume to "dampen" the effect of the chiller either over- or undercooling the chilled water. Another example, and perhaps most importantly, is where the load drops below the minimum capacity the chiller can operate. In this case, the chiller will cycle a compressor on and off to meet the load. If there is not enough fluid in the system, the compressor will incur too many starts, which will result in undue wear and tear on the chiller.

To evaluate whether there is sufficient volume in the system requires first estimating the amount of fluid in the system. To find the volume, the amount of fluid in the chiller evaporator, piping, and coils must be added together. Water volume can be approximated from data used for sizing the system compression tank (see Chapter 8).

The sufficient fluid volume, V_w, can be determined *via* Eq. (11-1) or (11-1a).

$$\text{For IP units, } V_w = \text{Ts}/500 \times \{[(\text{TD}/(H_1 - H_2)) + (\text{TD}/H_2)]\} \quad (11\text{-}1)$$

$$\text{For SI units, } V_w = \text{Ts}/250 \times \{[(\text{TD}/(H_1 - H_2)) + (\text{TD}/H_2)]\} \quad (11\text{-}1a)$$

where

V_w = fluid volume, gal. [L]

Ts = time from start to start, minutes

TABLE 11-9

Typical Chiller Controller Dead Bands

Chiller Compressor(s) Type	Ts = Time between Starts (Minutes)	Minimum Operating % Capacity of One Compressor	TD = Dead Band of Chiller Controller (°F) [°C]
Scroll	15	25	4 [2.2]
Reciprocating	15	25	
Rotary screw	30	25	
Centrifugal	30	10	

H_1 = minimum operating capacity of the chiller, Btu/h [kW]

H_2 = minimum applied load on the chiller, Btu/h [kW]

TD = dead band of chiller controller, °F [°C] as shown in Table 11-9

In general, for small systems that use scroll compressors, care must be taken to ensure that the system water volume is at least 7.5 gal./ton [8.1 L/kW] for typical air-conditioning applications and 10+ gal./ton [10.8+ L/kW] for process cooling applications or where the chiller serves only a single AHU (or other load). Buffer tanks must be baffled to ensure that there is no stratification and that fluid entering the tank is adequately mixed. If multiple tanks are required due to space limitations, it makes no difference whether the tanks are piped in parallel or series, though a series configuration results in more "steps" of baffling.

For typical comfort cooling applications, since the primary goal is to prevent compressor short cycling, the better location for the buffer tank is on the return side of the chiller(s). With a primary–secondary or primary-only variable flow piping configuration, the buffer tank needs to be located upstream of the chiller(s), between the bridge/bypass return connection and the pump(s) inlet. Also, in this configuration, the buffer tank can replace the air separator that is normally required, since flow velocity through the tank is very low and entrained air will naturally separate.

For process applications, or where chilled water supply temperature control is critical, the buffer tank should be located on the supply side, downstream of the chiller(s), since the tank volume will tend to maintain a more constant supply temperature as the compressor cycles. An added benefit is that even if the chiller fails, there is an extended "coast down" time created by the stored cooling capacity in the buffer tank. But, in this configuration, the buffer tank is located in the high-pressure region of the system and an air separator is still required on the suction side of the pumps.

11.4 Condensers and Cooling Towers

11.4.1 Air-Cooled Condensers

Refrigerant condensers are defined in terms of the *heat sink* to which the heat removed from the refrigerant is rejected:

1. Air-cooled condensers consist of a refrigerant-to-air coil and fan(s) for rejecting heat to the outdoor air.

2. Water-cooled condensers consist of a refrigerant-to-water heat exchanger. Refrigerant heat is transferred to the water, which may be a pond or lake, a river, or the water flow from a cooling tower.
3. Evaporatively cooled condensers are air-cooled condensers that use a supplemental water spray to reduce the outdoor temperature by adiabatic cooling.

In HVAC applications, condensers are typically a part of packaged equipment and their performance is included as part of the overall package performance ratings.

Air- and evaporatively cooled condensers must be installed in accordance with the manufacturer's requirements to provide adequate airflow and prevent warm discharge air for "short-circuiting" and entering the unit, reducing its capacity. Adequate clearance around the unit is a must, and if screening is used, it must have at least 50% open area and not extend above the top of the condenser.

11.4.2 Cooling Tower Fundamentals

In water-cooled HVAC systems, the cooling tower has the role of rejecting the heat collected during the space cooling process to the ambient atmosphere. In addition, the inefficiencies of the water chiller represent heat that is added to the condenser water circuit and must be rejected by the tower.

Once a design temperature range for the condenser water is established, the condenser water flow rate can be determined from Eq. (8-9) or (8-9a), based on the total heat rejection value determined as a function of the compressor power input, as shown in Table 11-10.

As condenser water flows through the tower, this heat is rejected to the ambient air and the condenser water is cooled, primarily, through evaporation of a small percentage of the total water flow. Evaporation is a process by which heat is absorbed by air and the remaining condenser water is cooled to the desired leaving temperature.

All of the data presented earlier are based on the use of electric-drive vapor compression cycle systems. With *absorption cycle chillers*, the amount of heat that is added to the cooling load, and that must be rejected through the cooling tower, is approximately 20%–50% greater than for an electric-drive chiller. Therefore, the condenser water flow rate, for the same range, will be approximately 20%–50% higher (or, for the same flow, the range must be approximately 20%–50% greater).

Every cooling tower, no matter how configured or constructed, must have the elements shown in Figure 11-13.

TABLE 11-10

Refrigeration System Total Heat of Rejection

Compressor Input Power (kW Input/Ton [kW Input/kW Output])	Approximate Heat of Compression (%), Including 2% for Condenser Water Pump(s)	Total Heat Rejection (Btu/h) [kW]
0.8 [0.22]	25	15,000 [4.4]
0.7 [0.20]	22	14,640 [4.3]
0.6 [0.17]	19	14,280 [4.2]
0.5 [0.14]	16	13,920 [4.1]
0.4 [0.11]	14	13,680 [4.0]

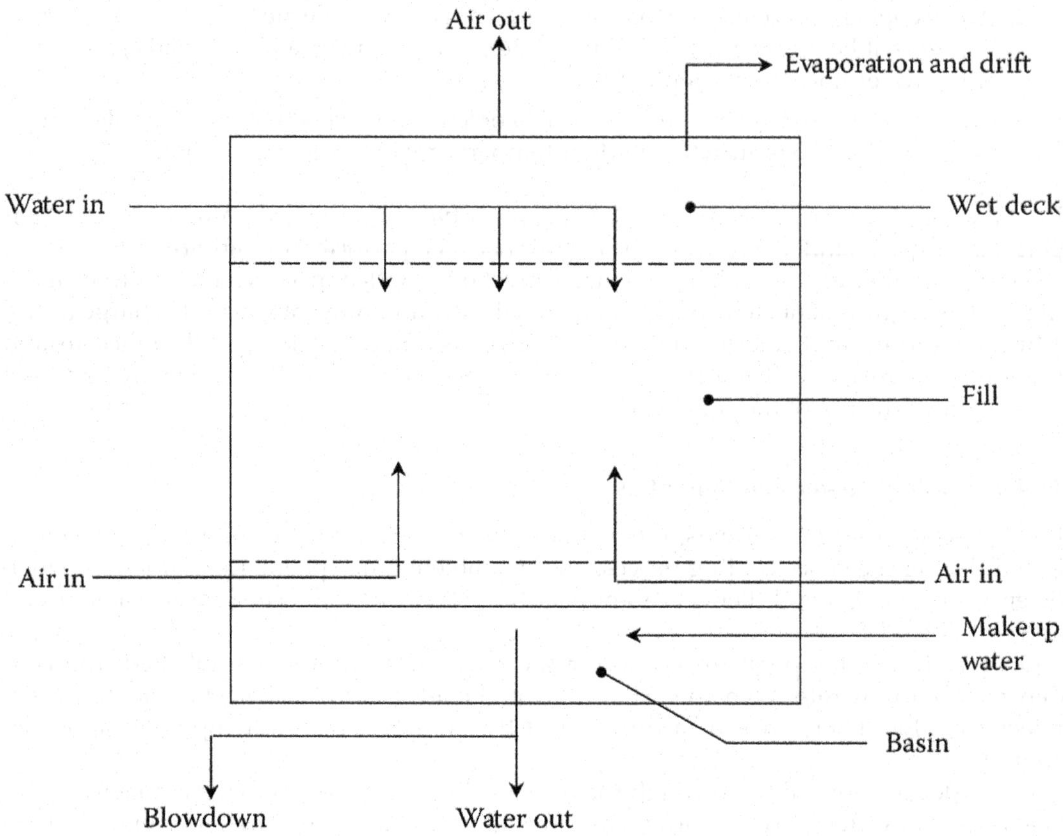

FIGURE 11-13
Cooling tower elements. [Fig. 9.1, *HVAC Water Chillers and Cooling Towers*, Herbert W. Stanford III, PE (CRC Press/Taylor & Francis, 2012).]

1. *Fill*: Heat transfer media in the cooling tower. Fill is designed to maximize the "contact" between the return condenser water and the ambient air. The better the contact, obviously the better evaporation and heat transfer occur.
2. *Hot water distribution ("wet deck")*: Pans or basins with metering outlets or spray nozzles designed to provide an even distribution of the return condenser water entering the fill
3. *Cold water basin*: The basin, either as an integral part of the tower or as a separate sump, collects the water passing through the tower for supply to the system by the condenser water pump. The basin must also be sized to contain enough water to supply the condenser water system until the pump returns water to the tower.
4. *Fan(s)*: All cooling towers used for HVAC applications are *mechanical draft* towers that use one or more fans to provide airflow through the tower.
5. *Inlet louvers and drift eliminators*: Inlet louvers act to force the air entering the tower into as straight and even flow pattern as possible, while the drift eliminators are designed to trap and remove any entrained water droplets that may be in the tower's leaving air.

Cooling towers and condenser water systems have their own vocabulary that must be mastered, as follows:

Approach: Difference between the condenser water supply temperature and the entering WB temperature

Casing: Exterior tower walls, excluding the louvers

Cell: Smallest tower subdivision that can operate independently. Usually, a tower cell is designed with independent water and airflow, and cells are controlled individually as a function of imposed overall cooling tower load.

Counterflow: Water flow path is configured at 180° to the airflow path; that is, the two flows are in opposite directions.

Crossflow: Water flow path is configured at 90° to the airflow path; that is, the two flows are perpendicular to each other.

Double flow: A crossflow tower where two opposed fill banks and air intakes are served by common fan and air plenum (called a "twin-pack" in the United Kingdom)

Drift: Unevaporated water droplets that are lost from the cooling tower

Eliminators: An assembly of baffles or other devices to remove entrained water droplets from the air leaving the cooling tower

Evaporation (loss): Condenser water that undergoes a phase change from liquid to vapor and exits the tower as part of an air–vapor mixture

Forced draft: Airflow is "pushed" through the cooling tower by one or more fans at the air inlet(s), resulting in the tower being under positive pressure.

Induced draft: Airflow is "pulled" through the cooling tower by one or more fans at the air outlet(s), resulting in the tower being under negative pressure.

Inlet louvers: Blades or other assemblies designed to (a) prevent water from splashing out of the tower and (b) promote uniform airflow through the fill.

Makeup (water): Water added to offset water lost to evaporation and drift

Nozzle: Device to control water flow through a cooling tower. Normally, nozzles are designed to produce a spray pattern by either pressure or gravity flow.

Range: Condenser water temperature increase (rise) in the condenser and decrease (drop) in the cooling tower

Recirculation: Air leaving the cooling tower that reenters the tower inlet, elevating the tower entering WB temperature

In HVAC applications rarely are "custom" cooling towers designed. Rather, manufacturers have designed a range of cooling towers whose characteristics are well defined. For any HVAC application, it is only necessary to use the specified flow rate, range, and ambient WB temperature to compute a required tower characteristic and then test that required characteristic against those available in any product line.

Manufacturers provide catalog data and selection software for the purpose of selecting a cooling tower for a specific application. *But since most HVAC designers and the purchasers of towers are not expert in the use of a particular manufacturer's catalog or selection software, it is always prudent to have the tower manufacturer perform any required tower selection. Then, if an error is made, the manufacturer will have to make good on the tower, not the designer or buyer.*

The selection of a cooling tower for a specific set of required performance parameters (condenser water flow rate, selected range, and ambient WB temperature) results in establishing a required *cooling tower characteristic*. But, what is the impact if the required performance changes?

As shown in Figure 11-14, there are three factors that define the requirements for a specific cooling tower characteristic:

1. Entering (ambient) air WB temperature
2. Condenser water flow rate
3. Approach

Once a tower is selected, that is, the cooling tower characteristic is established, changes in any of these three factors may necessitate a change in the cooling tower. Typically, the performance parameters that are subject to change include the following:

1. Increase in entering WB temperature. Too often, the wrong ambient WB temperature is selected for tower sizing.
2. Increase in rejected heat load. This may dictate increasing the condenser water flow rate and/or increasing the range.

A change in WB temperature and/or a change in range will result in a change in approach. Therefore, there are three performance impact relationships as shown in Figures 11-15–11-17.

Figure 11-15 illustrates the impact of changing the tower approach by either changing the design WB temperature or changing the condenser water range. Obviously, a relatively small increase in required approach will require a much higher cooling tower characteristic.

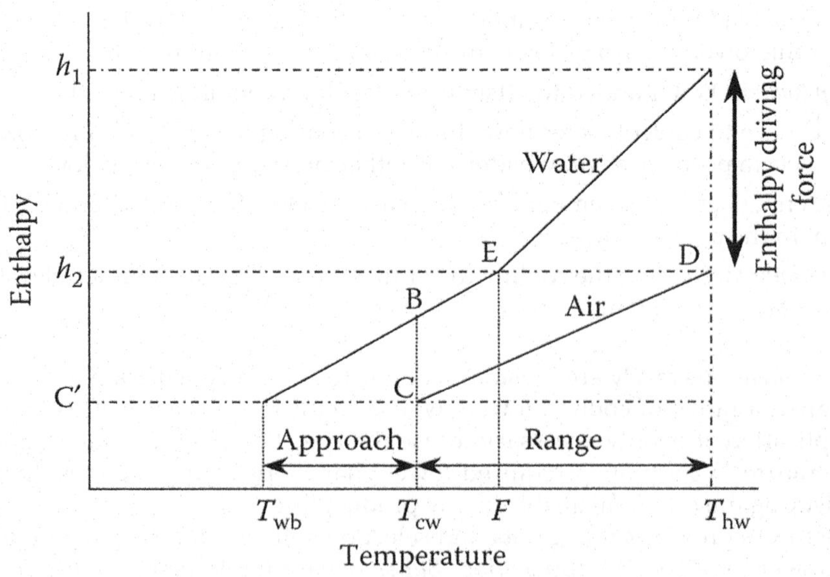

FIGURE 11-14
Graphical representation of cooling tower characteristics. [Fig. 9.3, *HVAC Water Chillers and Cooling Towers*, Herbert W. Stanford III, PE (CRC Press/Taylor & Francis, 2012).]

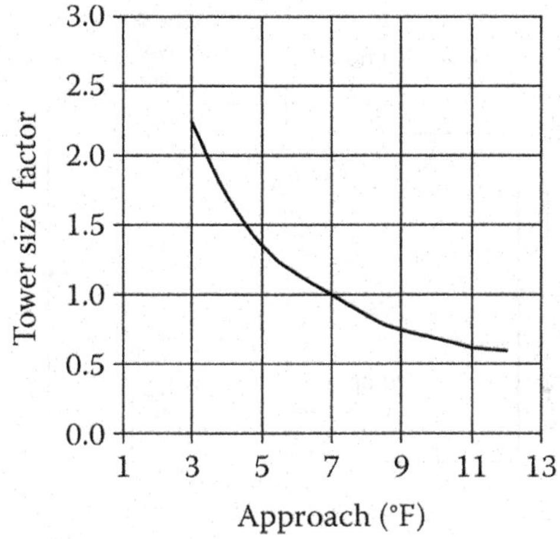

FIGURE 11-15
Cooling tower performance as a function of approach. [Fig. 9.5, *HVAC Water Chillers and Cooling Towers*, Herbert W. Stanford III, PE (CRC Press/Taylor & Francis, 2012).]

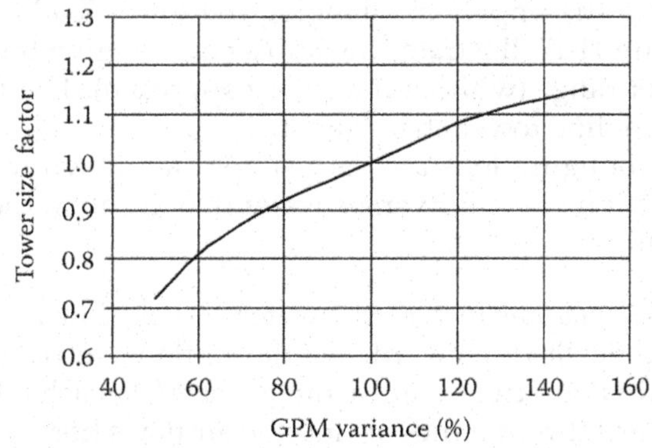

FIGURE 11-16
Cooling tower performance as a function of water flow. [Fig. 9.6, *HVAC Water Chillers and Cooling Towers*, Herbert W. Stanford III, PE (CRC Press/Taylor & Francis, 2012).]

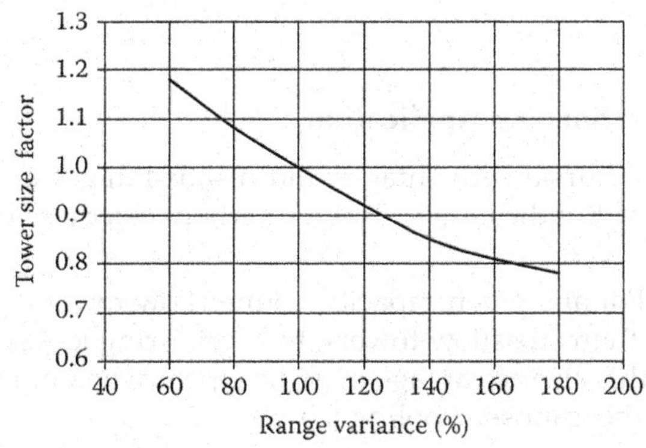

FIGURE 11-17
Cooling tower performance as a function of range. [Fig. 9.7, *HVAC Water Chillers and Cooling Towers*, Herbert W. Stanford III, PE (CRC Press/Taylor & Francis, 2012).]

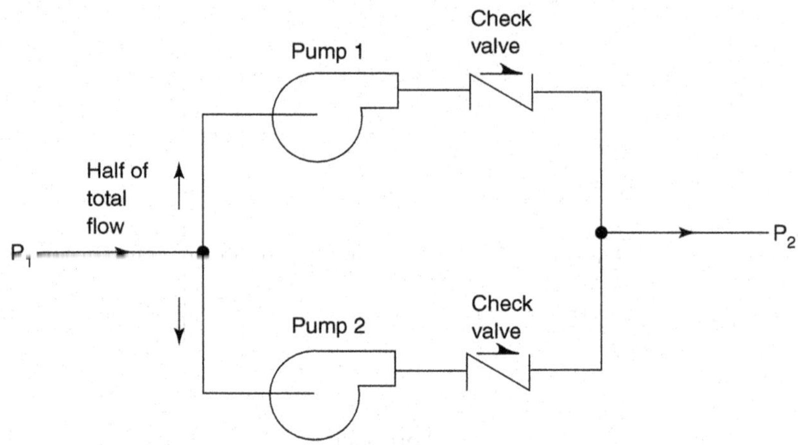

FIGURE 11-18
Cooling tower configurations. [Fig. 8.8, *HVAC Water Chillers and Cooling Towers*, Herbert W. Stanford III, PE (CRC Press/Taylor & Francis, 2012).]

Figure 11-16 illustrates the impact of changing the tower load by changing the water flow rate, whereas Figure 11-17 illustrates the impact of changing the tower load by changing the condenser water range (without impacting approach). These two variables have a smaller impact on the cooling tower characteristic.

Basic cooling tower configurations are very simple, as shown in Figure 11-18, and are dictated by (1) the direction of the air versus water flow through the tower fill and (2) the location of the tower fan(s):

1. *Air/water flow*: In *counterflow* towers, the water and air flow in opposite directions: the water flows vertically downward, and the air flows vertically upward. In *crossflow* towers, the two flow streams are arranged at 90° to each other: the water flows vertically downward through the fill, and the air flows horizontally through it.
2. *Fan location*: The tower is defined as *forced draft* when its fan is arranged to blow air through the tower. In this configuration, the fans are located on the entering airside of the tower and the fill is under positive pressure. In *induced draft* towers, the fan is located on the leaving airside of the tower and the fill is under negative pressure.

11.4.3 Tower Configuration and Application

There are waterside and airside advantages and disadvantages to both counterflow and crossflow cooling towers, but the largest differences between these types of cooling towers are as follows:

1. *Size/arrangement*: For any given capacity, counterflow towers will typically have a smaller footprint than crossflow towers, thus requiring less space for their installation. However, this size advantage may disappear when multiple tower cells are required to meet the imposed cooling load:
   a. Small counterflow towers may require air intake on only one side, and it is relatively simple to arrange multiple towers in a line, spaced just far enough apart for maintenance.

b. For larger counterflow towers, two cells can be joined, which effectively reduces the air entry condition for each cell from four sides to three sides. If more than two cells are required, the next one or two must be spaced at least one tower width away to ensure proper air entry.
c. Multiple crossflow tower cells can be joined side-by-side to form a tower as long as desired. However, it is better to limit this side-by-side arrangement to only three or four cells and to provide an airflow and maintenance separation of at least one tower width between each set of cells. When more than four cells are joined, there are often adverse airflow patterns at the middle cell(s).

2. *Fan horsepower/energy*: Counterflow cooling towers generally have higher airflow static pressure losses than crossflow cooling towers, requiring more fan horsepower and resulting in greater fan energy consumption. These higher losses are due primarily to counterflow towers being taller, with a greater fill height, and due to the water spray wet deck.
3. *Pump horsepower/energy*: The nozzles used in the counterflow cooling tower spray wet deck require, typically, 5 psig [175 kPa] residual water pressure to provide proper atomization. This, coupled with the fact that counterflow towers tend to be taller than crossflow towers, results in increased pump head, pump motor horsepower, and pump operating energy.

Crossflow towers are the better selection when it is desirable to minimize tower fan energy consumption, minimize pump size and pumping energy, and provide ease of maintenance. The crossflow tower is better with waterside economizer applications.

The counterflow tower is the better selection when the available space (footprint) is limited and/or where icing during winter operation is a concern. The counterflow tower may also be the better selection when very close approach is needed (less than 5°F [3°C]).

When there is only one chiller, the normal design approach is to select one cooling tower to serve that chiller's condenser. However, when there are multiple chillers, there are three options for designing the condenser water system, which are as follows:

1. Each chiller can have an individual, dedicated cooling tower. With this configuration, individual condenser water pumps and supply and return condenser water piping between each chiller and its tower are required. The cost of this piping is typically higher than the cost for a single (but larger) condenser water supply loop serving all chillers from all towers.
2. Multiple cooling towers or a large multicell tower can be connected to a common condenser water system to serve all of the chillers. Each chiller, however, must have a two-position isolation control valve. The advantage to this configuration is versatility, since any tower and any pump can serve any chiller.
3. Multiple cooling towers or large multicell towers are connected to a common condenser water system with an individual pump serving each chiller. This configuration eliminates the need for chiller isolation valves and *is the recommended configuration when the chillers have different capacities.* (The downside of this configuration is, of course, loss of the versatility provided by the common parallel pumping arrangement.)

The multicell approach has a number of operating advantages:

1. More than one cell can be used to serve a single chiller, providing a lower condenser water supply temperatures and reducing chiller energy consumption.
2. The failure of tower cell does not necessarily mean that any specific chiller cannot be used, since all chillers are served by all of the tower cells.
3. An additional, redundant tower cell can be included to provide "backup" in the event of a tower cell failure or to maintain the required condenser water supply temperature in periods of higher-than-expected WB temperature.
4. Only one tower makeup water control system is required, rather than one for each tower under the first option, reducing maintenance requirements.
5. One cell (or more) can be shut down for maintenance without interfering with a specific chiller's operation.

A disadvantage of the multicell approach is that automatically controlled two-position isolation valves are required on both the inlet and the outlet of each tower cell. These isolation valves eliminate water flow through cells that are not active.

The best place to locate any HVAC cooling tower is on the roof of the building it serves. This placement satisfies two conditions that are critical to proper tower performance, which are as follows:

1. The elevation in the tower's operating water level must be above the condenser water pump (and any piping between the pump suction and the tower basin).
2. Towers must have sufficient clearance from walls and other obstructions around them to prevent *recirculation* of discharge air back into the inlet. If there are multiple towers, they must be located to avoid the discharge of one tower entering any adjacent tower. Generally, a building roof is clear of obstructions that can result in recirculation, *provided that obstruction by aesthetic screening is not added.*

However, rooftop tower locations increase the difficult of maintenance, require additional structure to carry the tower and water weight, and create "aesthetic" issues for the designing architect. When a ground location is selected, the tower grillage must raise the basin level to ensure that the tower's operating water level is above the condenser water pump and any piping between the pump suction and the tower basin. Great care must be taken to ensure that a ground-level tower location will not result in air recirculation and/or noise reflection.

The airflow requirement of an HVAC cooling tower is approximately 270–300 cfm/ton [160–175 L/s·kW]. Thus, (1) sufficient free area around the tower for the unobstructed flow of large air quantities must be provided and (2) the tower must be located sufficiently clear of obstructions to prevent discharge air from reentering the tower (or entering an adjacent tower).

Ground-level placement considerations for a tower depend, initially, on the type of tower, crossflow or counterflow. Crossflow towers have an air intake on either one (*single-flow* towers) or two sides (*double-flow* towers), whereas counterflow towers have air intakes on all four sides. However, for both types of towers, the following recommendations apply:

1. *Adjacency*: Building walls, screen walls, trees, and so on form obstructions for cooling towers if the tower is located too close to them. While most tower manufacturers provide information about layout for their cooling towers, most of the

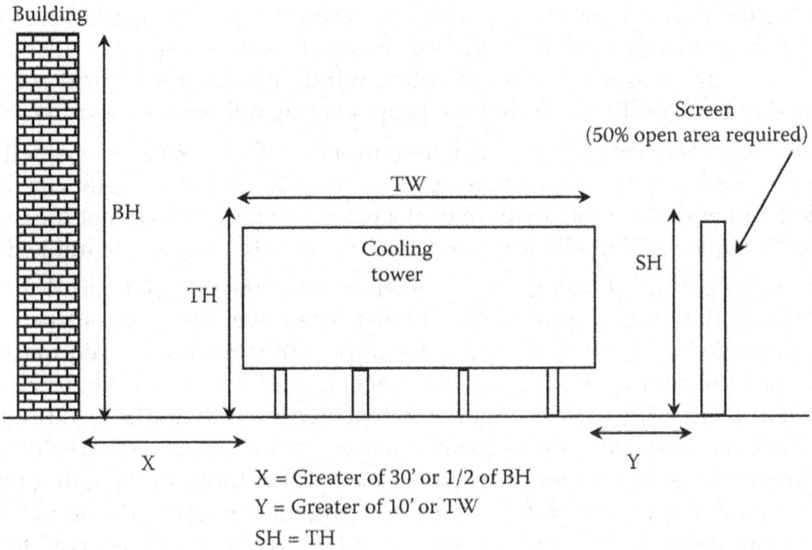

FIGURE 11-19
Cooling tower installation criteria. [Fig. 11.7, *HVAC Water Chillers and Cooling Towers*, Herbert W. Stanford III, PE (CRC Press/Taylor & Francis, 2012).]

recommended clearances are minimal and are not sufficient to always ensure proper tower performance. Therefore, the recommended clearance from adjacent obstructions is shown in Figure 11-19.

Cooling towers are rarely "pretty"; thus, architects will often specify that ornamental screening be installed around the tower. To prevent tower performance problems, the following design criteria for screening must be followed:

1. The screening must provide at least 50% open area on the air intake side(s) of the cooling tower, and the open area must be sufficient to allow for 600 fpm [3 m/s] or low airflow velocity through the screen.
2. The top of the screening must be no higher than the tower air discharge.
3. Provide clearance between the screening and tower air intake side(s) of the greater of 10 ft or the width of the tower. If there are multiple towers or multiple tower cells within a screened enclosure, these clearances must be increased by at least 50% (to 15 ft [4.5 m]).
4. Maintenance clearance between the screening and the non-active sides of the tower must be at least 4 ft [1.3 m] to provide adequate passage and maintenance space.

2. *Elevation*: Ideally, any HVAC cooling tower should be installed so that the operating water level is higher than all of the elements of the condenser water system. In every installation, the operating water level must be higher than the condenser water pump and all of the condenser water supply piping between the tower basin and the pump. Thus, the pump is maintained at a "flooded" condition and cannot lose prime.
3. *Prevailing wind*: Since counterflow towers typically have air intakes on all four sides, tower orientation relative to the prevailing summer wind is not important. Orient a crossflow cooling tower so that a blank side is facing the wind. Wind can

influence the performance of a double-inlet crossflow tower by creating an unbalanced airflow through the two sides. (If a crossflow tower must be installed with one inlet in the direction of the prevailing wind, the tower manufacturer must be apprised of this condition so that the proper sizing adjustments can be made.)

4. *Multiple towers or tower cells*: No more than four crossflow cooling tower cells should be close-coupled. For counterflow towers, usually only two cells can be close-coupled and even then the tower manufacturer must guarantee that the individual cell performance will not be impaired since each cell is losing one intake side.
5. *Dunnage and grillage*: *Dunnage* is the steel and/or concrete structural foundation upon which the tower support *grillage* rests. The tower grillage routinely consists of structural steel members arranged to support the tower at the required bearing points. On the roof of a building, dunnage should consist of structural column extensions so that the tower weight is transferred directly to the building foundation elements. *Never attempt to install a tower directly on the roof structural system!* Roofs are never designed for the types of loads introduced by a cooling tower and its water load. Even if the roof structure is designed to support the operating tower weight, vibration produced by the tower will assuredly be transferred to the roof structure and create serious noise problems. The bottom of the grillage should be 3–4 ft [1.0–1.3 m] above the roof to provide clearance for roofing maintenance. On the ground, dunnage may consist of concrete piers or columns, each with properly designed footings and reinforcement.

One critical part of every cooling tower system that is often incorrectly designed is the *cooling tower(s) bypass valve*. For any condenser water system, either single tower or multitower/cell, the initial system start-up can create a problem. If the system has been shut down overnight, as often happens in the spring and fall, the condenser water temperature in the tower basin may be relatively cold. If this cold water is circulated to the condenser, the chiller may be shut down by its safety operating controls due to "low condenser water temperature" or "low refrigerant temperature." In either case, the chiller will simply not start until the condenser water temperature is increased to above the safety set point temperature. To prevent this problem, a modulating three-way control valve should always be included with HVAC cooling tower systems.

The control valve is normally open to the bypass (normally closed to the tower). Thus, upon start-up, all condenser water flow bypasses the tower and simply circulates through the condenser. As the chiller loads, the condenser begins to reject heat to the water, raising its temperature. As the temperature rises, to set point, the control valve modulates the bypass port closed, diverting the water to the tower, upon which the tower capacity control scheme takes over.

The bypass control valve should be placed as close to the condenser as possible, but it must also be at an elevation at or below the tower basin operating water level if the bypass line is connected directly to the condenser water supply line. If not, when the chiller and its condenser water pump shut down at night, the bypass port reopens and the water in the bypass line will drain to the same level as the tower operating water level. This, in turn, creates an "air gap" or "bubble" in the bypass. Upon start-up the next morning, this air bubble will enter the pump, causing it to lose prime and all flow will stop… immediately causing the chiller to shut down!

If the valve cannot be installed below the tower operating water level, the bypass line must be piped to the tower basin. Thus, while the drain down will still occur, the resulting air bubble cannot enter the pump. (An alternative would be to install a check valve in the suction piping between the bypass connection and the tower basin. But, over time, the check valve will probably leak and the problem will return.) In this configuration, however, the basin water, at a low temperature, is supplied to the condenser, so chiller start-up problems may result.

The bypass control valve, ideally, would have linear flow characteristics. However, butterfly valves, which are normally used for bypass duty, have equal percentage flow characteristics. However, a butterfly valve will have a near-linear relationship between "% open" and "% flow" if it is selected to operate with a maximum 70° rotation and a 33% authority setting.

Another common error is failing to include and/or correctly size an equalizer line between multiple tower or cell basins that are used in a common condenser water system configuration. Where each tower/cell has an individual basin, the basins can be connected together with an equalizer line. No matter how closely matched the piping is between towers/cells, flow to each can never be *exactly* balanced. Without the equalizer line, one basin would probably overflow, whereas the other was pumped dry. The equalizer line allows water to flow between basins and compensate for the minor flow imbalances between the basins that will exist.

Based on field experience, the equalizer line should be sized to accommodate at least 15% of the flow rate of the largest tower/cell. On this basis, Table 11-11 summarizes the recommended equalizer line sizes relative to tower/cell flow rate.

With an equalizer line, the water operating level will be essentially the same in each tower/cell basin. Therefore, it is critical that each tower/cell be installed so that its recommended operating water level is at the same elevation as adjacent towers/cells.

11.4.4 Closed Circuit Evaporative Cooler

Closed circuit water source heat pumps require some method for the rejection of excess heat and the *closed circuit evaporative cooler*. These evaporative coolers are "first cousins" of the cooling tower, *the difference being the use of a closed water-to-air cooling coil in lieu of open fill.*

An open water system for heat rejection is created by using a water basin, a circulating pump, and a spray-type wet deck in a counterflow configuration to the airflow. By wetting the coil surface and allowing evaporative cooling to occur, the effective air temperature is

TABLE 11-11

Recommended Equalizer Line Size

Largest Tower/Cell Flow Rate (gpm) [L/s]	Equalizer Line Size (NPS in.) [DN]
Up to 240 [15]	4 [100]
241–627 [15–39]	6 [150]
628–1,167 [40–74]	8 [200]
1,168–1,925 [75–121]	10 [250]
1,926–2,820 [122–178]	12 [300]
2,821–3,465 [179–225]	14 [350]
3,466–3,850 [226–243]	2 @ 10 [250]
3,851–5,640 [244–356]	2 @ 12 [300]

reduced by 10°F–15°F [5°C–8°C]. Water flow to an evaporative cooler is typically between 1.0 and 1.25 gpm/ton [0.018–0.022 mL/J] simply because their range is higher (usually 25°F–30°F [14°C–17°C]). The evaporation rate, however, is the same as for a cooling tower, 0.1% per degree of range, and water treatment is required.

11.5 Evaporative Air Cooling

In hot, dry climates, the use of evaporative cooling can be implemented to cool the air circulated in the facility, reducing the energy burden imposed by mechanical refrigeration. The two types of evaporative coolers are as follows:

1. *Direct evaporative coolers*: They consist of an air handler section and direct evaporative cooling section. These units are operated in the fan-only mode when the outdoor air temperature is low enough to meet the cooling load requirements. When ventilation cooling is inadequate, a water pump circuit is energized to circulate water over a wetted media over which the outdoor air passes to provide evaporative cooling. The saturation effectiveness of direct evaporative coolers may range between 80% and 99%.
2. *Indirect evaporative coolers*: They cool air without adding moisture; therefore, they do not increase humidity. Indirect coolers include an air-to-air heat exchanger, evaporative media (such as the media described earlier), and a secondary (outdoor) air fan and outdoor air louvers. The secondary air is cooled as it flows through the evaporative media. The cooled air flows through a heat exchanger that indirectly cools the primary (supply) air to the space.
    a. Another type of indirect cooler incorporates cooling tower, which includes similar components except that the air-to-air heat exchanger is replaced with a water cooling coil. The cooling tower provides cooled water to a water cooling coil. Primary air flows through the cooling coil and is cooled without the addition of moisture. The primary benefit of indirect coolers is their capability to cool without adding moisture, that is, increasing the humidity of the primary air.
    b. The saturation effectiveness of indirect coolers may range between 50% and 75%.
3. *Indirect/direct coolers*: They combine direct and indirect coolers into multistage units as follows: the first stage of cooling is done by ventilation only, the second stage of cooling by indirect cooling, and the third stage of cooling by direct evaporative cooling. Indirect/direct evaporative coolers can also be provided with refrigeration coils to deliver cooling when the outdoor air humidity is too high for evaporative cooling.

11.6 Waterside Economizer Cycle

When an HVAC system cannot take advantage of an airside economizer cycle, the next best approach is to use a *waterside economizer cycle.*

The conventional waterside economizer cycle uses a cooling tower to produce cold water when the ambient WB temperature is 50°F [10°C] or lower. At 50°F [10°C] WB, a typical cooling tower will produce 40%–50% of its design capacity at an approach of about 5°F [3°C]. Thus, a 500 ton [1,750 KW] cooling tower will deliver approximately 250 tons [875 kW] of cooling with a 45°F [7°C] supply water temperature at 50°F [10°C] ambient WB. But, in most comfort cooling applications, the need for dehumidification is reduced during the heating season and a 5°F–10°F [3°C–7°C] increase in the chilled water supply temperature can be tolerated. Thus, the number of hours of available free cooling can be significantly increased.

Obviously, this cold water can be used to provide cooling without operating a water chiller. However, to separate the "dirty" tower water from the "clean" chilled water, a heat exchanger is typically used, at only a loss of 2°F–3°F [1°C–2°C] in supply water temperature.

Two types of heat exchangers may be used for waterside economizer applications, which are as follows:

1. *Plate-and-frame heat exchangers,* often called simply "frame heat exchangers," consist of thin corrugated plates separated by rubber or neoprene gaskets and mounted in a rigid frame. The two water flows are arranged counter to each other on opposite sides of each plate. Close temperature approach, 1°F–2°F [0.5°C–1.0°C], is obtained by having a very large heat transfer area in a relatively small package.
2. *High-efficiency shell-and-tube heat exchangers* provide high heat transfer area by (a) using a large number of small-diameter enhanced surface tubes and (b) increasing the tube length with a U-shaped configuration. An approach of 2°F [1°C] can be obtained with this type of heat exchanger.

Shell-and-tube heat exchangers have lower cost, less weight, and more versatile configurations than plate-and-frame heat exchangers. They are also easier to insulate and require less maintenance. Closer approach (higher efficiency) can be obtained with the plate-and-frame heat exchanger.

While the economizer can be applied to a single tower system, most energy conservation codes require that it be an "indirect" system, which allows the chiller and the economizer to operate concurrently. Since chillers typically will not operate at a 45°F [7°C] condenser water supply temperature, bypass temperature control becomes critical and often fails. Therefore, where multiple chillers and cooling towers are available, while it is preferred to pipe multiple cooling towers or tower cells in parallel as a common system, it is recommended that one tower be isolated for economizer duty, leaving at least one other chiller and tower available to supplement the economizer cooling capacity, as shown in Figure 11-20.

The energy consumption by a conventional waterside economizer for the pumps and cooling tower is only about 15%–20% of the energy consumption required by mechanical cooling.

An alternative to the conventional waterside economizer is to transfer heat between the cooling tower water and the chilled water inside a chiller through the use of *refrigerant migration,* also known as a *thermosyphon.* When the temperature of the water from the cooling tower is colder than the desired chilled water temperature, the compressor is turned off and automatic shutoff valves inside the chiller refrigerant circuit are opened. Because refrigerant vapor migrates to the area with the lowest temperature,

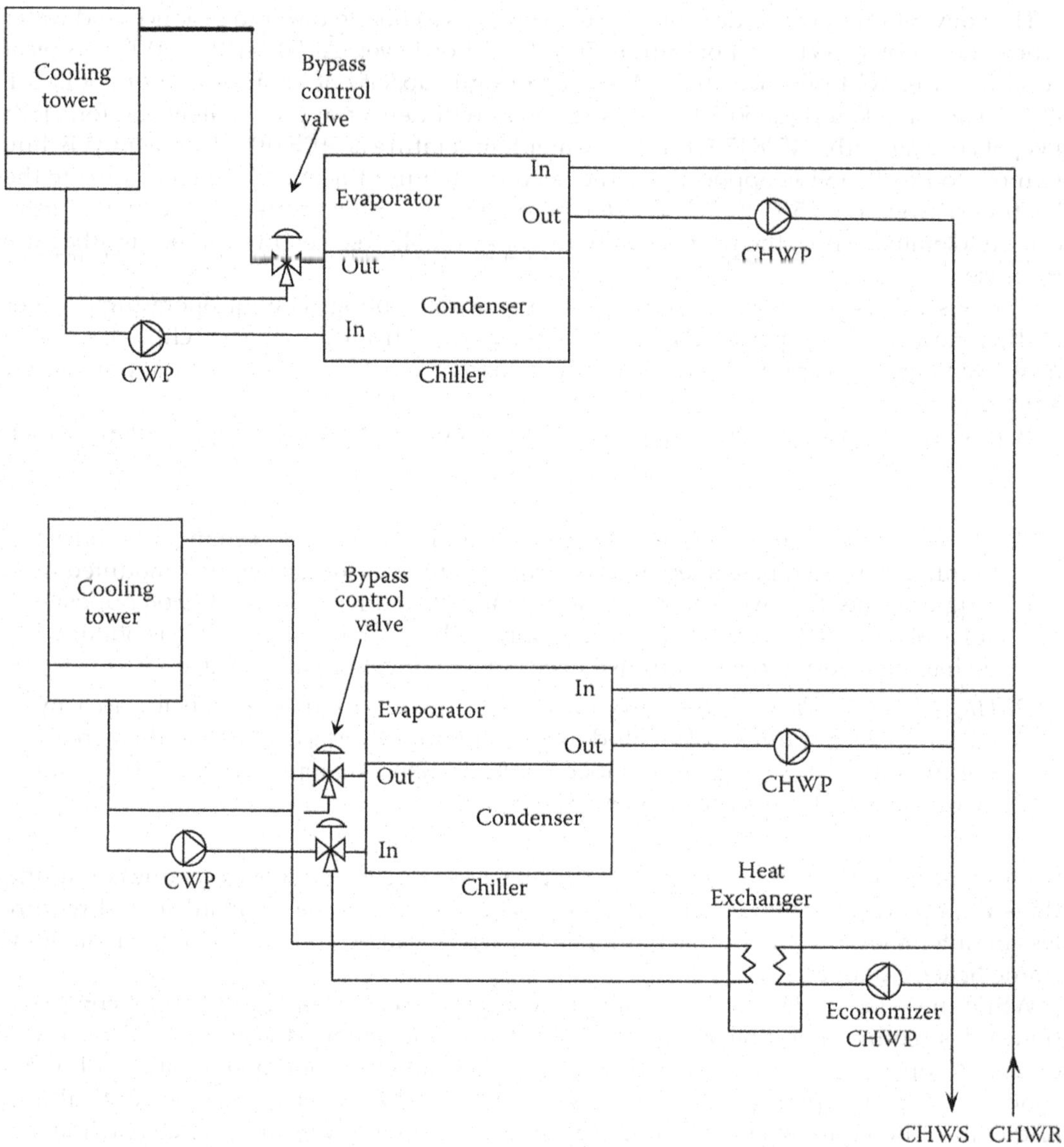

FIGURE 11-20
Multiple chiller waterside economizer cycle piping. [Fig. 14.3, *HVAC Water Chillers and Cooling Towers*, Herbert W. Stanford III, PE (CRC Press/Taylor & Francis, 2012).]

refrigerant "boils" in the evaporator and the vapor migrates to the cooler condenser. After the refrigerant condenses, it flows by gravity back to the evaporator. This allows refrigerant to circulate between the evaporator and the condenser without the need to operate the compressor(s).

Depending on the cooling load profile and the climate zone in which the chiller is installed, it is possible for refrigerant migration in a chiller to satisfy significant hours of cooling load without mechanical refrigeration, especially if the cooling loads can tolerate a warmer chilled water supply temperature.

If the free-cooling chiller is part of a multiple chiller installation piped in parallel, simultaneous operation is not possible without blending leaving chilled water temperatures. To avoid this issue, the free-cooling chiller should be piped the same way as the heat exchanger in Figure 11-20, thus acting as a "precooler" for the chilled water return.

Bibliography

HVAC Water Chillers and Cooling Towers, Herbert W. Stanford III, PE (CRC Press/Taylor & Francis, 2012).

Engineers Newsletter, Vol. 37-3, "Free Cooling Using Water Economizers", The Trane Co., La Crosse, Wisconsin.

ASHRAE Handbook-HVAC Systems and Equipment, American Society of Heating, Refrigerating, and Air-Conditioning Engineers, Inc. (Atlanta, GA, 2016).

12

Heating Systems and Components

12.1 Firing Fossil Fuels

With the exception of electric resistance and heat pump type heating systems, heating systems rely on the burning (more commonly called "firing") of fossil fuels. By definition, fossil fuels are solid, liquid, and gaseous substances that are carbon-based by-products of organic materials. The common fossil fuels utilized today in heating, ventilating, and air-conditioning (HVAC) applications are fuel oil (mostly No. 2), natural gas, and (to a limited extent) propane (liquefied petroleum gas or LPG). Coal, while widely applied in the past, is no longer a viable fossil fuel choice due its pollution levels.

Each of these fuel types can be broken down into subtypes, each with a different heat release rate or *heating value* (i.e., Btu [W]/unit of fuel). In the HVAC industry, the term *higher heating value* (HHV), which takes into account the latent heat of vaporization of water in the combustion products, is utilized to define the heat-release rate for a fuel. Table 12-1 is a summary of HHVs for common fuels.

12.1.1 Oil-Fired Systems

An oil burner is defined as "a mechanical device for preparing fuel oil to combine with air under controlled conditions for combustion." To burn oil, it is necessary to first convert the fuel from a liquid into a vapor for combustion, typically in two steps:

1. *Atomization*. The liquid fuel is introduced in a fine mist (with the smallest possible droplet size) to enhance the vaporization and, then, combustion processes.
2. *Vaporization*. The oil droplets are heated and change state from liquid to gas.

Air for combustion may be supplied by either *natural* or *mechanical draft* that utilizes a fan. Natural draft burners depend on the hot stack gases rising, creating a low-pressure region

TABLE 12-1

HVAC Fuels HHVs

Fuel	Higher Heating Value
No. 2 Fuel oil	140,000 Btu/gal [10,800 W/L]
No. 4 Fuel oil	144,000 Btu/gal [11,150 W/L]
No. 5 Fuel oil	150,000 Btu/gal [11,600 W/L]
No. 6 Fuel oil (low sulfur)	143,800 Btu/gal [11,130 W/L]
Natural gas	1,000 Btu/cf [10.3 W/L]
Propane (LPG)	91,600 Btu/gal [7,100 W/L]

that "sucks" combustion air in from the space. For mechanical draft burners, a fan "forces" or "induces" air into the combustion chamber. More complex gas–air mixing patterns are possible, and combustion efficiency is significantly improved over that of the atmospheric burner. Ignition of the air/fuel mixture is usually by electric spark.

An oil-firing system consists, in addition to the burner, of an oil storage tank, piping, and a pump to deliver the oil from the tank to the burner. No. 4, 5, and 6 fuel oils have high viscosities at room temperatures and require heating systems to keep them at a temperature high enough to be pumped and atomized. For No. 4 oil, this temperature is about 100°F [38°C] and about 120°F [50°C] and 160°F [70°C] for Nos. 5 and 6 oil, respectively. Generally, these "heavy" oils are used only in large boiler systems. No. 2 fuel oil requires no supplemental heating since its viscosity is relatively low at room temperature.

Commercial/industrial burners are typically atomizing types. They inject fuel oil into the combustion chamber as a fine, conical spray. The burner forces combustion air into the oil spray, causing a mixing of air and oil. An electrical spark ignites the mixture and sustained combustion takes place.

These burners are capable of almost complete burning of the fuel oil, without visible smoke, when they are operated with excess air as low as 20% (approximately 12% carbon dioxide in the flue gases). Atomizing oil burners are generally classified according to the method used for atomizing the oil: *pressure atomizing*, *air atomizing*, or *steam atomizing*.

Pressure-atomizing oil burners are utilized for most No. 2 oil firing. Atomizing is accomplished by pumping the oil at a pressure from 100 to 300 psig [690–2,070 kPa] through a burner nozzle (orifice) which, due to expansion into the much lower pressure region of the combustion chamber, breaks it into a fine mist that swirls into the combustion chamber as a cone-shaped spray. Combustion air from a fan is forced through the burner around the oil nozzle and is directed into the oil spray. Ignition for smaller capacity burners is usually accomplished by an electric spark applied near the discharge of the burner nozzle. For burners above 20 gph [76 L/h], a spark-ignited gas or oil igniter is used.

Pressure-atomizing burners are designated to be either forced draft or induced draft. A forced draft burner has a fan and motor combination with capacity enough to supply all the air for combustion into the combustion chamber at a pressure high enough to force the gases through the heat exchange equipment without the assistance of an induced draft fan or chimney draft.

Variation in firing rate, in response to load variations, is accomplished by simultaneously varying the oil pressure to the burner nozzle and regulating the airflow by a damper. Large burners (15 gph [50 L/h]+) are equipped with modulating controls, while smaller burners utilize "on-off" or "step" control.

Air-atomizing burners are similar in construction to pressure-atomizing burners. Compressed air and oil are supplied to individual parts within a nozzle that forces the oil to break up into small droplets as a result of the shear forces created by the atomizing air. Combustion air from a forced draft fan is routed through the burner throat, where it mixes with the oil spray inside the combustion chamber. The ignitor is similar to that used on pressure-atomizing burners.

This type of burner is well suited for heavy fuel oils, including No. 6, and has a wide "turndown" or load range of 3:1 for the smaller sizes and 6 or 8:1 for the larger sizes. Turndown is accomplished by simultaneously varying oil pressure, atomizing air pressure, and the flow rate of combustion air entering the burner. Air-atomizing burners use 2.5–8.0 cf of compressed air per gallon of fuel oil.

Steam-atomizing oil burners accomplish atomization by the impact and expansion of steam. Oil and steam flow in separate channels through the burner gun to the burner

TABLE 12-2

Oil Burner Selection Criteria

Type of Burner	Size Range (gph) [L/h]	Oil Type
Pressure atomizing	0.5–50 [2–200]	No. 2 or 4
Air atomizing	0.5–700 [2–2,600]	No. 2, 4, 5, or 6
Steam atomizing (register type)	80+ [300+]	No. 2, 4, 5, or 6

nozzle, where they mix before discharging through an orifice into the combustion chamber. Combustion air, supplied by a forced draft fan, passes through the directing vanes of the burner register, through the burner throat, and into the combustion chamber. These vanes give the air "pre-swirl" and the burner throat directs it into the oil spray, where mixing of air and oil takes place.

Full-load oil pressure at the burner inlet is typically 50 psig [345 kPa] and 10 [69 kPa] psig at minimum load. The turndown ratio is about 8:1, and this wide load range makes the steam-atomizing burner ideally suited to varying boiler loading and modulating control. Depending on the burner design, steam-atomizing burners use from 1 to 5 lb of steam to atomize a gallon of oil. Where no steam is available for startup, compressed air is often used for atomizing until the boiler begins to produce steam.

All oil burners are selected on the basis of their firing rate (input) and rated firing efficiency, typically 80%–84%, depending on the boiler configuration. Table 12-2 gives the typical selection criteria for oil burners.

The second component of the oil-firing system is the oil pump. Basically, the pump must be sized on the basis of the firing rate required (gph [L/h]) and the pressure required at the burner, plus piping losses and static head.

The third component is the fuel storage tank. Standard oil storage tanks range in the size from 50 gal [200 L] to over 60,000 gal [230,000 L]. Proper selection of the tank(s) size depends on the overall firing rates and on the available delivery lots. Typically, an oil supply of 14–30 days is considered adequate, depending on the location and the dependability of oil supplies. Since even heavy oil will store for 12–18 months with no problems, larger tank sizes invoke no operating penalties. Also, since oil purchased in larger delivery lots (up to about an 8,000 gal [30,000 L] tanker capacity) will generally result in lower costs, hence increasing storage capacity to allow for larger lots may be a positive factor.

Except for when the project location is not served by natural gas distribution lines, boilers (except, perhaps, for smaller, light commercial applications) should be designed for dual-fuel operation, typically fuel oil and natural gas.

Fuel oil storage tanks should be designed for *aboveground* installation, as shown by Figure 12-1, with fuel transfer pumps located within the boiler room.

Aboveground storage tank design and construction standards are addressed by:

1. Federal Water Pollution Control Act (40 CFR Part 112)
2. National Fire Protection Association *Standard 30* (NFPA 30) and *Standard 30A* (NFPA 30A)

Federal regulations do not actually use the terms "aboveground storage tank." Instead, the term "bulk storage container" is used and is defined as "…any container used to store oil. These containers are used for purposes including, but not limited to, the storage of oil prior to use, while being used, or prior to further distribution in commerce. Oil-filled electrical, operating, or manufacturing equipment is not a bulk storage container."

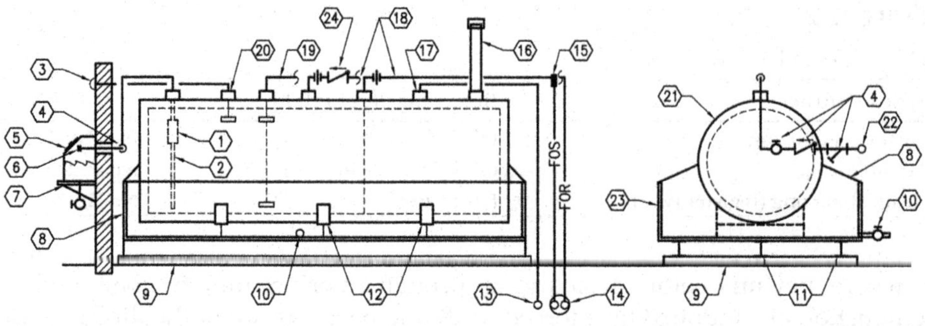

NOTES KEYED TO X/XXX:

1. OVERFILL PREVENTION VALVE.
2. ALUMINUM DROP TUBE.
3. LIGHT AND HORN.
4. STRAINER, CHECK AND BALL VALVE.
5. ACCESS DOOR.
6. FILL ADAPTER AND DUST CAP.
7. 15 GALLON CAPACITY SPILL CONTAINMENT WITH DRAIN AND LOCKING BALL VALVE. BOLT TO WALL.
8. CONTAINMENT DIKE WITH RAINSHEILDS.
9. 4" THICK CONCRETE PAD.
10. DIKE DRAIN WITH LOCKING BALL VALVE.
11. SUPPORT BASE CHANNELS.
12. SADDLES (TYPICAL).
13. 3/4" MONITORING CONDUIT.
14. 3/4" FOS AND FOR IN 4" U.G. CONDUIT WITH LEAK DETECTION.
15. ANTI–SYPHON DEVICE.
16. VENT
17. INTERSTITIAL MONITOR.
18. F.O.S. AND F.O.R.
19. TANK LEVEL GAUGE FOR REMOTE READING AT CONTROL PANEL AND CONNECT TO DDC CONTROL SYSTEM.
20. HIGH LEVEL ALARM WITH AUDIBLE HORN.
21. FUEL OIL STORAGE TANK.
22. TO FILL ADAPTOR.
23. STEEL SECONDARY CONTAINMENT DIKE.
24. CHECK VALVE.

FIGURE 12-1
Typical aboveground fuel oil storage tank installation. (SW Standard Detail O02.)

A "bulk" storage container is defined as one containing 55 gal [210L] or more liquid and all bulk storage container installations must be constructed so a secondary means of containment is provided for the entire capacity of the largest single container and sufficient freeboard to contain precipitation. Diked areas must be sufficiently impervious to contain discharged oil.

Each bulk storage container installation must be engineered to avoid accidental discharges and include the following devices:

1. High liquid level alarms with an audible or visual signal at a constantly attended operation or surveillance station (in smaller facilities an audible air vent may suffice.)
2. High liquid level pump cutoff devices set to stop flow at a predetermined container content level direct audible or code signal communication between the container gauge and the pumping station a fast response system for determining the liquid level of each bulk storage container. Direct vision gauges may be used, but a person must be present to monitor gauges and the overall filling of bulk storage containers.

12.1.2 Gas-Fired Systems

Gas burners are significantly simpler than oil-firing systems since (1) the fuel is already in a vapor state and (2), at least for natural gas, no storage system is required. Gas (natural or LPG) must be mixed with air and ignited in a gas burner.

Electronic ignition has generally replaced the gas pilot, and therefore, that portion of the train may not be required.

Gas burners are generally referred to as "atmospheric" or "power." *Atmospheric burners* operate with natural draft and are simple tube-type burners where the gas is introduced at the center nozzle and the combustion air is induced around it. This type of burner is often used in furnaces, smaller boilers, fired unit heaters, and so on.

Power burners utilize mechanical draft to supply and control combustion air. The power burner fan moves air through the burner and forces it through the heat exchanger. A stack is required to release the flue gas products at a height above a nuisance level and the combustion process occurs under pressure. Fuel gas is introduced into a controlled airstream designed to produce thorough gas–air mixing but still capable of maintaining a stable flame front.

Large conventional and all condensing boilers utilize fully modulating gas burners with at least a 5:1 turndown ratio.

12.2 Furnaces

A *furnace* is defined as a direct-combustion air-heating device. The air to be heated is circulated by a fan around the outside surface of a combustion chamber, which serves as the heat exchanger. Hot flue gases transfer their heat to the circulating air as illustrated in Figure 12-2.

Furnaces have their widest application in the residential sector but have specialized HVAC applications in the industrial/commercial sector, such as for make-up air heating, fired unit heaters, or packaged rooftop heating and cooling units.

Generally, since they operate at relatively low temperatures, furnaces typically take advantage of condensing combustion. See Section 12.3.1 for further discussion of condensing combustion.

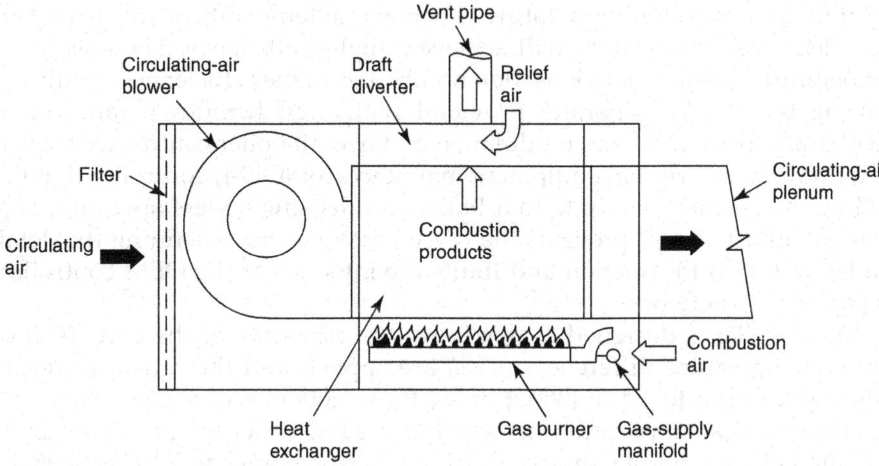

FIGURE 12-2
Typical noncondensing horizontal gas-fired furnace.

12.3 Boilers

A boiler is a pressure vessel designed to transfer the heat produced by combustion to a fluid. The definition has been expanded to include transfer of heat from electrical resistance elements to the fluid or by direct action of electrodes on the liquid. In boilers of interest to the HVAC industry, the fluid is usually water, either as liquid ("hot water") or as steam.

12.3.1 Boiler Types, Ratings, and Efficiency

Boilers are generally specified on the basis of working pressure and temperature, fuel used, shape and size, usage (such as for heating or process), steam or water, and so on. They may fire fuel oil, LPG, or natural gas or be equipped with dual fuel burners that may fire any combination of these fuels.

Conventional boilers operate without condensing within the flue gas leaving the boiler. This precaution is necessary to prevent corrosion of the metals used in the boiler construction and, thus, *noncondensing boilers must be operated at a minimum 140°F entering water temperature to prevent internal condensation from occurring and to minimize the potential for thermal shock within the boiler.*

Condensing boilers achieve higher efficiencies by condensing water vapor contained in flue gases. When fuel gas is combusted in a boiler, approximately 90% of the energy contained in the fuel is converted into sensible heat (heat that causes a change in temperature) and approximately 10% is converted into latent heat (heat involved in phase change) that is stored in the water vapor by-product of the combustion process.

In conventional boiler technology, the latent heat contained in the water vapor is allowed to escape with the flue gases. Condensing boilers reclaim that latent heat by condensing the water vapor and transferring that heat back into the return water. This process of condensing the water vapor can increase the steady-state efficiency of an average boiler from 78%–84% to between 86% and 98%, depending on boiler operation.

These potential efficiency gains can only be attained if the boiler is operating in condensing mode achieved when the temperature of the return water is below the dew-point temperature of the water vapor in the flue gas. For fuel gas boilers, the dew point is approximately 127°F [53°C], and generally, if the return water temperature is below that temperature, the water vapor will start to condense and the system will achieve a higher efficiency. The lower the return water temperature, the more condensation occurs and higher efficiencies result.

Condensing boilers are generally provided with high turndown ratio, modulating mechanical draft burners. Burner modulation and overall boiler control are typically provided by a proprietary original equipment manufacturer (OEM) controller that is capable of controlling the operation of up to four boilers condensing boilers operating in parallel. This situation almost always prevents the HVAC designer from defining the details as to how a boiler system is to function and limits the interface of the OEM controller with a direct digital control network.

The OEM controller is designed to maximize the efficiency of the boiler(s) it controls. And, if large temperature differences (TDs) are applied and the return water temperature is allowed to drop to 100°F [38°C] or so, these boilers will achieve 96%–98% firing efficiency. However, the applications in which these low water temperatures are satisfactory are limited and the normal expected firing efficiency range will be 89%–96%.

To allow condensing, the boiler must utilize stainless steel or cast aluminum internal construction for protection against acid corrosion and include a method of collecting and disposing of the acidic condensate produced. Condensate drain systems must be designed to drain the amount of condensate produced at full load, typically 0.008 gal/1,000 Btu/h [0.01 L/kW]. In multiple boiler operations, it is common to manifold the drains from each boiler. A condensate drain must be installed from the bottom of any vertical common flue. The pH of the condensate produced by condensing boilers ranges from 3.0 to 6.0 and the drain installation must comply with code requirements. Where necessary, a neutralizing tank may be required to ensure that the final discharge pH meets code limits.

With few exceptions, all boilers are constructed to meet the American Society of Mechanical Engineers (ASME) *Boiler and Pressure Vessel Code,* Section IV, for heating boilers (low-pressure boilers) and Section I for power boilers (medium- and high-pressure boilers).

Every boiler is rated at the maximum working pressure determined by the ASME *Boiler Code* (or other code) under which it is constructed and tested and is equipped with safety controls and pressure-relief devices mandated by code requirements. Low-pressure boilers are designed for maximum working pressures of 15 psig [103 kPa] steam or 160 psig [1,100 kPa] hot water.

12.3.2 Application Considerations

There are a number of design considerations in the application of boilers:

1. *Heating medium.* For most HVAC systems, economics and control versatility favor hot water as the heating medium. For large systems that must include significant distribution systems (college campuses and so on) or where temperatures in excess of 200°F [93°C] are required (hospitals, laboratories, and so on), steam is the choice.
2. *Boiler type.* It is usually dictated by the capacity requirement and design hot water supply temperature and TD. For simple heating applications where hot water supply temperature may be ≤127°F [55°C], hot water condensing boilers are more efficient and should be used. Where higher hot water supply temperatures are required, conventional boilers are applied and cast iron or flexible tube boilers are good choices up to about 100 BHP [980 kW] output. For larger boilers, three-pass fire-tube boilers provide good firing efficiency and reliable service at a reasonable cost. The highest efficiency conventional boilers are four-pass fire-tube types.

   There are many condensing boiler types and technologies available. *Low-mass (low volume) condensing boilers,* defined as having 20 gallons [76 L] per 1,000,000 Btu/h [300 kW] or less of water volume, often require constant flow, while *high-mass (high volume) boilers* with 50-gallons [189 L] per 1,000,000 Btu/h [300 kW] or more of water volume are generally low-flow tolerant and can be used in variable primary flow configured systems. Low mass designs typically have high water-side pressure drops necessitating a primary–secondary pumping system, while high-mass boilers have much lower pressure drops.

   For retrofit applications that require a higher hot water supply design temperature, but that may also have another load available that can operate at lower temperatures, several high-mass condensing boilers offer dual-return connections. Dual returns can condense when a small amount (generally 10% or greater) of return water is brought back to the boiler as a separate loop, with a temperature below 127°F [55°C]. When connected to the lower temperature return instead of mixing in the system piping, the coldest possible water is introduced to the boiler, thereby increasing the condensing potential and fuel efficiency of the system. When a higher temperature return connected at a warmer location of the heat exchanger combines with the low-temperature return located near the coolest flue gases, the boiler can achieve condensing performance even when a majority of the system requires operation at noncondensing temperatures.

3. *Load profile.* While peak load will dictate the overall boiler or plant capacity needed, the part-load profile will dictate the sizing of multiple boilers. For single boiler systems, define minimum required boiler capacity as the total system peak heat loss, plus 2%–5% for line losses. Single boiler systems should be used only for the smallest applications.

   For reliability and better load versus capacity control, multiple boiler systems are preferred. For two boilers, select each boiler's capacity to satisfy 60%–65% of the total system peak heat loss. For systems with three or more boilers, select each boiler's capacity on the basis of the total system peak heat loss divided by $(N-1)$, where N is the number of boilers provided. Thus, in the event of a boiler failure, the remaining boilers will satisfy the peak heat loss of the building.

   With conventional noncondensing boilers, due to limited turndown even with modulating burners, boiler start/stop can is typically based on keeping boilers operating at 50%–100% of design capacity.

   Condensing boilers are more efficient at lower load levels and boiler start/stop can is typically based on keeping boilers operating at 10%–50% of design capacity as much as possible.

12.3.3 Boiler/Furnace Venting

Boiler venting for noncondensing combustion should be designed in accordance with the American Society of Heating, Refrigerating, and Air-Conditioning Engineers (ASHRAE) *Handbook—HVAC Systems and Equipment*. Where multiple conventional boilers must be vented with a common breeching and stack, follow the stack and breeching guidelines for design and sizing criteria.

Boiler venting for condensing combustion requires the use of special vent material, typically Type AL 29-4C, a double-wall, stainless steel fabricated vent. Condensate from the second-stage heat exchanger and from the vent must be collected, trapped, and piped to a sanitary drain. In some cases, the acid nature (pH = 2.9–4.0) of the condensate may require acid neutralization before disposal.

Combustion air intakes should be designed in accordance with NFPA 54, *Natural Fuel Gas Code*, as follows:

1. Two-permanent-openings method: Two permanent openings, one "high" (commencing within 12 in. [300 mm] of the top of the exterior wall) and one "low" (commencing within 12 in. [300 mm] of the floor) are required. The openings must communicate directly with the outdoors. Each opening must have a minimum free area of 1 in.^{2} [645 mm^2] per 4,000 Btu/h [1,172 kW] of total fuel input rating of all fuel-firing equipment in the space.
2. One-permanent-opening method: One permanent opening, commencing within 12 in. [300 mm] of the top of the enclosure, must be provided. The opening must directly communicate with the outdoors and must have a minimum free area of (1) 1 in.^{2} [645 mm^2] per 3,000 Btu/h [880 kW] of the total input rating of all fuel-firing equipment located in the space and (2) not less than the sum of the areas of all vent connectors in the space.

12.4 Hydronic Heating Systems

In hydronic heating systems, water serves as both a heat transfer and heat distribution medium. Typically, two (or more) steps of heat transfer take place: (1) water is heated by the boiler(s) and (2) this hot water is circulated through finned-tube heating coils, where it gives up heat to the air. Generally, there is forced-air circulation of air over heating coils.

The heating rate for a hot water system, or any section of it, is dictated by the water flow rate and the temperature drop of the water through the heating coils and/or radiation and is determined by Eq. (8-8). In the design of a hydronic system, the heating requirement is known from the heat-loss calculations. Then a TD is selected, typically 20°F–40°F [11°C–23°C], and necessary flows are computed by Eq. (8-9) or (8-9a). The selection of TD depends primarily on the available heating water temperature, which is dictated by the type of heating required. For typical heating requirements, to reduce "air buoyancy" effects, heating coil leaving air temperatures are maintained ≤95°F [35°C]. Therefore, a hot water supply temperature not exceeding 140°F [54°C] is required, with a TD of 30°F [17°C]. Note that as the TD decreases, the flow rates, pipe sizes, and pumping energy increases. Thus, it is more cost-effective to select the largest TD practical.

For the typical hot water heating system, water is circulated from one or more boilers through one or more heating coils and heat is transferred to the supply airstream of an air-handling system(s), either at the air handler or at terminal units. Direct radiation heating systems, which can take advantage of higher leaving air temperatures, are rarely utilized in nonresidential construction. In cold climates, though, radiation heating may be utilized for supplemental perimeter heating.

Piping of multiple noncondensing boilers typically consists of a primary–secondary system. Thus, when in use, the flow through each boiler remains constant. Each boiler is provided with a primary pump that is started/stopped with the boiler, this meets the requirement imposed by energy conservation codes that the flow in the boiler plant can be automatically reduced, correspondingly, when a boiler is shut down. In addition, a three-way bypass valve is required to ensure that the return water temperature is never lower than 140°F [60°C] in order to prevent undesired internal condensation.

A primary–secondary pumping configuration must also be used with multiple low-mass, low-volume condensing boilers to protect them from thermal shock, low or no flow that can result in localized boiling and scaling, and/or excessive flow that can cause erosion. But, primary-only pumping typically may be used if multiple condensing boilers are the high-mass, high volume type that can deal with large variations in flow, have no minimum return water temperature requirements, and have a low water pressure drop. *If in doubt about which piping configuration to use, the designer must determine the limitations, if any, imposed by each boiler manufacturer for the specific type of boiler being considered for the application.*

There are basically two ways to incorporate primary-only variable flow: (1) utilize one set of variable flow primary pumps for the entire system or (2) utilize both variable flow primary and variable flow secondary pumps, which vary flow through their respective loops.

To provide that the flow in the boiler plant can be automatically reduced, correspondingly, when a boiler is shut down, one of the following control devices in each boiler supply line is required:

1. *Flow limiters.* These devices, installed right before the inlet to each boiler in the system, prevent any overpumping of the boiler. It is acceptable to vary flow through larger, flexible watertube condensing boilers. But each manufacturer has its own

set of specifications for minimum and maximum flows through each boiler, which must be observed so as not to damage the boiler. These flow limits override any variance in flow maintained for the sake of efficiency.

2. *Two-way isolation valves.* These valves stop flow through a boiler when it is not running. Each valve must be timed to open and close based on the boiler volume and the boiler control response time, as defined by the boiler manufacturer. These valves must also be interlocked with the primary flow pumps so they are never running unless at least one boiler isolation valve is open. The pumps need to run for 2–5 min after boiler is shutoff in order to remove the residual heat from the boilers, therefore, the valve must remain open for a short period even after the boiler stops.

Variable flow controls must be configured to:

1. Establish the minimum flow rate for each boiler at start-up in accordance with the manufacturer's specifications.
2. Operate the pumps based on the minimum system volume required for boiler start-up and staging to avoid short cycling or high limit temperature overrides. Maintain a pump minimum flow rate based on the pump manufacturer's recommendation.
3. Three-way valve(s) should be installed at the end of the system to maintain required minimum flow through the boiler (or the pumps, whichever has the greater minimum requirement). These valves maintain a pathway for the water to flow back to the boiler(s) when the coil's two-way control valves are closed. Also, a certain amount of flow is also necessary prior to boiler start-up and following boiler shutdown so that all the minimum flow requirements are met during these brief periods.

Under light load conditions, a low-mass boiler can easily heat the system's low water mass more quickly than the system can use it, which can cause "short cycling." An obvious solution is to simply add mass to the system by using a properly sized hot water *buffer tank*. The mass of water stored in a buffer tank will take longer to come to temperature as the boiler begins to fire and will store any excess heat until it is demanded by the system. This allows for longer run and standby times for the boiler, reducing boiler cycling.

In hot water systems with low-mass boilers, a buffer tank is required whenever the minimum boiler output is greater than the minimum imposed system load and the water volume in the system is too low to absorb the extra boiler output without the water temperature "spiking" and causing the boiler to cycle "off."

To properly size a buffer tank, the following must be known:

1. The minimum desired run time of the boiler (for either a single boiler or the last boiler firing in a multiple boiler sequence, typically a minimum 10-min run time for most low-mass, low-volume condensing boilers is required).
2. The start-up firing output of the boiler (typically 30% of the rated boiler capacity).
3. The minimum system load, based on the zone with the least demand.
4. The system ΔT (°F) [°C] at minimum system load, computed as the peak load design TD multiplied by the ratio of the minimum system load to the peak system load.

On this basis, the minimum required system volume V_w, in gallons [L], can be computed using Eq. (12-1) or (12-1a).

$$\text{For IP units,} \quad V_w = (Q_b - Q_s)/(50 \times \Delta T) \qquad (12\text{-}1)$$

$$\text{For SI units,} \quad V_w = (Q_b - Q_s)/(25 \times \Delta T) \qquad (12\text{-}1a)$$

where

Q_b = Start-up firing output, Btu/h [W]
Q_s = Minimum system load, Btu/h [W]

If the actual system volume, determined on the same basis as for expansion tank sizing in Chapter 8, is less than V_w, a buffer tank sized to make up the difference is required.

12.5 Steam Heating Systems

Steam boiler capacity is generally presented in terms of *boiler horsepower* (BHP) [kW], a measure of boiler energy output. One boiler horsepower is equivalent to 34.5 lbs/h (PPH) of steam production (33,472 Btu/h) [9.8 KW] and/or 139.5 ft^2 [13 m^2] of equivalent direct radiation (EDR).

12.5.1 Steam Quality

For general heating and humidification of noncritical spaces (offices, classrooms, manufacturing, and so on), "conventional" steam systems may be used. Blowdown control, typically coupled with a chemical-based water treatment program (see Chapter 14), will address removal of dissolved solids from the feedwater. And, the high temperatures of steam generation will eliminate the vast majority of potential organic contaminates.

But, in some applications, particularly for hospitals, research facilities, and so on, "higher quality" steam may be required as follows:

1. *High Quality Steam*: It is required for general health care humidification and other applications and can be produced by typical steam boilers if two measures are taken:
   a. No condensate piping corrosion treatment chemicals may be used unless they are USFDA (U.S. Food and Drug Administration) approved as *food grade*. Neutralizing or filming amines are typically used in steam systems to minimize corrosion in condensate piping. These amines "carryover" during steam production. The USFDA states that amines that have approval for use in steam that contacts food products are also acceptable when steam is used in medical sterilizers. That certainly makes them acceptable for humidification use!
   b. Line filters with 10–20 μm media and Type 316L stainless steel construction are included upstream of each humidifier to trap metallic impurities, and so on. Note that this introduces a maintenance requirement, and the design must include careful attention to providing access to the filter for change-out. Routinely, two filters, piped in parallel, are required, each with isolation valves so the filters can be changed without shutting off steam service.

2. *Higher Quality Steam*: Using individual or central steam-to-steam generators, hospitals typically require the use of higher quality steam for humidification of operating rooms, delivery rooms, intensive/acute care, and other critical care areas. Museums, libraries, laboratories, and so on normally require higher quality steam to prevent contamination.

   To produce this steam quality level, feedwater consists of 100% untreated domestic water introduced into a steam-to-water heat exchanger. The heat exchanger, then, produces lower pressure steam for humidification, leaving behind almost all of the impurities, including dissolved solids, that it contains.

   Because these heat exchangers rapidly collect the dissolved solids removed from untreated domestic water when producing steam, they must be routinely manually cleaned. Most manufacturers use stainless steel construction and provide ready access for cleaning.

3. *Clean Steam*: Using a direct-fired or steam-to-steam generator, "clean" steam is steam produced from feedwater made up of 100% domestic water that contains minimal contamination. This is accomplished by removing dissolved solids and other contaminants from the domestic supply before it enters a steam generator.

   There is no specific standard for "clean" steam, but the following criteria for steam generator feedwater quality represents a good guideline for design:

   a. <0.02 ppm free chlorine. This can be accomplished by installing activated carbon filtration.
   b. <0.05 ppm dissolved oxygen. Chemical oxygen scavenging and/or deaeration may be required.
   c. <1.0 ppm hardness. Requires a water softener.
   d. <15.0 ppm suspended solids. Requires a water softener
   e. <2,000 ppm dissolved solids. Requires conductivity-based automatic blowdown control.

      Chemical water treatment may be required to maintain some of these conditions. Any chemicals used must be FDA approved ("food grade"). Do not use corrosion control chemicals as part of water treatment, since this results in a highly corrosive water condition, the steam generator components in direct contact with the water must be constructed of Type 316L stainless steel. Any nonmetallic component must be fabricated from ethylene propylene diene monomer and/or fluoropolymer tetrafluoroethylene. Even with this construction, routine maintenance and cleaning of the steam generator is required, often involving chemical cleaning/etching and repassivation of the steel surfaces.

      Line filters with 3–15 μm media and Type 316L stainless steel construction must be included upstream of each humidifier. The design must include careful attention to providing access to the filter for change-out. Routinely, two filters, piped in parallel, are required, each with isolation valves so the filters can be changed without shutting off steam service.

4. *Pure Steam*: Using a direct-fired or steam-to-generator, "clean" steam is steam produced from feedwater made up of 100% distilled, reverse osmosis, or deionization-purified water.

   There is no specific standard for "clean" or "pure" steam for humidification applications. However, there are statutory requirements for regulating water purity defined by the U.S. Pharmacopeia. Thus, it makes sense to design pure steam systems for pharmaceutical or biotechnology projects so that heat exchanger feedwater meets the criteria for "Clean Steam" defined previously.

Most pharmaceutical and biotechnology facilities will have a purified water system that should be the source of feedwater for the humidification system(s). If there is no purified water system available, then the design must include one.

5. *Steam for Sterilizers*: In health care facilities and laboratories it is required to meet the requirements of the Centers for Disease Control (CDC's) *"Guideline for Disinfection and Sterilization in Healthcare Facilities."* This guideline has only minimal requirements for steam quality:
   a. The steam must be "saturated", that is, 97%–100% dry steam. Steam must not be superheated.
   b. Steam pressure must be high enough to maintain 270°F [132°C] within the sterilizer chamber. This temperature is equivalent to 60 psig [414 kPa] steam entering the sterilizer. (Higher pressure steam is not desirable since it will become superheated when expanded in the sterilizer and not properly condense.)

   Historically, steam for sterilizers has met the "High Quality Steam" requirement discussed previously. However, there is significant discussion about steam contaminants in the current literature, and most experts are now recommending the limits on sterilizer steam feedwater contaminants shown in Table 12-3.

   In Europe, health care sterilizers are required to operate using "Pure Steam." And, many major U.S. health care facilities have adopted this requirement to help fight rampant reinfection occurrences. Note, however, that some sterilizers contain types of steel and other materials that will corrode rapidly with pure steam and that Type 316L stainless steel is required for their construction.
6. *Steam for Processes*: Process steam is typically conventional steam. However, steam for direct use in the production of pharmaceuticals, food products, and so on have specific requirements that must be met. Each project must be carefully researched to determine steam quality requirements.

   Steam-to-steam heat exchangers are used to produce higher quality or pure steam using conventional or higher quality steam as the heating medium. These heat exchangers may be "stand-alone," producing steam for a number of imposed

TABLE 12-3

Sterilizer Steam Contaminant Limits

Contaminant Upper Limit	
Silica	1 mg/L
Iron	0.2 mg/L
Cadmium	0.005 mg/L
Lead	0.05 mg/L
Other heavy metals (total)	0.1 mg/L
Chloride	3 mg/L
Phosphate	0.5 mg/L
Conductivity (TOC)	15 μS/cm
pH	6.5–8
Hardness	0.02 mmol/L

loads (humidifiers, sterilizers, and so on), or be an integral part of a humidifier. For sizing, the load imposed on the heat exchanger must be determined based on the anticipated peak simultaneous steam demand of connected humidifiers, sterilizers, and so on. Size the heat exchanger for at least 150% of the computed peak load to account for instantaneous loads that may be imposed.

12.5.2 Steam Heat Transfer

In transferring heat steam to any heated medium, the steam gives up its enthalpy of vaporization, together with a small fraction of its sensible heat due to some subcooling of the condensate. The quantity of heat transferred from the steam is a function of the rate of flow, its specific heat, and the TD of the medium being heated. For all practical purposes, this must equal the heat transmitted to the conditioned medium by the heat exchanger (air heating coil or steam-to-hot water converter) and is expressed by Eq. (12-2).

$$Q_s = W_s\left(h_f - h_{fc}\right) \tag{12-2}$$

where
Q_s = heat transfer rate from steam, Btu/h [W]
h_f = enthalpy of the liquid of the steam, Btu/lb [W/kg]
h_{fc} = enthalpy of the liquid of the condensate, Btu/lb [W/kg]
W_s = weight of steam condensed, lb/h [kg/h]

In addition to meeting heating and/or humidification requirements, steam has significant advantages for heating energy distribution:

1. Steam heat content, per pound, is much higher than for water, reducing pipe sizes and initial distribution system cost.
2. Steam pressures provide for transport of the heating medium without the need for pumps.

Thus, steam is often used as the primary heating medium where relatively large-scale distribution is required for college campuses, military facilities, and so on. Steam is also the necessary choice when temperatures above 200°F–210°F [93°C–99°C] are required, particularly in industrial processes.

12.5.3 Steam-to-Water Heat Exchangers

In many cases where the primary distribution for heating energy is steam, this heat is transferred to a smaller, individual hot-water heating systems for local distribution since control of a hot water system is simpler and better than for the steam system. The heat transfer is accomplished by a simple steam-to-water heat exchanger, called a "converter" in the industry.

Steam-to-water heat exchangers are typically shell and tube type, consisting of a number of tubes mounted inside a cylindrical shell, with steam within the shell, flowing over the outside of the tubes, as the second fluid (typically water) flows through the tubes. Heat

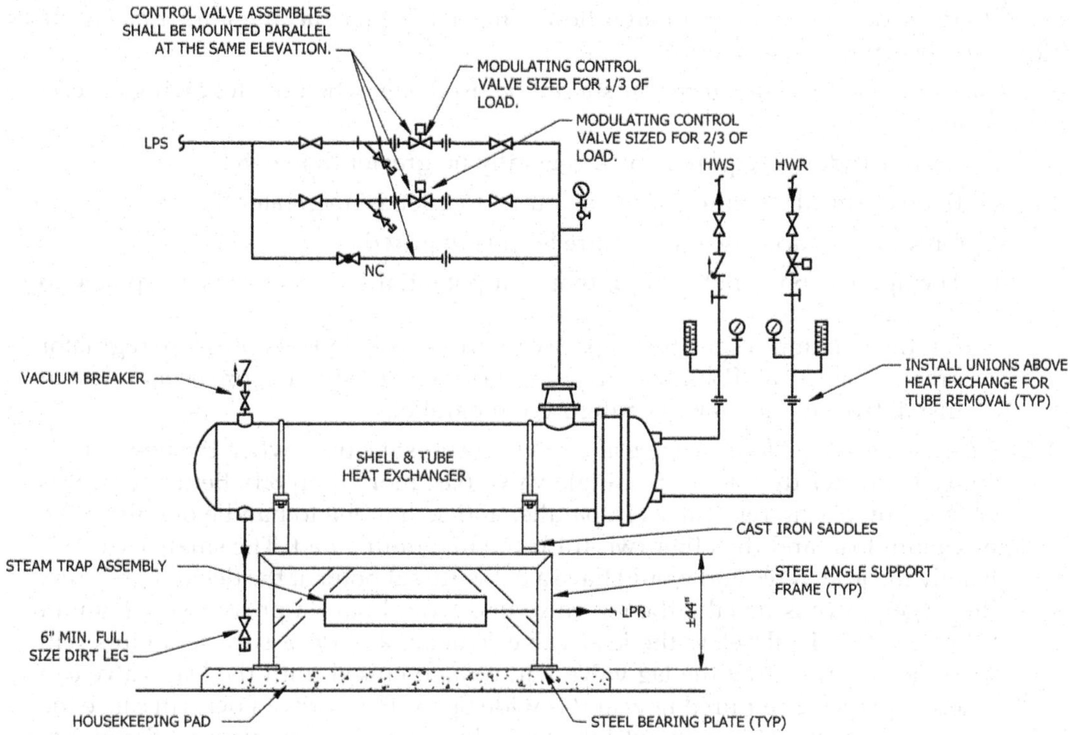

FIGURE 12-3
Steam-to-water heat exchanger piping. (SW Standard Detail HX01.)

transfer on the steam side is accomplished via phase change as the steam changes from gas to liquid state, creating a temperature increase on the water side. Heat transfer efficiency approaches 97%–98%.

Recommended heat exchanger piping is shown in Figure 12-3.

12.5.4 Feedwater System

For any steam boiler system, a feedwater system is required to provide a mix of returned condensate and makeup water to the boiler(s). Design system with multiple feedwater pumps, with a total capacity of twice the largest boiler steam generation, plus the sum of the remaining boilers steam generation rates. Size each feedwater pump for a flow rate equal to this total flow divided by (*N* to –1), where *N* is the number of pumps provided and select each feedwater pump to produce a discharge pressure equal to the boiler pressure, plus 20%. A *deaerator* is required as part of the feedwater system if the percentage of condensate return is less than 75% of the steam generation capacity.

12.5.5 Steam Pressure Reducing Valves

Typically, steam is produced at relatively high pressure (80–100 psig) [550–690 kPa] for high efficiency transport in the steam distribution system. However, in all or part of the system, this pressure is reduced, an adiabatic process, to a lower pressure to one better suited to the type of load and for better flow control. For this duty, *pressure reducing valves* (PRVs) are incorporated, as follows:

1. *One-stage single pressure regulator* should be used *only* when the following criteria is met:
    a. Load turndown requirement is generally no greater than 6–8:1.
    b. Ratio of specific volume of steam, outlet to inlet, is no greater than 3:1.
    c. Only one reduced steam pressure level is required.
    d. The application is not critical, that is, a potential PRV failure is acceptable to the user.

    If the maximum specified capacity requires selection of a pressure regulator greater than 12 in. [DN240] valve size, however, it will be more economical to install two smaller valves configured in parallel.
2. *One-stage station with parallel pressure regulators* must be used when the flow rangeability is greater than that of a single valve (i.e., more than 8:1). Better control is achieved by piping two valves in parallel and sizing one to handle one-third the maximum load and the other two-third the maximum load. The smaller valve is usually the lead valve and would have a pressure set point at the desired pressure. The larger valve is usually the lag valve and would have a pressure set point of 2–3 psig [14–20 kPa] below the lead valve. This offset of pressure set points will stage the valves so that the lag valve will remain closed until the lead valve can no longer pass the required flow and is wide open. This lack of flow will cause the set pressure to drop slightly until the lag valve opens and regulates at the higher demands of flow.
3. *Two-stage series pressure regulators,* as shown by Figure 12-4, provide pressure reduction by using two valves in series to reduce the pressure in stages.
    Depending on the volume of fluid required and pressure reduction, the second stage valve typically will be larger in size than the first stage valve. Unless a specific intermediate pressure of the fluid is required, this intermediate pressure is typically selected so as to keep the pressure turndown ratios of both valves as similar as possible to help equalize and maximize the service life of both valves.

    Typically, two-stage series PRVs are needed when an intermediate pressure steam supply is required. Also, to reduce PRV-generated noise, use a two-stage series station when the specific volume ratio, outlet to inlet, is greater than 3 to 1.

    In critical applications, such as hospitals, and research facilities, where reliable steam flow is a prime criteria, *two-stage parallel pressure regulators,* as shown by Figure 12-5, should be used.
4. *Do not oversize valves!* Oversizing will cause "hunting" and poor steam pressure control. Size PRVs so that capacity does not exceed the required maximum load by selecting valve C_v, as follows:

    Determine the choked or critical flow index:

    If $p2/p1 \geq 0.58$, then flow is "choked" (sometimes called "critical")

    If $p2/p1 < 0.58$, then flow is "non-choked" (sometimes called "noncritical" or "subcritical")

    where

    $p1$ = inlet absolute pressure (psia) [kPa]

    $p2$ = outlet absolute pressure (psia) [kPa]

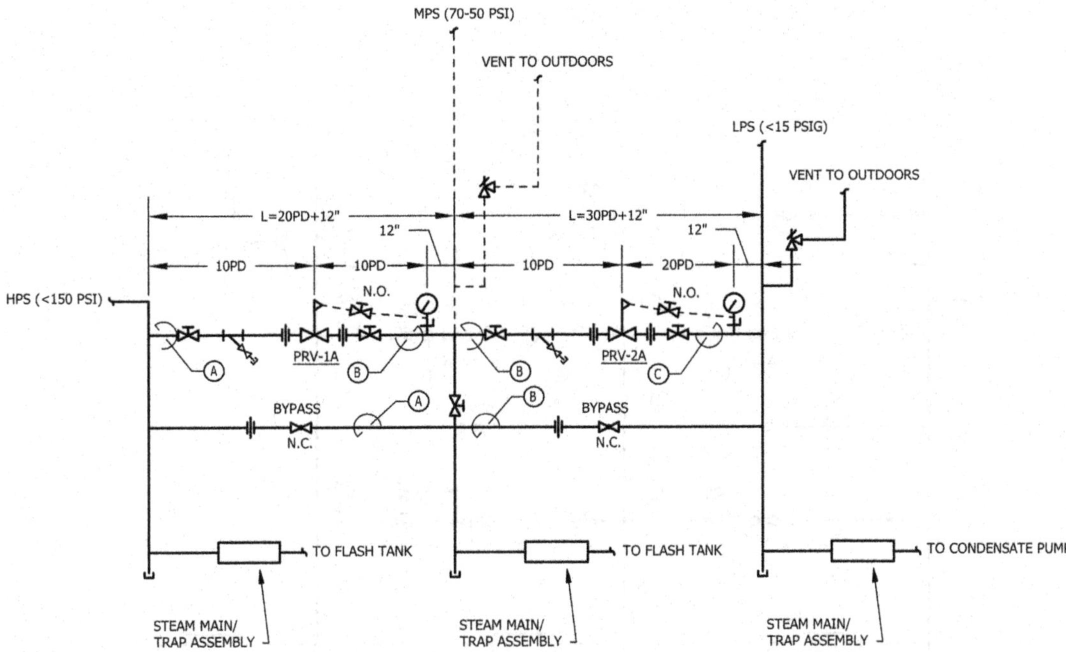

NOTES:

1. INSTALL STRAINERS HORIZONTALLY WITH BLOW DOWN ROTATED TO BOTTOM.
2. REDUCING STATION TO BE SIZED AND PIPED PER MANUFACTURER'S REQUIREMENTS.
3. PD = PIPE DIAMETERS, NPS OR DN (SEE TABLE 12-4)
4. UNIONS REQUIRED FOR PIPING 2" NPS AND SMALLER, OMIT ON PIPE WITH FLANGED CONNECTIONS.

FIGURE 12-4
Two-stage series steam pressure reducing station piping. (SW Standard Detail S01C.)

In IP units, for *choked flow*, the flow capacity (C_v) is computed by Eq. (12-3).

$$C_v = W/(1.71 \times p1) \tag{12-3}$$

and for *non-chocked flow*, the flow capacity (C_v) is computed by Eq. (12-4).

$$C_v = W/\left\{2.1 \times \left[(p1+p2)(p1-p2)\right]^{1/2}\right\} \tag{12-4}$$

where

$$W = \text{required maximum load}(\text{lb/h})$$

The C_v computed above represents the "minimum required" C_v. Actual C_v, based on available PRV size and configuration, will rarely match the required C_v exactly. *In selecting PRVs, the actual C_v should never be greater than 120% of the minimum required value.*

In SI units, the K_v value is the flow coefficient which corresponds to a water flow rate given in m³/h at a differential pressure of 1 bar and a water

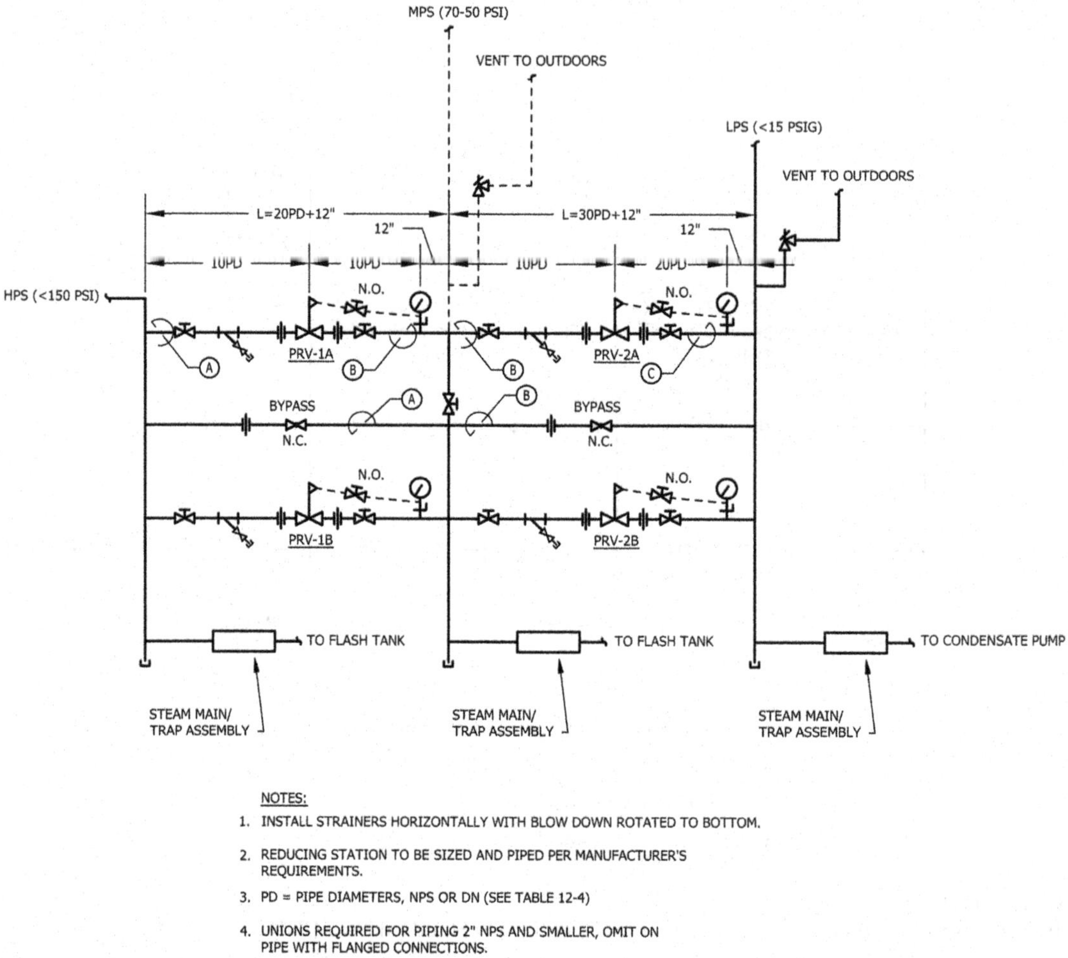

FIGURE 12-5
Two-stage parallel steam pressure reducing station piping. (SW Standard Detail S01A.)

temperature between 5°C and 30°C. The IP system uses the flow coefficient C_v which corresponds to water flow rate given in gal/min at a pressure difference of 1 psi and a water temperature of 60°F. The relationship between K_v and C_v is defined by Eq. (12-5).

$$K_v = 0.86 \times C_v \tag{12-5}$$

For *choked flow,* the flow capacity (K_v) is computed by Eq. (12-6).

$$K_v = (W/461)\left[(t1+273)/(p1-p2)p2\right]^{0.5} \tag{12-6}$$

and for *non-chocked flow,* the flow capacity (K_v) is computed by Eq. (12-7).

$$K_v = \left[W/(230+p1)\right](t1+273)^{0.5} \tag{12-7}$$

where

W = required maximum load, kg/h

$t1$ = inlet steam temperature, °C

A K_{vs} value provided by manufacturers is the K_v value at nominal valve lift for a specific series of PRVs. The K_{vs} value represents the maximum throughput to be calculated for a valve and should never be greater than 120% of the computed value.

5. Make sure that *dry* steam, with no entrained condensate, enters the valve. The design should include a drip leg and trap ahead of valve. In critical cases, to ensure dry steam enters the valve, an upstream steam moisture separator may be required.
6. Do not locate a pressure-reducing valve station close to the steam control valve(s) as the two elements will "fight" with each other, resulting in wide pressure swings that may be ultimately controlled on by the release of stream by the safety pressure relief valve.
7. Allow *ample* room for the *PRV station*. Provide a minimum of 10 pipe diameters in length upstream and make sure that the sensor connection is at least 20 pipe diameters downstream of the PRV.
8. Piping flow velocities both upstream and downstream of the PRV are recommended to be 4,000–6,000 fpm. On that basis, the Table 12-4 can be used to select PRV upstream and downstream piping sizes (indicated as "A", "B", and "C" in Figures 12-4 and 12-5), based for Sch. 40 pipe:

12.5.6 Steam Condensate Receivers and Pumps

Condensate receivers and pumps are required within any steam piping system to collect steam condensate and provide pumped return of that condensate to the boiler(s) feedwater system. The following are the basic design criteria for selecting and sizing condensate receivers and pumps:

TABLE 12-4

Recommended PRV Piping Sizes (NPS) [DN]

Flow Rate (PPH) [kW]	HPS (80–100 psig) [550–690 kPa]	MPS (30–70 psig) [200–500 kPa]	LPS (≤15 psig) [≤5 kPa]
≤200 [1,960]	3/4" [20]	1" [25]	1-1/4" [32]
300 [2,940]	3/4" [20]	1" [25]	1-1/2" [40]
400 [3,920]	1" [25]	1-1/4" [32]	2" [50]
500 [4,900]	1" [25]	1-1/4" [32]	2" [50]
600 [5,880]	1-1/4" [32]	1-1/2" [40]	2-1/2" [65]
800 [7,840]	1-1/4" [32]	2" [50]	3" [80]
1,000 [9,800]	1-1/2" [40]	2" [50]	3" [80]
1,500 [14,700]	2" [50]	2-1/2" [65]	3" [80]
2,000 [19,600]	2-1/2" [65]	3" [80]	4" [100]
3,000 [29,400]	3" [80]	4" [100]	6" [150]
4,000 [39,200]	3" [80]	4" [100]	6" [150]
5,000 [49,000]	4" [100]	6" [150]	6" [150]
6,000 [58,000]	4" [100]	6" [150]	8" [200]
8,000 [78,400]	4" [100]	6" [150]	8" [200]
10,000 [98,000]	6" [150]	6" [150]	10" [250]

1. *Always select duplex pump sets.* Thus, even if one pump or motor fails, condensate can be returned. Also, having two pumps that alternate use reduces the "starts per hour" imposed on each pump motor by 50%.
2. *Select condensate pump discharge pressure* on the basis of the pressure loss ("head") between pump discharge and boiler feedwater tank, including any lift required (since this is an "open" system), and the "back pressure" that will be present in the boiler feedwater tank.

   The pressure (head) that any pump must produce is the total pressure loss imposed through the highest pressure loss circuit under design (peak) flow conditions.

   Piping pressure loss can be quickly estimated by determining the total piping length and multiplying that length by the average design pressure loss per unit length (ft of water/100 ft of piping) as for any hydronic piping system. To account for fitting pressure losses, add between 30% (for long, straight piping runs) to 50% (for typical main/branch systems within buildings) of the straight pipe pressure loss.

   Condensate pumps must overcome *static lift*, in feet, represented by the height difference between the water level in the basin (receiver) and the water level of the boiler deaerator or feedwater tank, as applicable.
3. *Condensate receivers/pump units are typically cataloged according to square feet of EDR.* Each square foot of EDR equals 240 Btu/h [70 kW].
4. *Size pump flow rate* for three (3) times the anticipated peak condensate flow rate to allow time (minimum 2 min) for entrained air in the receiver to be vented. (This criteria also addresses the potential for higher than anticipated condensate flow rates during cold start-up periods.)
5. *Size receiver volume* for four (4) times the anticipated peak condensate flow rate since only about 75% of the total tank volume is usable.
6. *Clearly detail/specify* that a 4″ NPS [DN100] vent to atmosphere from each receiver is required. A size of vent will adequately vent any flash steam that may occur in the receiver when flash steam recovery is not implemented.

Bibliography

ASHRAE Handbook-HVAC Systems and Equipment, American Society of Heating, Refrigerating, and Air-Conditioning Engineers, Inc. (Atlanta, GA, 2016).

Cameron Hydraulic Data, Ingersoll-Rand (Woodcliff Lake, NJ, 1988).

Steam: Its Generation and Use (42nd Edition), (Babcock & Wilcox Enterprises, Inc., Charlotte, NC. 2015).

13

HVAC Controls

While almost all commercial, institutional, and industrial facilities control their heating, ventilating, and air-conditioning (HVAC) systems *via* a *direct digital control* (DDC) system, many HVAC engineers continue to find the ever-changing DDC technology difficult to understand and utilize. This is not surprising considering that most HVAC designers have a mechanical engineering background with no real expertise in electronics, computers, networks, communications protocols, and so on.

Fortunately, though, HVAC designers do not have to be computer experts to obtain good control of HVAC processes. The real key is to understand and then fully describe how the HVAC systems, subsystems, and individual components are intended to function by defining the required *sequences of operation*.

HVAC designers must treat controls design seriously... it is, after all, the key to intended and effective HVAC systems operation. There are several elements to this:

1. Controls design cannot be properly performed if put off until the end of the HVAC design process. An HVAC system's performance requirements and controls designs are interdependent. *Therefore, temperature controls design is an iterative process that must be performed as part of the overall HVAC design process, beginning in the Schematics Design phase.* Leaving the controls design to the end of the construction documents phase will greatly increase the chance that neither the HVAC system nor its controls will work well.
2. Detailed operational sequences and control point definitions are the most important components to control system design. It should be obvious that a controls contractor cannot reasonably be expected to extrapolate an HVAC system's intended operation sequence from the mechanical plans and specifications alone. This is important for two reasons:
   a. The cost of a DDC system is primarily determined by the control point quantities and types (analog points are much more expensive than digital points). While a controls contractor can reasonably extrapolate a list of points from a well-written operation sequence, making this effort part of the bid process increases risk that the winning contractor will not provide the intended system.
   b. The process of developing point definitions provides the engineer with an excellent check of the operation sequence and the rest of the control system design.

Without a well-engineered and well-specified control system design, the contractor cannot reasonably be expected to provide what is intended.

13.1 DDC Fundamentals

Every HVAC process is controlled by three elements: a *sensor* or other input device, a *controller*, and a controlled *output* device.

The sensor measures the controlled medium or other control input in an accurate and repeatable manner. Common HVAC sensors are used to measure temperature, pressure, relative humidity (RH), airflow state, and so on. Other variables may also be measured that impact controller logic. Examples include other temperatures, time-of-day, or the current electrical demand condition. Additional input information (sensed data) that influences the control logic may include the status of other parameters (airflow, water flow, current) or safety (fire, smoke, high/low temperature limit, or any number of other physical parameters). Sensors are the first, as well as the weakest, link in the chain of control.

The controller processes data that is input from the sensor and/or other devices, applies the logic of control, and causes an output action to be generated. This signal may be sent directly to the controlled device or to other logical control functions and ultimately to the controlled device. The controller's function is to compare its input(s) with a set of instructions, such as set point, throttling range, and action and then produce an appropriate output signal. How the controller functions is referred to as the *control response*, typically one the following:

1. *Two-position* control compares the value of an analog or variable input with the programmed instructions and generates a digital (two-position) output. The instructions involve the definition of an upper and lower limit for the measured variable. The output changes its value as the input crosses these limit values. There are no standards for defining these limits. The most common terminology used is *set point* and *differential or throttling range*. The set point, plus (or minus) half the differential, indicates the point where the output "pulls-in," "energizes," or is "true." The output changes back, or "drops-out," after the input value crosses through the value equal to the set point minus (or plus) half the differential.

   Two-position control can be used for simple control loops (temperature control and so on.) or limit control (such as "freezestats" that are designed to shut down an air-handling unit when the unit mixed air temperature is too low). The input can be any measured variable including temperature, RH, pressure, current, liquid level, and so on.

   Time can also be the input to a two-position control response. This control response provides a schedule for operations and functions such as an old fashioned time clock. The output "makes" when the time is in the defined "on" time and "breaks" during the defined "off" time.

   A common application of two-position control is the residential heating system. With a set point of 70°F [21°C] and a 4°F [2°C] differential, the thermostat (the "controller") will energize the heating system when the space temperature falls to 68°F [20°C] and turn it off when the temperature rises to 72°F [22°C] in the space. Because of thermal lag, the actual operating differential will be somewhat larger than the controller differential. The result of this control is shown by Figure 13-1.
2. *Floating* control is a control response that produces two possible digital outputs (DOs) based on a change in a variable input. One output increases the signal to the controlled device, while the other output decreases the signal to the

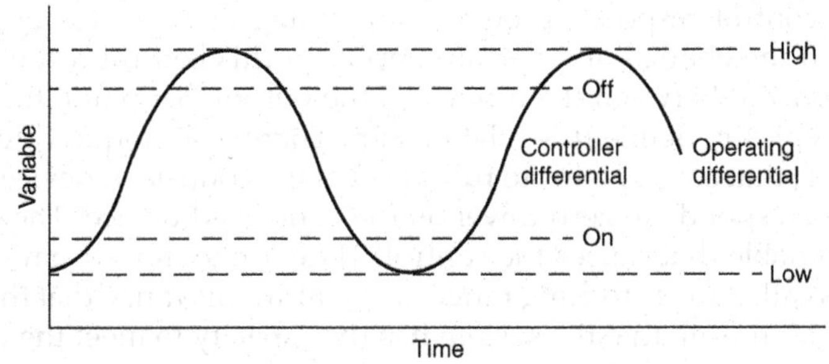

FIGURE 13-1
On/off control operation.

controlled device. This control response also involves an upper and lower limit with the output changing as the variable input crosses these limits. Again, there are no standards for defining these limits, but the terms set point and deadband are common. The set point sets a midpoint and the deadband sets the difference between the upper and lower limits.

When the measured variable is within the deadband (or neutral zone), neither output is energized and the controlled device does not change, it stays in its last position. For this control response to be stable, the sensor must sense the effect of the controlled device movement very rapidly. Floating control does not function well where there is significant thermodynamic lag in the control loop. Fast airside control loops respond well to floating control. An example of floating controls is shown by Figure 13-2.

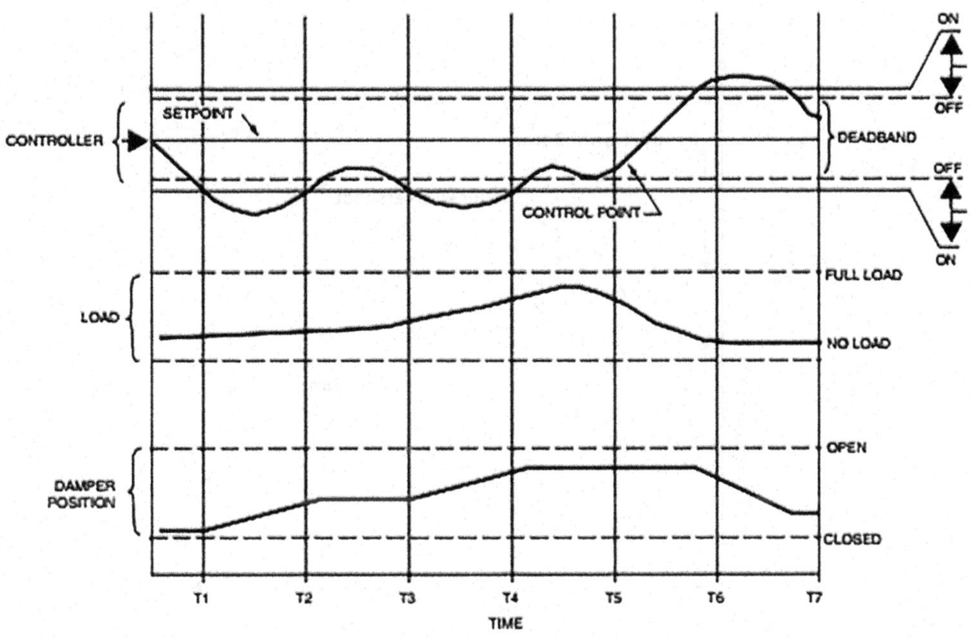

FIGURE 13-2
Floating control operation.

3. *Proportional* control response produces an analog or variable output change in proportion to the change of a varying input. In this control response, there is a linear relationship between the input and the output. Set point, throttling range, and action typically define this relationship. There is a unique value of the measured variable that corresponds to full travel of the controlled device and a unique value that corresponds to zero travel on the controlled device. The change in the measured variable that causes the controlled device to move from fully closed to fully open is called the throttling range. It is within this range that the control loop will control, assuming that the system has the capacity to meet the requirements.

   The action dictates the slope of the control response. In a direct acting proportional control response, the output will rise with an increase in the measured variable. In a reverse acting response, the output will decrease as the measured variable increases. The set point is an instruction to the control loop and corresponds to a specified value of the controlled device, usually half-travel, as shown by Figure 13-3.

   With proportional control, the final control element moves to a position proportional to the deviation of the value of the controlled variable from the set point, in a linear manner. The final control element is seldom in the middle of its range because of the linear relationship between the position of the final control element and the value of the controlled variable. The set point is typically in the middle of the throttling range, so there is usually an offset between the control point and the set point.

   In digital control logic, proportional control can be represented mathematically as Eq. (13-1).

$$V = (K \times E) + M \tag{13-1}$$

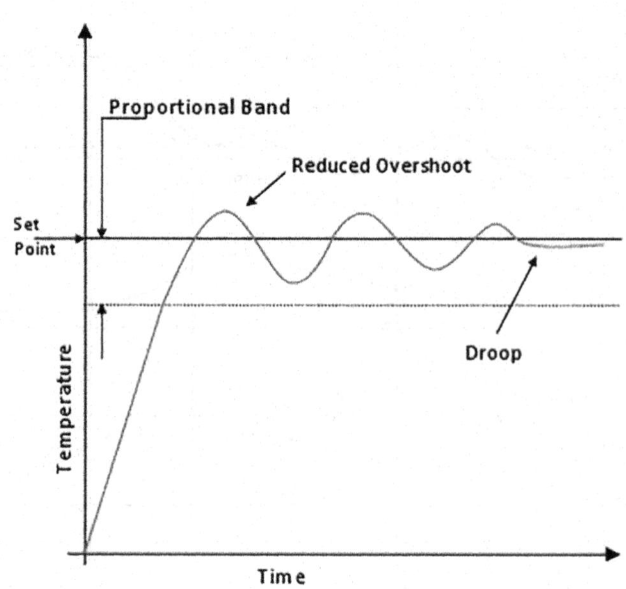

FIGURE 13-3
Proportional control operation.

where

V = Output (control) signal

K = Proportionality constant (gain) = sensor span/throttling range

E = Deviation (control point − set point)

M = Value of the output when the deviation is zero, usually the output value at the middle of the output range

4. *Proportional plus integral* (PI) control involves the measurement of the offset or "error" that can occur with proportional control over time. This error is integrated and a final adjustment is made to the output signal from the proportional part of this model. This type of control response will use the control loop to reduce the offset to zero. A well set-up PI control loop will operate in a narrow band close to the set point. It will not operate over the entire throttling range.

   PI control loops do not perform well when set points are dynamic, where sudden load changes occur, or if the throttling range is small.

   In digital control logic, PI control is represented by Eq. (13-2)

$$V = (K \times E) + M + (K \times T1)\int E\,dt \qquad (13\text{-}2)$$

where

V = Output (control) signal

K = Proportionality constant (gain) = sensor span/throttling range

E = Deviation = (control point − set point)

M = Value of the output when the deviation is zero

$T1$ = Reset time

$K/T1$ = Reset gain

dt = Differential of time (increment in time)

   Reset of the control point is not instantaneous. Whenever the load changes, the controlled variable changes, producing an offset. The proportional controller makes an immediate correction, which usually still leaves an offset. The integral function of the controller then makes control corrections over time to bring the control point back to set point.

   Integral windup can occur with PI controllers. This is an excessive "overshoot" condition caused by the integral function making continued correction while waiting for feedback on the effects of its prior correction(s). Integral windup can occur when the controlled system is off, the heating or cooling medium fails or is not available, or one control loop overrides or limits another. DDC systems generally must have control software to prevent integral windup.

5. *Proportional plus integral plus derivative* (PID) control adds a predictive element to the control response. In addition to the proportional and integral calculation, the derivative or slope of the control response will be computed. This calculation will have the effect of dampening a control response that is returning to set point so quickly that it will "overshoot" the set point.

   In digital logic, PID control is represented as follows:

$$V = (K \times E) + M + (K \times T1)\int E\,dt + (K \times \mathrm{TD} \times dE/dt) \qquad (13\text{-}3)$$

where

V = Output (control) signal

K = Proportionality constant (gain) = sensor span/throttling range

E = Deviation = (control point—set point)

M = Value of the output when the deviation is zero

$T1$ = Reset time

$K/T1$ = Reset gain

dt = Differential of time (increment in time)

TD = Rate time (time interval by which the derivative advances the effect of proportional action)

KTD = Rate gain constant

dE/dt = Derivative of the deviation with respect to time (error signal rate of change)

Adding the derivative function to create PID control is a labor-intensive and time-consuming process, requiring "tuning" and retuning of the control loop to eliminate instability. *Few (if any) HVAC processes require PID control and applying it for those that don't need it can create more problems than it solves… add the derivative function only after proportional, and then PI, control fail to work satisfactorily.*

Designers must evaluate the specific requirements associated with each control loop to select the proper control mode:

1. What degree of accuracy is required?
2. What amount of offset, if any, is acceptable?
3. What types of load changes are anticipated (size, rate, frequency, and duration)?
4. What are the system characteristics (time lag, reaction rate)?

Table 13-1 can be used as a general guide to selecting control modes for HVAC process.

TABLE 13-1

Recommended Control Modes for HVAC Control Applications

Control Application	Control Mode
Space temperature	P (floating control is acceptable only for simple comfort air-conditioning applications with a wide throttling range)
Mixed air temperature	PI
Coil discharge temperature	PI (cooling), P (heating)
Hot water supply temperature	P
Airflow or water flow	PI (with wide throttling range and fast reset rate) or PID if PI proves unstable
Fan static pressure	PI
Water pressure	P
Humidity	P (PI if throttling range is 5% or less)
Dew-point temperature	P (PI if throttling range is 2°F or less)

13.2 DDC Input/Output Points

13.2.1 Input/Output Basics

A *digital input* (DI) [sometimes called a binary input (BI)] typically consists of a power supply (voltage source), a switch, and a voltage-sensing device (analog-to-digital [A/D] converter). Depending on the switch's open/closed status, the sensing device detects a voltage or no voltage condition, which in turn generates a logical/binary 0 or 1, on or off, alarm or normal, or similar "either-or" defined state.

A DO typically consists of a switch (either mechanical as with a relay, or electronic as in a transistor or triac) that either opens or closes the circuit between two terminals depending on the binary state of the output.

An *analog input* (AI) is a measurable electrical signal with a defined range that is generated by a sensor and received by a controller. The AI varies continuously in a definable manner in relation to the measured property.

The analog signals generated by some types of sensors must be conditioned by converting to a higher level standard signal that can be transmitted over wires to the receiving controller. AIs are converted to digital signals by the A/D converter typically located at the controller. A/D conversion is limited to a small range of DC voltage, so that internal or external input circuitry must change the character of noncompatible signal types to a DC voltage range within the limits of the A/D converter.

There are basically three types of AI and output signals: voltage, current, and resistance. Common voltage signals used in the controls industry are 1–5 volts direct current (VDC), 2–10 VDC, 3–15 VDC, 0–5 VDC, 0–10 VDC, and 0–15 VDC. The 4–20 mA current signal has become the industry's most commonly used signal for use with analog and digital controllers. Resistance measurement is most commonly associated with direct inputs from temperature-sensing devices, such as *thermistors* and *resistance temperature detectors* (RTDs).

An *analog output* (AO) is a measurable electrical signal with a defined range that is generated by a controller and sent to a controlled device, such as a variable speed drive or actuator. Changes in the AO cause changes in the controlled device that result in changes in the controlled process. Controllers first produce a DO that is then converted to an analog signal. The analog circuitry is typically limited to a single range of voltage or current, such that output transducers are required to provide an output signal that is compatible with controlled devices using something other than the controller's standard signal.

Inputs and outputs can also be used in special configurations, such as accumulating points, pulse width modulation (PWM) signals, and tri-state or floating points.

Accumulating points are typically associated with inputs and are special in that during each scan the controller adds the input point value to the accumulated value. Accumulating points may have either AI or DI. Typically, accumulating points are also used to determine energy quantities, such as kWh from a power sensor and Btu [kW] from flow and temperature sensors.

PWM signals are based on the amount of time a DO circuit is closed over a fixed time base. This amount of time can range from 0% to 100% of the time base, providing an analog value for each time period that represents the time base of the signal.

A tri-state (or "triac") signal consists of two digital signals used together to provide three commands. This type of signal is commonly used to operate a damper or valve actuator in a modulating fashion but may also be used with a transducer to generate an analog signal.

If both DOs are "off," the actuator does not move. Output 1 "on" will cause movement in one direction; output 2 "on" will cause movement in the other direction. The fourth possible signal (both outputs "on") is not used in tri-state operation. The concept was initially developed to allow electric controls consisting of single-pole, double-throw switches with a center-off position to control actuators in a modulating fashion. Modulating operation is achieved by this action because the actuators being controlled drive slowly so the change in position is proportional to the amount of time the output remains energized.

13.2.2 DI Devices

DI devices consist of a switch, either two- (open/close) or three-position (close low/open/close high) in one form or another, used to complete a DI circuit.

Property-sensing (flow, level, and so on) switches are rated by parameters such as adjustment range, accuracy or repeatability, and deadband or differential. The range of a control switch is specified by upper and lower process values between which the switch has been designed to operate. The accuracy or repeatability of a control switch is a value typically measured in process units or percent of range that represents the expected maximum deviation from set point at which the switch will operate under test conditions. The switch differential or deadband is the change in process value required to cause the state of the switch to change. For example, a pressure switch that makes at 10 psig and breaks at 8 psig has a 2 psig differential.

Manual (or "hand") switches are used as DI devices and in hardwired electrical control circuits associated with DOs. Hand switches come in numerous sizes, shapes, and configurations. Common switch types include rotary, selector type, toggle, and push button. Selector and toggle switches are almost always maintained contact type. Push buttons may be momentary or maintained contact type. Selector switches can have key operators to prevent tampering.

Limit switches convert mechanical motion or proximity into a switching action. Limit switches are most commonly used in DDC control systems to provide position status feedback to the controller for valve and damper positions (an "end" switch).

Temperature switches (also called thermostats, aquastats, or freezestats depending on their application) are commonly used in DDC systems to provide a DI when a process medium temperature rises or falls to a set point temperature.

Flow switches are used to provide a DI to DDC controls systems when a fluid flow rate has risen above or fallen below the set point value. Common applications include safety air and water flow interlocks for electric heaters and humidifiers, chiller safety interlocks, and burner safety interlocks. Numerous technologies are available, but the most common types used in DDC systems are mechanical and differential pressure types.

Differential pressure type flow switches operate on the principle that a difference in pressure is always associated with fluid flow or the principle that the total pressure (TP) of a flowing fluid is always greater than the static pressure (SP). These differences in pressure can be accurately predicted for a given situation and related to the fluid flow rate.

Level switches are used in DDC control systems to provide a DI when the fluid level in a tank, vessel, or sump has reached a predetermined height. Common applications include cooling tower sump level control and monitoring, steam condensate tank level, storm water and sewage sump level monitoring and control, and thermal storage tank level monitoring. Numerous mechanical and analog technologies are currently available. Some analog technologies include capacitance, ultrasonic, and magnetostrictive-based devices in combination with solid-state electronics to provide a switching action based on the level.

More commonly used technologies include devices that employ the use of a float (integral, rod and float, submersible), conductivity probe, or differential pressure mechanism.

Pressure switches are used in DDC systems to provide status indication for fans, filters, and pumps and to provide flow and level status indication by virtue of the predicable relationships between pressure and these values. Typical mechanical pressure switches use a piston, bellows, bourdon tube or diaphragm, and a magnetic or mechanical linkage to convert the forces resulting from the measured pressure into repeatable motions used to operate one or more switches.

Vibration switches are used to provide a signal when vibration levels in rotating machinery, such as fans (particularly with induced draft cooling towers), reach unsafe levels. Vibration switches are commonly applied on large cooling towers and air-handling unit fans for safety reasons.

Moisture detecting switches are commonly used to detect moisture under raised floors, in piping and tank containment areas, and in the drain pans of air-handling units to alert system operators before damage or flooding occurs. Most moisture detecting switches are instruments of the float or conductivity type. Float types are adapted to actuate at very low fluid levels. Conductivity types may consist of point-sensitive probes located very close to the bottom of a low point or sump where water will collect, or they may be ribbons or strips with wires separated by a nonconductive material, such that when any portion of the ribbon is exposed to liquid moisture, the electrical circuit is completed and the switch mechanism activates.

Current sensing relays are used in DDC systems to monitor the status of electrical devices. The devices typically have one or more adjustable current set points. Common applications include fan and pump on/off status feedback. Current switches can detect broken fan belts if properly adjusted. Current relays can also be used for phase monitoring.

DI devices can be used as direct DDC inputs and the control logic written to define the function of each device in relation to other devices in the control of a specific equipment item. For example, the input from a freezestat must indicate "closed" before a start relay will close to start a fan. This is called *software interlock*. However, with software interlock, a failure of the DDC system means that the HVAC equipment control circuit is no longer interlocked or protected by its safety devices. Therefore, it is more common to used hardwired interlock for safety or limit controls.

13.2.3 AI Devices

There are numerous AI devices utilized in HVAC control. The main categories of devices include (but are not limited to) the measurement of temperature, humidity, dew point, pressure, flow (liquid, air), liquid level, light level, electrical attributes (voltage, current, phasing, power), energy, occupancy, position, and gas concentration.

One of the most common properties measured in the HVAC control world is temperature, a measurement necessary in a host of HVAC control strategies, and several temperature measurement technologies exist for use with DDC control systems:

1. RTDs operate on the principle that the electrical resistance of a metal changes predictably and in an essentially linear and repeatable manner with changes in temperature. The resistance of the element at a base temperature is proportional to the length of the element and the inverse of the cross-sectional area. RTDs are commonly used in sensing air and liquid temperatures in pipes and ducts and as room temperature sensors. DDC systems may accept RTD inputs directly, or a transmitter with voltage or current output may be used.

The accuracy of an RTD sensor is typically expressed in percent of nominal resistance at 0°C [32°F]. RTDs are relatively accurate when compared to other sensing devices and have good stability characteristics. They can be made from many materials, some of which include platinum, tungsten, silver, copper, nickel, nickel alloys, and iron. Currently, most RTDs used in the HVAC field are constructed in film configurations with platinum, nickel, or nickel iron.

Since the resistance of the sensor is the property being measured, the resistance of all elements of the circuit, including the sensor leads, affects the measurement. With RTDs, and particularly those with lower base resistance values, the resistance of long leads can amount to several percent or more of the sensor circuit. This can result in significant error, and correcting this problem is to include a separate transmitter with the sensor.

2. *Thermistors* are commonly used for sensing air and liquid temperatures in pipes and ducts and as room temperature sensors. The term "thermistor" evolved from the phrase "thermally sensitive resistor." Thermistors are temperature-sensitive semiconductors that exhibit a large change in resistance over a relatively small range of temperature. There are two main types of thermistors: *positive temperature coefficient* and *negative temperature coefficient* (NTC). NTC thermistors are commonly used for HVAC temperature measurement.

   Unlike RTDs, the temperature-resistance characteristic of a thermistor is nonlinear and cannot be characterized by a single coefficient. Manufacturers commonly provide resistance-temperature data in curves, tables, or polynomial expressions. Linearizing the resistance-temperature correlation may be accomplished with analog circuitry or by the application of mathematics using digital computation. The lead resistance of most thermistors is very small in comparison to sensor resistance.

   While some designers prefer RTDs for all temperature sensing, for comfort applications it is acceptable to use precision thermistors (10,000 ohms) for room temperature sensing and 1,000 ohm platinum RTDs for HVAC unit temperature sensing. Thermistors are not quite as accurate or stable as RTDs, but they are easier to wire, cost less, and almost all DDC systems accept them directly. Typically, occupant set point adjustment of space temperature set point is not desired. However, if the owner insists on allowing occupants to adjust set points, this must be clearly indicated on the control drawings. Occupant adjustments are limited to +/–2 °F from set point(s) defined by the sequences of operation.

3. *Thermocouples* are available for space, pipe, and duct application. Thermocouples operate on the principle that when two dissimilar metals are joined at both ends and one of the ends is at a different temperature, a voltage that is proportional to the temperature of the junction is produced. This principle requires that the leads be made of the same metals in order to achieve reasonable measurement accuracy. The signal level from a thermocouple is in the millivolt range such that transmitters are often used to overcome the effect of the leads. Although in widespread laboratory and industrial use, thermocouples are not widely used in commercial HVAC control applications.

Table 13-2 compares the most common temperature measurement technologies applicable to DDC control systems for HVAC. The comparisons made are general in nature and not intended to be all inclusive for each sensor type.

TABLE 13-2

Comparison of Temperature Measurement Technologies

	Thermistor	RTD	Thermocouple
Advantages	Large resistance change with temperature. Rapid response time. High resistance eliminates problems with lead resistance. Low cost. Good stability. Interchangeable.	Linear resistance with temperature. Good stability Wide range of operating temperature. Interchangeable over a wide temperature range.	Widest operating range. Simple and rugged. Low cost. No external power supply required.
Disadvantages	Nonlinear. Limited operating temperature range. Interchangeable over only narrow temperature ranges. Subject to inaccuracy due to self-heating. Current source required.	Small resistance change with temperature. Slower response time. Subject to self-heating. Transmitter or three- or four-wire leads required for lead resistance compensation. Some types easily damaged by shock or vibration. External circuit power source required.	Nonlinear. Lower stability. Reference junction temperature compensation required. Radio frequency or electronic noise can affect low signal level.

RTDs, thermocouples, and thermistors are all small devices with similar mounting techniques used for all of the types. Sensors for pipe and duct mounting are commonly sheathed in a stainless steel sheath of 1/8″–1/4″ diameter (larger and smaller diameters are available). Sensors for liquid piping systems may be mounted with direct immersion into the fluid or installed in a *thermowell* to allow removal without draining the piping system and to reduce the likelihood of leaks. Sensors installed in wells should be installed with a heat transfer compound filling the space between the sensor and the well to insure good thermal contact between the measured fluid and the sensor.

In measuring the temperature of air in large ducts, it is typical to use an averaging element because the air temperature may vary over the cross section of the duct (particularly at fittings just downstream of heating or cooling coils). RTD and thermistor sensors have been developed that accomplish this using multiple sensors installed in a single flexible tubular element. The element is typically arranged in a serpentine fashion to obtain representative measurements over the entire cross-sectional area of the duct. Very large ducts or air-handling unit casings often require multiple sensors that are customarily wired in parallel-series arrangements. Averaging elements are commonly applied downstream of mixing dampers and large heating or cooling coil banks.

Sensors for outdoor air applications should be located in normally shaded areas to prevent the heating effects of solar radiation. These sensors are usually provided with a shield or hood to reduce the effects if exposed to direct sunlight and prevent direct contact with precipitation.

After temperature, the second most common measurement in the HVAC industry is RH. RH sensors are used in DDC control systems to measure humidity in spaces and ducts. Commonly applied sensor types include thin-film capacitance, bulk polymer resistance, or the integrated circuit type that combines a sensor (commonly of the capacitance type) and some of the signal conditioning circuitry to form a solid-state device.

Thin film capacitance sensors operate on the principle that changes in RH cause the capacitance of a sensor made by laminating a substrate, electrodes, and a thin film of hygroscopic polymer material to change in a detectable and repeatable fashion. Because of the nature of the measurement, capacitance humidity sensors are combined with a transmitter to produce a higher level voltage or current signal.

Capacitance-type RH sensor/transmitters are capable of measurement from 0% to 100% RH with application temperatures from –40°F to 200°F [–40°C to 93°C]. Capacitance sensors are affected by temperature such that accuracy decreases as temperature deviates from the calibration temperature. Sensors are available that are interchangeable within plus or minus 3% without calibration. Sensors with long-term stability of <±1% per year are available.

Bulk polymer resistance sensors use the principle that resistance change across a polymer element varies with RH and is measurable and repeatable. As with capacitance humidity sensors, polymer resistance sensors are combined with transmitters to produce a higher level voltage or current signal.

Bulk polymer resistance humidity sensor/transmitters are commonly capable of measurement from 0% to 100% RH with application temperatures from –20°F to 140°F [–29°C to 60°C]. Resistance sensors are affected by temperature such that accuracy decreases as the temperature deviates from the calibration temperature. Bulk polymer resistance humidity sensors are not commonly interchangeable. Sensors with long-term stability of <±1% drift per year are available.

It is a common practice when measuring RH to combine a temperature sensor and transmitter into the same device as the humidity sensor. Adding a microprocessor makes it possible to calculate and transmit dew-point temperature or enthalpy. Often, these devices can be configured to output calculated humidity ratio, wet-bulb temperature, and absolute humidity as well as dew point.

Pressure is measured in DDC systems in order to control the operation and monitor the status of fans and pumps. Space "barometric" pressure is sometimes measured and used for control. Pressure is also the basis of many flow and level measurements. Pressure sensors commonly used with DDC systems include capacitance and variable resistance (piezoelectric and strain gage) types.

Capacitance pressure sensors typically use a capacitance cell consisting of a diaphragm exposed to the pressure medium separated from another plate by a fill fluid. When the applied pressure deflects the diaphragm, the capacitance characteristic of the sensing element changes. The capacitance cell is excited by a high frequency source. The frequency changes as the capacitance of the cell changes. This frequency shift is converted to the output signal by the transmitter electronics. Capacitance transmitters are available configured for either differential or gauge pressure measurement. Usual outputs are voltage or current.

Capacitance transmitters are available with ranges from a few inches water column (in. wg) to thousands of pounds per square inch (psi).

The major considerations for the installation of a pressure element in a fluid system should include the following:

1. Sensor location (pipe mounted, tank mounted, remote).
2. Isolation of the sensing element from undesirable and potentially damaging transient pressures, such as those resulting from water hammer and turbulence.
3. Temporary isolation from the pressure source for maintenance and release of trapped pressure when removing the sensor for maintenance or for setting zero during calibration.

4. Over-range protection for differential pressure instruments.
5. Protection from occurrence of temperature outside of the range of the sensor application.
6. Venting trapped, non-condensable gases in liquid-sensing piping.
7. Draining trapped liquids from gas.

Pressure snubbers or dampeners are used to reduce the magnitude of pressure transients. These can be a sintered metal element with small openings, a small orifice fitting, a high-pressure drop valve (such as a needle valve), or a pressurized gas-filled container mounted on the sensing piping.

The designer must incorporate valves to provide isolation, venting, drain, and pressure relief for pressure instruments installed in piping systems.

Flow measuring devices are used in DDC systems to monitor air and liquid flow rates. Typically, airflow-measuring devices are used to monitor and control the output of fans, dampers, and associated equipment used to control outside airflow, terminal unit airflow, and space pressures. Liquid flow is commonly measured to maintain required flows in boilers, chillers, and heat exchangers and to control and monitor energy production and use (requires temperature measurement also).

Flow rate is typically obtained by measuring a velocity of a fluid in a duct or pipe and multiplying them by the known cross-sectional area (at the point of measurement) of that duct or pipe. Common methods for measuring airflow include hot wire anemometers, differential pressure measurement systems, and vortex shedding sensors:

1. "Hot Wire" or *thermal anemometers* operate on the principle that the amount of heat removed from a heated temperature sensor by a flowing fluid can be related to the velocity of that fluid. Most sensors of this type are constructed with a second, unheated temperature sensor to compensate the instrument for variations in the temperature of the air. Hot wire sensors are available as single point instruments for test purposes, or in multipoint arrays for fixed installation. Hot wire type sensors are better at low airflow measurements than differential pressure types and are commonly applied to air velocities as low as 50 ft/min.
2. *Differential pressure flow devices* in common use in HVAC systems include Pitot tubes and various types of proprietary velocity pressure (VP)-sensing tubes, grids, and other arrays. All of these sensing elements are combined with a low differential pressure transmitter to produce a signal that is related to flow velocity.

   The Pitot tube measures both SP and TP. The difference between these two values is the VP. VP is a function of the air velocity that can be computed using Eq. (7-10) or (7-10a).

   The need to sense multiple points in the cross section of a duct gave rise to averaging type sensors with arrays of pressure-sensing points. This is called an airflow monitoring station, as shown in the following figure, and is commonly used in HVAC applications.

   Some differential pressure based flow stations include transmitters that have the capability to electronically extract the square root of the measured pressure and provide an analog signal that is linear with respect to velocity, whereas others provide an analog signal that is proportional to measured pressure and depend upon the DDC system to calculate the square root and, therefore, velocity.

Velocity range is limited by the range and resolution of the pressure transmitter used. Most differential pressure-type stations are limited to a minimum velocity in the range of 400–600 fpm [2–3 m/s]. Maximum velocity is only limited by the durability of the sensor.

In recent years, flow arrays have been developed for installation in the inlets of centrifugal fans and across outdoor intakes (exterior and interior), making total airflow measurement much more convenient.

All airflow sensors work best in sections of ducts that have uniform, fully developed flow. All airflow-sensing devices should be installed in accordance with the manufacturers recommended straight runs of upstream and downstream duct in order to provide reliable measurement. A number of manufacturers offer flow straightening elements that can be installed upstream of the sensing array to improve undesirable flow conditions. These should be considered when conditions do not permit installation with the required straight runs of duct upstream and downstream from the sensor.

Common methods used to measure liquid flow include differential pressure measurement systems, vortex shedding sensors, positive displacement flow sensors, turbine-based flow sensors, magnetic flow sensors, ultrasonic flow sensors, and "target" flow sensors. For water flow measurements, *differential pressure flow devices* in common use in HVAC systems operate either by measuring VP (insertion tube type) or by measuring the drop in pressure across a restriction of known characteristic (orifice, flow nozzle, Venturi):

1. *Insertion tube* type flow sensors are usually constructed of a round or proprietary-shaped tube with multiple openings across the width of the flow stream to provide an average of the velocity differential across the tube and an internal baffle between upstream and downstream openings to obtain a differential pressure. Insertion tube type meters have a low permanent pressure loss and, with proper installation and associated pressure instruments, are satisfactory for many common applications. Insertion tube flow sensors are available that can be installed and removed through a full port valve so that installation and service are possible without draining the section of piping in which they are installed.
2. A *concentric orifice* plate meter is the simplest and least expensive of the differential pressure type meters. The orifice plate constricts the flow of a fluid to produce a differential pressure across the plate. The result is a high pressure upstream and a low pressure downstream that is proportional to the square of the flow velocity. An orifice plate usually produces a greater overall pressure loss than other flow elements. An advantage of this device is that cost does not increase significantly with pipe size.
3. *Venturi tube* meters exhibit a very low pressure loss compared to other differential pressure meters, but they are also the largest and most costly. They operate by gradually narrowing the diameter of the pipe and measuring the resultant drop in pressure. An expanding section of the meter then returns the flow to very near its original pressure. As with the orifice plate, the differential pressure measurement is converted into a corresponding flow rate. Venturi tube applications are generally restricted to those requiring a low pressure drop and a high accuracy reading. They are widely used in large diameter pipes.
4. *Flow nozzle* meters may be thought of as a variation on the Venturi tube. The nozzle opening is an elliptical restriction in the flow but with no outlet area for pressure

recovery. Pressure taps are located approximately 1/2 pipe diameter downstream and 1 pipe diameter upstream. The flow nozzle is a high velocity flow meter used where turbulence is high (Reynolds numbers above 50,000) such as in steam flow at high temperatures. The pressure drop of a flow nozzle falls between that of the Venturi tube and the orifice plate (30%–95%).

5. *Vortex shedding* flow meters operate on the principle that when a fluid flows around an obstruction in the flow stream, vortices are shed from alternating sides of the obstruction in a repeating and continuous fashion. The frequency at which the shedding alternates is proportional to the velocity of the flowing fluid. Vortex flow meters provide a highly accurate flow measurement when operated within the appropriate range of flow. Vortex meters are commonly applied where high quality water, gas, or steam flow measurement is desired.

Positive displacement flow meters are used where high accuracy at high turndown is required and reasonable-to-high permanent pressure loss will not result in excessive energy consumption. Applications include water metering such as for potable water service, cooling tower and boiler make-up, steam condensate, and hydronic system make-up. Positive displacement meters are also used for fuel metering for both liquid and gaseous fuels. Common types of positive displacement flow meters include lobed and gear-type meters, nutating disk meters, and oscillating piston type meters. These meters are typically constructed of metals such as brass, bronze, cast, and ductile iron but may be constructed of engineered plastic, depending on service.

Due to the close tolerance required between moving parts of positive displacement flow meters, they are sometimes subject to mechanical problems resulting from debris or suspended solids in the measured flow stream. Positive displacement meters are available with flow indicators and totalizers that can be read manually. When used with DDC systems, the basic meter output is usually a pulse that occurs at whatever time interval is required for a fixed volume of fluid to pass through the meter. Pulses may be accepted directly by the DDC controller and converted to flow rate, or total volume points, or a separate pulse-to-analog transducer may be used. Positive displacement flow meters are one of the more costly meter types available.

Turbine and propeller type meters operate on the principle that fluid flowing through the turbine or propeller will induce a rotational speed that can be related to the fluid velocity. Turbine and propeller type flow meters are available in full-bore, line-mounted versions and insertion types where only a portion of the flow being measured passes over the rotating element. Full bore turbine and propeller meters generally offer medium to high accuracy and turndown capability at reasonable permanent pressure loss. Turbine flow meters are commonly used where good accuracy is required for critical flow control or measurement for energy computations. Insertion types are used for less critical applications. Insertion types are often easier to maintain and inspect because they can be removed for inspection and repair without disturbing the main piping. Some types can be installed through hot tapping equipment and do not require draining of the associated piping for removal and inspection.

Magnetic flow meters operate on the basis of Faraday's Law of electromagnetic induction, which states that a voltage will be induced in a conductor moving through a magnetic field. The magnitude of the induced voltage *E* is directly proportional to the velocity of the conductor *V*, conductor width *D*, and the strength of the magnetic field *B*. As shown in the following figure, magnetic field coils are placed on opposite sides of a pipe to generate a magnetic field.

As a conductive liquid moves through the field with average velocity *V*, electrodes sense the induced voltage. The distance between electrodes represents the width of the conductor. An insulating liner prevents the signal from shorting to the pipe wall. The only variable in this application of Faraday's law is the velocity of the conductive liquid *V* because field strength is controlled constant and electrode spacing is fixed. Therefore, the output voltage *E* is directly proportional to liquid velocity, resulting in linear output. Magnetic flow meters are used to measure the flow rate of conducting liquids (including water) where a high-quality, low-maintenance measurement system is desired. The cost of magnetic flow meters is high relative to many other meter types.

Ultrasonic flow sensors measure the velocity of sound waves propagating through a fluid between two points on the length of a pipe. The velocity of the sound wave is dependent upon the velocity of the fluid such that a sound wave traveling upstream from one point to the other is slower than the velocity of the same wave in the fluid at rest. The downstream velocity of the sound wave between the points is greater than that of the same wave in a fluid at rest. This is due to the Doppler effect. The flow of the fluid can be measured as a function of the difference in time travel between the upstream wave and the downstream wave. Ultrasonic flow sensors are nonintrusive and are available at moderate cost. Many models are designed to clamp on to existing pipe.

A *target* meter consists of a disc or a "target" which is centered in a pipe. The target surface is positioned at a right angle to the fluid flow. A direct measurement of the fluid flow rate results from the force of the fluid acting against the target. Useful for dirty or corrosive fluids, target meters require no external connections, seals, or purge systems. Target flow meters are commonly used for liquid flow measurement and less commonly applied to steam and gas flow.

As with airflow, all liquid flow sensors work best when fully developed, uniform flow is measured. To attain fully developed, uniform flow sensors should be installed in accordance with the manufacturers recommended straight runs of upstream and downstream pipe in order to provide the most reliable measurements.

With most liquid flows measured for HVAC applications, density changes with pressure and temperature are relatively small and most often ignored due to their insignificant effect on flow measurements. When measuring the flow of steam or fuel gases, unless temperature and pressure are constant, ignoring the effect density changes with varying temperature and pressure will often result in significant or gross errors. For this reason, it is common to measure the temperature and pressure, in addition to the flow, and electronically correct the result for the fluid density. This correction may be done using an integral or remote microprocessor-based "flow computer" or it may be made in the DDC controller with suitable programming.

Selection of fluid flow meters is based on four factors: pressure drop, accuracy, turndown, and cost. Turndown defines the flow range over which the meter is accurate. For example, a turndown of four indicates that the minimum flow at which accuracy is maintained is 25% of the meter's range or rated maximum flow. At a turndown of 30:1, the meter is accurate down to about 3% of its flow range. Table 13-3 summarizes these factors for each flow measurement technology.

The accuracy of most meters is defined in terms of percent of full range. Therefore, it is important that designers select flow meters that have a maximum rated flow relatively close to the anticipated HVAC process flow. Often, this introduces a pressure drop trade-off that must be evaluated.

TABLE 13-3

Attributes of Flow Measurement Technologies

Technology	Fluid(s)	Relative PD	Typical Accuracy	Turndown	Relative Cost
Hot wire	Air	Very low	1%	50–12,000 fpm velocity	Moderate
Pitot tube	Air/liquids	Low	5%	4:1	Moderate
Orifice plate	Liquids/steam	High	3%	4:1	Low
Venturi tube	Liquids/steam	Low	3%	4:1	High
Flow nozzle	Liquids/steam	Medium	3%	4:1	High
Vortex shedding	Air/steam/ liquids	Medium	0.5%	20:1 air/steam 30:1 liquids	High
Positive displacement	Liquids	High	0.1%	100:1	Low to moderate
Turbine or propeller	Steam/liquids	Low	1%	30:1	Turbine: high prop: low
Magnetic	Liquids	Very low	1%	30:1	High
Ultrasonic	Dirty liquids	Very low	1%–5% depending on liquid	20:1	Moderate
Target	Liquids/steam	Medium	1%	20:1	Low

The turndown of differential pressure meters (Pitot tube, orifice plate, Venturi tube, and flow nozzle) can be improved to between 10:1 and 16:1 by using dual transmitters, one for the low range of flow and one for the high range.

Liquid-level measurements are typically used in DDC systems to monitor and control levels in thermal storage tanks, cooling tower sumps, water system tanks, pressurized tanks, and so on. Numerous sensing technologies are available. Common technologies applicable to HVAC system requirements are based on hydrostatic pressure, ultrasonic, capacitance and magnetostrictive-based measurement systems.

Measurement by hydrostatic pressure is based on the principle that the hydrostatic pressure difference between the top and bottom of a column of liquid is related to the density of the liquid and the height of the column. For open tanks and sumps, it is only necessary to measure the gauge pressure at the lowest monitored level. For pressurized tanks, it is necessary to take the reference pressure above the highest monitored liquid level. Pressure transmitters are available that are configured for level monitoring applications. Pressure instruments may also be remotely located; however, this makes it necessary to field calibrate the transmitter to compensate for elevation difference between the sensor and the level being measured.

Ultrasonic level sensors emit sound waves and operate on the principle that liquid surfaces reflect the sound waves back to the source and that the transit time is proportional to the distance between the liquid surface and the transmitter. One advantage of the ultrasonic technology is that it is noncontact and does not require immersion of any element into the sensed liquid. Sensors are available that can detect levels up to 200 ft from the sensor. Accuracy from 1% to 0.25% of distance and resolution of 1/8″ is commonly available.

Capacitance level transmitters operate on the principle that a capacitive circuit can be formed between a probe and a vessel wall. The capacitance of the circuit will change with a change in fluid level because all common liquids have dielectric constant higher than that of air. This change is then related proportionally to an analog signal suitable for DDC AIs. Resolution of 1/8″ and accuracy of 1%–0.25% of span are available.

Magnetostrictive level transmitters operate on the principle that an external magnetic field can be used to cause the reflection of an electromagnetic wave in a waveguide constructed of magnetostrictive material. The probe is composed of three concentric members. The outermost member is a protective, product-compatible outer pipe. Inside the outer pipe is a waveguide, which is a formed element constructed of a proprietary magnetostrictive material. A low-current "interrogation" pulse is generated in the transmitter electronics and transmitted down the waveguide creating an electromagnetic field along the length of the waveguide. When this magnetic field interacts with the permanent magnetic field of a magnet mounted inside the float, a torsional strain pulse, or waveguide twist, results. This waveguide twist is detected as a return pulse. The time between the initiation of the interrogation pulse and the detection of the return pulse is used to determine the level of measurement with a high degree of accuracy and reliability. Accuracy and resolution of 1/16″ or better are available from some manufacturers.

Ambient light-level sensors used in DDC systems are typically used to turn on night lighting when light level drops below a set point level and/or to turn off indoor or outdoor lighting when ambient levels are sufficient. Light-level sensors can be used to control the output of dimmable fluorescent lighting to set point level. An accuracy of ±1% reading is common.

Monitoring of electrical power is performed by DDC systems to determine status or condition of HVAC system components, determine power and energy consumption of various components, and implement usage and demand control strategies to reduce building energy costs. The two most common electrical measuring devices used for DDC are current transducers and power measuring devices:

Current transformers are used to monitor current flow to motors, heaters, or electrical distribution systems. Their input may be used for demand limiting purposes, control, or energy accounting. The sensing element of a current transducer is typically a current transformer. It transforms the current being monitored into a higher voltage, lower current. Additional circuitry reduces this voltage to the desired level. Current transducers may have line and load terminals for the monitored current, or they may be arranged as a coil that the current carrying conductor passes through. With this arrangement, the load conductor induces the current in the transformer *via* the electromagnetic field surrounding the conductor. Current transformers and transducers are available with solid or split cores. The split core device may be installed without disconnecting the power conductor provided that there is sufficient slack in the conductor and room in the enclosure. Accuracy of ±0.5% of full scale is readily available.

With the increased interest in indoor air quality and the need to monitor potentially dangerous gases, *gas concentration* measurements have become increasingly more prevalent in DDC system design. Many devices are currently available for use in HVAC applications:

1. *Carbon monoxide* detectors are used to operate ventilation equipment to prevent carbon monoxide levels from becoming unsafe. They are also used to warn facility owners and occupants of unsafe levels in garages, loading docks, tunnels, and other areas where vehicles are operated. Solid-state sensing technology is most commonly used. Single or multiple sensing point versions are available that can provide contact closures at one or more set levels and/or analog signals that are proportional to carbon monoxide concentration.

2. *Carbon dioxide* concentration inside of buildings has been related to general ventilation adequacy and is often recommended for application in DDC systems to control outdoor air fans and dampers to keep the concentration below set point levels, that is, *demand control ventilation. However, there are well-documented issues with the accuracy and reliability of carbon dioxide sensors and use of these devices should be avoided (see Chapter 10).*
3. *Refrigerant gas* detectors are used in the control of emergency ventilation systems to evacuate hazardous concentrations of refrigerant gas in refrigeration machinery rooms and other enclosed areas. Gas-specific detectors are available to detect individual refrigerant types. The most commonly used sensor types are infrared, photo-acoustic, and solid-state sensing technologies. Single or multiple sensing point versions are available that can provide contact closures at one or more set levels and/or analog signals that are proportional to refrigerant concentration.

13.2.4 DO Devices

DO devices are used to provide two-position control (open/close, on/off, and so on) of valves, dampers, electric motors, lighting and external signaling devices, such as alarm bells and indicator lights. The most common DO devices are relays, contactors, starters, and two-position valve and damper actuators. DOs may also be used to control analog devices using PWM. PWM is accomplished by monitoring a timed closure of a set of contacts. The amount of time the contacts are closed is proportional to a level of performance for the controlled device.

A *relay* is a device where the power applied to a coil or input terminal causes the path between pairs of separate, additional terminals to either allow the electrical current flow or stop the current flow. Contactors and starters are essentially large relays designed for interrupting and applying power to loads (i.e., integral horsepower motors) and significant resistance loads (i.e., lighting and heaters). The most common types of relays are standard instantaneous control, latching, and timing. Contactors and starters can be considered common types of heavier duty relays with and without load protection.

Standard *control relays* may be electromechanical or solid state. Electromechanical control relays use a magnetic coil and armature to cause contacts to open or close when current is applied to the coil. Solid-state relays use semiconducting devices that become electrically conductive between output terminals when a voltage is applied to the input.

Latching relays are a variation of the standard instantaneous control relay where the contacts change position when initially energized but do not revert to the normal state (when the input signal is removed) until a separate reset signal is applied. Latching relays may have mechanical latches using a set and reset coil, or they may latch magnetically. Latching relays are also available with manual reset latches.

Time delay relays are a variation of the standard instantaneous control and latching relay where a fixed or adjustable time delay must occur following a change in the control signal before the switching action occurs. Common time delay relay configurations include on delay, off delay, and on/off delay.

Two-position actuators are used to control the linear or rotary motion of a final control element to one of the two positions, usually open or closed. The two most common types of two-position actuators are the *solenoid type* and the *rotary type.*

13.2.5 AO Devices

AO devices are used to provide modulating control of sequencers, final control element actuators, electric motors (through variable speed drives or PWM circuits), and silicon controlled rectifiers (SCRs).

The sequencing of multiple on–off devices based on a single AO from a control loop is required for cooling towers with multiple two-speed fans, multistage electric heaters, and multistage refrigeration systems. This sequencing can be accomplished within the DDC controller or it may be accomplished externally using a discrete sequencing device. These devices have two or more relay or DOs that are adjusted to spread the signal range that they turn on and off. For example, a two-stage sequencer might be adjusted so the stage one relay turns on at 37.5% analog signal level and off at 12.5%. The stage two relay would be adjusted to turn on at 87.5% analog signal and off at 67.5%. More advanced sequencers may incorporate adjustable inter-stage time delays, minimum on and off times, and so on.

Variable frequency drives (VFDs) are typically used to vary the speed of HVAC motors (see Chapter 1) in order to control the output of driven equipment. SCRs are used to regulate an AC power supply to a resistive electrical load, such as an electric heater, to provide continuously variable output (see Chapter 6).

Actuators (also called "operators") are one of the most important components of DDC systems today. Actuators impart movement to the final control elements that are the key to control of HVAC processes.

Valve and damper "action" is defined as "normally open" or "normally closed"... the position of the valve or damper when its actuator is exerting 0 torque due to absence of a control signal. In most applications, the "failure mode" of a coil or heat exchanger is defined by the "normal" valve position, open or closed. With valves, heating applications typically utilize normally open control valves so that heat is applied even if the control system fails. For cooling, the valve mode is optional: in more northerly climates, normally closed cooling valves are typically utilized, while normally open valves are used in warmer climates.

To have a valve or damper return to its normal position in the event of a control signal or power failure, the use of a *spring return* type actuator is required. Most electronic/electric actuators are not supplied with spring return and designers must specify the need for spring return (at a higher cost) when this type of action is needed for electric/electronic actuators.

Modulating electric/electronic actuators must have integral *drive motors* rated for at least 1,200 starts/hour and 60,000 full stroke cycles to help ensure long life. A positioning circuit accepts an analog control signal (typically 0–10 V or 4–20 mA). The actuator then interprets this control signal as the valve position between the two limit switches (maximum closed position and maximum open position). To achieve this, the actuator has a position sensor (usually a potentiometer), which feeds the actual rotation position back to the positioning circuit. In this way, the actuator can be positioned along its stroke in proportion to the control signal.

Other common problems with electric/electronic actuators include the following:

1. Torque ratings are limited for spring return type actuators.
2. Speed of movement can be as low as 4 s/mm that in rapidly varying control applications may be too slow. (This is particularly true for cooling tower bypass valves.)

3. Their "plastic" enclosures and mounting assemblies are not particularly robust and tend to fail, requiring actuator replacement. The designer must pay careful attention to the operator construction at the time of vendor submittals and *always request a sample of each type and size of actuator to be used.*

One of the most common applications in which the limitations of electric/electronic actuators can lead to operating difficulties is control of a condenser water bypass valve. The problem tends to appear at the start-up during the fall, winter, and spring when tower basins are full of relatively cold (typically, 40°F–60°F) water. With large chillers and very short bypass circuits, the heat rejected by the chiller quickly overwhelms the relatively small thermal flywheel represented by the recirculation loop. Once set point temperature is achieved, if the valve cannot position itself to redirect flow over the tower faster than the chiller can raise the temperature in the recirculation loop, the condenser temperature will spiral out of control, and the chiller will trip off on high head pressure. Complicating matter is that water temperature tends to step up in increments, rather than ramp up, as the system warms up. There is not much opportunity for water to mix in the recirculation loop; so with each pass through the chiller, the water in the loop tends to ratchet up by the condenser temperature rise. This is a frustrating and difficult problem and is best addressed during design.

Large three-way control valves are formed by mounting two two-way butterfly valves on a piping tee. Butterfly valves have equal percentage flow characteristics and work well when used to control flow through coils and heat exchangers. However, when coupled together to form a three-way condenser water bypass control valve, individual valve operators are required for each of the two butterfly valves so the two valves can be modulated independently to overcome the nonlinearity, percent flow versus percent open, for each valve. Therefore, for this duty, the two butterfly valves must be modulated independently in accordance with Table 13-4 to maintain more balance flows, while maintaining constant condenser water flow rate.

Valve actuators, no matter the type, must be selected for *tight shutoff* against the maximum system pressure with which that can be developed to prevent "valve lift" due to system pressure, resulting in unintended flow through the coil or heat exchanger. For water systems, the maximum possible system pressure is equal to the pump "cut-off" pressure, the pressure produced by the pump at (or near) a zero flow condition. For steam systems, the maximum possible pressure is equal to the maximum anticipated steam pressure at the valve.

TABLE 13-4

Condenser Water Bypass Butterfly Valves Modulation

Percent (%) Open	
To Cooling Tower (NO)	**To Bypass (NC)**
100	0
80	85
60	90
40	93
20	96
0	100

13.3 Final Control Elements

The manipulated variable in most HVAC systems is usually a fluid (air, water, and steam, typically), whose flow rate is changed in response to a change in the controlled variable's condition. The devices that manipulate these fluids are steam and water (or other liquid) control valves and air dampers, which are called the final control elements.

The importance of the correct selection of control valves and dampers to proper control system operation cannot be overemphasized.

13.3.1 Control Valves

There are four valve characteristics that must be selected by the designer: required *valve configuration and operating mode*, the *type of valve* required for the application, the *valve rangeability*, and the *flow coefficient* (C_v [K_v]).

Hydronic control valves may be three- or two-way configuration. Three-way valves are utilized when the water system is designed for constant flow, while two-way valves are used in variable flow systems. A standard globe or butterfly valve is inherently a two-way valve. A three-way globe valve simply has a second port added while three-way action using butterfly valves requires two valves and a pipe tee.

Three-way control valves may be the "mixing" or "diverting" type. With mixing valves, the valve has two inlets and one outlet, allowing two input flows to "mix" to create a single output flow, while diverting valves have one inlet and two outlets, splitting the single input flow to two output flows. Diverting valves, due to the pressures involved, require more robust construction and are more expensive than mixing valves, which are typically preferred for HVAC applications.

The type of control valve selected for water flow control applications may be a globe valve or a butterfly valve. The actuator torque required to operate a globe-type control valve is fairly constant from the full closed to full open positions. But, for butterfly valves, the total valve operating range must be limited to a range from 0° (closed) to 70° (80% open). The dynamic torque required to move a butterfly valve starts fairly high as it necessary to "unseat" the wafer from the annular seal. The torque requirement falls off quickly as the valve approaches 25° of rotation (about 30% open), increases very rapidly to peak at 75°–80° rotation, and then drops suddenly at approximately 85° rotation. This makes flow control unstable beyond 70° of stem rotation.

Water valves are available with three flow characteristics: "quick opening," "linear," and "equal percentage":

1. Quick opening control valves are selected for two-position (open–closed) control. These valves allow as much as 90% of their design flow rate when only 20% open, so there is little or no delay in flow as the valve opens or closes.
2. Linear control valves have flow rates that are in direct proportion to the stem travel of percentage open. These valves are used only to control loads with little variation and where very "tight" control is required.
3. Equal percentage control valves are designed to be "slow opening," allowing only about 10% flow when 50% open. This characteristic is exactly counter to the capacity versus flow relationship for typical coils and heat exchangers, where 10% flow produces 50% heat transfer capacity. Therefore, coupling an equal percentage

control valve with a typical coil or heat exchanger results in an essentially linear control function, as shown by Figure 6-2, which is ideal for proportional temperature control.

Rangeablity (also called "turndown") is the measure of the ratio of the maximum controllable flow rate through the valve to the minimum controllable flow rate. The rangeability should be as large as possible and never less than 30:1. Large rangeability reduces potential control problems under light load conditions.

The C_v, [K_v] commonly called the *valve coefficient*, is the flow rate through the valve resulting in a 1 psig pressure drop and is used to define the valve size to obtain the desired design pressure drop at the design (maximum) flow rate.

Water valves for modulating control duty must be selected to provide a wide-open pressure drop equal to or greater than the full-load pressure drop through the coil or heat exchanger being controlled. Thus, the pressure drop through the valve, not through the coil, will control the actual flow rate (the valve with have authority over the coil). Once the required pressure drop is established, the required water control valve flow coefficient can computed in IP units using Eq. (13-4)

$$C_v = Q/\left(\text{PD}^{0.5}\right) \tag{13-4}$$

where

Q = The coil or heat exchanger peak flow rate (gpm)

PD = Pressure drop required through the wide-open control valve (psig), equal to or greater than the coil or heat exchanger full-flow pressure drop

In SI units, the value of K_v can be calculated on the basis of Eq. (13-5).

$$K_v = Q/\left(\rho/\text{PD}\right) \tag{13-5}$$

where

Q = Flow rate (m^3/h)

PD = Pressure drop (bar)

ρ = Density (kg/m^3)

For applications where the C_v is greater than 160 [K_v > 140], the rangeability of a 70° rotation butterfly valve will be better than the rangeability of a globe valve with the same C_v. Since butterfly valves have very low wide-open pressure drop, valves in throttling duty will almost always be smaller than line size and must be installed with concentric reducers to equalize the water pressure on the valve wafer.

Water valves for two-position control duty, no matter the design, may be line size or (more typically) one size smaller. Butterfly valves for two-position duty use their full 90° of rotation.

Steam control valves are always globe type. For modulating control duty, valves should be selected for a wide-open pressure drop equal to 80% of the pressure differential (in psig) between the steam supply and the condensate return to establish authority. The required flow coefficient in IP units can then be computed using Eq. (13-6)

$$C_v = \left(Q \times V^{0.5}\right)/63.5 \times \text{PD}^{0.5} \tag{13-6}$$

where

Q = The coil or heat exchanger peak steam flow rate (lb/h)

PD = Pressure drop required through the wide-open control valve (psig)

V = Specific volume of the steam at the supply pressure (cf/lb)

In SI units, the valve coefficient K_v is defined as the flow rate in cubic meters per hour [m³/h] of water at a temperature of 16°C, with a pressure drop across the valve of 1 bar and can be computed using Eq. (13-7) or (13-8), as applicable.

$$\text{If } P2 > P1/2,\ K_v = Q/\left[22.4 \times (\text{DP} \times P2)^{0.5}\right] \tag{13-7}$$

$$\text{If } P2 < P1/2,\ K_v = Q/(11.2 \times P1) \tag{13-8}$$

where

Q = Steam flow rate (kg/h)

DP = Differential pressure (bar)

$P2$ = Pressure downstream of the valve (absolute bar)

$P1$ = Pressure upstream of the valve (absolute bar)

Steam valves for two-position control duty may be line size or one size smaller.

13.3.2 Control Dampers

A modulating control damper can be thought of as an "air control valve" and for the inherent damper characteristics (i.e., airflow vs. percent open) to be nearly linear. This typically requires that the damper be an opposed blade type, sized so that its wide-open pressure drop is 8%–10% of the total system pressure drop.

The pressure drop through a wide-open opposed blade damper is very low, as shown in Figure 13-4.

HVAC control dampers typically have a free area ratio of 0.70–0.80 (unless they are very small where the free area ratio can be reduced to 0.50–0.60).

The most common application of control dampers in air systems involves dampers for controlling air mixing for an AHU's economizer cycle. To size these dampers, the first step is to evaluate the "system" in which the damper is installed. For an economizer cycle application, each damper has its own system defined as the outdoor air, return air, and relief airflow paths and each damper is selected based on the individual system or path pressure condition. Figure 13-5 shows the effect of variable pressure drop on opposed blade flow characteristics when sizing dampers to comply with American Society of Heating, Refrigerating, and Air-Conditioning Engineers (ASHRAE) *Guideline 16*.

With an airside economizer cycle, mixing of the outdoor and return airstreams to maintain uniform air temperature and eliminate the potential for coil freezing due to "stratification" is very important. The following approaches help to ensure complete mixing of the return and outdoor airstreams:

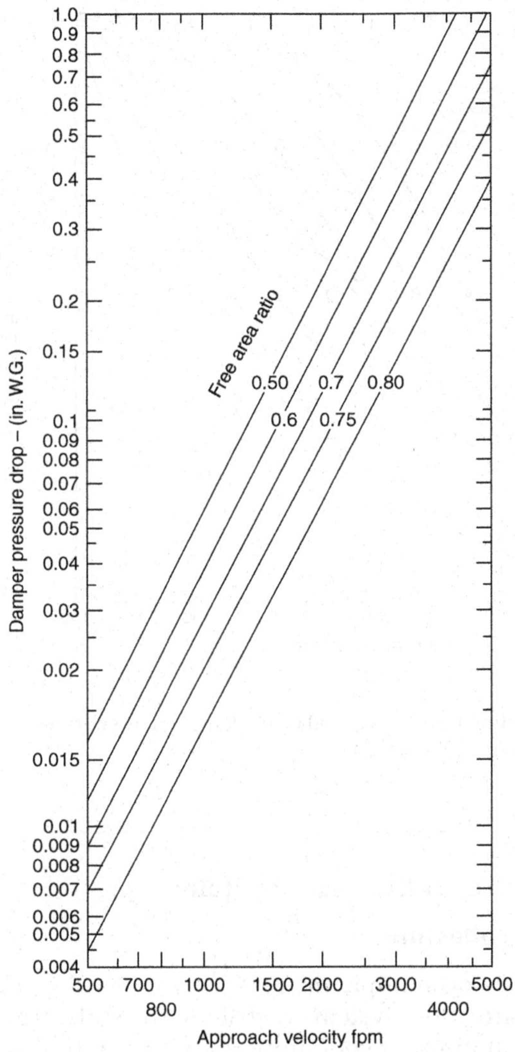

FIGURE 13-4
Opposed blade control damper pressure loss. (Source: Johnson Controls, Inc., *Engineering Data Book*, Section Db, Fig. 1.)

1. Arrange the intersection of the return and outdoor air paths 180° from each other so that the two airstreams meet "head on" to promote mixing.
2. If the two airstreams must meet at 90° from each other, utilize parallel blade dampers, arranged to provide near "head on" airflow paths. For systems with low damper velocities, additional mixing baffles may also be required.
3. If the two airstreams enter the mixing plenum parallel to each other on the same side, install baffles to force the two airsteams to mix and/or utilize parallel blade dampers installed vertically to provide near "head on" airflow paths.

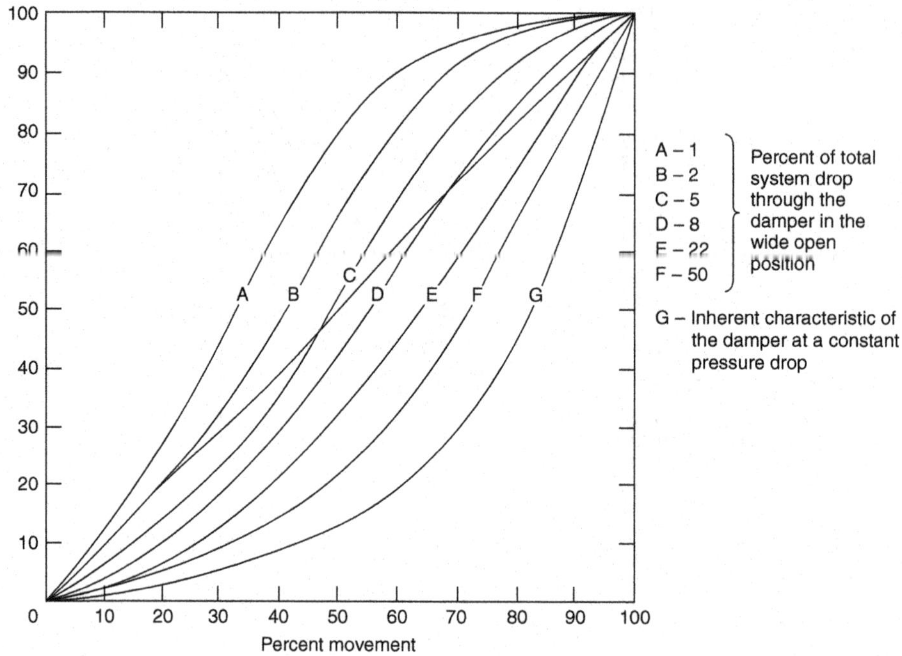

FIGURE 13-5
Effect of total air system pressure drop on opposed blade damper flow characteristic. (Source: Johnson Controls, Inc., *Engineering Data Book*, Section Db, Fig. 2.)

13.4 Direct Digital Controllers and Systems

13.4.1 DDC System Architecture

A DDC "system" consists of multiple microprocessor-based controllers with the control logic dictated by software. Most systems distribute the software to remote controllers to eliminate the need for continuous communication capability, that is, each controller can "stand-alone."

The typical DDC system network architecture consists of three network levels:

1. System management Level
2. Building level
3. Field level

The top level of an HVAC control system is typically the building management system that monitors, supervises, controls, and reports on building control systems. Its basic functions are as follows:

1. Providing information on those supervised building functions including, but not limited to, current status, archived historical information, summaries, analysis, displays, and reports on control and management functions.
2. Detecting, annunciating, and managing alarms and other conditions.

3. Diagnostic monitoring and reporting of system functions, nodes, devices, and communication networks.
4. Interfacing or integrating between individual building control systems.

The building level controllers are *networked* and communicate with the management level. These controllers can manage HVAC equipment directly (typically, major components such as air-handling units) or indirectly through lower level controllers that are networked downstream. System level controllers handle the operations of all downstream field-level controllers, collect and maintain data, and can operate as stand-alone units if communication is lost to the management level. System controllers have a peer-to-peer relationship with other controllers.

Field-level controllers serve building floors, specific areas, applications, and HVAC components. Field-level controllers communicate with one or more building-level controllers.

Many DDC systems use commercially available local area network (LAN) technologies (e.g., ethernet) for system communication. These LAN technologies are also popular for use in office or factory automation communication, sometimes referred to as an "intranet." Therefore, the designer has the option of specifying a dedicated LAN for the DDC system or allowing the DDC system to share the facility LAN with other systems.

13.4.2 Controllers and Control Loops

The key element in any DDC system is the microprocessor-based digital controller.

The controller microprocessor performs several functions. First, it accepts the input signals, either binary or analog, that define the condition or the value associated with the measured variable. Then, based on the control logic defined by the stored program(s), computes an output condition or value for the manipulated variable. AI and AO signals must be converted to and from digital values by "converters" that are part of the controller electronics. Other functions, such as communications with other controllers, time-keeping, and data storage are also performed by the controller microprocessor.

Building-level controllers typically have the following features:

1. Real-time accurate clock function
2. Full software compliment
3. Large total point capacity
4. Support for global strategies
5. Buffer for alarms/messages/trend and runtime data
6. Freeform programming
7. Downloadable database
8. High analog/digital converter resolution
9. Built-in communication interface for PC connection.

Custom application controllers (CACs) are digital controllers with capacity to connect directly to numerous input and output devices. They are fully programmable to accommodate building and equipment-customized operating logic and sequences of operation and include the ability to communicate with other CAC and building-level controllers. CACs typically applied to control of AHUs, central plant subsystems, and so on.

Application specific controllers (ASCs) are limited function field-level controllers designed for control of smaller HVAC components such as direct expansion (DX) packaged units, fan coil units, and so on. The logic (programming) for ASCs is developed by the ASC manufacturer, incorporated as "firmware," and typically allows only limited project-specific programming to be included. Most ASCs are low in cost but may have limited functionality for a specific project. Their use must be carefully evaluated by the designer. *It is recommended that ASCs be used in lieu of fully programmable secondary controllers only where it fully complies with the required sequence of operation for the HVAC application.* ASCs should incorporate user-defined password protection so that local set point change, and so on. is possible only after password verification.

Control "loops" are at the heart of a DDC system's control logic. A "closed" control loop must have an output (typically an AO or DO point), an input (typically an AI or DI point), a set point, and a defined control sequence of operation ("logic") resident in the controller. For any HVAC process control, there is a *controlled variable* (temperature, humidity, pressure, and so on) and a corresponding *manipulated variable* (fan speed, water flow, airflow, and so on) The control loop then, based on the variation between the set point and the actual condition resulting from a "disturbance" (based on the "feedback" input provided by a sensor), provides an output signal to the final control element that changes the manipulated variable to return the controlled variable to its set point condition... very simple in concept, sometimes not so simple in execution.

A control loop may be "open." In an open control loop, the output has no effect on the input. For example, while a boiler (the output) is turned on when the outside air temperature (the input) is below 55°F (the set point), the boiler's operation clearly has no effect on the outside air temperature. An open control loop often can use a DI point or even a software variable as its input. An example is AHU start/stop (the output) based on a schedule (where the time of day is the input and the start time is the set point).

13.4.3 Networks and Communication

HVAC designers, for the most part lacking any background in electronics or *information technology* (IT), have some difficulty understanding DDC networks and how DDC system components communicate with each other and their operators. For a practical matter, detailed understanding of these elements is not necessary in order to design and specify a DDC system. However, a basic understanding is necessary.

A network is the "infrastructure" over which the connected devices are connected to each other. These connections involve equipment such as routers, switches, bridges, and hubs using cables (copper, fiber, and so on) or wireless technologies (Wi-Fi).

A network communications protocol consists of a standard set of rules that define characteristics for transporting data between network devices (see Section 6).

Networks are defined by separate specific functions (called "layers"), as follows:

1. *Physical*: Cable or media standards. Electrical characteristics, length limitations, how computers or other devices mechanically and electrically connect.
2. *Data link*: Format of data on the network and how it flows. Standards describe how data are packaged into "packets," including things such as maximum length and what type of address is included for the sender and receiver. Other standards describe how a device can gain access to the physical layer, what to do if there is contention, how to tell if there is a transmission error, how to mark packets to keep them in sequence, and so on.

3. *Network*: Provide routing and related functions that enable multiple physical network segments to be combined into an inter-network. This layer provides the standards for logical naming and addressing of devices so you can route to them even if they are not physically connected to the same network.
4. *Transport*: Provide reliable process-to-process communication. This layer implements "connections," which require that data flow in sequence, that errors be detected and corrected, and that data transmissions be acknowledged, if desired—some applications such as Web servers do not check if you actually got the data. At this level, the network also provides "addresses" for different types of services to make sure that data received by a computer goes to the correct application process.
5. *Session*: Concept of tying together multiple transport streams into a single "session."
6. *Presentation*: Defines data format conversion, compression, encryption, and so on.
7. *Application*: What the user is trying to accomplish, for example, email, file transfer, and Web browsing.

Today, *Transmission Control Protocol/Internet Protocol* (TCP/IP) is by far the most dominant suite of networking protocols since it is the basic communication protocol of the Internet.

TCP/IP is a two-layer program. The higher layer, TCP, manages the assembling of each message or file into smaller packets that are transmitted over the Internet and received by a TCP layer that reassembles the packets into the original message. The lower layer, IP, handles the address part of each packet so that it gets to the right destination. Each device communicating over the network checks this address to see where to forward the message. Even though some packets from the same message are routed differently than others, they'll be reassembled at the destination.

The Internet uses additional higher layer application protocols that utilize TCP/IP. These include the World Wide Web's (www) *Hypertext Transfer Protocol* (http), the *File Transfer Protocol* (ftp), and so on, which lets you log on to remote computers, and the *Simple Mail Transfer Protocol* (smtp) that handles email. These and other protocols are often packaged together with TCP/IP as a "suite."

For wired networks, Ethernet is a popular standard that defines a number of wiring and signaling standards for the physical layer of a network. Ethernet has been standardized as IEEE 802.3. Its twisted pair wiring form has become the most widespread LAN technology in use.

Three types of networks are typically used in DDC systems. At the highest level of communication, a client-server network is usually utilized. This may be a stand-alone network utilized solely by the DDC system or may be a site network (LAN or WAN) that serves as an "Intranet." In the second case, the control applications share use of the network with the normal IT functions. This type of network is typically used as the primary LAN in most DDC systems.

A client-server network consists of one or more computers that act as servers, managing the network and the data traffic over it. Each device connected to this network, then, is a client that makes use of the network for communications with other clients. In the control environment, the client may be a controller or a communications interface device.

For smaller DDC systems, a peer-to-peer network may be used as the primary network. This type may also be used as a "subnetwork" on larger systems. On this type of network each device can share information with any other device on the LAN without going through a server or communications manager, as indicated in the following figure:

A peer-to-peer network can be enhanced to utilize the IP, allowing any device on the network to be connected to and communicate over the Internet.

Operator interface devices can be connected to the primary LAN through a communications interface. But, a Web server and appropriate software can be used to allow each interface device to communicate the Internet, using a conventional Web browser. This type of connection allows the DDC system to broadcast alarms or other information as email or text messages to cell phones, and so on.

When systems become larger than the capacity of a single subnetwork, such as at a university campus, a medical center, or even a large K-12 school district, a higher level of architecture is added to allow the use of multiple subnetworks.

The site LAN or WAN may be an existing network used as an Intranet or it may be a dedicated controls network. The site LAN/WAN is used to connect multiple subnetworks and site computers. Multiple subnetworks can be connected to a single-site LAN/WAN that allows information sharing between devices on different subnetworks, which could be individual buildings on a common site or different sites. Consideration should be given to the pros and cons of sharing the facility LAN/WAN versus using a dedicated LAN:

Pros: The major benefit of using the facility LAN is that the network infrastructure, including routers and interconnecting cabling, is already in place. This eliminates a significant material and installation expense. There is typically a network administrator or IT department responsible for managing and maintaining the network. Because the network is already extended to all users in the facility, any connected computer will have access to the DDC data (depending on the format of the data, proprietary software may be required to view it).

Cons: There are security concerns involved with deploying the DDC system over a network shared by other users. Since many systems use Windows-based technology, they are susceptible to viruses, worms, or hacking over the network. Educational facilities are notorious for having connected users with the time and ability to improperly use the network. Fortunately, most IT departments are savvy to these risks, but for smaller facilities, this could be a concern. Also, when the facility LAN is down for maintenance, there is no access to DDC data. In some cases, remote access to the facility LAN may not be possible or allowed; this eliminates the possibility for remote monitoring and troubleshooting. For new construction projects, the installation and startup of the facility LAN may be on a schedule that does not allow for timely installation and start-up of the DDC system. Another concern is that of available bandwidth. Excessive network traffic can reduce the available bandwidth, resulting in slower communication, delayed alarm reporting, or possibly lost data. Although security issues imposed by the DDC system on the network are unlikely and bandwidth use by DDC systems is typically very low, some IT managers are hesitant to allow DDC systems to share the LAN.

The designer should weigh these pros and cons based on the project-specific parameters. If the choice is made to utilize the facility LAN, early coordination with IT management personnel is critical.

For the HVAC designer, two aspects are important: (1) all networks and the communications over them must meet open standards so that any device meeting that standard can be connected to it and (2) the network must be reasonably "robust" to support the level of data traffic required for HVAC control.

Communication between two different devices on a common network requires a common protocol, a common communication speed, and a method of data formatting. Vendors build their devices around these criteria, so communication between devices by the same manufacturer is routine. The ideal condition for "interconnectivity" is for any third party to provide a component controller that is "native" to the overall building control system.

For a building DDC system to communicate with other DDC systems using a different communications protocol, there is a need for an interface or gateway to translate between the two systems. The proper operation of the gateway is dependent on the continued use of the specific revised levels of software on both systems. It typically requires the support of the manufacturer at the corporate level to implement and cooperation between manufacturers, a highly improbable state of affairs!

In the past, it has been a particular irritant to designers and owners that no DDC system provided by one controls manufacturer could be modified or extended by any other controls manufacturer because the protocols used by each manufacturer were both unique and proprietary, what is called a *proprietary protocol*.

Today, designers require the use of an *open protocol* that complies with a specific standard protocol and allows components from different manufacturers to coexist on the same network. These components would not need a *gateway* to communicate with one another and would not require a manufacturer-specific workstation to visualize data. This would allow more than one vendor's product to meet a specific application requirement.

Unfortunately, the sole use of an open or standard protocol does not guarantee that a DDC system will be an "open" system. A manufacturer has the ability to use open or standard protocols, yet create a closed system, thus generally continuing a building owner's dependence on a single manufacturer. The job of policing compliance with a standard protocol by the manufacturer falls to the designer.

Today, the most widely accepted and applied standard protocol used for commercial controls is BACnet, an open protocol published by ASHRAE, defined by ASHRAE Standard 135, *BACnet—A Data Communication Protocol for Building Automation and Control Networks*.

BACnet has seen rapid acceptance because it can be "scaled" to any size application. BACnet can be easily applied at the device level to create an interoperable subnetwork of intelligent systems. Another key factor to BACnet's growth is the implementation of reliable compatibility testing to ensure that BACnet devices are truly fully compliant with the standard. The BACnet Manufacturers Association (BMA) operates BACnet Testing Laboratories (BTL), which tests the BACnet functionality of any product submitted by a manufacturer in accordance with the ASHRAE Standard 135.1, *Method of Test for Conformance to BACnet*. Products that pass these tests are "listed" by BTL.

The primary communication protocol for BACnet is BACnet/IP with operates over an Ethernet TCP/IP physical network and allows BACnet devices to communicate directly with each other, over the Internet, or over an owner's Intranet. Web servers, connected to the Internet and the BACnet/IP LAN, are configured to receive information from BACnet devices and subnetworks and present them in a form that can be viewed and operated from standard Web browsers.

13.5 HVAC Sequences of Operation

There are about as many sequences of operation for each type of HVAC system and its components as there are design engineers. But, since about 2010, there has been an attempt to create "standard" sequences that can be applied by designers. The first of these efforts is Ref. 1, and at the time this book is being written, ASHRAE is working to finalize this draft. However, the draft guideline can be used as a good starting point for the designer to develop sequences for a specific project.

Historically, details of controls have been defined on the drawings produced as part of the construction documents for a project. However, since about 2015, the trend has been to include control sequences as part of the specifications and the Construction Specifications Institute as assigned Section 23 09 93 for that role.

Control sequences of operation are typically presented in "narrative" format since experience has proven that this format is better understood by HVAC designers to describe their design intent, for owners and contractors to understand how each HVAC system is to operate, and for control vendors to determine DDC system requirements. They are presented as individual sequences for each HVAC system "element" and, thus, it is required that the HVAC designer assemble these "element" control sequences to define the complete control requirements for each specific HVAC system.

Interface to original equipment manufacturer (OEM) controls is often required as part of a sequence of operation, mostly with packaged DX systems, chillers, and boilers. For packaged DX systems, interface can typically be accomplished at one of four levels:

1. For small units (5 tons or less) that are CAV type without an airside economizer, a thermostatic interface is preferred since that provides low cost, almost seamless connectivity to the DDC system.
2. As the units of 7.5–20 tons capacity become complicated with variable air volume (VAV) operation and/or an airside economizer, some variety of a basic OEM controller is usually provided. The typical basic OEM controller (e.g., Trane's "ReliaTel") can handle most comfort applications well, though it may have some limitations where high limit humidity control or more sophisticated ventilation airflow control is required. More ever, this level of controller typically does not work satisfactorily in critical environments such as hospitals and research facilities. Interface with these controllers is limited and must be carefully evaluated by the designer.
3. For large units (25+ tons) serving higher quality commercial facilities or hospitals, research laboratories, and so on, a fully configurable OEM controller is required. Since these types of applications require higher quality, higher performance packaged systems (Trane's "Intellipak," for example), the typical OEM controller provided will normally meet most design requirements. These controllers are available as native BACnet devices, so interface with the DDC system is typically not difficult if the designer takes care to require I/O point types that are compatible with the controller.
4. For units required to deal with 30%–100% outdoor air for ventilation, a fully configurable OEM controller is required, no matter what its capacity is. Since DX equipment has limitations relative to normal SHRs that can be address, the operation of these units with high outdoor airflow is very complex and must use the OEM controller's operating sequence.

Interface with chiller OEM controls today is less difficult than in the past. A fully configurable OEM controller is required and typical OEM controller provided will normally meet most design requirements. These controllers are available as native BACnet devices, so interface with the DDC system is typically not difficult if the designer takes care to require I/O point types that are compatible with the controller.

Interface for control of conventional boilers is not a problem and can generally be addressed through use of a fully programmable controller provided with the DDC system. Condensing boilers are more difficult to interface (see Chapter 12), simply because most manufacturers provide OEM controllers with very specialized firmware operating with ModBus communications. *Thus, the DDC designer's only workable option is to simply interface with the limited number of I/O points made available by the OEM and leave it to the OEM controller to control operation of the boiler(s) based on the design piping configuration.*

Finally, it is recommended that the designer include a "schematic" of each HVAC system and subsystem on the drawings and show the location of each required I/O point and final control device in order to define not only the point itself but where each input or output device is located within each HVAC system.

13.6 DDC System Security

Up until about 2000, all DDC systems were essentially "stand-alone" systems, the access isolated to the internal network supporting their communication. But, now, most DDC networks are connected to the Internet.

While this connection offers many benefits, such as ease of access and use by the facility staff, it opens the network to "hacking" and other potential security issues. This has already happened, with systems being maliciously hacked and locked by "ransomware."

In recent years, Wi-Fi, the wireless LAN standardized by IEEE 802.11, has been used instead of (or in addition to) Ethernet in many installations. *However, the use of Wi-Fi (utilizing so-called "wireless" sensors and actuators) as part of the DDC system network expands this risk geometrically and should be avoided.*

Guidelines for DDC system security are not yet well developed, though ASHRAE has been routinely updating ASHRAE *Standard 135* to improve the security of BACnet networks. Thus, it falls to the HVAC designer to review with DDC system vendors (and with the owner's IT department if the DDC systems connects to the facility WAN) to ensure that these systems are protected as well as possible in accordance with the general guidelines provided here.

Where the DDC system uses a Web interface, that interface can be combined with either a *virtual private network* (VPN) or a remote desktop solution. The combination of either approach adds the protection of data hiding through encryption and robust security through a commercial security solution.

Combining a VPN with a Web server interface provides the benefit of restricting the applications that can interact with the DDC system. The VPN provides a secure connection to the Web server taking away the reliance on the Web server's security implementation for user authentication. Having the Web server installed on the only computer that has access to the DDC system network ensures that no other programs are using the VPN to access the control system. This reduces the risk that rogue programs on the user's computer can attack the DDC system.

Combining a remote desktop solution with a Web server does not provide any more protection than using desktop solution with a non-Web-based interface. In either case, the effort to secure the DDC system is reduced through the single computer being allowed access to the control system. Is the IT network providing security? Or, is the control network open to anyone who can get inside the building and make a physical connection?

If the DDC network is segregated from the IT network, there are still threats. Are there any computers directly connected to the control network? Are there any networked controllers that are in the user space (networked thermostats, supervisory controllers in unlocked closets, and so on?) Is there any equipment that is outside the building, such as roof-top units? Are there network controllers attached to the exposed equipment?

If an attacker can get access to a controller, the attacker can probably take over the controller's connection to the network.

Segregate the control network from the IT network. Segregation from the IT network reduces the attack points for a control network. To start, it means that users on the IT network do not have carte blanche access to the control network. It also ensures that slipups in security on the IT network have a reduced chance of impacting the control network.

Segregation can be achieved through IT technology (virtual local area networks and managed switches) or through physical separation. While some costs can be reduced by using the same infrastructure, using a completely separate network also has advantages. The controls industry seems to be split when it comes to whether to share IT infrastructure or to provide a completely separate physical network. *From a security perspective, providing completely separate infrastructure can be the better choice.*

With seemingly constant changes to IT networks, each change is a source for the introduction of security holes. It is not uncommon for IT personnel to forget that there is even a control network installed (until their equipment loses its AC) and that changes they make can cripple a control system network. Having a completely separate infrastructure reduces the probability that the day-to-day changes made to the IT network will negatively impact the control network.

Once segregated, the connections between the corporate network and the control network can be limited and protected with firewalls.

If there are connection points to the control network that are not used, or which are provided solely for service tools, ensure they are turned off when not in use. Do not make it easy for an attacker to connect into the control network by leaving a spot on the wall he can plug into.

Use simple sensor networks for controllers in the user space or apply network security on sensor networks. Any equipment that is in the user space (i.e., in the occupied portion of the building instead of in mechanical rooms or locked in cabinets) should be separated from the main control network.

The most common ASC found in the user space is a *smart thermostat*. An attacker can pull it off the wall and plug in a computer to gain access to the control network. The simplest method to protect against stealing a smart sensor's network connection is to use a simple sensor network instead of more powerful network or use "dumb" sensors in the space and leverage the intelligence of supervisory controllers.

Remove default usernames and passwords: Most control system products come with default usernames and passwords to allow for initial system configuration. These default login credentials should be removed.

Ensure the use of strong passwords: Since the IT department is not usually responsible for software on the control network, frequently IT password policies are not enforced on control network. While no users enjoy having the change in their passwords, or having to remember yet another new password, if a strict password policy is required to protect the IT network, why would it not also be in place to protect the control network?

Ensure that control system security patches are up to date: Any control system product reachable from the IT network should have security patches installed. Many control system products are based on common operating systems, Web server applications, and so on. Security patches for these products should be field tested and then installed.

Install tamper alarms on all equipment cases: For the really paranoid who have to protect against malicious insiders with access to roam about the facility and into mechanical rooms, the installation of tamper alarms on all equipment cases can provide an indication not only when someone might be physically damaging equipment, the tamper alarms also provide a warning that someone might be attempting to take over the network connection of a device.

13.7 DDC System Design Checklist

DDC system design is a relatively straightforward process:

1. Review the HVAC design to initially judge if it can meet the building's/owner's requirements for controllability and function, for example, zoning, part- and low-load operation, and areas of varying operating schedules. *Most commonly, poor control results not from poor controls, but from poor HVAC system design.*
2. Review the HVAC equipment specifications to determine what controls/safeties are specified with each piece of HVAC equipment (AHUs, packaged units, chillers, VFDs, packaged pumping sets, and so on). It may be obvious that some of these controls or safeties should be eliminated and instead provided by the DDC system. OEM or vendor controls may require integration into the DDC system or simply be supplanted by the DDC system. OEM controllers or gateways must be carefully evaluated at the submittal review stage of the project!
3. Develop control point definitions and operational sequences for each project in a top-down fashion. It is recommended that designers never use control sequences and schematics from a previous project so as not to repeat past errors or omissions.

   Determine all HVAC systems, subsystems, and components to be controlled. Determine the owner's specific intent for systems operation, set points, and so on. Keep the input and output points to the minimum number required for HVAC control and efficient building operation… follow the keep it simple, stupid (*KISS*) principle.

   Create a schematic control diagram of each system that includes all components to be controlled (e.g., fan/pump motors, and valves/dampers).

   As part of the sequences of operation, define the modes of operation for each component (e.g., on/off, occupied/unoccupied, heating/cooling, and dehumidification). Alternatively, if an operation mode affects multiple components directly, then these modes may be defined as separate items.

   Define failure modes. Failure can occur from a global aspect such as building power loss and control system failure or can be more localized due to the failure of as specific component such as a fan, pump, and chiller. For global failures, there is very little to be done except to ensure that the HVAC systems restart automatically when power or control is restored. For localized failures, failure mode definition can be complex. For example, if the chiller fails and cooling is not available, what should a VAV fan speed controller do? In this same case, should VAV terminal units fail open, fail closed, or simply remain where they are? What should happen if the campus steam supply fails during the winter? Are there specific control modes required for smoke or fume venting to be initiated by an alarm condition?

4. Include the following elements in the contract documents on the drawings:
   a. Define the overall DDC system architecture and clearly indicate whether an independent DDC LAN or owner's LAN/WAN is utilized to be for communications.
   b. For each HVAC system... air-handling system, chilled water system, steam system, hot water system, and so on... there must be a system schematic or diagram showing all system components and the location of each control sensor and actuator that is required.
   c. On the HVAC floor plans, show the location of key control system elements such as panels, workstations, major cable and raceway routes, and the sources of power. Locate duct SP sensors, piping differential pressure sensors, flow sensors, flow monitors, and so on.

Bibliography

Guideline 36P, High Performance Sequences of Operation for HVAC Systems, BSR/ASHRAE (Atlanta, GA, June 2016).

"Securing a Control Systems Network," Carl Nielson (*BacNET Today and the Smart Grid, a Supplement to the ASHRAE Journal*, November 2013, Pgs. B18–B22).

Standard 135, BACnet - A Data Communication Protocol for Building Automation and Control Networks, ASHRAE (Atlanta, GA, 2016).

Guideline 13, Specifying Building Automation Systems, ASHRAE (Atlanta, GA, 2015).

Section IV

Special Considerations

14

Special HVAC Design Considerations

14.1 Indoor Air Quality

Trying to define *indoor air quality* (IAQ) [sometimes referred to as *indoor environmental quality*] is somewhat akin to the Supreme Court defining pornography…we know what it isn't, even if we don't know exactly what it is. American Society of Heating, Refrigerating, and Air-Conditioning Engineers (ASHRAE) *Standard 62.1* defines "acceptable indoor air quality" as "air in which there are no known contaminants at harmful concentrations as determined by cognizant authorities and with which a substantial majority (80% or more) of the people exposed do not express dissatisfaction." The problem with this definition is simply who wants to live with the 20% of the building occupants who do express dissatisfaction!

In broad terms, experience shows that "good" IAQ results from the following:

1. *Temperature and humidity control* by the building heating, ventilating, and air-conditioning (HVAC) systems within the ranges of 70°F–77°F [21°C–25°C] and 30%–60% relative humidity (RH) (see Chapters 3, 4, 6, and 10). If building occupants are too hot or too cold, they are "uncomfortable" and may become more sensitive to other IAQ issues. Therefore, it is important that indoor temperatures be maintained with a comfort range, as defined by ASHRAE *Standard 55*. Unfortunately, this standard is complex and requires that design engineers take into account occupant clothing, air currents, and other factors that are far outside of the normal HVAC designer's realm.

   Thus, we fall back on experience to tell us that indoor temperatures of 70°F–73°F [21°C–23°C] in winter and 73°F–77°F [23°C–25°C] in summer, assuming acceptable air motion and humidity condition, will result in "comfort" for most occupants. Experience shows that women tend to be more comfortable at the upper ends of these rangers, while men seem to like the lower temperatures.

   Based on numerous studies, we know that when the indoor humidity falls below 30% RH or rises to above 60% RH, each for extended periods, the probability of occupant issues with viruses, fungi, respiratory infections, allergic rhinitis, asthma, static electricity (on the low end of this range), and so on increase significantly.

2. *Adequate dilution ventilation* with outdoor air to maintain occupant effluents (CO_2 and odor) to acceptable levels (see Chapter 10).

   IAQ is dependent on outdoor air quality since outdoor air is assumed to be "clean" and acceptable for dilution ventilation. However, in some locations, during at least part of the year (usually in summer), ambient air conditions are seriously degraded by vehicles, power plants, and/or adverse atmospheric pressure conditions. For industrial buildings, or buildings located in industrial areas, the ambient outdoor air may be contaminated with particulates and/or gaseous pollutants. This condition can exist even outside of industrial areas if buildings are located close to vehicle traffic, have loading docks, or even use combustion engine powered lawn equipment...all of which produce a wide range of pollutants that can be "sucked in" through outdoor air intakes. Finally, with climate change, global ambient CO_2 levels are increasing and have risen already to 407 ppm as of mid-2017 (from 300 ppm in 1950). All of these factors must be considered by the HVAC designer to determine if air cleaning of the outdoor air is required entering the HVAC system in order to ensure good IAQ.

3. *Adequate lighting.* Poor lighting can be an irritant that exacerbates other IAQ problems and can produce its own health risks in the form of headaches and eye strain. While lighting design is beyond the scope of this text, the electrical designer must pay careful attention to the brightness, glare, and quality of light as required for good lighting.
4. *Low noise and vibration levels.*
5. *Ongoing monitoring and maintenance to avoid introducing pollutants or contaminants* into the indoor air from either indoor or outdoor sources.

Poor IAQ, real or perceived, can be the result of many factors:

1. As energy costs have risen and the emphasis on energy cost savings have assumed importance with building owners and managers, HVAC systems have been modified to reduce ventilation airflows under certain conditions, to eliminate the use of reheat for humidity control, to be started as late as possible and shutdown as soon as possible, to broaden temperature control ranges, and so on. All of these factors can degrade IAQ.
2. Many buildings are now designed with better envelope air barriers that reduce infiltration airflows and the resulting *secondary ventilation*; "least first cost" based design and/or poor construction has cut corners, resulting in increased water intrusion, poor lighting, and noise or vibration problems, along with HVAC systems that cannot satisfy both temperature and humidity control needs, particularly in hot, humid climates; and the use of more and more synthetic or manufactured building materials that have potential for releasing pollutants and contaminants as they age.
3. Buildings change in response to changing business or occupant needs, or simply need to be updated as they age, and renovation of existing buildings is common. This renovation means that new building products and new building systems can regularly introduce new (or exacerbate existing) IAQ problems.

People factors also enter into the IAQ equation.

1. Building occupants are regularly exposed to news coverage on indoor environmental issues, including asbestos, polychlorinated biphenyls, lead paint, and mold. Problem buildings, such as the Environmental Protection Agency (EPA) headquarters building in Washington, DC, a state office building in California, and dormitories at North Carolina Central University, Durham, NC, resulted in

well-publicized IAQ problems that raised public awareness. This awareness, as one author states, "…has raised the expectations of occupants regarding the quality of the air in their personal space."

2. And, there are some individuals who are hypersensitive to fungi, volatile organic compounds, and other contaminants or pollutants that require very special indoor conditions control and addressing these individuals' concerns may require implementation of management solutions in lieu of technical ones.

14.2 Antiterrorism Design for HVAC Systems

14.2.1 Introduction

The term *terrorism* is defined in the United States by the Code of Federal Regulations as "...the unlawful use of force and violence against persons or property to intimidate or coerce a government, the civilian population, or any segment thereof, in furtherance of political or social objectives."

The FBI further describes terrorism as either domestic or international, depending on the origin, base, and objectives of the terrorists:

1. Domestic terrorism is the unlawful use, or threatened use, of force or violence by a group or individual based and operating entirely within the United States or its territories without foreign direction, committed against persons or property to intimidate or coerce a government, the civilian population, or any segment thereof, in furtherance of political or social objectives.
2. International terrorism involves violent acts dangerous to human life that are a violation of the criminal laws of the United States or any state or that would be a criminal violation if committed within the jurisdiction of the United States or any state. International terrorist acts occur outside the United States or transcend national boundaries in terms of the means by which they are accomplished, the persons they appear intended to coerce or intimidate, or the locale in which the perpetrators operate or seek asylum.

The U.S. State Department defines terrorism as "premeditated, politically motivated violence perpetrated against noncombatant targets by subnational groups or clandestine agents, usually intended to influence an audience."

Terrorists can be "domestic" or "international" and the FBI considers "ecoterrorism" and "animal terrorism" the country's leading domestic terror threats. For the better part of a decade, the Earth Liberation Front (ELF) and Animal Liberation Front (ALF) were seen not as antisocial thugs but idealistic young kids. After all, the public generally supported their goals: a clean environment and compassion toward animals. *Then came September 11, 2001.* For the ELF and ALF, that was the day they took joint credit for firebombing a McDonald's in Tucson, Arizona three days before.

More widely recognized domestic terrorists include right-wing anti-government, radical religious, and white supremacist organizations and their "lone wolf" followers that spawned the bombers of the Federal office building in Oklahoma City in 1995, the "Unibomber," and school shootings in more recent years.

International terrorists are recognized primarily because of the World Trade Center and Pentagon attacks, also on September 11, 2001, by anti-American Muslim fundamentalists. The 9/11 report issued in July 2004 clearly defines Islamist extremists as the primary international terrorists targeting the United States.

Without a doubt, the architects and engineers who designed the Federal office building in Oklahoma City, the World Trade Center, the Pentagon, and certainly the McDonald's in Tucson did not envision that these facilities would be attacked by terrorists and did not include "antiterrorism" as a building design criteria. *Today, however, terrorism directed toward buildings is common enough that every designer must consider design elements that help mitigate these attacks.*

The facilities that are "desirable" targets for terrorist attack include airports and other transportation centers, government buildings (especially those with high traffic and symbolic recognition, including courthouses, jails, police stations, and post offices), schools, stadiums and arenas, shopping centers and restaurants, and symbolic structures (e.g., Statue of Liberty, Golden Gate Bridge).

14.2.2 Terror Threats against Buildings

Buildings are susceptible to two primary types of terrorist threats, *explosives* and *chemical, biological, and/or radiological* (CBR) weapons. Comprehensive protection against the full range of possible threats is prohibitively costly and probably not even possible. However, a level of protection that reduces the risk of mass casualties resulting from terrorist attacks can be provided for all occupants of the buildings we design.

Explosive materials are designed to release a large amount of energy in a very short time. Part of the energy is released as heat and part as shock waves that travel through the air and the ground. The shock wave (air blast) radiates at supersonic speed in all directions from the explosive source, diminishing in intensity as the distance from the source increases. These waves reflect off the ground, adjacent structures, and other surfaces, reinforcing the intensity of the blast's effects. These reflections are most pronounced in dense urban environments where neighboring structures can create a canyon effect or where pockets of blast energy can become trapped in reentrant (concave) corners. Buildings experience the effects of explosions in several stages. The initial blast wave typically shatters windows and causes other damage to the building facade. Then, the blast wave enters the building and exerts pressure on the structure. Air-blast pressures within a building can actually increase as the pressure waves reflect from surfaces and can cause injuries to the occupants directly by means of physical translation, ear, lung, and other organ damage, or debris from building elements and contents. Finally, the building frame is loaded globally and responds as it would to a short-duration, high-intensity earthquake.

People exposed to explosions can be killed or injured by the intense heat and pressure generated at the site of the detonation, where temperatures can range up to 4,000°C [7,200°F] and pressures to several hundred times atmospheric pressure.

Once the blast wave enters the occupied spaces or, in an indoor explosion, the pressure wave is reflected off walls, floors, and ceiling, forming in effect a series of pressure pulses. The response of the ears and lungs to repetitive pulses is similar to the response to a single long pulse. Because the damaging effects of blast pressure indoors often exceed those of an unconfined explosion of similar size, it is crucial to minimize the opportunity for blast pressures from outdoor explosions to enter occupied spaces and to protect such spaces from even small explosive devices.

For people within structures subjected to blast effects, penetration by glass fragments and impact by other blast-induced debris have been consistent causes of death and serious injury. People are also subject to blunt trauma from furniture, accessories, and nonstructural building components such as overhead lighting and ductwork that become detached from their moorings. Smoke and inhalation of dust also cause blast-induced injury.

CBR threats consist of CBR weapons that can be introduced into a building by any one of the following three scenarios for their delivery.

1. Large-scale, external, airborne release targeting a broad area
2. External release close to and targeted at a specific building
3. Internal release, typically delivered by mail or parcel service

14.2.3 HVAC Design to Minimize Explosive Threats

In the event of an explosion directed at a high occupancy building, the primary design objective is to protect the people by preventing building collapse and, for the HVAC designer, limiting injuries due to flying debris and the direct effects of the blast pressure wave entering the building. Next, the HVAC design should facilitate building evacuation and rescue efforts.

The key concepts for providing secure and effective mechanical systems in buildings is (1) separation, (2) hardening, and (3) redundancy:

1. *Separation* begins with keeping critical mechanical systems as far away from high-threat areas of the building, including lobbies, loading docks, mail rooms, garages, and retail spaces…anywhere an explosive device can be delivered and detonated easily. To help meet this goal, mechanical equipment rooms should be constructed of solid masonry walls (or at least reinforced concrete masonry unit properly anchored to the structural system) and have secured access.

   Provide positive pressurization of stairwells and vestibules to aid occupant emergency egress.

   Air systems should be designed to operate in a "smoke evacuation mode." This mode can be initiated by a signal from the fire alarm system, but the design should also include an emergency manual selector switch so that authorities can operate the systems in this mode or simply shut them down. In the smoke evacuation mode, each system's outdoor damper and relief air damper must open fully and the return damper close. If the system is the variable air volume type, the fans should be indexed to provide maximum airflow. All building exhaust systems should continue to operate while in this mode.
2. *Physical "hardening"* or protection of critical piping, ductwork, electrical power, and control wiring conduit, and so on will help ensure the survival of these systems from direct blast effects. Route piping, conduit, and ductwork so they are not located in or on exterior walls and avoid close proximity to high-threat areas of the building. Support piping, conduit, and ductwork to meet the seismic design requirements for Seismic Design Category D.

   Outdoor air intakes should be designed to eliminate a direct, linear path between the outdoors and critical HVAC mechanical equipment to mitigate blast pressure wave effects. This can be accomplished by designing (1) outdoor intake plenums that are constructed of concrete masonry and (2) connections to these plenums at 90° from the intake airflow direction.

3. *Redundant critical systems and controls,* installed in widely separated locations, increase the likelihood that these systems will remain in operation during the rescue period.

14.2.4 HVAC Design to Minimize CBR Threats

HVAC systems design elements required to address CBR threats fall into three categories: (1) reduce the potential that airborne CBR agents can enter the building, (2) zoning and HVAC systems control to limit the air distribution of CBR agents that may be released internally, and (3) air cleaning to reduce CBR concentration levels:

1. Airborne CBR agents can enter the building from the outdoors by being released into the air close to a building entrance or into the outdoor air intake of an HVAC system. To reduce the potential for these types of attacks, the HVAC design should be such that (a) the building is maintained at a positive pressure and (b) outdoor air intakes are protected. High-threat areas in a building (lobbies, loading docks, mail rooms, garages, and retail spaces) should be maintained at a negative pressure relative to the rest of the building, but at a positive pressure relative to the outdoors, to prevent the spread of internally released CBR agents. Return air intakes represent another potential access point for introducing CBR agents.

   Outdoor air intakes, as illustrated by Figure 14-1, should be located on a secure roof or high sidewall. The lowest edge of the outdoor air intakes should be placed at least 12 ft [4 m] above the ground or above any nearby accessible level (i.e., adjacent retaining walls, loading docks, handrails, and so on). Also, the entrance to the intake should be covered with a sloped metal mesh or louver to reduce the threat of objects being tossed into the intake. A minimum slope of 45° is generally adequate.

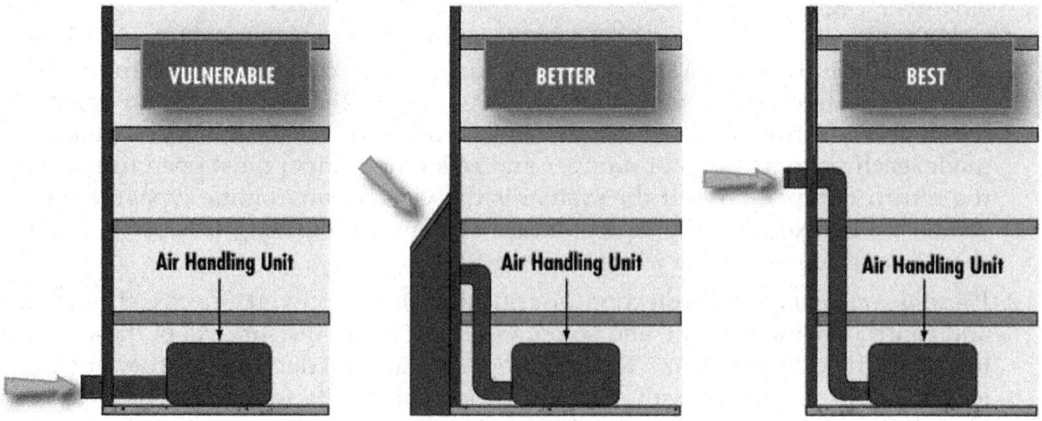

FIGURE 14-1
Outdoor air intake protection against terrorist attack. [Source: Public Domain (Federal Emergency Management Administration).]

HVAC return air grilles should always be located in areas that are inaccessible or are under security surveillance. Return air plenums in public areas or high-threat areas should not be interconnected with return plenums in the remainder of the building.

Ducted return systems offer limited access points for introducing CBR agents, while any point in the ceiling of a ceiling return plenum can be used for access. *Therefore, ducted return systems are recommended.* Even, then, the return system should be designed so that it is not shared by multiple air-handling systems.

2. Certain areas within a building are more likely to be the entry points for CBR agents, including lobbies, mailrooms, and other receiving areas, along with public spaces, including retail. These areas should have HVAC air systems that are totally independent from the rest of the building and from other high-threat areas.

   Theoretically, automatic CBR agent detectors can be used to initiate protective actions such as shutdown of ventilation systems, closing outside air intakes, or turning on filtration systems. Current biological detection technology requires a minimum delay of approximately 15 min to detect the presence of biological agents. Practical application of chemical detection is limited by shortcomings in response time, false alarms, broad-spectrum capability, maintenance requirements, cost, and the quantity of sensors needed at air intake locations. *Thus, application of detectors for terrorist threats should be limited to the following uses: first entry determination by first responders, monitoring casualties before medical treatment, determining the extent of the hazard, and determining when protective measures are no longer required.*

3. Particulate filtration, as discussed in Chapter 10, is the first step in CBR agent cleaning. High-efficiency particulate air (HEPA) filters are capable of removing from 99.97% (minimum efficiency reporting value (MERV) 17) to 99.9999% (MERV 20) of particle sizes between 0.3 and 1.0 μm.

   Under normal circumstances, biological agents, except for viruses, and toxins can be collected using appropriately selected particulate filters.

   For radiological agents, air-cleaning devices would be ineffective at stopping any blast and radiation itself; however, they would be useful in collecting the material from which the radiation is being emitted. Micrometer-sized aerosols from a radiological event are effectively removed from airstreams by HEPA filters. This collection could prevent distribution throughout a building; however, subsequent decontamination of the HVAC system would be required.

   For chemical agents, including industrial toxic chemicals, gas-phase air cleaning is required. Blister and nerve agents are strongly adsorbed by activated carbon. Blood and choking agents are not strongly retained by activated carbon, but additives, such as metal oxides and other reactants found in the U.S. military carbon ASZM-TEDA, may be used in the sorbent to degrade the hazard.

   Before selecting an air-cleaning strategy, the designer must evaluate potential CBR threats, ventilation/filtration requirements, and IAQ. The answers to these questions will guide in making decisions about what types of filters and sorbents (if any) should be installed in an HVAC system, how efficient those filters and/or sorbents must be, and what procedures must be developed to maintain them after installation.

14.3 HVAC Water Treatment

The HVAC designer must become knowledgeable of the water treatment problems and the requirements for water treatment programs to solve these problems. This knowledge forms the basis of design for water treatment systems.

14.3.1 Hot and Chilled Water Systems Chemical Treatment

Chilled and hot water systems are *closed recirculating water* systems. For water treatment, closed systems have many advantages:

1. There is no loss of water in the system (except when a leak occurs) and, thus, no need for makeup water. Therefore, deposition or scaling is not a problem.
2. Once filled and entrained air removed, the closed system creates an anaerobic environment that eliminates biological fouling as a problem.
3. Closed systems reduce corrosion problems because the water is not continuously saturated with oxygen (an oxidizer) as in open systems. The low temperatures common to HVAC systems further reduces the potential for corrosion.

Water treatment to eliminate the potential for corrosion is required only when the water system is initially filled with water or when it is drained and refilled due to maintenance, repair, or modification.

Corrosion in a closed system can occur due to oxygen pitting, galvanic action, and/or crevice attack. To prevent these conditions, the "shot feed" method of chemical treatment is used. With this method, a bypass chemical feeder is used to add treatment chemicals to the system in a one-time "shot" just after the system is filled with water. For the mixed metallurgy (steel and copper) systems typical for HVAC water systems, a molybdate corrosion inhibitor is best, with treatment limits of 200–300 ppm recommended.

14.3.2 Evaporative Cooling Water Systems Chemical Treatment

Evaporative cooling water systems are open to atmospheric contamination and the water treatment issue becomes much more involved. *A specialized water treatment program is required to control deposition (scaling), corrosion, and microbiological fouling of evaporative coolers (cooling towers, closed circuit industrial coolers, or other spray coil systems), along with associated heat exchangers and piping.*

As a portion of the evaporative cooling water flow is lost by evaporation (approximately 0.1% of the flow rate per degree of temperature range) through the evaporative cooler, the concentration of dissolved solids increases since solids are left behind as the liquid evaporates. Makeup water, which is added to the evaporative cooling water system to offset the evaporation losses, will add dissolved minerals at a lower concentration level than the evaporative cooling water and, thus, some equilibrium concentration level will be maintained. However, if this equilibrium concentration level is high enough that deposition can occur, a program to control solids' concentration at a lower level is required.

The concentration of dissolved solids can be reduced by adding more makeup water, which has a lower solids concentration, to the evaporative cooling water, which has a higher concentration. However, to add water to the system, an equal amount must be removed from the system by *blowdown*, the intentional "dumping" of evaporative cooling water to drain.

For corrosion control, evaporative cooling water pH should be maintained between 4 and 10, and in most evaporative cooler water treatment programs, the desirable range for pH is between 8 and 9 in order to maintain water alkalinity to a reasonable level (400 ppm or less). However, in sheetmetal evaporative coolers, a pH of 7.0–8.0 is preferred to help prevent white rust corrosion. Therefore, at least for most metal evaporative coolers, the ideal pH range is 7.5–8.5.

Water hardness and alkalinity are a function of the hardness and alkalinity of makeup water, the amount of evaporation and drift loss from the evaporative cooler operation, and the blowdown proposed to yield the desirable pH to prevent both deposition and corrosion. The term *cycles of concentration* defines the ratio of the desired concentration of dissolved solids in the evaporative cooling water to the concentration of dissolved solids in the makeup water. Cycles (of concentration) is equal to the concentration of dissolved solids (ppm) in the basin water, that is, the point of blowdown, divided by the concentration of dissolved solids (ppm) in the makeup water. This relation can be expressed in terms of water flow by Eq. (14-1)

$$\text{Cycles}(\text{of concentration}) = \text{MU}/\text{BD} \tag{14-1}$$

where

MU = total makeup water flow, which is the sum of evaporation + blowdown, gpm [L/s]

BD = blowdown flow, gpm [L/s]

MU in this equation can be replaced with the value (E+BD), where E is the evaporation rate, and rearranged to yield Eq. (14-2).

$$\text{BD} = E/(\text{Cycles} - 1) \tag{14-2}$$

Since drift water loss is not included in these calculations, the actual required BD flow can be reduced by the amount of drift loss from the tower (about 0.5%–1.0%)

There are two ways of controlling blowdown in an HVAC evaporative cooler system:

1. *Constant blowdown* with manual adjustment based on periodic water hardness analysis is the simplest method. However, since the amount of blowdown is constant, the loss in water and water treatment chemicals is high and this really represents the most expensive approach.
2. *Controlled blowdown* based on continuous monitoring of the water hardness as indicated by its conductivity. Automatic control minimizes the waste of water and water treatment chemicals and is the preferred method.

The amount of makeup water is reduced significantly as the number of cycles is increased from 2 to 6. However, there is only a further 5% reduction as the cycles are increased from 6 to 10, and only a further 2% reduction as cycles are increased to 20. *Therefore, in most evaporative cooler applications, cycles of concentration is maintained between 5 and 10 and deposition inhibitors are added as necessary.* While lower cycles represent loss of more water and treatment chemicals, the amount of treatment chemicals required tends to go down with cycles, and 5–10 cycles usually represent a good balance point.

In these cases, chemical treatment, that is, the addition of deposition inhibitors to the water system, becomes necessary to augment blowdown. The most common scale inhibitors used in evaporative cooling water systems are *phosphonates,* which are organic phosphate compounds, such as high-density polyethylene, which function by adsorption on the crystals as they form and prevent them from attaching to metal. Thus, these crystals precipitate out of solution, usually in the tower basin.

Corrosion is the loss of metal; it literally "dissolves." *General corrosion* is widespread and is caused, usually, by impurities in the metal or characteristics of the metal or its environment that results in an overall fouling of the metal surface. *Localized corrosion* results, mostly, from scratches, stress, or localized environment and is the most common reason for "metal failure." If dissimilar metals with different electrical potentials are used in an evaporative cooling water system, galvanic corrosion is enhanced and the metals simply corrode faster, particularly at and near the point(s) of contact between the metals.

The first step in corrosion control is to minimize the contact between water and mild steel materials. All primary wetted surfaces…wet decks for induced draft towers and all cooler basins… should be constructed of stainless steel. This typically increases tower costs but is generally a worthwhile investment. Use plastics or fiberglass for the tower casing, wet deck covers, intake louvers, drift eliminators, and fill. Finally, if mild steel is used for the tower structural frame, it should be galvanized *and coated with an epoxy or polymer final protective coating.*

However, the piping in most evaporative cooling water systems will be steel and must be protected from corrosion. This is accomplished by using one or more treatment programs, as follows:

1. *Passivating (Anodic) inhibitors.* These chemicals form a protective oxide film on the metal surface which is not only tough but, when damaged, quickly repairs itself. *Othophosphates, a common passivating inhibitor, should not be used in evaporative cooling water systems containing stainless steel since it will make the metal brittle over time.*
2. *Precipitating inhibitors.* These form complexes that are insoluble at the higher pH and, thus, precipitate out of the water. Zinc is a good precipitating inhibitor. Molybdate will also act as a precipitating inhibitor and, thus, can serve as a corrosion inhibitor using two mechanisms.
3. *Adsorption inhibitors.* These are organic compounds containing nitrogen, such as amines, or sulfur or hydroxyl groups. Due to the shape, size, orientation, and electrical charge of the molecule, they will attach to the surface of the metal, preventing corrosion. Their drawback is that they form thick, oily surface films that reduce heat transfer capability.

Biological fouling results from bacteria, fungi, zooplankton, and phytoplankton or algae introduced through makeup water or filtered from the air passing through an HVAC evaporative cooler. "Fouling" results when these micoorganisms grow in open systems rich in oxygen (an aerobic process) and form slime on the surfaces of the tower, piping, and heat transfer surfaces of the evaporative cooling water system. Slime is an aggregate of both biological and nonbiological materials. The biological component, called the *biofilm,* consists of microbial cells and their byproducts. The nonbiological components consist of organic and/or inorganic debris in the water that has become adsorbed or imbedded in the biofilm layer.

The number one method of controlling biological fouling is to keep evaporative coolers clean. At least twice during the cooling season, the basin should be drained, scrub cleaned, and allowed to fully air-dry before refilling. Then, the use of chemical treatment will complete the control chore.

There are two kinds of antimicrobial chemicals or *biocides* used in evaporative cooler water treatment programs to control biological fouling: oxidizing and nonoxidizing.

1. *Oxidizing chemicals* include chlorine, bromine, and ozone that *oxidize* or accept electrons from other chemical compounds. (**Warning!** *Oxidizing chemicals, particularly chlorine, can react with steel, including stainless steel, and cause rapid corrosion. To prevent this, concentrations of these chemicals must be kept low, ideally to less than 0.7 ppm. Oxidizing chemicals must be introduced into the evaporative cooling water system in a way to be rapidly dispersed to prevent localized high concentrations.*)
2. *Nonoxidizing antimicrobials* attack cells and damage the cell membrane or the biochemical production or the use of energy by the cell, resulting in its death, and are sometimes referred to as "surface-active" biocides. Typical nonoxidizing biocides include isothiazolinones, glutaraldehyde, mechanical biological treatment, and polyquat.

Microbials in evaporative cooling water systems can become resistant to a single method of attack, or some microbials may be more or less immune to one type of attack. Therefore, it is recommended that both types of treatment chemicals be used (oxidizing and nonoxidizing), either blended together or in alternating treatment patterns, as indicated by periodic water testing results. The key to a successful biological treatment program is maintaining adequate chemical treatment levels at all times via continuous feed of antimicrobials into the evaporative cooling water system.

HVAC evaporative coolers are excellent "air washers." Pollen, dust, microbes, leaves, and other debris in the air are readily trapped and removed by the water and deposited in the basin to form "sludge," which can foul heat transfer surfaces. In the majority of cases, regular basin cleaning will control the problem, but in many urban and/or industrial areas, the amount of sludge formed during normal tower operation may be so great that specific control measures must be installed.

The most common general fouling control method is the use of *side-steam filtration*. Here, a portion of the evaporative cooling water flow is diverted through a filter for removal of dirt and suspended solids. Typically, filters are sized so that the entire water volume is filtered each hour. Thus, it necessary to determine the total system water volume contained in the tower basin and wet deck, evaporative cooling water piping, condenser(s), and so on and divide that volume, in gallons, by 60 min to establish the required filter flow rate, in gpm.

The most common (an economical) type of sidestream filter is the sand filter with backwash, much as used for swimming pool applications. These filters will remove suspended solids of 50 μm and larger particle size.

Two other water treatment concerns relative to evaporative coolers are *white rust* and *Legionella* (see Section 14.5.5 below). Since the late 1970s, there has been a significant increase in the use of galvanized steel evaporative coolers, particularly in the HVAC market. With the advent of new evaporative cooler water treatment polymers and corrosion inhibitors, most evaporative cooling water treatment programs are now utilizing little or no acid addition and operating pH levels have increased from 6.5–7.5 to as high as 9.5 (with the new alkaline polymer approach). As the technology of the water treatment industry has changed, the corrosion of galvanized steel has now become a major concern.

The term *white rust* refers to the premature, rapid loss of galvanized coating on evaporative cooler galvanized metal surfaces. White rust is evidenced by a white, waxy, nonprotective zinc corrosion deposit on wetted galvanized surfaces. This rapid loss of the galvanizing results in the corrosion of the underlying steel and, instead of tower systems that will last 20–25 years, equipment will have drastically shortened life spans. Based on current information, it is well established that white rust can form if the following conditions exist:

1. The galvanized coating is not properly "passivated" when the tower is placed in service. Passivation is a process that allows the zinc coating to develop a natural nonporous surface of basic zinc carbonate. This chemical barrier prevents rapid corrosion of the zinc coating from the environment, as well as from normal evaporative cooler operation. The basic zinc carbonate barrier will form on galvanized surfaces within 8 weeks of tower operation with water of neutral pH (6.5–8.0), calcium hardness of 100–300 ppm, and alkalinity of 100–300 ppm.
2. Evaporative cooling water is maintained at pH above 8.0.
3. High evaporative cooling water alkalinity (above 300 ppm).
4. Low evaporative cooling water calcium hardness level (below 100 ppm).
5. The lack of phosphate-based corrosion inhibitor in the evaporative cooling water treatment program.

For most galvanized metal HVAC evaporative coolers, and so on, white rust will occur if not prevented by the designer incorporating the following requirements in project design and/or specifications:

1. A secondary barrier coating on all wetted surfaces, such as epoxy or polymer finish. An even better approach is to specify new towers to have wetted surfaces such as basins and wet decks to be constructed of stainless steel. This option is normally available for only a 15%–20% cost increase.
2. Require that each galvanized metal cooler be properly passivated upon system start-up.
3. Require that a phosphate-based corrosion inhibitor be incorporated into the water treatment program, along proper dispersants.

14.3.3 Evaporative Cooling Water Systems Nonchemical Treatment

Since about 2000, with the increased emphasis on sustainable building design, owners and designers have been searching for nonchemical water treatment systems that perform as effectively as chemical-based systems. Results so far have been mixed.

Chemical-based evaporative cooling water treatment has been widely applied, researched, and documented for decades and is an effective way of treating all conditions associated with evaporative cooling water chemistry. Each of the major nonchemical water treatment methodologies, therefore, must be considered as only partially effective, primarily only for a specific type of water condition control and additional chemical treatment for other conditions will typically be required.

There are several nonchemical water system technologies that are available in the HVAC market, the most common being sidestream filtration, ozone generation, ultraviolet treatment, magnetic systems (both fixed magnets and electromagnets), pulse power (PP; sometimes called alternating current induction coils), and hydrodynamic cavitation (HDC)and ultrasonic cavitation (USC). Ozone and ultraviolet treatments are essentially limited to controlling microbiologic fouling and may be used to replace chemical biocides such as chlorine and bromine in the water system.

1. *Ozone* is an evaporative cooling water antimicrobial that eliminates the need for chemical treatment for microbiological fouling. Ozone (O_3) is an unstable form of oxygen (O_2) that has a relatively short half-life, usually less than 10 min. Ozone is a powerful biocide and virus deactivator and will oxidize many organic and inorganic compounds. Ozone treatment cannot be used with water with excessive hardness (500 ppm or higher calcium carbonate) or with sulfates greater than 100 ppm. Also, because of its short life, ozone should not be used in large systems or systems that have long piping runs that would require long residence times to get complete coverage.
2. *Ultraviolet light* is also used effectively as an antimicrobial treatment in evaporative cooling water systems.
3. Every few years, salesmen make the rounds to call on owners and designers and talk about the "wonders" of *magnetic treatment for deposition control* in evaporative cooling water systems. Their product consists of strap-on magnets installed around evaporative cooling water piping. Promoters of these devices claim that dissolved solids, which dissociate into charged ionic salts in solution, can be easily removed by allowing them to pass through a magnetic field. *Tests by independent sources have proved conclusively that magnetic treatment has no effect on deposition rates in evaporative cooling water systems.* Magnetic treatment is simply a scam and, when encountered, should be immediately rejected as a viable water treatment method.
4. *Sidestream filtration,* unless reverse osmosis filters are used, is limited to capturing larger particles that enter the water system from the atmosphere, including dust, pollen, and so on. The need for sidestream filtration can be readily identified by build-up of sludge or sand in the tower or cooler basin. Sidestream filtration is commonly used to augment nonchemical water treatment systems.
5. PP systems use induction coils powered by high-frequency AC currents that create a variable magnetic and electric fields in and around the water passing through. With PP systems, the primary method for deposition control remains blowdown, controlled by monitoring the water conductivity and maintaining it at a moderately low level. Blowdown is also used as the means of controlling pH of the water.

   While the science behind PP systems has not been well defined, manufacturers purport that dissolved particulates in the cooling water are stripped of their natural static charge by these fields and become "seed cores," the preferred attachment site for calcium carbonate dissolved in the water, completing treatment for deposition control.

   Sidestream filtration is generally required to capture suspended particles, both the larger particles that may enter the water basin from the outdoors and the ones precipitated during the PP process.

   Numerous comparison studies indicate that PP systems effectiveness for deposition control can be equivalent to chemical treatment systems. However, control of corrosion and biologic fouling has been found to vary widely between tests.

Some comparison studies indicate that the overall amount of deposition by PP systems is not less than the deposition within comparative chemical-treated systems. However, the PP system deposition is a "soft" material and easily cleaned by pressure washing, while chemical system deposition tends to be "hard" material, requiring more rigorous cleaning methods.

No peer-reviewed, scientific study of PP system effectiveness against corrosion has been conducted. While some comparison tests report that PP systems do inhibit corrosion, others indicate little or no effect on corrosion and, in fact, report the increased occurrence of "white rust." Thus, to be safe, PP systems should include supplemental treatment to address corrosion, primarily by control of pH via increased blowdown (i.e., reducing cycles of concentration) or by chemical means.

6. HDC and USC systems use flow velocity energy or high-frequency sound waves to create high and low pressure regions that result in water cavitation, that is, water "flashes" in to a gaseous state.

   HDC/USC systems also utilize blowdown to reduce water hardness (dissolved solids) and control pH. Sidestream filtration is generally required to capture suspended particles, both the larger particles that may enter the water basin from the outdoors and the ones precipitated during the HDC/USC process.

   HDC systems consist of a pressure equalizing chamber and a cavitation chamber. Water is pumped into the pressure equalizing chamber from both ends and then is channeled into the cavitation chamber through nozzles that both increase its velocity and imparts rotation to the water stream. A low pressure is created at the core of the rotating water stream that results in some of the flow "flashing" from a liquid to a gas state (steam). When the two water streams "collide" in the middle of the chamber, these steam bubbles collapse. This process causes calcium carbonate and other dissolved solids to precipitate out as water flashes into steam.

   This same process "flashing" process is produced by USC systems using high-strength ultrasound waves to propagate pressure waves within a cavitation chamber.

   No peer-reviewed, scientific study of HDC/USC system effectiveness against corrosion has been conducted. While some comparison tests report that HDC/USC systems do inhibit corrosion, others indicate little or no effect on corrosion and, in fact, reported increased occurrence of "white rust." Thus, to be safe, these systems should include supplemental treatment to address corrosion, primarily by control of pH via increased blowdown (i.e., reducing cycles of concentration) or by chemical means.

14.3.4 Steam and Condensate Systems Chemical Treatment

Lack of quality chemical-based water treatment can cause a wide range of boiler system failures:

1. Poor heat transfer and/or overheating and rupture due to deposits on tubes.
2. Corrosion failures due to oxygen pitting, chelant corrosion due to excess concentration of sodium salt over a period of time, caustic attack due to tube deposits in phosphate-treated boilers caused by caustic dissolving magnetite film, and/or acid attack due to poor pH control.

To avoid these potential problems in steam boiler systems, water treatment systems and programs must be implemented, as follows:

1. *Makeup Water Pretreatment.* The most common method of reducing water hardness and alkalinity is to "soften" the incoming make-water to reduce the concentration levels of dissolved sodium salts via a *zeolite water softening process.* This is an *ion exchange process* that uses "strong acid cation" resin to exchange calcium and magnesium ions for sodium ions. Regeneration of resin is done by treating with sodium chloride solution, then rinsing.

   Deaeration (DA), the next step in pretreatment, is designed to remove dissolved oxygen from makeup water producing localized corrosion (pitting) in piping and boiler tubes. Dissolved gases can be removed by lowering the pressure in the atmosphere contacting the liquid by "vacuum deaeration." However, this process is relatively inefficient so "pressure deaeration" is normally used for boiler feedwater pretreatment. With this process, feedwater is sprayed into a low pressure (typically 5 psig [34 kPa]) steam chamber where contact with steam heats it to within just a few degrees of the saturation (flash) temperature. As feedwater temperature is elevated, the oxygen solubility is reduced by 97%–98% and the dissolved oxygen dissociates. Oxygen and a small amount of steam are vented to the atmosphere. DA is normally required when more that 15% makeup is required (i.e., less than 85% condensate recovery). Most units are rated at 0.005 cm^3/l (7 ppb) of oxygen in feedwater. However, periodic feedwater testing should be done to ensure proper DA operation (2–4 times per year).
2. *Deposition Control.* Just as for evaporative coolers, *blowdown* is the primary method of deposition control for boilers. The amount of blowdown is a function of the quantity and quality (hardness) of makeup water.

   The blowdown requirement is determined by measuring boiler water electrical "conductance," a measure of the amount of conductive solids in the water (pure water has zero conductance). The recommended boiler water conductance is 3,500 M℧/cm or less (where, electrically, a "mho" is the conductive equivalent to an "ohm" of resistance).

   Two methods are used for boiler blowdown:

   a. Manual "bottom" blowdown consists of opening blowdown valves in accordance with an operating schedule dictated by periodic boiler water testing. This method removes both dissolved solids and sludge, but to be effective, frequent short blowdown periods are required. The boiler operator must monitor boiler water level during blowdown to prevent boiler operating problems.

   b. Automatic "top" blowdown can be intermittent or continuous, as determined by conductance monitoring and the water treatment control system. With this method, blowdown is taken from the highest water level, where dissolved solids concentrations tend to be higher, resulting in a more efficient process. Heat recovery from the wasted boiler water is possible and is recommended. Note that with automatic blowdown, some manual blowdown may still be required for sludge removal.

   The amount of boiler blowdown required is computed with Eq. (14-3).

$$\text{Required blowdown}(\%) = 100 \times (A/B) \tag{14-3}$$

TABLE 14-1

Desired Dissolved Solids Concentrations

Boiler Operating Pressure (psig)	Total Dissolved Solids (ppm)	Total Alkalinity (ppm)
0–50	2,500	500
51–300	3,500	700
301–345	3,000	600

where

A = actual dissolved solids concentration in the feedwater (mix of makeup and condensate).

B = desired dissolved solids concentration in the boiler water are recommended Table 14-1.

Additional deposition control can be provided with chemical treatment by phosphate or phosphate/polymer, chelant, or chelant/polymer additives if blowdown fails to provide the required levels of total dissolved solids and/or alkalinity in the boiler water.

3. *Corrosion Control.* Corrosion in boiler systems occurs as the result of up to three mechanisms:
   a. Galvanic corrosion occurs when a metal is coupled with another metal of different electrical potential (valence). This condition induces an electrical current and loss of electrons at the cation. Galvanic corrosion can be caused by metallic scale deposits, surface pitting, or scratches, and so on exposing different materials within the metal. This process is accelerated when two different metals are used, such as steel and copper.
   b. Caustic corrosion occurs in the presence of a concentration of caustic (such as NaOH) due to steam blanketing that allows salts to concentrate on surfaces and/or localized boiling caused by porous deposits on tube surfaces. The caustic dissolves the protective magnetite layer, causing a loss of the underlying metal.
   c. Acidic corrosion is caused by low feedwater pH and results in metal "thinning" (general corrosion) and/or local corrosion at bolts and other stress points.

   Corrosion in boiler systems is generally controlled by a combination of pH control (8.5–12.7 maximum range) and oxygen control (5–7 ppb) via DA and/or oxygen scavenging chemicals.
4. *Condensate System Corrosion.* Corrosion in condensate piping occurs due to two conditions:
   a. Oxygen "pitting," which results in localized loss of steel, occurs due to contact with air in atmospheric pressure pumped return systems. This problem can be treated by injection of an oxygen scavenger chemical into the condensate but is a difficult and expensive method of treatment.
   b. Acid corrosion occurs when carbon dioxide in air reacts with water to form carbonic acid that attacks steel. This corrosion is enhanced by decomposition of feedwater alkalinity that produces carbon dioxide. This problem can be treated by adding amines to feedwater to neutralize acids. Amines are introduced into the boiler water and "carry-over" into the condensate system with the steam.
5. *Boiler Fireside Problems.* Deposition and corrosion can occur on the fireside of boilers. Deposition occurs as deposits of fuel ash components on surfaces and may

require treatment by fuel additives to dilute deposits. Corrosive fuel ash components may deposit on the boiler surfaces. These deposits have low melting points and, when in a liquid state, attack metal surfaces. Treatment additives to boiler fuels can raise melting points of deposits (but make deposits harder to remove). The best way to avoid this problem is to burn "clean" fuels (natural gas, light oils, low-sulfur/low-ash heavy oils).

14.3.5 *Legionella* Risk Management in HVAC Water Systems

In 1976, 34 attendees at an American Legion Convention in Philadelphia died from a pneumonia-like disease that was later traced to the then-unknown bacteria that we now call *Legionella*. Since this initial outbreak (which was really preceded by earlier events in Austin, Minnesota, in 1957 and Washington, DC in 1964), the recognition of *Legionella* as a serious problem has grown significantly. In July 2015, a major Legionnaires' disease outbreak erupted in the Bronx borough of New York City, NY. In that outbreak, 12 people died and over 100 residents were diagnosed with the disease. According to the U.S. Centers for Disease Control, Legionnaires' disease infects approximately 25,000 people in the United States each year and 10%–15% of these cases are typically fatal.

Legionella is a bacteria that is common in surface waters, including lakes, rivers, and so on. The bacteria survive routine water treatment, and low concentrations are introduced into most potable water supplies. *Legionella* thrives in water temperatures between 68°F [20°C] and 122°F [50°C], with optimal growth occurring between 95°F [35°C] and 115°F [46°C]. Low pH and high levels of aquatic growth (microbiota, amoebae, algal slime, and so on) enhance bacteria growth. Water temperatures above 135°F–140°F [57°C–60°C] kill the bacteria.

At these temperatures, the ideal habitats for *Legionella* include evaporative cooler and evaporative condenser systems, where with temperatures typically range from 85°F–100°F [29°C–38°C]. Other potential breeding grounds for *Legionella* include domestic hot water systems in schools and hospitals, humidifiers, spas or whirlpools, and even vegetable misters in supermarkets.

The major mechanism for infection by *Legionella* is via inhalation of aerosolized water droplets or particles containing the bacteria. Evaporative cooler or evaporative condenser sprays introduce aerosolized water droplets and, therefore, represent prime mechanisms for infecting humans. There is no evidence that drinking water with the bacteria in it will cause disease nor can the disease be passed by human-to-human contact.

There are two types of disease caused by *Legionella*: "Legionnaires' Disease" is a severe form of pneumonia that may be fatal, while "Pontiac Fever" is a nonfatal flu-like illness. Legionnaires' disease symptoms can vary from a cough and low fever to rapidly progressive pneumonia, coma, and death. Symptoms occur typically within 3–9 days after exposure.

To monitor and control *Legionella*, the following steps are recommended:

a. Test for *Legionella* in evaporative cooler water. While there is an academic debate over the cost-versus-benefit of routinely testing for *Legionella*, it is dumb (from both an ethical and a liability point of view) to ignore any potentially life-threatening condition in a facility. Testing must be specific for *Legionella*: "total bacteria" tests promoted by some water treatment companies are inadequate since there is no correlation between total bacteria and *Legionella* concentrations.

b. If *Legionella* is found, reduce or eliminate the bacterial concentration. Evaporative coolers (and evaporative condensers) can be decontaminated by slug chlorination with 50 ppm of free residual chlorine, along with a dispersant. Then maintain 10 ppm of free residual chlorine for a least 24 h, while maintaining the water pH at 7.5–8.0. Finally, drain the system and repeat the process.

c. Evaporative cooler systems should be drained and flushed twice each year. All surfaces should be cleaned and allowed to air-dry before reuse. Continuous feed water treatment systems for biocides are required to maintain consistent concentration levels.

d. Keep up with the latest information on *Legionella*. ASHRAE has published *Guideline 12, Minimizing the Risk of Legionellosis Associated with Building Water Systems*. Then, ASHRAE *Standard 188, Legionellosis: Risk Management for Building Water Systems*, significantly expands on *Guideline 12* in code language and has been codified by some municipalities. This standard is an excellent design guide for reducing the potential for *Legionellosis* in HVAC systems.

14.4 Vibration and Noise Mitigation

14.4.1 Noise Definition and Design Criteria

Sound is the hearing sensation caused by a physical disturbance in the air. *Noise* is simply objectionable sound.

Sound waves in air arise from variations in pressure above and below the ambient pressure and can be caused by vibration of a building surface, repetitive pulsation in an airstream (due to rotating fan wheels or blades), turbulence caused by air movement impacting on surfaces in ductwork and/or air distribution devices, turbulence in water or steam piping systems, and the impact of one mass on another (such as water hammer in a piping system). Some can hear sound in the frequency range of 20–20,000 Hz, but in reality, few can hear above 8,000–12,000 Hz.

The human ear is not equally sensitive to all frequencies of sound and noise can be perceived as "hissy" when the noise is concentrated in the higher frequencies, "rumbly" when concentrated in the lower frequencies, and "neutral" when concentrated in the mid-range of frequencies.

To be acceptable, the "loudness" of noise from mechanical systems must not prevent people from hearing sound they want to hear. Additionally, the "quality" of mechanical noise must not be intrusive or annoying. Therefore, designers must be concerned about both the level of mechanical noise and its quality. To establish a design criterion, acoustic engineers have developed the *room (noise) criteria* (RC method) to replace the older *noise criteria* (NC method).

The NC method of rating noise established specific limits of sound pressure (loudness), in decibels (dB), in each of eight frequency ranges (referred to as *octave bands*) as shown in Table 14-2. Using this criterion, designers could specify noise limits for plumbing and HVAC equipment using a simple, single number.

The NC method does not address the quality of mechanical noise, since sound sources with equal NC levels may actually sound markedly different. To improve the criteria, the RC method reduces the maximum octave band dB levels in the lower and upper octave bands, resulting in sound that has a more neutral quality.

TABLE 14-2

NC and RC for Acceptable Noise

		Maximum Allowable Sound Pressure (dB) to Meet Specified Criterion					
Octave Band/ Frequency	**Method**	**25**	**30**	**35**	**40**	**45**	**50**
1	NC	(No criteria established)					
16 Hz	RC	-	-	65	70	75	80
2	NC	(No criteria established)					
31.5 Hz	RC	-	55	60	65	70	75
3	NC	54	57	60	64	67	71
63 Hz	RC	45	50	55	60	65	70
4	NC	44	48	52	56	60	64
125 Hz	RC	40	45	50	55	60	65
5	NC	37	41	45	50	54	58
250 Hz	RC	35	40	45	50	55	60
6	NC	31	35	40	45	49	54
500 Hz	RC	30	35	40	45	50	55
7	NC	27	31	36	41	46	51
1,000 Hz	RC	25	30	35	40	45	50
8	NC	24	29	34	39	44	49
2,000 Hz	RC	20	25	30	35	40	45
9	NC	22	28	33	38	43	48
4,000 Hz	RC	15	20	25	30	35	40
10	NC	21	27	32	37	42	47
8,000 Hz	RC	(No criteria established)					
11	NC	(No criteria established)					
16,000 Hz	RC	(No criteria established)					

Table 14-2 summarizes the RC method dB limit in each octave band. With the RC method, designers can still specify a single number noise limit criterion, but the resulting noise will be more neutral and acceptable to the listener. Table 14-3 indicates the recommended RC method design criterion for various types of occupancy.

Designing for a criterion below RC-25 requires specialized expertise and the architect should retain an acoustic consultant to work with the design team. The need for such low RC criterion should be carefully evaluated, since these low sound levels are very difficult (and expensive!) to obtain. Also, while the background noise in industrial occupancies may allow higher than RC-50 criterion, designing mechanical systems for a higher criterion is not recommended.

14.4.2 Indoor Noise Control in HVAC Systems

The noise produced by HVAC systems falls in three general categories:

1. Noise from air systems.
2. Noise from piping systems.
3. Noise due to vibration from mechanical equipment being transferred through the building structure.

TABLE 14-3

Recommended Room Criterion

Maximum RC	Environment	Typical Occupancy
25	Extremely quiet environment, suppressed speech is audible, suitable for acute pickup of all sounds.	Broadcast or recording studios, concert halls, music rooms, bedrooms, special classrooms for the very young or hearing impaired, foreign language instruction classrooms, and so on.
30	Very quiet, suitable for large conferences; telephone use satisfactory.	Residences, general classrooms, theaters, libraries, executive offices, directors' rooms, large conference rooms, and so on.
35	Quiet, suitable for conference at 15 ft. table; normal voice 10–30 ft; telephone use satisfactory.	Private offices, school cafeterias, court rooms, churches, small conference rooms, and so on.
40	Satisfactory for conferences at 6–8 ft table; normal voice 6–12 ft; telephone use satisfactory	General (open) offices, school corridors, laboratories, restaurant and entertainment facilities, and so on.
45	Satisfactory for conferences at 4–5 ft table; normal voice 3–6 ft, raised voice 6–12 ft; telephone use is somewhat difficult.	Retail stores, cafeterias and fast food dining, lobby or public areas, and so on.
50	Unsatisfactory for conferences; normal voice 1–2 ft., raised voice 3–6 ft.; telephone use is difficult.	Workshops, machine rooms, industrial process areas, and so on.

The noise from air systems can be held to acceptable limits careful attention to equipment selection, vibration isolation, and ductwork and air distribution design by the HVAC engineer.

Fan selection is the first step in HVAC air system noise control. Only airfoil blade or backward-inclined centrifugal fans or vaneaxial fans should be used in air supply systems since these are the quietest types of fans. It is important that the HVAC engineer select fans for operation in their most efficient operating range to maximize the fan performance. Air-handling units should be "blow through" configuration and, for noise-sensitive applications, provide an acoustic (sound attenuating) double-walled discharge plenum.

Fan discharge transitions must be kept "smooth", with no abrupt change in direction or cross-section contraction near the fan discharge and air velocities in ducts must be kept within ASHRAE recommended limits based on the room noise criterion established for the areas over which the ducts pass.

Air distribution devices must be selected based on the room noise criterion established for the area served. All balancing dampers must be located as far upstream of air distribution devices as possible and *never* be installed in the throat of the device.

The use of interior duct liner and/or sound attenuators should be avoided. Sound attenuators are expensive, introduce significant airflow pressure losses, and often don't perform as advertised. Sound-attenuating double-walled duct (with a perforated inner wall) may be used for the first 20–40 ft [6–12 m] of supply ductwork to provide for good fan noise reduction.

Piping noise problems parallel those of plumbing systems and the same measures to control that noise must be designed by the HVAC engineer:

1. Water flow and piping characteristics to reduce flow noise
2. Vibration isolation of piping from the structure
3. Pump vibration isolation
4. Water-hammer control

Steam systems may produce noise due to water hammer, control valve "chattering", and/or trap operation. Again, careful attention to the piping arrangement, slope, drip legs, and control valve sizing by the HVAC engineer should minimize these potential problems.

The problem of mechanical vibration/noise being transferred to the building structure and creating low frequency (rumbly) noise problems requires attention by both the architect and the HVAC engineer. In Chapter 4, the size requirements for mechanical equipment rooms (fan/air-handler rooms, boiler rooms, and chiller rooms) were addressed. However, the *arrangement and configuration* of these rooms, along with the choice of construction materials, can have significant impact on noise transfer. To keep noise from mechanical equipment rooms from becoming a problem, the following architectural and structural design guidelines are recommended:

1. Locate mechanical equipment rooms as far as possible from noise-sensitive areas.
2. Floors of mechanical equipment rooms must be concrete, with the following minimum thickness:

   Boiler and Chiller Rooms 8" [200 mm]

   AHU/Fan Rooms 6" [150 mm]
   Mechanical Closets 4" [100 mm]
3. Each piece of mechanical equipment should be installed on a 4" [100 mm] high concrete *housekeeping pad* for additional mass.
4. Walls of mechanical equipment rooms should be solid, high-density construction to reduce transmission of airborne noise. Where adjacent areas are noise-sensitive, (a) use double walls and cut the floor between the walls to eliminate floor transmission and/or (b) provide sound absorptive material to the inside of the walls.
5. Doors to mechanical equipment rooms are often serious "sound leaks". If possible, design each mechanical equipment room with a vestibule. At the very least, door/jamb assemblies should be tight fitting, with both head and threshold seals.
6. Coordinate each mechanical equipment room configuration with the HVAC engineer to allow ample separation between HVAC equipment and the room walls in order to reduce the potential for radiated noise transfer.
7. A "floating floor" for the mechanical equipment room may be necessary to prevent floor transmission to adjacent noise-sensitive areas, particularly if the mechanical equipment room is above the noise-sensitive area.
8. All mechanical room floor, ceiling, and wall penetrations must be arranged to prevent the transfer of vibration from piping and duct to the building.
9. Vertical shafts must provide sufficient clearance between the duct and the walls to prevent transfer of noise. The minimum separation is 10% of the duct dimension in each direction or 6" [150 mm], whichever is greater. Walls of duct shafts should be solid, high-density construction to reduce transmission of airborne noise.

10. Do not locate terminal units above the ceiling of spaces that require RC-35 or lower design room noise criterion.
11. Do not locate fan powered terminal units, fan coil units, water source heat pump units, and so on above the ceiling of spaces that require RC-40 or lower design room noise criterion.

In commercial buildings, the roof structural systems tend to be fairly "light" in order to keep first costs to a minimum, with the use of long span bar joists and light metal roof deck being common. Typically, packaged rooftop-mounted HVAC equipment is selected for this type of building. Often, vibration from a rooftop unit is transferred to the roof assembly that, like a drum skin, amplifies the vibration and creates a serious noise problem. Roof penetrations by the supply and return ducts are usually poorly installed and, again, noise and vibration is transferred to the roof.

To reduce the potential of rooftop packaged unit noise problems; the following architectural and structural design guidelines should be followed:

1. Never locate a rooftop unit over a noise-sensitive area (offices, conference rooms, and so on)
2. Locate rooftop units in areas of maximum roof structural stiffness to reduce the "roof drum" effect. Locating units directly over columns is best, while locating them over beams close to a column is almost as good. If these locations are impossible, add structural beams and/or an inertia pad under the unit to stiffen the roof construction. The best installation method is to extend the columns through the roof to support a separate steel framework for the rooftop unit, totally eliminating direct roof contract.
3. Provide for adequate vibration isolation within the rooftop unit. Spring vibration isolators must support each fan/motor/drive assembly within the unit. Duct connections at the unit must be made with flexible connectors to eliminate vibration transfer. Electrical connections must be made with flexible raceway.
4. Install high-density backing (minimum 1.5 lb/sf)[7.3 kg/m^2], such as ½" [13 mm] gypsum wallboard (GWB), on suspended acoustical ceiling tiles installed below a rooftop unit. (Make sure the ceiling suspension system is designed for this additional loading.)

14.4.3 Outdoor Noise Issues in HVAC Systems

Numerous governments have property line sound limits that must be met. Under this type of regulation, *maximum sound pressure* levels (dBA), measured at adjacent property lines, are established. Typical maximum allowable sound pressure values are 50–60 dBA in residential areas and 60–70 dBA in industrial areas.

More stringent standards have been developed by the World Health Organization (WHO). The WHO standards are being promoted by numerous anti-noise groups and their adoption by at least some states and municipalities is considered but a matter of time. WHO makes the case that the effect on the health and safety of the listener by a combination of noise "events" is related to the combined sound energy of those events. Thus, they introduce the term "LA_{eq},T" to define A-scale sound pressure level (dBA) averaged over a time period T. As shown in Table 14-4, these standards establish maximum LA_{eq},T limits at or within various types of occupancies or environments.

TABLE 14-4

World Health Organization Community Noise Limit Guidelines

Environment	Critical Health Effect(s)	LA_{eq}, dBA	T, H
Outdoor living area	Serious annoyance.	55	16
	Moderate annoyance.	50	16
Dwelling (living areas, inside)	Speech intelligibility and moderate annoyance	35	16
Bedrooms (inside)	Sleep disturbance	30	8
Bedrooms (outside, window open)	Sleep disturbance	45	8
Schools (classrooms, inside)	Speech intelligibility, disturbance of information extraction, message communication	35	During class
Schools (playground, outdoors)	Annoyance	55	During play
Hospital (patient rooms, inside)	Sleep disturbance	30	8 (night) 16 (day/ evenings)
Industrial, commercial, shopping, and traffic areas (inside and outside)	Hearing impairment	70	24
Outdoors in parkland and conservation areas	Disruption of tranquility	Minimize disruption	

Outdoor HVAC equipment must be selected, located, and configured to ensure that the source sound power levels are adequately attenuated to meet the property line sound limitations required.

There are two caveats relative to manufacturer-provided sound data:

1. The noise data provided is typically based on a single piece of equipment. Where multiple units are utilized, the manufacturer must be queried for cumulative effect sound data.
2. The noise emission from most outdoor HVAC equipment will, most likely, not be uniform in all directions. The manufacturer data must define the equipment orientation for each noise emission value.

The equipment location and orientation can have a pronounced effect on the sound condition. Always orient a quiet side of the equipment toward any sound reflecting wall or side of a building or toward adjacent property. Locate equipment at least 20 ft away from any sound reflecting surface.

Sound dissipates over distance. As shown by Table 14-5, the amount of sound attenuation varies with the multiples of distance from the point of sound measurement (distance from the rated noise measuring point). At a distance 10 times greater than the sound rating distance from the tower, approximately 20 dBA attenuation is expected. Thus, equipment with a rated noise level of 86 dBA at 5 ft [1.5 m] away will have an anticipated noise level of 65 dBA at 50 ft [15 m] away, a 21 dBA reduction. Note that the dissipation is not linear and an increase to 25 times the sound rating distance from the tower only reduces the sound level by approximately 25 dBA.

TABLE 14-5

Effect of Distance on Sound Level

Multiple of Sound Rating Distance	Approximate Sound Reduction (dBA)
5x	13
10x	20
15x	24
20x	26
25x	29

To avoid a community noise problem, locate outdoor HVAC equipment as far from adjacent property lines as possible. Since HVAC cooling towers and air-cooled chillers or condensing units may have sound levels of 65–85+ dBA, the WHO criteria require that they not be located within 50–100 ft [15–30 m] of property lines without taking additional attenuation measures:

1. Select/approve the quietest equipment possible for the application. For cooling towers, use low rpm axial fans with the greatest number of blades for the application.
2. Orient a quiet side of the equipment toward any sound reflecting wall or side of a building and toward adjacent property.
3. Reduce radiated noise with a solid wall sound barrier between the equipment and the adjacent property. Sound barrier noise attenuation, however, is a function of the wall height and the distances between the tower, the barrier, and the property line. An acoustical consultant should be retained to design any sound barriers that may be required.
4. Only as a last resort should sound attenuators or sound enclosures be utilized. These devices are expensive, can seriously impact the performance of the equipment, and create significant maintenance problems.

Sound barriers applied to reduce the sound emitting from outdoor HVAC equipment generally have a maximum practical barrier attenuation of 10–15 dBA. Barriers should be constructed of solid, high-density nonporous materials. The barrier material should have a Sound Transmission Coefficient at least 10 dB higher than the desired barrier attenuation goal. Common barrier materials include steel, precast concrete, brick or other masonry units, heavy wood, and composition boards. No matter what materials are used, design life for the barrier should be 30 years or longer. General design guidelines for sound barriers are as follows:

1. The "line of sight" between the noise source and the receiver must be cut off completely by the barrier.
2. Barriers should be at least 5 times wider than they are high.
3. Barriers should be built as close as possible to either the receiver of the source to maximize the "sound shadow."
4. Sound barriers must be designed to meet wind loading and other structural requirements and not interfere with site drainage.

Barriers are less effective at reducing low frequency noise than mid and high frequency noise. Hence, to calculate the effectiveness of a noise barrier, it is necessary to know the source noise levels in octave or one-third octave bands. Hedges or single trees (or rows of trees) do not themselves make effective noise barriers.

14.5 HVAC Systems Start-Up and Commissioning

14.5.1 HVAC Systems Operation Prior to Commissioning

Even before a project's HVAC systems may be completely installed and commissioned, HVAC contractors are increasingly being asked by general contractors to perform early start-up of HVAC systems to provide temporary heating, cooling, or dehumidification during the construction process. *This practice can pose a negative significant impact to the integrity of the HVAC equipment and HVAC distribution system and should not be allowed by the designer.* Specifications should include specific requirements that contractors shall provide *temporary* heating and/or cooling to support their construction activities, especially the drying of concrete and the application of interior finishes, including gypsum wall board.

Sheet Metal and Air Conditioning Contractors' National Association (SMACNA) has published a paper entitled *Temporary Uses of HVAC Systems in Building Construction Projects* that outlines the realities and risks of using building HVAC systems to provide temporary heating, cooling, and/or dehumidification, as follows:

1. Misuse of permanent HVAC systems to heat ("dry out") open areas under construction because such operations exceed design specifications. The filter systems, even with the addition of construction or prefilters, are also incapable of providing the dust-holding capacity required to protect the permanent HVAC equipment and duct system. Construction filters cannot sufficiently protect the permanent HVAC system from excessive amounts of construction dust, particularly the most common source, dust created from sanding of GWB.
2. Use of the permanent HVAC system in an attempt to dry wet surfaces, such as drying recently poured concrete floors to permit or expedite the installation of carpet or wooden floors. Permanent HVAC systems are not designed or constructed to perform in such a manner. Indeed, such activities may result in subsequent IAQ problems associated with mold and other related airborne contaminants.
3. Initiation of the HVAC equipment warranty period when the equipment is started. Early startup of the permanent HVAC system for temporary heating, cooling, dehumidification, or for other reasons may also void the warranty on that system's equipment.
4. Early start-up of the permanently installed HVAC system will result in reduced equipment life, operating efficiencies, and potential equipment damage.
   a. Motors typically used in HVAC applications have open windings and the accumulation of construction dust raises the operating temperature and leaches oil away from bearings.
   b. Coils are manufactured under very clean conditions but have residual oil on the heat exchanger surfaces, which cause dust not captured by filters to tightly

adhere to the surface. This reduces the efficiency of the energy exchange, especially when cooling coils condense moisture causing certain types of dust (from GWB and plaster) to harden.

5. The potential of increased tenant complaints and claims. Dust and particulates in HVAC ductwork is increased exponentially when the permanently installed HVAC system is used for temporary heating, cooling, or dehumidification during the construction process. In such circumstances, the stage is set for potential mold-related conditions and consequent tenant complaints.
6. Total energy costs will generally be higher than the cost to use temporary heating, cooling, and dehumidification equipment readily available in the market place. For a permanent HVAC system to have any beneficial effect in heating or cooling, a construction site requires continuous operation at maximum capacity. In contrast temporary heating and moisture removal equipment use energy directed to exactly where it is needed and the total energy costs are usually less.

And, for HVAC systems designers, there is a potential liability risk for allowing incomplete, unbalanced, and only partially controlled HVAC to be utilized by the general contractor to aid in his or her construction process. The detrimental effects, both direct and indirect, from early system use can result in a claim by the owner when problems develop. At the very least, an owner can usually make a good case that early use of the systems shortened their life and increased their life-cycle costs.

14.5.2 The Commissioning Process

Once installation is completed, *HVAC systems must be properly commissioned* to ensure that they function in accordance with the design intent and provide the intended level of temperature and humidity control, air cleaning, ventilation, and proper air distribution.

ASHRAE *Guideline 0* defines commissioning as "a quality-oriented process for achieving, verifying, and documenting that the performance of facilities, systems, and assemblies meets defined objectives and criteria."

Commissioning is, therefore, an "umbrella" process for all the planning, delivery, verification, and managing risks to critical functions performed in, or by, facilities. Commissioning uncovers deficiencies in design or installation using peer review and field verification. Commissioning also accomplishes higher energy efficiency, environmental health, and occupant safety and improves IAQ.

Commissioning is a quality assurance-based process that delivers preventive and predictive maintenance plans, tailored operating manuals, and training procedures. Essentially, the commissioning process formalizes review and integration of all project expectations during planning, design, construction, and occupancy phases by inspection and functional performance testing, and oversight of operator training and record documentation.

Commissioning is often misinterpreted to focus solely on testing during the end of the construction phase. But commissioning is actually a collaborative process for planning, delivering, and operating buildings that work as intended. ASHRAE defines commissioning as "...the process of ensuring that systems are designed, installed, functionally tested, and capable of being operated and maintained to perform in conformity with the design intent... Commissioning begins with planning and includes design, construction, start-up, acceptance, and training and can be applied throughout the life of the building." This definition accurately depicts commissioning as a holistic process that spans from predesign planning to post-construction operation and can be thought of as a checks-and-balances system.

There are seven basics elements that must be included in a specification by the designer for proper HVAC systems commissioning:

1. *Establish the Commissioning Team.* HVAC systems commissioning must be performed by a commissioning "team," consisting of the installing personnel, the controls subcontractor, and the *testing, adjusting, and balancing* (TAB) subcontractor(s) and must include the following:
   a. Systematically evaluating all pieces of HVAC equipment, subsystems, and systems to ensure that they are working according to specifications and the design intent. This includes measuring temperatures and flow rates from all HVAC devices and calibrating all sensors to a known standard.
   b. Reviewing the sequence of operations to verify that the controls are providing the correct interaction between equipment, subsystems, and systems.
   c. Performing verification procedures, functional performance tests, and validation with appropriate documentation, to demonstrate proper HVAC systems operation.
   d. Providing operations and maintenance (O & M) documentation and owner training.
2. *O&M Documentation.* Maintenance and Operating (O & M) Manuals, in both hard copy and electronic format, must be furnished to the owner. Manuals should be prepared in accordance with ASHRAE *Guideline 4.*
3. *Owner Instruction and Training.* After installation has been completed, equipment has been tested, systems have been placed in permanent operation, and all adjustments have been made, a competent start-up technician must be provided as required to instruct the owner's designated operating personnel in the operation and maintenance of the equipment. Arrange for each installer of equipment that requires regular maintenance to meet with the owner's designated personnel to provide proper operation and maintenance. Instructions and training must, as a minimum, consist of classroom instruction and "hands on" instruction of approximately equal duration, addressing the following elements: theory of operation (usually provided by the designer), proper start-up and shutdown procedures, proper handling of emergency situations, safety considerations, controls, routine maintenance procedures, and recommended preventive maintenance procedures.
4. *HVAC Design Intent and Basis of Design.* The *designer* must provide a detailed description of the design intent and basis of design for the HVAC systems, including the objectives of each HVAC system and its functional use, sequences of operations, control set points and operating parameters, and performance criteria required of each HVAC component.
5. *Systems Start-up.* Start-up checklists for each type of equipment and system must be developed by the designer and/or the commissioning team.
6. *TAB.* The extent of TAB work must include all HVAC components, subsystems, and systems installed as part of this project. After systems have been started up and initially adjusted, the TAB subcontractor must perform tests and accomplish the balancing, in accordance with the ASHRAE *Handbook-Applications* and ASHRAE

Standard 111, necessary to provide the air and water flows within –5% to +10% of those indicated on the contract documents.

Reports must be certified by the TAB agency and must be a certified proof that the systems have been tested, adjusted, and balanced in accordance with the referenced standards; accurately represent how the systems have been installed; define how the systems are operating at completion of the TAB procedures; and record all final quantities measured to establish normal operating values of the systems. Instruments used for testing and balancing of HVAC systems and components must have been calibrated within a period of 6 months prior to balancing and tested for accuracy prior to start of work.

7. *Functional Performance Tests and Certification*. The designer must specify functional performance tests to prove all modes of the sequences of operation and to verify all other relevant contract requirements. Tests must begin with equipment or components and must progress through subsystems to complete systems. Upon failure of any functional performance test checklist item, the contractor must correct all deficiencies in accordance with the applicable contract requirements. *The checklist must then be repeated until it has been completed with no errors.*

   A major part of these tests includes testing of the controls system(s) operation. To the maximum extent possible, all potential operating modes must be tested and evaluated.

   Based on these functional performance test checklists, the commissioning team must prepare standardized reporting forms for each item of equipment, subsystem, and system to document the required functional performance tests. Each test must be certified with the signature and date of signing by each member of the commissioning team.

14.6 Correcting HVAC Performance for Altitude and Temperature

The U.S. National Institute of Standards and Technology uses a temperature of 20°C (293.15 K, 68°F) and an absolute pressure of 1 atm (14.696 psia) [101.325 kPa]. This standard is also called *normal temperature and pressure* (abbreviated as NTP) or *standard temperature and pressure*.

Based on Boyle's law, we know that air volume varies inversely with pressure and, based on Charles' law, we know that air volume varies directly with temperature, as shown by Eq. (14-4).

$$V_2/V_1 = (P_1/P_2)\cdot(T_2/T_1) \tag{14-4}$$

where

$V_{1,2}$ = Volume at states 1 and 2 (cf) [m³]

$P_{1,2}$ = Absolute pressure at states 1 and 2 (psia) [kPa]

$T_{1,2}$ = Absolute temperature (°R = °F + 460)[°K = °C + 273]

Pressure is obviously defined by elevation, while temperature is defined by the psychrometrics of the air-conditioning process involved. Table 14-6 lists barometric pressures for

TABLE 14-6

Atmospheric Pressure as a Function of Elevation

Elevation (ft)	Pressure (psia)	Elevation [m]	Pressure [kPa]
0	14.7	0	101.3
2,000	13.7	1,000	89.9
3,000	13.2	2,000	79.5
4,000	12.7	3,000	70.1
5,000	12.2	4,000	61.6
6,000	11.8	5,000	54.0
7,000	11.3	6,000	47.2
8,000	10.9	7,000	41.1
9,000	10.5	8,000	35.0
10,000	10.1	9,000	30.7

elevations up to 10,000 ft [9,000 m] above sea level within the range of air temperatures normally encountered in HVAC applications. *In practice, an elevation of less than 2,000 ft [600 m] does not warrant correction of air volume from standard conditions.*

Normally, heating/cooling loads and required airflows are determined on the basis of standard air conditions. Actual airflow rate (cfm)[m^3/s] is defined by the actual absolute temperature, T_a (°R)[°K], and pressure, P_a (psia)[kPa], compared to the normal or standard absolute temperature, T_s (°R)[°K], and pressure, P_s (psia)[kPa], as shown by Eq. (14-5).

$$\text{actual airflow rate} = \text{standard airflow rate} \times (P_s/P_a)\cdot(T_a/T_s) \qquad (14\text{-}5)$$

Since duct and equipment static pressure losses are normally computed on the basis of *standard airflow,* these losses must also be adjusted for the increased *actual airflow rate.* Based on the affinity laws discussed in Chapter 9, both static pressure losses and the fan brakehorsepower requirement must be recomputed using the (*actual airflow/standard airflow*)^2 to determine actual static pressure loss and the factor (*actual airflow/standard airflow*)^3 to determine actual fan brakehorsepower.

HVAC heating and cooling equipment operating at higher elevations produces less than its catalog rating based standard air-conditions. Only the calculated actual operating parameters (capacity, static pressure, cooling/heating loads, and so on) required at the higher elevation should be tabulated in equipment schedules on the design drawings or in the technical specifications. A statement, such as "*All capacities shown or indicated are at xxxx feet [xxxx m] above sea level elevation,*" should be included as part of each applicable equipment schedule. *Thus, it becomes the responsibility of the equipment manufacturer to make needed adjustments to enable their equipment to meet the requirements scheduled at the indicated elevation.* (Note that the equipment manufacturer must provide submittals to document proposed equipment performance.)

Bibliography

ASHRAE Handbook-HVAC Systems and Equipment, American Society of Heating, Refrigerating, and Air-Conditioning Engineers, Inc. (Atlanta, GA, 2016).

FEMA 427, Primer for Design of Commercial Buildings to Mitigate Terrorist Attacks, Federal Emergency Management Agency (Washington, DC, 2003).
FEMA 426, *Reference Manual to Mitigate Potential Terrorist Attacks Against Buildings*, Federal Emergency Management Agency (Washington, DC, 2003).
The Nalco Water Handbook, 4th Edition, NALCO, An Ecolab Company (McGraw-Hill Education, 2017).
Practical Guide to Noise and Vibration Control for HVAC Systems, 2nd Edition, ASHRAE (Atlanta, GA, 2005).
Temporary Uses (Early Start-Up) of HVAC Systems In Building Construction Projects, SMACNA (www.smacna.org/docs/default-source/project.../summary-document-pdf.pdf, accessed 2018).

15

Engineering Economics and Design Decision-Making

15.1 Introduction

Engineering economics is a subset of economics concerned with the use and"...application of economic principles" in the analysis of engineering decisions. As a discipline, it is focused on the branch of economics known as microeconomics in that it studies the behavior of individuals and firms in making decisions regarding the allocation of limited resources. Thus, it focuses on the decision-making process, its context, and environment. It is pragmatic by nature, integrating economic theory with engineering practice. But, it is also a simplified application of microeconomic theory in that it avoids a number of microeconomic concepts such as price determination, competition, and demand/supply. It draws upon the logical framework of economics but adds to that the analytical power of mathematics and statistics.

Engineers seek solutions to problems, and the economic viability of each potential solution is normally considered along with the technical aspects. Fundamentally, engineering economics involves formulating, estimating, and evaluating the economic outcomes when there are alternatives to accomplish a defined purpose.

Considering the time value of money is central to most engineering economic analyses. Cash flows are discounted using an interest rate, except in the most basic economic studies.

For each problem, there are usually many possible alternatives. The opportunity cost of making one choice over another must also be considered. There are also noneconomic factors to be considered, such as color, style, and public image; such factors are termed attributes.

Costs as well as revenues are considered, for each alternative, for an analysis period that is either a fixed number of years or the estimated life of the project. The salvage value is often forgotten, but is important, and is either the net cost or revenue for decommissioning the project.

Some other topics that may be addressed in engineering economics are inflation, uncertainty, replacements, depreciation, resource depletion, taxes, tax credits, accounting, cost estimations, or capital financing. All these topics are primary skills and knowledge areas in the field of cost engineering.

The seven-step procedures used to assist decision-making are as follows:

1. The recognition, definition, and evaluation of the problem.
2. Search for potential as well as feasible alternatives.
3. Incorporating the basic cash flow approach.

4. Decision should serve the long-term interest of the organization.
5. Analyzing the economic aspects of the engineering problem.
6. The preferred alternative is based on the total effort.
7. Attention to ensure feedback for improvement of operation.

The economics of heating, ventilating, and air-conditioning (HVAC) systems include capital costs and operating costs. The operating cost comprises energy consumption, maintenance and repair, recommissioning, replacement, and asset preservation. Life-cycle analysis factors in escalation and time cost of money over the study's period of time.

Often, engineers are asked to determine the best option between one or more HVAC system options. There are several ways to represent the results when comparing the life-cycle costs of two or more options. Simple payback, net savings, savings-to-investment ratio (SIR), adjusted internal rate of return (IROR), and discounted payback are all methods to measure an HVAC option's economic performance over time.

The economics of HVAC systems can be broken into two distinct buckets: capital expenditures and operating expenditures. Building owners have varying expectations regarding the overall economic impact of HVAC systems. It is safe to say all owners care about the capital investment; however, not all owners understand the impact of capital investment on operating costs.

There is a balance between capital costs and operating costs for each project. Owners often have a threshold for spending capital to gain operating cost improvements. The engineer will need to understand this balance and help the owner make decisions that align with the basic project requirements.

Life-cycle costing is a methodology by which a designer can evaluate the total cost of a building and its individual components over the entire period that the building or a component is expected to be utilized.

This type of analysis is used to compare alternative design options so that designers can select the most cost-effective alternative. Alternative design elements that (1) have different first costs and/or (2) will impact operating and/or maintenance costs differently must be compared utilizing a common basis, as shown by Eq. (15-1).

$$LCC = I + E + M + R \tag{15-1}$$

where

LCC = Life-cycle cost of the alternative

I = Investment or first cost of the alternative

E = Present value of energy costs associated with the alternative over its performance life

M = Present value of the recurring maintenance and repair costs associated with the alternative over its performance life

R = Present value of the nonrecurring replacement cost(s) associated with the alternative over its performance life

Life-cycle cost analysis (LCCA), then, proceeds in five basic steps for each design alternative:

1. Compute the first cost associated with the alternative.
2. Establish the economic life, in years, for the alternative.

3. Determine the annual operating and maintenance costs for the alternative over its economic life.
4. Compute replacement costs for the alternative if major renovations or replacement is required with the economic life period selected.
5. Finally, utilizing a consistent calculation method, compute the total cost to purchase and operate each alternative over a common economic period or life cycle.

The design alternative with the lowest total cost to purchase and own (lowest life-cycle cost) is the alternative that should be selected.

While LCCA is the most common method of comparing alternative HVAC design decisions, other comparison measures may also be applied:

1. *Simple payback* (years) is computed by dividing the capital cost difference between two alternatives and dividing that result by the annual operating cost difference. This does not factor in the time cost of money or maintenance/repair, replacement, or salvage value.
2. *Life-cycle payback* (years) is computed by dividing the capital cost difference between two alternatives and dividing that result by the life-cycle cost difference. This metric is rarely used, but it does account for the time cost of money and maintenance/repair, replacement, and salvage values.
3. *Net life-cycle cost savings* equals the difference between the life-cycle cost of one alternative, less the life-cycle cost of another alternative.
4. *IROR* for each design alternative is computed by dividing the average annual operating cost savings by the initial capital cost. This method accounts for time cost of money and maintenance/repair, replacement, and salvage value.

15.2 Defining HVAC Design Alternatives

The key to meeting the requirement for LCCA of facilities is to apply life-cycle costing to *all* of the significant design decisions that must be made.

But, trying to create a comprehensive list of the design alternatives that may require analysis is sort of like trying to count the number of stars in the universe … after a while the task becomes counterproductive. Suffice it to say that there a number of design areas within which the HVAC designer will be called upon to identify and analyze potential design alternatives. These basic areas include the following:

1. Alternatives that reduce (or even eliminate) heating/cooling loads imposed on secondary HVAC systems. Thus, every heating and/or cooling load component becomes a candidate for improvement. Simple things such as alternative building construction to reduce thermal heat transfer; revised fenestration elements to reduce solar heat gains in the summer, yet maximize the benefits of solar heat gain in the winter; minimizing ventilation air needs while eliminating infiltration loads; lighting alternatives, including the use of daylighting, to meet visual requirements while minimizing imposed heat gains, and so on. And, the potential

array of alternatives includes even basic architectural design decisions, such as overall building shape, orientation, and interior space configuration.

2. Alternative secondary HVAC systems that are capable of meeting the owner's needs while imposing the least loads on the primary HVAC systems that serve them.
3. Alternative primary HVAC systems, and their configurations, that are capable of meeting the loads imposed by the secondary systems they serve, while utilizing the most cost effective method(s) of input energy consumption (electricity, fossil fuel, and so on) use.

Alternatives must meet program, performance, and budget constraints. Obviously, larger more general alternatives will have a greater impact on total life costs than more detailed alternatives involving specific building components or subsystems.

Overall, determining alternatives to evaluate is a daunting task. For most design projects that have limited design fees associated with them, the list must quickly be reduced to a finite number of design alternatives that fall within the scope of the project, its construction budget, and the owner's expectations. Otherwise, project analysis will last forever and the project will never be built!

Stanford University has done one of the better jobs of defining the scope of design alternatives to be considered by the design team. During the Schematic Design and Design Development phases of the project design, designers make increasingly detailed decisions about the final design for the building, including mechanical, electrical, structural, telecommunications, and plumbing systems. During this period, LCCA and comparison between design alternatives should be done.

Stanford has defined potential life-cycle cost comparisons in six general categories: Energy Systems, Mechanical Systems, Electrical Systems, Building Envelope, Siting/Massing, and Structural Systems. Within each category, specific comparisons involve options for satisfying the same needs. The 14 comparison areas are defined, as follows, with examples of options that might be considered in each:

1. Energy Systems
   a. Central plant-connected versus stand-alone systems (steam and chilled water (CHW))
   b. Alternative energy systems (e.g., solar photovoltaics, solar thermal, fuel cells)
   c. Equipment options for cooling systems (e.g., air-cooled chillers vs. water-cooled chillers vs. refrigerant-based DX units)
2. Mechanical Systems
   d. Secondary HVAC systems (e.g., multiple single-zone systems vs. variable air volume [VAV] multiple zone systems; water source heat pumps vs. air source heat pumps, and so on)
   e. Water distribution systems (e.g., various piping systems and pumping options)
3. Electrical Systems
   f. Indoor lighting sources and controls
   g. Outdoor lighting sources and controls
   h. Distribution (e.g., transformers, buss ducts, cable trays)

4. Building Envelope
   i. Skin and insulation options
   j. Roofing systems (various materials and insulation methods)
   k. Glazing, daylighting, and shading options
5. Siting/Massing
   l. Orientation, floor-to-floor height, overall building height, and interior space configuration
   m. Landscape, irrigation, and hardscape options
6. Structural Systems
   n. Systems/materials selection (e.g., wood vs. steel vs. concrete, cast-in-place vs. precast, and so on)

The design team must determine which of the six categories of studies and the 14 comparative analyses have the highest potential life-cycle cost benefit for the project. The design team and owner then must decide together to determine the details of each analysis. A "base case" will be established and the team will then draw upon its collective experience to identify potential alternatives to the base case for analysis.

15.3 Estimating Capital Requirements

15.3.1 Capital Cost Estimating

Capital costs are the sum of the design, construction, and financing costs necessary to provide a finished building or building component ready for use. Sometimes called the initial investment cost and includes land costs, legal fees, surveys, soil borings, environmental assessments, site demolition, and relocation of existing structures and utilities. Design fees including all consultants. Construction costs, including costs of labor, material, equipment, general conditions, insurance, performance and payment bonds, contractor's overhead, and profit. Other owner costs, including project administration/construction management fees, material's testing, contingencies, permits, and regulatory fees.

Construction costs, as estimated by architects and engineers, are often poorly estimated and even more poorly documented, particularly in the early stages of design. Estimating building costs on a "per square foot" basis, or mechanical costs on "per ton" or "per cfm" basis, often leads to unpleasant surprises at later stages of design or upon receipt of bids. Obviously, the first step in the calculation of accurate life-cycle costs is the accurate estimation of construction costs.

Cost estimating data are readily available to designers from R. S. Means Co., Inc., Construction Plaza, 63 Smiths Lane, Kingston, MA, 02364-0800, 800/334-3509 (Phone), 800/632-6732 (Fax). Means publishes annual construction cost data in several different categories, including building construction, mechanical, electrical, repair and remodeling, interiors, residential and light commercial, site work, and concrete and masonry. Also published are special cost data on square foot costs and assemblies' costs. Most of these data are also available on CD for use with cost-estimating software.

Most construction cost-estimating methods are based on the "quantity take-off" method whereby the estimator determines the quantity of each item of material and labor required and multiplies these by "unit costs" to obtain a final, total cost. The construction costs for most design element alternatives must be estimated in this manner, particularly if the quantities of materials change between alternatives. For example, to compare alternative fenestration, the amount of glazing, and mullions must be known in order to compute a cost of construction. Likewise, alternative HVAC systems must be compared utilizing accurate estimates of equipment quantities and capacities, including ductwork and piping.

It is only necessary to develop the scope of each alternative design element sufficiently to make valid quantity estimates, determine the current unit costs, and multiply the two to produce a construction cost estimate.

There are, however, classes of design elements that do not lend themselves to detailed quantity-based estimates. These include the basic architectural elements such as building shape, orientation, and space layout, plus, in some cases, the fenestration and building envelope. For these design elements, a relatively accurate, yet simple, method of producing cost estimates must be used, such as the "Assemblies Cost Data" developed by R.S. Means Co., Inc.

Finally, the construction estimate must include, as applicable, a number of costs that are often overlooked by designers:

1. Special equipment and/or rigging.
2. Demolition.
3. Additional architectural and/or structural requirements associated with HVAC alternatives.
4. Additional HVAC requirements associated with architectural alternatives.
5. Contractor overhead (insurance, bonds, taxes, and special conditions), typically 15%–20%.
6. Contractor profit, typically 5%–20%.

Other costs that are included in the capital requirement are design fees, which may increase or decrease as a function of the selected alternative; special consultants' fees; mock-ups; special testing; and so on.

All investment costs are to occur at the beginning of Year 1 of the analysis period (economic life).

15.3.2 Time/Money Relationships

There are two "time" factors that impact economic analysis of design alternatives:

1. Inflation is defined as the decrease in the value of money due to rising prices over time. As the value of money decreases, future costs will be higher, in today's terms, than current costs so future costs have to be adjusted to current values to provide valid comparison. This is accomplished by applying *present value factors* whereby all future costs are summed to yield their current value. If there was no such thing as inflation or if money could be borrowed without interest expense, the present value of all future costs would simply be the arithmetic sum of all

those costs. Adding this value to the initial capital cost would yield the total cost to buy and own the alternative over its defined analysis life.

Inflation factors may be difficult for designers to determine, but each April, the National Institute of Standards and Technology (NIST) of the U.S. Department of Commerce publishes a supplement entitled *Energy Price Indices and Discount Factors for Life-Cycle Cost Analysis* to their *Handbook 135*. This supplement provides regional present value factors that may be used for LCCA. The supplement for the current year (2017) can be accessed at https://www.nist.gov/publications/energy-price-indices-and-discount-factors-life-cycle-cost-analysis-150-2017.

Each of the present value factors is presented on the basis of varying *discount rates*. The rate to be used for any analysis must be defined by the owner since it represents an investor's opportunity cost of money over time, meaning that an investor wants to achieve a return at least as high as that of her next best investment. Hence, the discount rate represents the investor's minimum acceptable rate of return.

For building and HVAC design alternatives, that rarely is an alternative that says "let's not build anything, but rather go buy a bond that yields x% interest." Therefore, the discounting can minimize the value of future energy, repair, and replacement costs, skewing the ranking of the alternatives by giving greater weight to capital costs. *Thus, in most cases, the discount rate should be 0%*. However, for some private clients, where the alternatives may actually address different investments (such as, do we build a high rise condominium or do we build a high rise office building?), the discount rate does impact the decision-making and must be defined by the investor.

2. Interest rates and resulting costs for loans (private sector) or bonds (public sector) increase are part of most facility's operating costs.

Other terms associated with the time/money relationships include the following:

Amortization—The gradual reduction of the balance in account according to a specified schedule of time and amounts. Usually the provision for extinguishing a debt, including interest, by means of regular monthly or annual payments.

Cash flow—The stream of monetary (dollar) values, costs, and benefits, resulting from a project investment.

Economic life—That period of time over which an investment is considered to be the least-cost alternative for meeting a particular objective. Most typically, this is the functional life of the alternative component(s).

Operating cost—The expenses incurred during the normal operation of a building or a building system or component, including labor, materials, utilities, and other related costs.

Sensitivity analysis—A procedure used as a result of uncertainty as to the actual value of a parameter or parameters included in an analysis. The procedure is to vary the value of the parameter or parameters in question and examine the extent to which these changes affect the results of the analysis. For example, if an analysis indicates that Alternative A is preferable to Alternative B, sensitivity analysis might be performed by increasing the estimated capital cost of Alternative A and then examining the results of the analysis under this change.

Uncertainty—State of knowledge about outcomes in a decision which is such that it is not possible to assign probabilities in advance. Ignorance about the order or value of things.

15.4 Recurring Costs Determination

15.4.1 Utilities Costs

Once an alternative's energy consumption for each energy source is determined, the annual cost of these sources must be determined. For fossil fuels, cost determination is fairly simple. For natural gas, the servicing utility will provide the applicable rate that must be applied. Gas rates tend to be straightforward, having only three to six cost tiers and limited seasonal factors (summer gas tends to be cheaper than winter gas). For fuel oil and LPG, unit costs are provided by the vender, again with some seasonal cost adjustments.

Electricity pricing (sometimes referred to as electricity tariff or the price of electricity) varies widely from country to country and may vary significantly from locality to locality within a particular country. The price of power generation depends largely on the type and market price of the fuel used, government subsidies, government and industry regulation, and local weather patterns. In many countries outside the United States, electricity is a government monopoly and pricing is based on politics and taxation criteria. However, within the United States and many other countries, the following general definitions apply to basic commercial and industrial electric rate schedules:

kWh	Kilowatthour, or the amount of electricity consumed over a given period of time, typically, 1 month.
kW	Kilowatt, or the amount of instantaneous power utilized by the facility. A kW of load operated for 1 h consumes 1 kWh of electrical energy.
Demand	The *maximum* amount of power utilized in a given period of time. Typically, this is the average (or "integrated") kW over a specified time period (often 15–30 min, but some are as low as 5 min). In all demand-based electric rates, a new billing demand is measured for each month.
Ratchet	A methodology defined in the rate schedule for establishing the monthly "billing" demand as some function of the "actual" demand over the previous months. (e.g., a rate may establish the billing demand to be the highest monthly demand in the previous 12 months.)
Floor	This is the minimum billing demand and is usually a function of the contract demand (the maximum KW contracted for), the actual demand, and some fixed demand. The maximum of these values, excluding the ratchet, establishes the minimum billing demand.
Demand Charge	The rate at which the billing demand is charged to the consumer, per kW of peak demand for the month.
Energy Charge	The rate at which the kWh consumption is charged to the consumer.
Extra Facilities	Fixed monthly charges by the utility to amortize their expense of installing special equipment for the benefit of the consumer, such as installing a totalizer meter on a plant that had more than one delivery point for electricity.

Time	Some rates charge more for both consumption and demand that occurs in the utility's high demand periods (summer over winter, mornings, and afternoons overnight time). These periods are usually defined as "on-peak," with lower rates during "off peak" times.

Utilizing these common definitions, the utilities will often have significantly different approaches to pricing their power. For example, some utilities utilize the "Hopkinson Rate Structure" for their general service rates. This structure sets a unit demand charge, along with a two- or three-step consumption rate. These types of rates are straightforward and require, each month, only a calculation of the billing demand times the unit demand charge plus the consumption times the energy charge. The only aspect of these rates that is complicated is the application of the ratchet to properly determine the monthly billing demand.

Other utility rates are significantly more complicated, such as the "Modified Wright Rate Structure" that applies the concept of "hours use of demand" and is, thus, a load factor rate. On the average, there are 730 h in each month. If a consumer had exactly the same load for all 730 h, he would have a 100% load factor and 730 h use of demand since his constant load would be his demand. Rarely does this happen. Most facilities have a demand that lasts only 40%–60% of the hours of utilization (that, in themselves, are usually less than 730 h per month.) The energy rates in the first block are higher than those in the second block, which are lower than those in the third block. Thus, penalizes the low load-factor consumers since more of their cost would be computed in the first rate block.

Of obvious concern is the impact of electrical demand imposed by a design alternative on overall electrical costs. While this varies with each utility, the following rules generally apply:

1. For the low load-factor consumer, for example, less than 200 h use of demand per month, the unit cost for electricity will be high.
2. For facilities that have several months in which the billing demand significantly exceeds the actual demand (usually caused by high seasonal cooling or heating requirements), significant savings would accrue from reducing the impact of these short demand loads, changing rate schedules, or selecting an alternative energy source for cooling or heating. (Thermal storage and cogeneration also offer relief from high ratchet demand charges.)

Thus, to accurately compute the electrical energy cost for each alternative, the designer must determine the applicable rate schedule(s) that may apply and use these schedules to compute costs.

15.4.2 Maintenance Costs

Annual recurring or "uniform" maintenance is a difficult cost element to estimate since (1) most designers have little experience in following up on their designs to determine long-term costs and (2) actual costs vary widely based on the owner's maintenance philosophy and/or the training and expertise of the maintenance staff.

The Association of Higher Education Facilities Officers defines the quality of maintenance in accordance with Table 15-1.

TABLE 15-1

Maintenance Levels

Maintenance Level	Description
1	Showpiece facility
2	Comprehensive stewardship
3	Managed care
4	Reactive maintenance
5	Crisis response

For most facilities, the minimum acceptable maintenance level corresponds to Level 2 ("Comprehensive Stewardship"), which includes the following:

> Maintenance activities appear organized with direction. Equipment and building components are usually functional and in operating condition. Service and maintenance calls are responded to in a timely manner. Buildings and equipment are regularly upgraded, keeping them current with modern standards and usage.

Another impact on maintenance costs is the increasing degradation of quality of workmanship in the construction trades and the wider use of new and relatively untried materials and methods. Higher than normal maintenance costs can result from poor detailing at the design stage, including insufficient allowance for expansion and contraction, incorrect choice or misuse of materials, poor joint conditions between different materials or components, and so on.

Designers can contribute significantly to reducing maintenance costs by asking four questions when designing each component or part of the building:

1. How can it be reached?
2. How can it be cleaned?
3. How long will it last?
4. How can it be replaced?

The analysis of some alternatives will require that the designer estimate the current costs of major repairs or even component replacement that will be required in the future. *These costs are to be estimated at current dollar values.* The future cost is computed by multiplying the current cost by the appropriate year's escalation factor.

15.5 Component Service Life and Replacement

Every building is assembled from a great many individual components … materials, systems, subsystems, devices, equipment, and so on. But, *every one of these building components will ultimately fail and must be replaced or renovated when it does.* Historically, the life cycle of building components has been represented by the "bathtub curve" of Figure 15-1.

However, the bathtub curve does not define the point at which "normal aging" failures finally require that the component be replaced or renovated.

Maximum "performance life" of a building component is the time over which the component serves its anticipated function over the range from 100% (when installed or initially

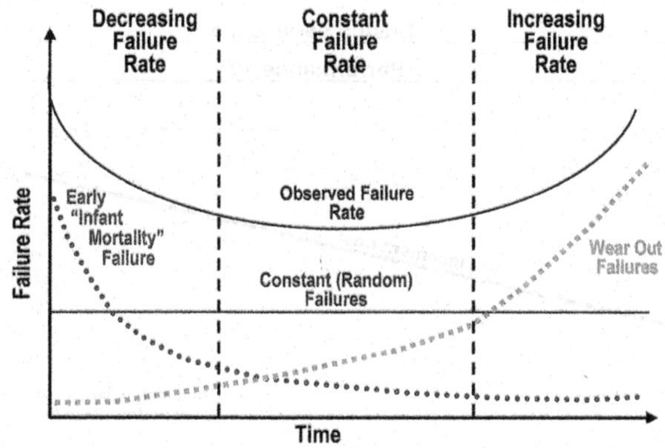

FIGURE 15-1
Component failure as a function of time. (Public Domain National Aeronautics and Space Administration.)

placed in service) to 0% (when it fails and the only option is to replace it). However, once the level of performance falls below some minimum (just how many roof leaks are acceptable?) *and* the cost of continuing to maintain a failing component threatens to exceed the cost of replacement, then the component has reached the end of its "design service life" (often referred to as its "economic life").

Figure 15-2, an enhancement of the "bathtub curve," ignores "infant mortality" failures associated with new construction and shows a more definitive relationship between

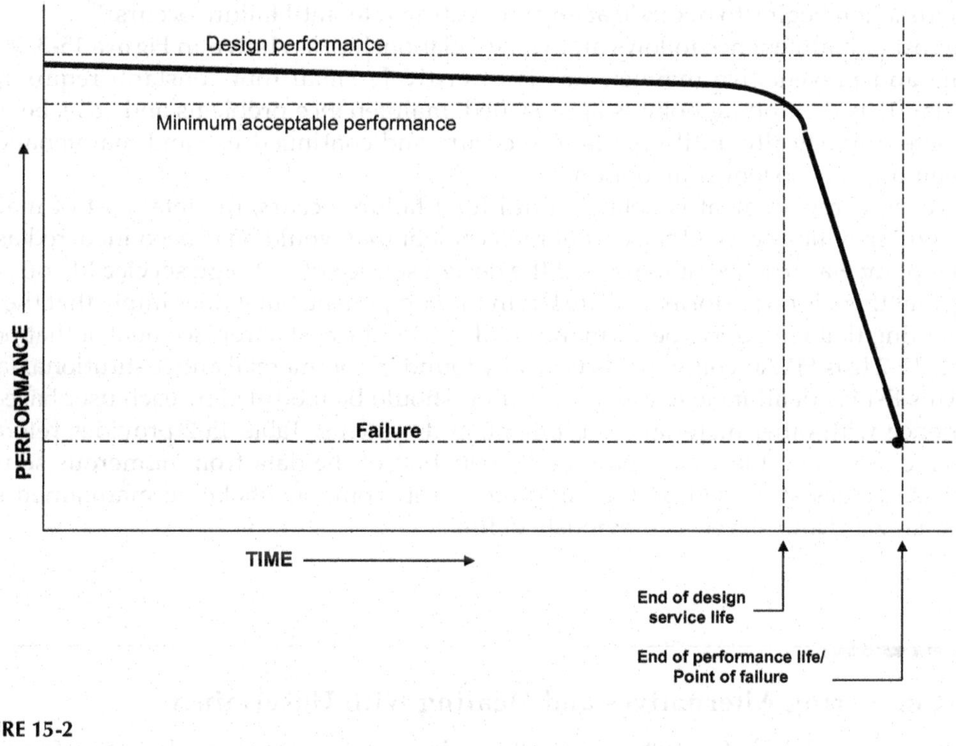

FIGURE 15-2
Component performance as a function of time. (Figure 2-2, *Effective Building Maintenance: Protection of Capital Assets,* Herbert W. Stanford III, PE The Fairmont Press/CRC Press/Taylor& Francis, 2010.)

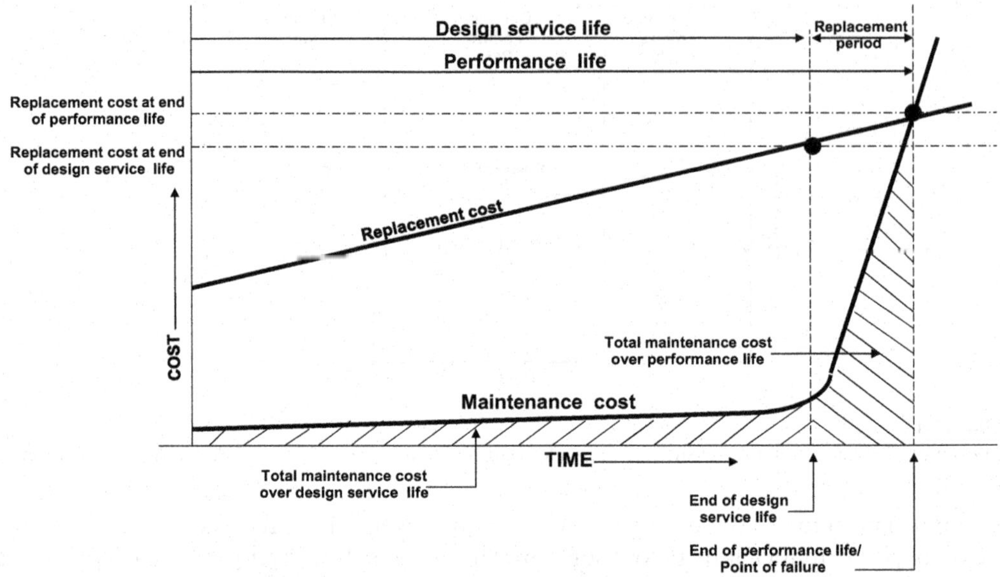

FIGURE 15-3
Maintenance/replacement cost as a function of time. (Figure 2-3, *Effective Building Maintenance: Protection of Capital Assets*, Herbert W. Stanford III, PE The Fairmont Press/CRC Press/Taylor& Francis, 2010.)

component performance and time. Over any component's performance life, function degrades slowly, even with adequate maintenance, until a "tipping point" is reached where function begins to degrade at an accelerating rate until failure occurs.

The cost of maintenance follows an inverse relationship, as shown in Figure 15-3. While routine and preventative maintenance is a relatively small (but constant) requirement from day 1, as the tipping point is approached, maintenance problems and costs begin to accelerate dramatically until final failure occurs and continued use and maintenance of the component is no longer an option.

However, if replacement is delayed until final failure occurs, the total cost of maintenance will inevitably exceed the cost of replacement that would have been incurred as the tipping point was reached at the end of the design service life. Design service life does not imply that there is no performance life left in the component, but it does imply that the cost of obtaining that remaining performance will exceed the cost of replacement at that point.

Table 15-2 lists HVAC components typically found in commercial and institutional facilities. This list is not all-inclusive and, over time, should be modified by each user based on experience with other materials, equipment, and systems. Table 15-2 provides reference economic or service life values based on evaluation of the data from numerous sources, and these values are based on the building and its components being maintained to at least a "stewardship" level, as previously defined.

15.6 Comparing Alternatives and Dealing with Uncertainty

After the life-cycle cost for each design alternative is computed, the relative performance of the alternatives may require further evaluation before a final selection is made.

TABLE 15-2

Typical Service Life Estimate for HVAC Elements

Elements	Sub-elements	Service Life (years)
1. Air-conditioning unit, packaged or split system	Window unit	7
	Packaged or split system, ≤10 tons	15
	>10 tons	17
	Packaged terminal air conditioner or package terminal heat pump	10
	Ductless split system	15
	Computer room unit	15
2. Heat pumps	Packaged or split system, ≤7.5 tons	15
	>10 tons	17
	Water source units	19
3. Air-handling units	Indoor	25
	Outdoor, rooftop	15
	Fan coil units	20
4.1 Boilers, conventional, non-condensing	Steel firetube, steam	25
	Cast iron, steam or hot water	25
	Electric, hot water	15
	Pressure burner, gas, oil, or dual fuel	21
	Atmospheric burner, gas	12
4.2 Boilers, condensing	Gas-fired, hot water type	25
5. Furnaces	Gas- or oil-fired	18
6. Unit heaters	Gas- or oil-fired	13
	Hot water or steam	20
	Electric	10
7. Radiant heaters	Gas-fired infrared heaters	15
8. Air terminals	Constant air volume, VAV boxes with or without reheat	20
	Fan-powered boxes	17
9. Ductwork	Duct work	30
	Diffusers, grilles, and registers	25
	Humidifiers	15
10. Fans (other than part of air-handling units)	Centrifugal	25
	Axial	20
	Propeller	15
	Exhaust, roof- or sidewall-mounted	20
11. Coils (other than part of air-handing units)	DX, water, or steam	20
	Electric	15
12. Heat exchangers	Shell-and-tube, steam, or HW	24
	Plate and frame	20
13. Chillers	Air-cooled, electric drive	15–17
	Water-cooled, electric drive	23
	Absorption, two-stage direct-fired	15
14. Cooling towers	Galvanized	10–15
	Galvanized w/stainless basin/deck	25
	Stainless steel	30+
	Ceramic	34

(Continued)

TABLE 15-2 (*Continued*)

Typical Service Life Estimate for HVAC Elements

Elements	Sub-elements	Service Life (years)
15. Condensers	Air cooled	20
	Evaporative	20
16. Pumps	Base-mounted	20
	Line-mounted	10–15
	Sump pump	10
	Condensate pump, indoor, above grade	15
	Condensate pump, outdoor, below grade	10–12
17. Controls	Pneumatic	20
	Electric	16
	Direct digital control	8–12
18. Piping, valves, and specialties	Water piping and valves	35+
	Steam piping and valves	35+
	Condensate piping and valves	15
	Water specialties: expansion tanks	20
	Flex connectors	15
	Strainers	25
	Air separators	25
	Pressure relief valves (PRVs)	10
	Backflow preventor (BFP) (reduced pressure principal/ reduced pressure zone)	15
	BFP (Double check)	15
	Steam specialties: PRVs	15
	PRVs	10
19. Insulation, pipe/duct	Fiberglass	15
	Calcium silicate	15
	Cellular glass	25
20. HVAC electric motors		15–30
21. Heating distribution	Piping and valves: Hot water	35
	Steam	20
	Condensate	12
	Manholes	35+
	Exposed insulation w/weather jacket	10
22. Cooling distribution	Piping and valves: Condenser water, CHW	35+
	Exposed insulation w/weather jacket	10
23. Fuel distribution	Gas piping: Above grade, painted	25
	Below grade, protected	30
	Below grade, plastic	40
	Oil piping: Above grade, painted	25
	Below grade, protected	30
	Below grade, plastic	40

15.6.1 Alternatives with Different Economic Lives

Sometimes, two alternatives for a design element will have different economic lives. For example, alternative building heating and cooling systems may have economic lives of 20 and 10 years, respectively. In this case, the "replacement" of the shorter lived system must be considered by assuming that this alternative system must be replaced once during the life of the other alternative. At the end of 20 years, both alternatives would require replacement, and thus, the analysis period is 20 years.

If the life of the shorter lived alternative were 15 years and the longer lived alternative were 20 years, the analysis becomes somewhat more complex. If we include a replacement cost for the first alternative at 15 years, then 10 years of life will remain at the end of the second alternative's life, and the cost comparison is not really valid. To solve this type of problem, the use of "equivalent uniform annual cost" is required.

The technique of *equivalent uniform annual cost* consists of converting each alternative into an equivalent hypothetical alternative having uniform recurring costs.

The total life-cycle cost for each alternative is computed *using the individual alternative's economic life*. The total life-cycle cost is divided by the alternative's economic life to obtain a *uniform annual cost*. The alternatives, then, can be compared on the basis of uniform annual cost, as illustrated by the following example:

Example: Compare Alternative No. 1, which has a 15-year life and a total life-cycle cost over that period of $175,000, with Alternative No. 2, which has a 20-year life and a life-cycle cost over that period of $200,000.

Solution: The uniform annual cost for each alternative is computed as follows:

Alternative No. 1: $175,000/15 years=$11,667/year

Alternative No. 2: $200,000/20 years=$10,000/year

Thus, Alternative No. 2 would be selected for this design element even though it has the apparent "higher" life-cycle cost.

Another approach is to make allowance for the associated residual service worth of the alternative with the longer life. This calculation involves identifying the future residual value at the end of the analysis period, then discounting the amount back to the present. The future residual value can be approximated by multiplying the future investment value (less future salvage value at the end of its service life) by the proportion of time remaining in the analysis period, compared to its service life.

15.6.2 Sensitivity Analysis

Performing LCCA requires that the designer makes certain "assumptions" upon which calculations will be based. As a professional, the architect or engineer is expected to make his assumptions based on experience and professional judgment after investigations. Obviously, if alternative life-cycle costs are computed on the basis of alternative assumptions, different assumptions probably would alter the outcome of the calculations. Thus, for certain design elements, sensitivity analysis is required.

Sensitivity analysis, very simply, consists of asking the question "what happens if this assumption is wrong by 10%, or 25%, or even 100%?" As professionals, HVAC designers should be able to make assumptions that will be reasonably close to actual conditions.

However, in some cases, particularly with the more innovative design elements such as passive solar systems, thermal storage, and daylighting, experience and detailed knowledge may be insufficient to overcome the uncertainty factor. In these types of design elements, or in any design element that the designer is not "comfortable," sensitivity analysis is required.

Sensitivity analysis consists of modifying the questionable assumptions so as to create a "family" of life-cycle costs and evaluating the resulting "sensitivity" of the analysis to the assumptions.

A good indicator that sensitivity analysis is required is (1) total life-cycle costs between alternatives that are very close and/or (2) when one alternative has a very fast simple payback over another alternative.

The outcome of an economic analysis can be sensitive to more than one type of assumption. For example, the construction cost for the alternative may be uncertain or the maintenance requirements may be "guesstimates." In these cases, alternative sets of assumptions will produce alternative life-cycle costs and the results can be plotted graphically. The resulting graphs may indicate a range of "required assumptions," that is, to produce the lower life-cycle cost, the assumptions must be within a specific range of values. The designer then can research his assumptions in order to establish the potential validity of these ranges.

15.6.3 Selecting Alternatives within the Project Budget

Design alternatives may be mutually exclusive or may be independent. Therefore, *the design with the lowest life-cycle cost* may include a number of alternative design elements resulting in a total project cost that exceeds the construction budget. In this case, the alternatives must be "ranked" on the basis of "cost effectiveness" so that the final design represents the lowest life-cycle cost *within the available budget.*

For mutually exclusive alternatives, the alternative with the lowest life-cycle cost is always selected. However, for independent alternatives, a comparative method of ranking cost effectiveness must be used.

The simplest comparative method is to evaluate the SIR between the alternatives and rank the alternatives in SIR order. Design alternatives resulting in a project cost within the budget are selected and all others, no matter how desirable, must be rejected.

To compare alternatives using SIR, first define the alternative with the lowest capital investment as the *base alternative.* Then, one at a time, all other alternatives are compared to the base alternative and ranked (ordered) on the basis of the computed SIR. The "savings" is the difference between the operating costs of the two alternatives. The "investment" is the capital cost difference between the two alternatives. If a design alternative has both a lower operating cost *and* a lower investment cost, there is no need to compute SIR since this alternative would be automatically selected. The more common case is that an alternative with a higher capital cost will have a lower operating cost. If the SIR is equal to or greater than 1.0, the savings equal or exceed the additional investment cost and the alternative is desirable.

For an alternative to be considered for selection, the SIR must be at least equal to 1.0, indicating that the operating cost savings at least equals all of the additional investment costs. The greater the value of the SIR, the better the economic performance of the alternative. *Thus, to select mutually exclusive alternatives for inclusion in the project, the alternatives with the highest SIRs are selected until the project budget limit is reached.*

15.7 Overview of LCCA

15.7.1 Basic Practices for LCCA

The following practices are typically applied when conducting life-cycle cost analyses for building design.

1. Design alternatives must be compared against a baseline reference alternate that is the lowest first cost of the alternatives being considered. The baseline alternate must offer a viable system, using state-of-the-art design features, and be in compliance with all project requirements. Where existing conditions form part of the baseline alternate, the analysis must not only include intended project work but also the additional costs necessary to achieve code compliance and reliable operation over the analysis period.
2. The analysis period should be chosen to fully represent all costs. When optimizing the design of a single system, all compared alternatives must be considered over the same analysis period. Where possible, the analysis period should be the smallest whole multiple of the service lives for the major systems involved in the analysis. Service lives of HVAC equipment can be found in Table 15-1.
3. Costs that have already been incurred or must be incurred, regardless of the chosen alternative, are excluded from the analysis.
4. Baseline and alternative first costs are typically those estimated for the construction award date. The LCCA can assume that the award date can be considered the zero point in time for the analysis period, with all other event times referenced to the construction award date. For greater simplicity, the year of design decision can also be considered as the zero point in time, and it can be assumed that the construction award will occur in that year.
5. Salvage values for alternatives are typically zero. However, in those cases where scrap values could impact decisions, the present value is calculated as its future value (scrap value) discounted back to the present.
6. Future one-time costs, such as replacement costs, are established by escalating a known today's value (using real growth rate) to its future value in the year it occurs, then discounting that value back to its present value (using a real discount rate).
7. *Due to possible margins of error in estimating costs, alternatives with a life-cycle cost differential of less than 10% should be judged inconclusive.*

15.7.2 Calculations and Analysis Tools

While there are equations for computing present value, and so on. based on a given inflation and discount rates, most computations can be more easily done using *factor values* rather the formulas. Factor values are pre-calculated values that correspond to a specific equation, a rate for interest and/or inflation, a discount rate (as applicable), and the number of years over which the analysis applies. Looking up these values can be faster and, perhaps, more accurate than using the equations. Appendix X contains factor tables commonly used for LCCA.

But, the use of computer programs can considerably reduce the time and effort spent on formulating the LCCA, performing the computations, and documenting the study. Building Life-Cycle Cost (BLCC) Program is an economic analysis tool developed by the NIST for the U.S. Department of Energy Federal Energy Management Program (FEMP). BLCC may be accessed at https://energy.gov/eere/femp/building-life-cycle-cost-programs.

Bibliography

Life Cycle Cost Analysis For State Facilities, State Construction Office, State of North Carolina (Raleigh, NC, 2001)

Engineering Economy (11th Ed.), William G. Sullivan, James A. Bontadelli, Elin M. Wicks (Prentice Hall, Upper Saddle River, NJ, 1999)

Guide To Effective Building Maintenance, Herbert W. Stanford III, PE, (CRC/Fairmont Press, 2010)

Handbook 135, National Institute of Standards and Technology (NIST), 1995

"Engineering Economics," John M. Watts, Jr. and Robert E. Chapman (Chapter 7, Section 5, NFPA *Handbook of Fire Protection Engineering*, 2002)

16

Building Information Modeling

16.1 Introduction

Historically, building design has been largely reliant upon two-dimensional technical drawings (plans, elevations, sections, and so on). Up until the 1990s, these drawings were produced via manual drafting techniques, augmented with photocopies materials. Computer-aided drafting (CAD) then revolutionized the industry with software-based drafting that also included some limited "attributes" and allowed three-dimensional (3D) representations to be easily made. Building information modeling (BIM) extends technical drawings beyond 3D, augmenting the three primary spatial dimensions (width, height, and depth) with time as the fourth dimension (4D) and cost as the fifth (5D). BIM, therefore, covers more than just geometry. It also covers spatial relationships, systems analysis, geographic information, and quantities and properties of building components (e.g., manufacturers' details). A BIM allows multiple designers to create a comprehensive design model, contractors and subcontractors to then add to the model with specific elements and information, and finally provides the owner with a detailed database of the building design and its construction, ready to integrate into a facility maintenance and management program.

An interdisciplinary group sponsored by McGraw-Hill Construction and the AIA California Council defined Building BIM as a "data rich," object-oriented and intelligent digital representation of the facility from which views and data can be extracted and analyzed to generate information that can be used to make decisions and improve the process of delivering and operating the facility.

BIM involves representing a design as combinations of "objects," some vague and undefined, some generic or product-specific, solid shapes, or void-space oriented (such as the shape of a room), and so on that carry their individual geometry, relations, and attributes. BIM design tools allow extraction of different views from a building model for drawing production and other uses. These different views are automatically consistent, being based on a single definition of each object instance. In BIM software, objects are defined in terms of parameters and relations to other objects, so that if a related object is amended, dependent ones will automatically also be revised. Each model element can carry attributes for selecting and ordering them automatically, providing cost estimates as well as material tracking and ordering.

For the professionals involved in a project, BIM enables a virtual information model to be handed from the design team to the various contractors and subcontractors and then on to the owner/operator; each professional adds discipline-specific data to the single shared model. This reduces information losses that traditionally occur when a new team takes "ownership" of the project and provides more extensive information to owners of complex structures.

16.2 BIM Advantages

Use of BIM goes beyond the planning and design phase of the project, extending throughout the building life-cycle, supporting processes including cost management, construction management, project management, and facility operation.

Building information models span the whole concept-to-occupation time span. To ensure efficient management of information processes throughout this span, a BIM manager might be appointed. The BIM manager is retained by a design build team on the client's behalf from the predesign phase onward to develop and track the object-oriented BIM against predicted and measured performance objectives, supporting multidisciplinary building information models that drive analysis, schedules, takeoff, and logistics. Companies are also now considering developing BIMs in various levels of detail, since depending on the application of BIM, more or less detail is needed, and there is varying modeling effort associated with generating building information models at different levels of detail.

Participants in the building process are constantly challenged to deliver successful projects despite tight budgets, limited manpower, accelerated schedules, and limited or conflicting information. The significant designer disciplines should be well coordinated, as two things can not take place at the same place and time. BIM aids in collision detection at the initial stage, identifying the exact location of discrepancies.

The BIM concept envisages virtual construction of a facility prior to its actual physical construction, in order to reduce uncertainty, improve safety, work out problems, and simulate and analyze potential impacts. Subcontractors from every trade can input critical information into the model before beginning construction, with opportunities to prefabricate or pre-assemble some systems off-site. Waste can be minimized on-site and products delivered on a just-in-time basis rather than being stock-piled on-site.

Quantities and shared properties of materials can be extracted easily. Scopes of work can be isolated and defined. Systems, assemblies, and sequences can be shown in a relative scale with the entire facility or group of facilities. BIM also reduces errors by enabling conflict or "clash detection" whereby the computer model visually highlights to the team where parts of the building (e.g., structural frame and building services pipes or ducts) may wrongly intersect.

Heating, ventilating, and air-conditioning (HVAC) engineers who implemented BIM early found themselves ahead of the curve. The use of BIM offered marketing opportunities for architects, so they began the migration to BIM before engineers and quickly found themselves without engineering partners capable of using the same design tools. Thus, early implementation of BIM allowed HVAC engineers to market the use of BIM as "additional services" to architects. Teaming with architects (and other engineers) utilizing BIM allowed design teams to market BIM services as an advantage to owners when pursuing new work.

In today's market, however, the use of BIM is essentially a standard service for all engineers. In the same way, CAD replaced drafting boards in the 1990s, BIM has replaced CAD in the 2010s. Without the marketing advantage and the ability to bill additional services to offset the higher costs of using BIM, what benefit does BIM really provide engineers?

One major advantage of BIM is coordination. When modeling the elements three dimensionally, it is possible to visually identify trouble areas between different building elements. Many software packages also allow different trades to communicate together. For example, when an HVAC engineer provides the correct horsepower for a pump, not only will the software populate the pump schedule to appear on the drawings but the electrical engineer can get this information from the HVAC engineers modeled element and automatically populate electrical circuits and panelboards.

Utilizing BIM improves integrated project approach to design. Using 3D views of mechanical equipment allows members of the team who are not versed in mechanical design see the impacts of decisions made faster (Figures 16-1 and 16-2).

Using BIM as a tool allows the user to communicate across all engineering disciplines, the architect, the owner, and the code officials more efficiently.

Another advantage is the software's ability to perform calculations for the HVAC designer. When the building elements are modeled correctly, the software is capable of summing airflows and water flows, performing pressure drop calculations, and much more. The accuracy of modeling is important. If you utilize plastic pipe in a model instead of copper, the computed pressure drop will reflect the differences in size and friction factor between the two types of piping.

But, like any tool, it is dangerous to solely rely on BIM. Anything generated by the software should be checked by the design engineer. This includes coordination, calculations, and anything else performed by the software. The software output is only as good as the data provided to it and the quality of the software itself. *Simply put, no program developer is going to pay for any error resulting from use of the software.*

FIGURE 16-1
Typical AHU 3D graphic created via BIM.

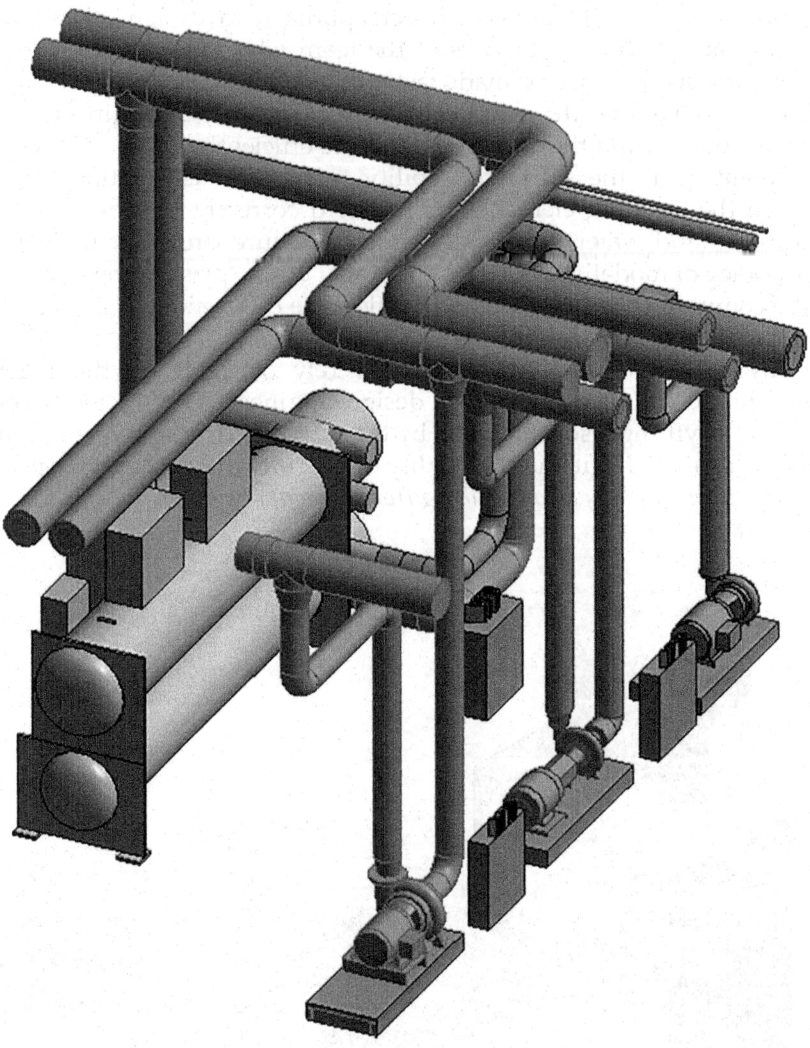

FIGURE 16-2
Typical piping 3D graphic created via BIM.

16.3 Applying BIM

Since BIM tools give users the ability to design their structure in 3D, a BIM platform will likely have the ability to visualize outside elements such as how sunlight will hit a building or how energy efficient the building is, as well as detect any potential clashes in the system or structure. This knowledge ensures the structure you are creating runs at peak performance.

Each phase of the construction process is planned out with models and even sub-models. BIM software enables users to sequence out these steps with the specific materials and crews needed for every individual phase of the process. Some BIM tools even allow you to animate the steps of your building process.

When it comes to the design and construction of a building or large structure, different teams will contribute various documents to the project. This might include drawings, customer requirements, phasing plans, and more. BIM solutions should be able to store all this information within the platform and allow multiple users to access it.

There is significant potential issues with the use of BIM:

1. **Incompatibility with construction partners:** BIM is not yet the universal standard in the construction industry, and a number of other tools can be used to create building models. As such, there is a possibility that contractors may not utilize BIM software to create buildings and therefore would not be able to use designer models.
2. **Lack of expertise:** BIM is a relatively new technology that is developing quickly. For that reason, there are a limited number of BIM software experts. Implementing BIM as the design tool on a project might require investment in additional training and support.
3. **Legal ramifications:** It is currently unclear where the ownership of BIM data falls. Whether it needs to be protected through copyright laws is also yet to be determined. When someone other than the original creator of a model contributes to the work, licensing issues may come up. And as there are many people involved in the design and construction of a building or large structure, this can potentially cause legal conflict.
4. **Cost:** Moving from CAD to BIM software is not a cheap undertaking. Not only does each designer has to pay for the software itself, but he/she will likely need to invest in training or consulting in order to learn the techniques of BIM. While switching will likely pay off in the long run, the upfront cost may be a significant hit, especially for small firms (which includes most HVAC design firms).

16.3.1 BIM Use by HVAC Designers

BIM software normally includes specific modules of design of major elements of any project, including HVAC systems. Starting with the special elements of the model created by the design architect, the HVAC designer then proceed to add HVAC components, piping and duct system layouts, details, and so on to be incorporated. At the same time, other design team members, such as structural, plumbing, electrical, and fire protection engineers are adding their respective elements to them model, automatically updating the model as evaluating each element to coordinate between the different design disciplines.

Frequent communication with the project designers is a must. Every BIM tool must have a feature that enables collaboration among colleagues. Collaboration features may include functionality such as real-time notifications and cross-platform communications.

The goal of communication is to maximize coordination (or lack thereof) between disciplines during the design process without exceeding the level of development (LOD). Early coordination can be done without the modeling software. Utilizing a piece of paper, a pencil, and the space requirements defined in Chapter 4, an HVAC engineer can make coordination easier prior to entering information into the model. This early coordination can minimize the time it takes to model the building elements. Once the information is modeled in the BIM software, each building element can be coordinated graphically with all building system.

Using the BIM software intelligently is important. "Modeling" each element in the design takes more time than creating basic two-dimensional plans. Because of this time investment, it is important to maximize the return from this investment of additional time and money.

BIM software is capable of doing many calculations for the HVAC designers. For the HVAC designer, sizing piping and ductwork, calculating total air pressure drop for a particular fan, to summing airflow of all terminal units, to calculating solar energy from a photovoltaic panel on a roof, and so on can be done by the software. But these calculations are only as good as the information entered into the program, and utilizing a specific software program does not remove the engineer's responsibility to verify the accuracy of the program outputs…*garbage in garbage out* is still a valid concern!

16.3.2 BIM Use by Contractors/Subcontractors

The construction contract(s) between the owner and the contractor(s) may require additional LODs to address construction sequencing and scheduling, documentation of equipment, creating of maintenance and operation requirements (that must be carefully developed based on design specification requirements and manufacturer requirements), and so on as components of the BIM.

16.3.3 BIM Use by Owners

Larger corporate and institutional owners often use facility management and/or facility maintenance software and want to incorporate the building model, especially 4D elements incorporated by the contractor(s). One drawback to this is that the older facility management/maintenance software typically is incapable of using BIM data and owners must be aware of potential problems before investing in 5D or 6D software.

16.4 AIA Contracts

Many owners consider the BIM "model" generated by the engineer as a "product." This is a misunderstanding that must be clarified early in the design contract process. Utilizing AIA Document B101, *Standard Form of Agreement Between Owner and Architect*, BIM modeling is an "Instrument of Service." AIA Document B101 defines "Instruments of Service" as representations, in any medium of expression now known or later developed, of the tangible and intangible creative work performed by each design discipline under their respective professional services agreement. The difference in risk between selling a "product" and a "service" is significant.

Historically, skilled contractors and tradesman utilized two-dimensional plans, a few details, and their own skill to construction buildings. Today, the skilled labor shortage in the United States has caused contractors to place more burden on the engineer. Today, contractors often anticipate an engineer's BIM model to be provided with a higher level of detail than was historically required by construction documents. AIA Document G202 *Project Building Information Modeling Protocol Form* will help project the engineer from this coordination. This document defines the LOD, or the specific minimum content requirements and associated authorized uses for each model element at five progressively detailed levels of completeness, as shown in Table 16-1.

TABLE 16-1

AIA Levels of Development for BIM

LOD	Description
100	Each element may be graphically represented with a symbol of other generic representation but does not rise to satisfy the requirements of LOD200. Information related to the element, such as cost and capacity may be derived from other elements. Each element may be analyzed on the basis of volume, area, orientation, and so on.
200	Each element is graphically represented as a generic system, object, or assembly with approximated quantities, size, shape, location, and orientation. Non-graphic information may also be attached to an element. Each element may be analyzed for performance of selected systems by application of generalized performance criteria assigned.
300	Each element is graphically represent as a specific system, object, or assembly in terms of quantity, size, shape, location, and orientation. Non-graphic information may also be attached to an element. Each element may be analyzed for performance of selected systems by application of specific performance criteria assigned.
400	Each element is graphically represented as a specific system, object, or assembly in terms of quantity, size, shape, location, and orientation, with detailing, fabrication, assembly, and installation. Non-graphic information may also be attached to an element. Each element may be analyzed for performance of selected systems by application of specific performance criteria assigned.
500	Each element is field-verified to confirm quantity, size, shape, location, and orientation. Non-graphic information may also be attached to an element.

Engineers' construction documents generally fall into LOD300, the level that requires the model to provide quantity, size, shape, location, and orientation of building elements. Additional information, typically incorporated by the contractors, vendors, suppliers, and so on may be incorporated for the BIM to rise to the level of LOD400 and LOD500.

Bibliography

Building Information Modelling: BIM Current and Future Practice, Karen Kensek and Douglas Nobel (Wiley, 2014).

Getting to Grips With Bim, James Harty (Routledge, 2015).

BIM Handbook, Chuck Eastman, Paul Teicholz, Rafael Sacks, and Kathleen Liston (Wiley, 3rd Edition, 2018).

17

Construction Contract Administration

17.1 Introduction

The vast majority of construction projects utilize American Institute of Architects (AIA) standard documents or owner-prepared "standard" documents. *Before designing any project it is the designer's responsibility to carefully review these documents and edit them as needed for each project.* For large projects or projects with special scope (such as sustainable designs), consultation with the design firm's attorney and/or professional liability insurance carrier may be warranted.

Typically, under terms of the *Owner–Designer Agreement*, services associated with construction *contract administration* (CA) are the responsibility of the designer. These services typically commence upon receipt of bids or negotiated construction prices and end, hopefully, on the date the designer executes the final certification for payment to the contractor(s) by the owner.

CA services are focused on the administrative aspects of construction instead of design. But, design issues and (all too often) issues having little to do with the design can surface during the construction period that may morph into litigation. While CA is a necessary, integral, and even helpful part of the overall construction process, careful attention by the designer to the potential litigation issues while performing CA is required to minimize claims risks. Finally, a CA "Practices and Procedures Checklist" is included in this chapter and should be utilized by the designer for each project.

The CA process is "successful" when the project is completed in accordance with the design intent, the owner is happy with the result, and nobody sues anybody. But, successful CA depends on the following:

1. *A clear set of expectations by the owner, designer, and contractor.* These expectations are often misunderstood and therefore either misstated or unstated. The owner–designer contract, the General Conditions of the Contract for Construction (generally AIA A201, but sometimes General Conditions are provided by the owner), sometimes the Supplementary General Conditions define critical construction administration requirements and describe them in sufficient detail to make the designer's scope of work clear to all who bother to read it…though some "education" of the owner and contractor is typically required.
2. *Good quality construction documents.* If the design issues have to be "fixed" during construction, then both the process and the results will be far from satisfactory to all parties.

Further, good communication during the construction period is critical. Mobile phones and computers with Internet access and email capability are standard equipment these days, but the designer's communication effectiveness is vastly increased by the following:

1. *Timely response to inquiries.* While the designer may not have the answer immediately, he should nevertheless respond as quickly as possible. Trust and confidence in the ultimate ability of the designer to resolve problems effectively as they arise is furthered by the designer's acknowledgment of receipt and a brief outline of the steps he will take to resolve such problems (including, where appropriate, a time frame for resolution).
2. *Use of standard forms.* The designer must provide standard forms for his own use and use by the contractor (requests for information [RFIs], submittals, change orders, substitution requests, payment applications, and so on). Communications in a format readily recognizable to all can speed and enhance the effectiveness of the designer's response. (Note that some clients may have their own standard forms that must be utilized.)

17.2 Design Period Considerations

Even before the project is released for bid or pricing, there are project elements that will impart significant requirements and potential liability for the designer during construction:

1. *Contract Time.* Usually, the "contract time" is the time for "final completion" of the work established by the owner. But, it is seldom in the owner's best interest to create an artificial demand for unrealistic early completion, and it is up to the designer to address this issue before the project is released. If, however, it is an economic or practical necessity to have the project completed by a certain time, then the construction process must be carefully evaluated by the designer and the project scope be revised, "phased," or otherwise adjusted to reasonably ensure completion within the owner's contract time.

   During the construction period, however, the contract time may be extended by scope changes that impact the "critical path" of the construction schedule. For any change to the project scope, a time change (if any) must be agreed upon at the same time that the scope change and associated price adjustment is defined.

   However, some owner's excessive pressure to maintain the completion date can lead to unfair decisions relative to contractor's requests for time extensions for excusable delay and on the time element of change orders. If the proposed contact time is inviolable to the owner, then the owner must actively cooperate by suppressing all changes that will affect the critical path of the project schedule and by timely performance of all contractual requirements. This would entail primarily making payments when due and making all decisions when needed. Undoubtedly, the designer will be forced into helping this process by fairly adjudicating time changes and dealing with the owner's concerns.

   Regardless of the owner's and contractor's best efforts to control construction delay, there will nearly always be some causes beyond anyone's reasonable control.

In addition to adverse weather, there may also be possible labor disruptions and actions of government. *One normally unacceptable delay is due to the contractor's claim of material or equipment unavailability or poor performance by subcontractors.* Both of these result from the contractor's "means and methods" of construction and are neither the designer's nor the owner's responsibility.

2. *Liquidated Damages.* When the completion date is of significant economic and/or functional importance to the owner, additional pressure can be applied by charging the contractor for overrunning the established by inclusion of a "liquidated damages" clause in the construction contract. A liquidated damages clause requires the contractor to forfeit a stipulated sum to the owner for each day of delayed completion. The intent is to provide an effective financial motivation to the contractor to complete on time.

   When liquidated damages are included as part of the construction contract, it is vitally important that the designer clearly define the date that construction actually starts by (typically the date of the Notice to Proceed), the number of calendar days of construction time allowed, and the target completion date. Construction time should be expressed in calendar days or by specific calendar dates.

   The actual damages likely to be incurred must be realistically estimated by the owner. The amount of liquidated damages should approximate the magnitude of actual monetary loss the owner would be likely to suffer if the project is not completed and ready for use at the designated time. *Amounts considerably greater than actual monetary losses would be considered punitive and should not be specified as courts and arbitrators are equally reluctant to enforce such penalties.*

   If the required project completion time is realistic and fair, project completion should not become an issue and no liquidated damages need be assessed. If the project time is unrealistic, some contractors will be opposed to liquidated damage provisions, considering them a penalty for failure to achieve the impossible, and will simply not bid or negotiate for the work. Other contractors may provide a price, but it will include the expected costs that will accrue from completing the project within a more realistic time. In that case, the contractors may include an allowance for potential liquidated damages in their breakdown of estimated costs. If all competitive contractors make such an allowance, the owner in effect will pay the liquidated damages if the project runs late and the contractor will gain a windfall if completion is on time or early. This would be equivalent to a bonus and would serve as a positive incentive for timely completion. A contractor who does not allow for liquidated damages in his price could end up with the project and the potential for conflicts over the application of liquidated damages escalates.

3. *Allowances, Unit Prices, and Alternate Bids.* Allowances may be included within a construction contract for a myriad of purposes, the most common of which is to allocate specific funds to offset the additional costs of dealing with uncovered and/or unforeseen conditions. Unit prices may be required for excavation (especially for rock removal) or other project elements where the quantity of work is difficult (if not impossible) to define. For most heating, ventilating, and air-conditioning (HVAC) projects, the use of allowances or unit prices is less common, but the use of alternate bid is routine.

   Alternate bids, when required, should always be defined in "additive" terms, that is, a cost increase to the base bid if an alternate bid is accepted. Deducting alternate bids are almost always priced to the contractor's advantage and should be avoided.

The scope of each alternate bid must be carefully defined on the drawings and by description in the specifications. If an alternate cannot be fully defined, then the use of an allowance or even unit pricing should be considered.

A special type of alternate is the "preferred product" alternate. Each of these alternates defines one or more manufacturers and/or equipment, used as the basis of design, which is preferred by the owner. The contractor, responding to this alternate bid requirement, will include the amount (if any) for providing the preferred manufacturer and/or equipment in lieu of any other manufacturer and/or equipment that may meet the requirements of the contract documents. In almost all cases, this will be an additive or "no cost" alternate, but it must be structured to include the potential of a cost deduction also.

4. *Special Inspections.* Seismic criteria may "trigger" the need for seismic special inspections and the HVAC designer must consult with the project structural designer to determine if the project meets the requirements for special inspections. Also, special inspections are typically required for all smoke control and smoke exhaust systems.

Special inspections are provided by neither the contractor nor the designer but by a qualified third party.

17.3 Preconstruction Procedures and Practices

CA services begin upon receipt of bids or contractor price proposals and include the following three steps:

1. Review and evaluation of the bids or proposed prices relative to the project budget, review, and evaluation of any alternative bids and/or unit prices, and so on.
2. Review and evaluation of the successful contractor's qualifications, including bonds and licensing. Minimum contractor requirements are defined in the Invitation to Bid, Instructions to Bidders, and General Conditions of the Construction Contract. Any contractor who cannot meet these qualification requirements must be rejected. Typically, the construction documents contain the following minimum qualification requirements:
   a. Contractor's licensing, based on project scope within the state in which the project is sited, demonstrates minimum knowledge and experience to perform the work.
   b. Payment and performance bonds verify financial capacity.
   c. Builder's risk insurance and other required insurance that shows that an insurance company is willing to assume certain contractor risks.
3. Based on the two preceding steps, issue a "Letter of Recommendation to Award" to the owner, and upon the owner's authorization, prepare (or help an architect prepare) the Contract for Construction to be executed by the contractor and owner.

Once the Contract for Construction is executed, the preconstruction period begins and the following steps are required before construction actually gets underway:

1. *Preconstruction Meeting.* The designer should take this opportunity to make sure the contractor has a clear understanding of the design intent. If this conference produces questions, it can still save many hours of labor (both design and construction) and thousands of dollars to address any issues raised before construction actually begins.

   This is also a good time to remind the contractor (in writing) of the importance of time deadlines, the negative impact of "dumping" submittals, and the prohibition of using submittals as vehicles for substitution or to offer contractor variations from the design.

   The agenda for this meeting must be developed and transmitted to all anticipated attendees *well in advance* of the meeting date to allow the contractor time to prepare the informational submittals by the meeting time.

   The designer must provide a report of the preconstruction meeting, documenting attendance, items discussed, and so on, utilizing a meeting minutes standard form and the Preconstruction Conference Agenda as an organizational outline.
2. *Submittals.* The contractor is required to immediately submit a list of proposed major subcontractors and material suppliers, proposed construction schedule, a proposed schedule for submittal submissions, and a proposed schedule of values. Each of these must be reviewed carefully by the designer and, as necessary, needed adjustments defined in writing for the contractor.
   a. The proposed construction schedule must confirm the required contract time, phasing, and so on. Details of the construction schedule are defined in the General Conditions to the Contractor for Construction.
   b. The proposed schedule for submittal submissions must be based on the master submittal list that defines the type(s) of submittals required by each specification section within each division. The proposed submittal schedule should be coordinated with the proposed construction schedule and include the following:
      1. The earliest date at which the product or element will need to be delivered for installation.
      2. The vendor's defined (*most pessimistic*) delivery time.
      3. The lead time required for the supplier and contractor to prepare the submittal.
      4. The time required for the contractor to review the submittal and process it to forward to the designer.
      5. The time for designer review (typically 10–14 calendar days from the date of receipt).
      6. Additional time required to correct the submittal if rejected, resubmit, and obtain the designer's review. (*Given the often poor quality of submittals from contractors and suppliers,* ***always*** *assume that at least one resubmission will be required.*)
   c. The proposed Schedule of Values has great importance. This schedule represents the contractors' allocation of costs to the various elements of the

construction project and must be evaluated in terms of both fairness and measurability. "Fairness" defines that the contractor has fairly allocated payment elements in a manner truly representative of cost elements, for example, not "frontend loaded" payments in order to maximize income during the early stages of the project. "Measurability" means that construction progress, in terms of percent of completion, for each construction element can truly be evaluated by the designer as construction proceeds...*you can't evaluate something you can't measure.*

The contractor should include a "general conditions" line item to be treated as equal to the contractor's project operation cost over the period of the contract time to be allocated on a monthly basis in each payment application. Additionally, line items for "Owner Training" and "Project Closeout" are typically required.

17.4 Construction Period Procedures and Practices

The construction period begins upon issuance of the Notice to Proceed to the contractor, which "starts the clock" for the allowable construction period defined in the construction contract. The contractor must submit complete lists or schedules of all proposed subcontractors and material suppliers and of all proposed construction materials and equipment immediately following the Notice to Proceed. Processing of the second and subsequent Certificate for Payment by the designer should be withheld until these lists have been submitted.

1. *Construction Progress Meetings.* Each meeting with the contractor provides additional opportunities to ensure that the contractor understands the design and is carrying out the design intent. If there is a true flaw or issue with the design (as opposed to a contractor's identification of a "change order" opportunity), then the meeting gives the designer yet another opportunity to provide a practical, cost-effective design solution.

   The designer must provide a report of project meeting, documenting attendance, items discussed, and so on, utilizing the meeting minutes standard form.
2. *Submittals Review.* The contractor may say he understands the construction documents but talk is cheap. Submittals are the designer's way of making sure the contractor "walks the walk."

   It is the contractor's responsibility to provide submittals that demonstrate compliance with the contract documents and the design intent. Thus, the submittal process helps the designer determine if the contractor actually understands the design intent and is prepared to construct. The purpose of the review is checking for conformance with the information given and the design concept expressed in the construction documents. It is not to determine accuracy or completion of dimensions, quantities, and installation and performance of equipment or systems, nor does it constitute an "approval" of construction means, methods, techniques, sequences, or processes.

   Note that "voluntary" submittals from the contractor, that is, submittals not required by the contract documents, should not be accepted for review as this review may result in the designer assuming additional, unintended liability.

Because submittals are often the subject of delay damages by contractors, the designer should be vigilant for contractor failures to adhere to the submittal schedule. Failures to provide timely submittals or attempts at "dumping" submittals on the designer in a fashion preordained to cause late response should be documented early in the process.

HVAC submittals must be in accordance with the requirements of the General Conditions, Division 01 specifications, and Division 23 specifications and demonstrate compliance with technical requirements by reference to each subsection of the Division 23 specification.

Material and equipment schedules, catalog cuts, manufacturers' data and shop drawings, and field working drawings as required by individual sections of the specifications must be provided. Shop drawings, technical data, and other such submittals required by individual sections also must be provided.

Equipment drawings, manufacturer's installation instructions as shipped with the equipment, field working and location drawings, wiring diagrams, and coordination drawings must be provided by the contractor for items of equipment, sleeves, foundations, curbs, wiring, ductwork, piping, and so on, as necessary for information and coordination of all trades. These drawings must be provided sufficiently in advance of installation to avoid delays and removal and reworking of installed work, so as to provide information to other trades when and as required.

Submittals are required by the General Conditions and Division 01 specifications to be checked before submission by technically qualified employees of contractor for accuracy, completeness, and compliance with contract requirement.

Finally, and most importantly, submittals must demonstrate compliance with technical requirements by reference to each subsection of the specifications. *Where a submitted item does not comply fully with each and every requirement of the specifications, the submittal must clearly indicate such deviations by being marked "**NON-COMPLYING FEATURE**."* Additional information that might assist the designer in product evaluation may be included with the submittal. This information should indicate how a specific noncomplying feature is believed by the contractor to meet the intent of the specification.

It is the contractor's responsibility to demonstrate compliance with the specifications and to clearly indicate any features that do not meet the specifications. It is not the designer's responsibility to identify noncompliance. Substantial noncompliance, as determined by the designer, is grounds for rejection of the submittal. Discovery of noncomplying features that have not been properly identified as such on submittals may require, at any stage of construction, the removal and replacement of the noncomplying item(s).

The designer is required to review submittals and a careful and complete review by the designer is as important as the original design itself. *This review is only for general conformance with the design concept of the project and general compliance with the information given in the contract documents.* Corrections or comments made on the shop drawings during this review do not relieve the contractor from compliance with the requirements of the plans and specifications. The contractor is responsible for dimensions to be confirmed and correlated at the jobsite; information that pertains solely to the fabrication processes or to the means, methods, techniques, sequences, and procedures of construction; coordination of his or her work with that of all other trades; and for performing all work in a safe and satisfactory manner. Finally, the contractor is responsible for any delay caused by his failure to observe submittals requirements and the time for completion of his contract will not be extended because of such delays.

The submittal review comment categories are defined in the guide specifications as follows:

1. *Reviewed.* Fabrication and installation or erection may be undertaken.
2. *Exceptions Indicated, Revise, and Proceed.* Fabrication and installation of erection may be undertaken. However, the contractor must comply with all notes or corrections indicated.
3. *Exceptions Indicated, Revise, and Resubmit.* Neither fabrication, installation, nor erection may be undertaken. Resubmit corrected copies for review. Corrections shall be limited to items marked, except that changes required in order to coordinate the corrections indicated shall be made.
4. *Rejected, Resubmit.* Neither fabrication, installation, nor erection may be undertaken. The contractor must revise the entire submission to comply with the information given in the contract documents and resubmit.

Submittals returned to the contractor with a "reviewed" or "exceptions indicated, revise and proceed" comment need not be resubmitted, except that corrected copies of "exceptions indicated, revise and proceed" submittals must be furnished for record when requested.

Submittals returned to the contractor with the "revise and resubmit" or "rejected, resubmit" comment must be corrected to comply with contract documents' requirements and resubmitted for review.

Shop drawings of work that involves more than one subcontractor must be coordinated by the contractor and submitted under one cover. No items may be fabricated, nor any portion thereof shipped to site, prior to receipt by the contractor of all applicable submittals, including manufacturer's data, samples, and shop drawings bearing a "reviewed" or "exceptions indicated" comment only.

3. *RFIs.* If used properly by the contractor, each RFI will seek clarification concerning some element of the contract documents or the design intent. All too often, however, an RFI will mask either the contractor's desperate plea to tell him "how to build it" or to setup for a change order premised on an alleged design error or omission (typically by issuing dozens of RFIs). The designer must be on guard for and resist either of these attempts.

   Each RFI must be handled promptly and professionally. It should answer the question "what is the design?" not "how am I supposed to build it?" The designer should also be thinking about defending the design from error or omission change order: Do the plans and specifications taken together describe the designer's intent so that the average, reasonably prudent contractor would be in a position to fabricate and install the subject element? Is this an attempt by the contractor to make up monies left out of the bid for the specified item?

4. *Changes to Work.* No set of contract documents are perfect, and it is routine for changes to the work to correct flaws in the contract documents and/or to address uncovered or unanticipated conditions. Changes represent modifications to the construction contract and may address the scope of the work, the construction contact amount, and/or the contract time.

a. A "Construction Change Directive" may be issued by the designer, as approved by the owner, to address the need for a change to the work that does not impact the construction contract amount and/or the contract time.

b. A "Change Order," as approved by the owner, may be issued for more substantive changes that impact the construction contract amount and/or the contract time. To initiate the change order process, the designer must define the required change, including revising and/or supplementing the contract documents, and issue a request for a change proposal from the contractor. Once the contractor's change proposal is received, the designer and owner must review it, negotiate any required modifications to it, and issue the change order for execution by all parties.

c. Project specifications should include a specified allowance for "uncovered and unforeseen conditions." When the need for changes to the scope of work, based on uncovered or unforeseen circumstances, arise during construction, and with the owner's approval, the designer may issue a Construction Change Directive authorizing the change and providing that agreed-upon contractor costs be deducted from the allowance. At the project completion, then, a single change order can be issued to summarize all debits to the allowance and provide for final deduction from the contract amount for any funds remaining.

5. *Contractor Payment Applications.* Typically, the designer is required to review and "certify" payment applications from the contractor, most commonly submitted on AIA forms G702 and G703.

   There are two major caveats for the designer when reviewing payment applications:

   a. Each payment application must truly reflect the work that has been accomplished and be in strict compliance with the Schedule of Values.

   b. The payment application process is taking on added significance as owners are using it as a vehicle to hold the designer accountable for payments to contractors for work done poorly or not at all, yet for which the designer certified payment. *The designer should certify payment only for work that has been completed in accordance with the contract documents.*

6. *Site Visits and Construction Observation.* Typically, the designer is required to visit the site at intervals "appropriate to the stage of construction." More often, the timing of these site visits will be dictated by the owner–designer agreement. These site visits may be scheduled weekly, biweekly, or monthly as dictated by the project scope and the owner's requirements. During these site visits, the designer must observe the quantity and quality of the work that has been done and prepare a "Field Report" delineating these observations, *particularly observed deviations for the requirements of the contract documents.*

   Contractors supervise construction while the designer observes construction to determine if the work is being done in accordance with the contract documents. This distinction is critical...the designer must not become so involved with the construction process that he/she can be deemed as "supervising" the work, a condition that imposes potentially significant and inappropriate legal consequences. Thus, designers do not "inspect" the work, rather they "observe" the work with the intent of identifying noncompliance with the contract documents. The means and methods utilized by the contractor are not the responsibility of the designer (though there is an exception if the designer sees that these means and methods create an immediate hazard to jobsite personnel).

Adequate and proper observation by the designer is a necessity. If a contractor fails to perform work in accordance with the contract documents and the designer certifies the Application for Payment for that work, then he/she may become liable for damages due to breach of duty to the owner by failing to observe unacceptable work. *Thus, observation must address both the progress and the quality of the work so as to keep the owner informed through the construction period.* Observe the work on a regular basis and do not be bashful about rejecting any work that is deemed noncompliant with the contract documents in the next Application for Payment.

The designer has the authority to reject work that does not comply with the contract documents or is otherwise "defective," for example, does not meet code requirements and demonstrates poor workmanship. This action by the designer must be fully defined in a field report provided to both the owner and the contractor. If the rejected work is significant in scope, the designer should issue a "Nonconforming Work Notice." Note that the designer does not have the authority to stop the work, only the owner or an authority having jurisdiction has that authority. But, when nonconforming work is found to be egregious, the designer may recommend to the owner that the work be stopped until all defects are determined and corrected.

7. *Contract Time.* The designer must take care not to be responsible for holding up construction progress for which the owner will have to pay the contractor delay costs. Nor will the contractor look kindly on construction delays caused by the designer's procrastination or lack of diligence in observation, submittal review, processing of payment applications, or decision-making, especially if liquidated damages are part of the contract requirements. These can lead to delay claims against the designer by the owner, the contractor, or both.

17.5 Completion and Closeout Procedures and Practices

As the construction period nears the end of the contract time, the designer should begin the process of accumulating required closeout documents, reports, and so on from the contractor and schedule inspection dates for substantial and final completion.

1. *Project Completion.* Perhaps the last (and most telling) opportunity to confirm the design's successful articulation are the requirements for determination of "substantial completion" and "final completion." During this period, the designer is obliged to determine that the work is available, functionally and aesthetically, for the owner's intended use and purpose. Everything that impedes a determination of substantial or final completion detracts from the project meeting the design intent. These inspections should be made with this in mind as well as the legal and financial consequences of completion. The safety net of a "punchlist" (the designer's list of construction elements that remain to be completed or corrected by the contractor) will be scant consolation if the list is so long or involved that everyone loses enthusiasm prior to rushing off to the next project. Care should be taken to distinguish a "punchlist" before final completion from a list of problems

or unfinished items which should have impeded the designer's determination of substantial completion in the first instance.

2. *Project Closeout*. The General Conditions of the Construction Contract and Division 01 of the project specifications typically define the actions and documentation required of the contractor prior to final payment. Closeout submittals must be reviewed by the designer. *Closeout can be a fairly lengthy process, and therefore, it is critical that the designer initiate the closeout process as early as feasible.*

17.6 Avoiding Litigation Pitfalls

There are three "pillars" that support properly executing the design professional's CA services and avoiding litigation that may result from these services.

1. *Control*. While it is anticipated that all designers want to perform their services in the best possible manner, it is imperative that there be some semblance of control of their client's expectations when it comes to construction administration. Whether the designer wants to believe it or not, in a majority of construction projects, it is the owner's expectation that the designer will use the utmost care and diligence in protecting the owner from deficiencies in the construction of the project by the contractor. *In other words, regardless of the terms of the contract and whether the design professional has minimal construction administration services responsibilities, the owners will typically hold the design professional to a higher standard, more akin to a construction manager.* In fact, it is fairly standard in litigation, especially when dealing with unsophisticated owners, that the owner will rely on the designer to see to it that the project is a success, and *any failure*, whether it is during or after the project, regardless of whether it is actually the result of a failure on the part of the designer, may lead to a professional liability claim.

   Educating the owner regarding the definition of construction site observations as well as the specific role of the designer during construction should lead to fewer or diminished exposure claims. The shifting of the owner claimed responsibility for construction management from the designer to the owner will make the owner more vigilant when dealing with issues that may arise during construction with the contractor. Additionally, it would be more difficult for an owner (who will hopefully by this time be more involved in the project) to simply state that they were relying completely on the designer in all of the decision-making. Stories abound in professional liability lawsuits where the owner claims that every decision they made was based on representations or decisions of the designer. Understanding that this is how owners view the role of designers during the construction phase should put every designer on their toes when providing site observation services.

   In the event of litigation against an designer by an owner, the owner will retain a consultant "expert" to give opinions as to the level of service provided by the designer and whether the designer "fell below the standard of care" in their work. It is generally accepted by those consultants that the project designer performing their site observation services are not required to make exhaustive inspections. However, those consultants are generally of the opinion that the designer is required to identify defects in the workmanship of the contractor. Additionally, those consultants would also agree that construction site observations are simply to ensure that the construction by the contractor being performed is in "general conformance" with the plans and specifications. As lawyers often see in lawsuits, however, there can be a wide variety of opinions as to the level of observance or "inspections" that a designer needs to perform pursuant to its construction site observations.

The designer who provides CA services should keep in mind that an owner's consultant has the benefit of hindsight and gets to play "Monday morning quarterback." Whether this is fair or not is irrelevant as to whether or not a claim will be brought.

Controlling the client's expectations is key to how the owner will feel during and after the completion of the project. Sometimes, how an owner "feels" about the services provided by the designer could be the flame that stokes the fire of litigation. If the owner's expectations are controlled during the construction process and there is proper communication by the designer (see below), the owner will feel more satisfied as to the services performed by the designer pursuant to the agreed upon scope. Conversely, if these expectations are not controlled, the owner will first look to the person holding a professional license as the culpable party for the failures of the project.

2. *Competence.* Most construction administration liability claims against designers involving their site observation services can be avoided by having the designer perform its work competently. The term "competence," however, is often described differently by owner's consultants in litigation. Whether the designer's representative is a technician, designer, or designer in training, there is a general level of experience and knowledge that is required when evaluating the work of the contractor. Attention to detail as to the construction documents is paramount in properly evaluating the contractor's work. Furthermore, an understanding of proper construction techniques and practices, as well as materials used, will help the designer in evaluating the work of the contractor.

   Is the contractor performing his work at a reasonable pace pursuant to the construction schedule? Does it appear that the contractor has skilled workers as well as appropriately credentialed subcontractors? While a successful project is a collaborative effort by the owner, designer, other consultants, and contractors, simply notifying the contractor's superintendent/foreman is not sufficient to notify the superintendent/foreman of sloppy workmanship and/or defective construction and does not satisfy the designer's responsibilities in terms of evaluating the work.

   Observing discrepancies in the construction should alert the designer to more thoroughly evaluating the contractor's work (even if this means spending a little more time at the site) and, as always, communicating to the owner. Knowledge of the materials being used by the contractor and their performance requirements, as well as their installation instructions, are critical requirements of a designer when performing their evaluations of the work. This is especially true when dealing with a new product that has not been around in the marketplace for very long. The designer is the professional with the license and eventually the buck will stop with the person holding the most credentials.

   While it is generally accepted that the designer performing construction administration site observations is not required to inspect the work, it cannot be understated that many litigation consultants analyze the work of designers during the project designer's construction administration services and routinely attempt to hold the designer up to that higher standard of investigation.

While the contractor is responsible for the means and methods of construction, invariably issues will come up during the construction process where the contractor will request information from the designer as to how to build a specific component. For most projects, comes in the form of an RFI, documenting the requested information.

However, it is not uncommon that during site observation work, the contractor will simply ask a question of the designer's representative who will, in turn, agree to a proposal set forth by the contractor. Prudent designers and their representative at the site should not automatically assume that the contractor has the knowledge to properly construct the building component in question. Therefore, the designer's representative must use all resources available to make the determination of whether the recommendation by the contractor is, in fact, the proper course of construction. Additionally, should the change be substantive in changing the character, materials, or look of the project, the change should be vetted through the owner. Again, the buck will stop with the designer and not the contractor should the proposed change result in defective workmanship and/or unacceptable deviations from the owner-approved plans.

Another common theme in designer liability claims are instances where the designer and the contractor have worked on several projects together and there is a general familiarity with one another. While this is often useful in the success of a project, it can also have some downside. For example, while the two companies may have worked together, it does not automatically assume that those persons involved in the actual day-to-day operations of the project are familiar with each other nor how things have been accomplished on other projects. What one superintendent may have understood the plans to mean on one project may be completely different from a superintendent on the next project. Additionally, the designer's representative on one project may assume that the contractor knows how to install a specific building component based on the history of the other project when, in fact, the contractor's superintendent and workers were not involved in that other project. The level of attention to detail during site observations and evaluations of the contractor's work should not lessen just because the designer is familiar with the contractor.

Most of the work by the designer is completed by this stage and most of the money has been paid from the owner to the designer. However, it is imperative the designer not lose focus during the site observation and evaluations of work stage of the project. The designer should see the project to the finish line and avoid the claim.

3. *Communication.* Designers should not underestimate the value of properly communicating and, when necessary, documenting those communications with the owner's representative. At a minimum, the designer representative should be noting discussions where decisions are made in field notes that are, in turn, copied to both the owner and the contractor memorializing the communications and decisions. This will help combat the owner from later taking an opposite position or conveniently forgetting the conversation and its substance. *Communicating clearly and effectively can reduce the chance of a liability claim in the first instance.*

While there is a working partnership with the contractor in attempting to get the project completed, it should be remembered that the designer is working directly for the owner and the law will treat the designer as the owner's agent. *In other words, the designer is required to "have the owner's back."* The duties of the designer flow to the owner and not to the contractor. Therefore, a prudent designer will be vigilant in their advocacy for the owner in any situation where there appears to be conflict between the owner and contractor.

When providing CA services, it is imperative that the designer not only communicate what is observed at the site but also to detail it in writing to both the owner as well as the contractor if he/she observes any problems of concern or deficiencies in the work of the contractor. While this may appear to be a documentation that would be self-serving and cover the designer, that is exactly the point. Keep in mind that when owners hire consultants during the litigation process to evaluate the work of the designer, the first thing they are looking at is for some basic factual background of the project and the work and services performed by the designer. Without proper documentation, the consultant is simply left to their own imagination and determination of the level of services provided by the designer. Do not give them that opportunity.

Documenting instances of construction deficiencies and advising the owner or the contractor of those deficiencies puts them on notice of the issue and should hopefully rectify the deficient construction. It is key that when the designer notices these deficiencies that there is some follow-up observation to ensure that the contractor is back on track for performing that specific construction pursuant to the contract documents. By documenting the deficiency, it shows that the designer is vigilantly performing its evaluation of the contractor's work and advising the contractor and the owner of those particular issues. Without properly documenting the deficiency in writing, it opens up the designer for a claim by the owner's consultant during litigation that the designer should have noticed the defect but did not. Regardless of whether or not the designer verbally communicated the issue to the owner and contractor, it is important to keep in mind that memories fade over time and professional liability claims against designers may take several years before they are alleged. Additionally, those individuals that the designer may have had verbal communication with may no longer be affiliated with the project in any way, and therefore, the consultant for the owner is again left to use their own imagination and speculation as to the level of services performed by the designer.

Another critically important issue in communicating with the owner is when a change order may be required which creates additional costs. The reason for the change order, as well as the decision of whether to accept or decline the change order, is vitally important to be put down in written form; written documentation is at the crux of many professional liability claims against a designer.

17.7 Construction Administration Checklist

The following checklist is based on projects that are delivered through the "design-bid-build" process. For negotiated price projects or "design-build" projects, some changes may be required.

17.7.1 Construction Administration Practices and Procedures Checklist

Pre-Bid Period. Consult with owner to define construction contract requirements to be incorporated in the project specifications, including the following:

1. Determine owner's required *Contract Time* or *Date of Completion* (but **not both**).
2. When the allowable construction time appears to be unrealistic, revise project scope, require phasing, or otherwise adjust the construction process.
3. Determine required *Liquidated Damages* (if any), ensuring that damage amounts are fair and reasonable.
4. Define *Allowances.*
5. Define *Unit Prices.*
6. Define *Alternate Bids* (especially for owner-preferred materials and/or manufacturers).
7. Define areas for contractor parking, laydown, and storage, and so on on the drawings.
8. Define use of existing utilities and/or sanitary facilities by contractor's personnel (as applicable).
9. Obtain owner's work rules (if any) and include as part of the project specifications.

Contract Award Period. Receive bids or proposals from contractors, evaluate contractor submissions, and prepare construction contract.

1. Review and evaluate the bids relative to the project budget, review and evaluate any alternative bids and/or unit prices, and so on.
2. Review and evaluate the low bidder's qualifications, including bonds and licensing.
3. If the low bidder's qualifications are unacceptable or the low bid must be rejected for procedural failures, reject the low bid and repeat steps 1 and 2 for the next low bidder.
4. Based on steps 1 and 2, issue a *Letter of Recommendation to Award* to the owner.
5. Upon the owner's authorization, prepare the *Contract for Construction* to be executed by the contractor and owner.

Preconstruction Period. Issue Notice to Proceed to the contractor and ensure that the contractor understands the project requirements and procedures.

1. Once the *Contract for Construction* is fully executed, issue the Notice to Proceed to the contractor.
2. Obtain contractor's list of major subcontractors and material suppliers and, with the owner, review and take other appropriate action.
3. Prepare and maintain the project directory (contacts list).
4. Preconstruction conference:
    a. Prepare and transmit a preconstruction meeting agenda well in advance of the conference date. Edit this agenda based on the specific project requirements. Delete references to specification sections that are not included in the contract documents.

   b. Prepare and distribute preconstruction meeting minutes as soon as possible after the conference date.
5. Obtain and review contractor's construction schedule to confirm contract time, phasing, and so on.
6. Obtain contractor's submittal schedule and review and take other appropriate action.
7. Obtain contractor's schedule of values and review and take other appropriate action.
8. Obtain contractor's submittals for site storage, lay down, and parking plans; fire protection plan, moisture and mold control plan, dust and containment plan, and waste management plan; review or take other appropriate action.
9. If required, notify the owner to submit applications for permanent gas, electric, water, telephone, and other services required for the project.

Construction Period. Periodically review the work by the contactor and ensure that the requirements of the contract documents are met.

1. Receive and respond to each submittal required by the technical sections of the specifications.
2. Periodically visit the site and observe the work.
   a. Review the *Owner–Designer Agreement* and schedule site visits in accordance with that agreement.
   b. Prepare and distribute a "Field Report" for each visit to the site.
   c. Prepare and distribute a "Nonconforming Work Notice" to address *significant* work that does not comply with the contract documents or is otherwise defective.
3. Keep the owner informed on the progress of the work.
4. At each scheduled construction progress meeting:
   a. Obtain and review the contractor's updated progress schedule and advise the owner of potential revisions to anticipated date of completion.
   b. Review status of submittals and concurrence with submittal schedule.
   c. Review status of any construction change directives, change orders, and so on.
   d. Prior to the first application for payment, receive, review, and approve, if appropriate, contractor's schedule of values.
   e. Record minutes of each meeting and distribute all attendees (ideally within 1 week of the meeting).
5. Receive and review the contractor's applications for payment; respond appropriately.
6. Verify requirements, if any, for reduction in retainage and have contractor submit consent of surety documentation.
7. Obtain and review required test reports (if any).
8. Receive submittals:
   a. Maintain submittal log and routinely check log against submittal schedule. Advise contactor of discrepancies.

  b. *Reject any "voluntary" submittal made by the contractor.*
  c. Review or take other appropriate action for each required submittal and return to the contractor.
9. Receive and respond to each RFI with a written clarification, a Construction Change Directive, or a request for a change proposal from the contractor. ***Reject*** *RFIs that raise trivial questions and/or are obviously self-serving by the contractor.*
10. For each potential change to the work:
    a. Issue "Construction Change Directive," as approved by the owner, to address the need for a change to the work that does not impact the construction contract amount and/or the contract time.
    b. Issue a "Change Order," as approved by the owner, for more substantive changes that impact the construction contract amount and/or the contract time.
       1. Define the required change, including revising and/or supplementing the contract documents.
       2. Issue a request for a change proposal from the contractor.
       3. Once the contractor's change proposal is received, the designer and owner must review it, negotiate any required modifications to it, and issue the change order for execution by all parties.

Substantial Completion. Ensure that the work is complete enough for use by the owner.

1. Receive from the contractor notification of substantial completion and list of items to be completed or corrected prior to final completion.
2. Obtain and review the completion documents required, including O&M manuals; contractor's documentation concerning owner training; and certificate of occupancy, occupancy permit, and/or final inspection approval(s) by authorities having jurisdiction.
3. Inspect the project to confirm substantial completion and prepare punch list of remaining work to be corrected or completed.
4. When the project is judged to be substantially complete, prepare Certificate of Substantial Completion.
5. If reproducible record drawings (including electronic files) are required, provide the contractor with appropriate media.
6. Request that the contractor submit project closeout documents (see below).

Final Completion. Review the contractor's request for final inspection and conduct a field site visit to confirm completion.

1. If the work is complete, prepare *Certificate of Final Completion.*
2. If the work is not complete, prepare a punch list of remaining work to be corrected or completed and define the date for reinspection.

Project Closeout. Ensure that final documentation required by the contract documents is submitted by the contractor.

1. Obtain and review the contractor's application for final payment, including required attachments such as waivers of lien and consent of surety documentation.
2. Review closeout documents required by the General Conditions, Division 01, and/or Division 23 of the specifications.
3. Obtain, review, and certify contractors' final payment request, including release of retainage.

Eleventh Month Inspection. Prior to expiration of the 1-year period of corrections, obtain the owner's authorization to conduct a site visit to determine if any (additional) warranty work is required by the contractor to remedy defects.

Index

A

Absorption, *see* Refrigeration; Water chillers
Acoustic Design, 400–407
  indoor noise, 401–404
  Noise Criteria, (NC) 159, 400
  outdoor noise, 404–407
  Room Criteria, (RC) 400
ADPI, *see* Air Diffusion Performance Index
Affinity laws, 243
AIA, *see* American Institute of Architects
Air cleaners
  final, 55, 56, 251, 255
  gas phase, 255–256
  particulate, 71, 253
  prefilter, 55, 247, 407
  ultraviolet, 155–157
Air-conditioning
  definition, 3
  processes, 33–52
Air-conditioning systems, 53–82
  dual duct, 59–62
  heat pumps, 72–82
  multizone, 62
  single zone, 56–58
  unitary, 68–72
  variable air volume, 58–59
  variable air volume/variable temperature, 62
  variable refrigerant flow, 70–71
Air Diffusion Performance Index, 147
Air distribution, 147–157
  air diffusion, 147–148
  displacement ventilation, 155
  underfloor, 155–157
Air-handling units, 249–262
Altitude correction, 410–411
American Institute of Architects, 431, 436–437, 439, 447
American National Standards Institute, 21, 71, 91, 187, 287
American Society of Heating, Refrigeration, and Air-Conditioning Engineers, Inc., 71, 72, 74, 77, 88, 90–91, 93, 95–96, 98, 99, 108, 114, 116, 118, 119, 130–131, 138, 141, 143, 148, 151, 155, 170, 173, 196, 205, 233, 255, 259, 267–271, 273, 277, 287, 290, 332, 368, 375, 377, 383, 400, 408, 409
Antiterrorism, 385–390
ASHRAE, *see* American Society of Heating, Refrigeration, and Air-Conditioning Engineers, Inc.
American Society of Mechanical Engineers, 90, 190, 331
ANSI, *see* American National Standards Institute
ASME, *see* American Society of Mechanical Engineers

B

BIM, *see* Building information modeling
Blower-coil unit, 249
Boilers, 330–335
  types/applications, 330–332
  venting, 332–335
Building codes/standards, 88–91
Building envelope, 7–8
Building information modeling, 431–437
  advantages, 432–434
  application, 434–436
  design contracts, 436–437

C

CAV, *see* Constant air volume
CDC, *see* Centers for Disease Control
Centers for Disease Control, 337
Chilled beams, 65–67
Chilled water
  approach temperature, 140–141
  buffer tank, 308
  configurations, 300–307
  storage, (*see* Thermal energy storage)
  supply temperature, 289
  temperature range, 289–290
  water treatment, (*see* Water treatment)
Chiller, *see* Water chillers
Climate Zones, 8–9, 74, 87, 89, 104, 146, 202, 259
Coefficient of performance, 74–76, 79, 80, 103, 274, 282, 283, 285, 287, 290
Comfort, 3, 5, 26, 64, 75, 107, 108, 114–115, 138, 148, 151–152, 250, 255, 263, 294, 308, 321, 354, 376, 383

Commissioning, 407–410
Condensers/cooling towers, 308–320
Constant air volume, 57, 59, 61–63, 67–69, 163, 164, 175
Construction contract administration, 439–454
  checklist, 453–454
  construction phase, 444–448
  design phase, 490–442
  litigation, 449–452
  preconstruction phase, 442–444
  project completion, 448
  project closeout, 448–449
Construction documents, 93, 110, 112, 436–437, 439, 442, 444, 450
Controls, direct digital, 345–380
  controllers and control loops, 370–378
  cybersecurity, 377–379
  design checklist, 379–380
  final control elements, 366–368
  fundamentals, 346–351
  input/output points, 351–364
  sequences of operation, 375–377
Cooling coils
  chilled water, 140–141
  direct expansion, (DX), 138–140
  heat transfer, 50–51
Cooling loads, 41–43
  latent cooling, 42–43
  sensible cooling, 41–42
Cooling Technology Institute, 91
Cooling towers
  configuration/application, 314–319
  evaporative cooling, 319–320
  fundamentals, 309–314
COP, *see* Coefficient of performance
Correcting for altitude/temperature, 410–411
Costs, 417–422
CTI, *see* Cooling Technology Institute

D

Dedicated outdoor air systems, 274–275
Desiccant cooling/dehumidification, 146–147
Design decisions, 413–430
DOAS, *see* Dedicated outdoor air systems
Duct
  air balancing, 164
  construction pressure classifications, 173–174
  design, 163–183
  layout and sizing, 171–174
  pressure loss calculation, 166–171
  sealing, 174
  support, 10

E

Electrical design
  basics, 12–25, 90
  devices/enclosures, 18, 23
  motor starters, 15–16
  motor/equipment drives, 21–23
  motors, 18–21
  VFDs, 16–18, 20
Electric heating coils, 141–143
Electric rates, 420–421
Energy conservation, 95–105
Engineering economics, 413–430
Evaporative air cooling, 320

F

Fancoil units, 67–68
FCUs, *see* Fancoil units
Filters, *see* Air cleaners
Final control devices
  dampers, 368–370
  control valves, 366–368
Fossil fuel firing, 325–329

H

Heat gain/loss calculation, 113–133
  benchmark loads, 132–133
  design conditions, 114–115
  fenestration, 118–122
  floors, 122–124
  heat balance/radiant time series, 128–130
  infiltration, 125–127
  internal, 124–125
  roofs, 115–118
  total equivalent temperature difference/time averaging, 130–131
  transfer function, 131
  walls, 115–118
Heat pumps
  advanced technology, 81
  air-source, 74–76
  concept, 72–74
  gas-fired engine driven, 80
  heat recovery, 80–81
  water-source, 76–80

Heat recovery
air-to-air, 99–102
internal source, 102
Heating coils
electric, 141–143
hot water, 141
preheat applications, 143–146
steam, 141
Heating systems
heat pumps, 72–81
hydronic, 333–335
steam, 335–344
Humidification, 262

I

Incremental HVAC units, 68–69
Indoor air quality, 383–385

L

Legionella, 399–400
Life-cycle costing analysis, 429–430

M

Maintainability, 91–94
Manufacturers Standardization Society, 10–11, 91
MSS, *see* Manufacturers Standardization Society

N

National Electrical Code, 12, 14
National Electrical Equipment Manufacturers Association, 16–18, 20, 23, 91
National Fire Protection Association, 88, 90, 265, 327, 332
NEC, *see* National Electrical Code
NEMA, *see* National Electrical Equipment Manufacturers Association
Net zero energy buildings, 108–110
NFPA, *see* National Fire Protection Association
Noise/vibration control, 400–407

P

Packaged HVAC units, 69
Piping
basics, 185–191
cooling coil condensate, 261–262
expansion control, 198–199, 209–210
freeze protection, 201–202
hydronic, (chilled water/hot water), 191–202
insulation, 215–218
refrigerant, 210–215
steam and condensate, 202–210
Properties of air, *see* Psychometrics
Psychrometrics
air mixing, 45–46
atmospheric air, 34–35
chart, 38–41
dry-bulb temperature, 34–35
processes, 36, 41–52
terminology, 35–36
water vapor, 35
wet-bulb temperature, 36–37
Pumps
characteristics/application, 221–225
configuration, 225–230
flow control, 230–231
head, 196–198, 221–225
static head, 224

R

Refrigeration, 277–323
absorption cycle, 282–285
chilled water, 288–307
direct expansion, 285–288
refrigerants, 277
vapor-compression cycle, 277–282

S

Seismic design, 12, 387, 442
Split HVAC systems, 69
Structural supports, 9–12
Sustainable design, 105–112

T

Terminal systems
space requirements, 245–249
types/applications, 245
Terminal units
single duct VAV, 157–160
fan-powered, 160
dual duct, 160–161
Terrorism/terror threats, 385–390
Thermal energy storage, 103–105
Thermodynamic systems, 5–9

U

UFAD, *see* Underfloor air distribution
Underfloor air distribution, 155–157
Unit ventilators, 67–68

V

Variable Air Volume, 57–59, 62–64, 69, 87, 139, 161, 242, 250 251
Variable frequency drives, 16–18, 20, 230–231, 280, 293, 295, 298, 364
VAV, *see* Variable Air Volume
Ventilation
  dedicated outdoor air systems, 274–275
  demand control, 271–274
  minimum outdoor airflow, 267–271
  outdoor air intake, 265–267
VFD, *see* Variable frequency drives
Variable flow refrigeration systems, 70–72
Vibration/noise control, 400–407

W

Water chillers, 288–308
Water treatment, 390–400
  chilled/hot water, 390
  condenser water, 390–396
  steam/condensate, 396–399
  white rust, 391, 393–394
Waterside economizer cycle, 320–323

Z

Zoning, 85, 86, 154,395,404